Lecture Notes in Computer Science 10438

Commenced Publication in 1973
Founding and Former Series Editors:
Gerhard Goos, Juris Hartmanis, and Jan van Leeuwen

More information about this series at http://www.springer.com/series/7409

Djamal Benslimane · Ernesto Damiani
William I. Grosky · Abdelkader Hameurlain
Amit Sheth · Roland R. Wagner (Eds.)

Database and Expert Systems Applications

28th International Conference, DEXA 2017
Lyon, France, August 28–31, 2017
Proceedings, Part I

 Springer

Editors
Djamal Benslimane (ID)
University of Lyon
Villeurbanne
France

Ernesto Damiani
University of Milan
Milan
Italy

William I. Grosky (ID)
University of Michigan
Dearborn, MI
USA

Abdelkader Hameurlain
Paul Sabatier University
Toulouse
France

Amit Sheth
Wright State University
Dayton, OH
USA

Roland R. Wagner
Johannes Kepler University
Linz
Austria

ISSN 0302-9743   ISSN 1611-3349 (electronic)
Lecture Notes in Computer Science
ISBN 978-3-319-64467-7   ISBN 978-3-319-64468-4 (eBook)
DOI 10.1007/978-3-319-64468-4

Library of Congress Control Number: 2017947505

LNCS Sublibrary: SL3 – Information Systems and Applications, incl. Internet/Web, and HCI

Printed on acid-free paper

This Springer imprint is published by Springer Nature
The registered company is Springer International Publishing AG
The registered company address is: Gewerbestrasse 11, 6330 Cham, Switzerland

Preface

The well-established International Conference on Database and Expert Systems Applications – DEXA — provides a forum to bring together researchers and practitioners who are actively engaged both in theoretical and practical aspects of database, information, and knowledge systems. It allows participants to exchange ideas, up-to-date information, and experiences in database and knowledge systems and to debate issues and directions for further research and development.

This volume contains papers selected for presentation at the 28th International Conference on Database and Expert Systems Applications (DEXA 2017), which took place in Lyon, France, during August 28–31, 2017.

DEXA 2017 attracted 166 submissions from all over the world. Decision on acceptance or rejection was based on at least three reviews for each submitted paper. After a thorough review process by the Program Committee members, to whom we owe our acknowledgment and special thanks for the time and effort they invested in reviewing papers, the DEXA 2017 chairs accepted 37 full research papers and 40 short research papers yielding an acceptance rate of 22% and 24%, respectively. Full papers were given a maximum of 15 pages in this volume and short papers were given an eight-page limit. Authors of selected papers presented at the conference will be invited to submit extended versions of their papers for publication in the Springer journal *Transactions on Large-Scale Data- and Knowledge-Centered Systems* (TLDKS). The submitted extended versions will undergo a further review process.

Two high-quality keynote presentations on "Structural and Semantic Summarization of RDF Graphs" given by Ioana Manolescu, Senior Researcher, Inria Saclay and Ecole Polytechnique, France, and "Omnipresent Multimedia – Pain and Gain of the Always Connected Paradigm" given by Gabriele Anderst-Kotsis, Johannes Kepler University Linz, Austria, were also featured in the scientific program of DEXA 2017.

This edition of DEXA also featured five international workshops covering a variety of specialized topics:

- AICTSS 2017: First International Workshop on Advanced ICT Technologies for Secure Societies
- BDMICS 2017: Second International Workshop on Big Data Management in Cloud Systems
- BIOKDD 2017: 8th International Workshop on Biological Knowledge Discovery from Data
- TIR 2017: 14th International Workshop on Technologies for Information Retrieval
- UCC 2017: First International Workshop on Uncertainty in Cloud Computing

The success of DEXA 2017 would not have been possible without the hard work and dedication of many people including Gabriela Wagner as manager of the DEXA organization for her highly skillful management and efficient assistance, Chirine Ghedira and Mahmoud Barhamgi as local Organizing Committee chairs for tackling different aspects of the local organization and their dedication and commitment to this event, Karim Benouaret, Caroline Wintergerst, Christophe Gravier, Omar Boussaid, Fadila Bentayeb, Nadia Kabachi, Nabila Benharkat, Nadia Bennani, Faty Berkaï, and Claire Petrel as local Organizing Committee members for supporting us all the way through, and Vladimir Marik as publication chair for the preparation of the proceedings volumes. Our special thanks and gratitude also go to the general chairs for their continuous encouragement and great support: Abdelkader Hameurlain (IRIT, Paul Sabatier University, Toulouse, France), Amit Sheth (Kno.e.sis - Wright State University, USA), and Roland R. Wagner (Johannes Kepler University, Linz, Austria).

DEXA 2017 received support from the following institutions: Lyon 1 University, Lyon 2 University, Lyon 3 University, University of Lyon, INSA of Lyon, LIRIS Lab, ERIC Lab, CNRS, FIL (Fédération Informatique Lyonnaise), AMIES Labex, and FAW in Austria. We gratefully thank them for their commitment to supporting this scientific event.

Last but not least, we want to thank the international community, including all the authors, the Program Committee members, and the external reviewers, for making and keeping DEXA a nice avenue and a well-established conference in its domain.

For readers of this volume, we hope you will find it both interesting and informative. We also hope it will inspire and embolden you to greater achievement and to look further into the challenges that are still ahead in our digital society.

June 2017

Djamal Benslimane
Ernesto Damiani
William I. Grosky

Organization

General Chair

Abdelkader Hameurlain   IRIT, Paul Sabatier University Toulouse, France
Amit Sheth   Kno.e.sis - Wright State University, USA
Roland R. Wagner   Johannes Kepler University Linz, Austria

Program Committee Co-chairs

Djamal Benslimane   University of Lyon 1, France
Ernesto Damiani   University of Milan, Italy
William I. Grosky   University of Michigan, USA

Publication Chair

Vladimir Marik   Czech Technical University, Czech Republic

Program Committee

Slim Abdennadher   German University, Cairo, Egypt
Witold Abramowicz   The Poznan University of Economics, Poland
Hamideh Afsarmanesh   University of Amsterdam, The Netherlands
Riccardo Albertoni   Institute of Applied Mathematics and Information
                        Technologies - Italian National Council of Research,
                        Italy
Idir Amine Amarouche   University Houari Boumediene, Algiers, Algeria
Rachid Anane   Coventry University, UK
Annalisa Appice   Università degli Studi di Bari, Italy
Mustafa Atay   Winston-Salem State University, USA
Faten Atigui   CNAM, France
Spiridon Bakiras   Hamad bin Khalifa University, Qatar
Zhifeng Bao   National University of Singapore, Singapore
Ladjel Bellatreche   ENSMA, France
Nadia Bennani   INSA Lyon, France
Karim Benouaret   Université Claude Bernard Lyon 1, France
Morad Benyoucef   University of Ottawa, Canada
Catherine Berrut   Grenoble University, France
Athman Bouguettaya   University of Sydney, Australia
Omar Boussaid   University of Lyon, France
Stephane Bressan   National University of Singapore, Singapore
David Camacho   Autonomous University of Madrid, Spain

Luis M. Camarinha-Matos   Universidade Nova de Lisboa + Uninova, Portugal
Barbara Catania   DISI, University of Genoa, Italy
Michelangelo Ceci   University of Bari, Italy
Richard Chbeir   UPPA University, France
Cindy Chen   University of Massachusetts Lowell, USA
Phoebe Chen   La Trobe University, Australia
Shu-Ching Chen   Florida International University, USA
Max Chevalier   IRIT, SIG, Université de Toulouse, France
Byron Choi   Hong Kong Baptist University, Hong Kong,
     SAR China
Henning Christiansen   Roskilde University, Denmark
Soon Ae Chun   City University of New York, USA
Alfredo Cuzzocrea   University of Trieste, Italy
Deborah Dahl   Conversational Technologies, USA
Jérôme Darmont   Université de Lyon, ERIC Lyon 2, France
Andre de Carvalho   University of Sao Paulo, Brazil
Roberto De Virgilio   Università Roma Tre, Italy
Zhi-Hong Deng   Peking University, China
Vincenzo Deufemia   Università degli Studi di Salerno, Italy
Gayo Diallo   Bordeaux University, France
Juliette Dibie-Barthélemy   AgroParisTech, France
Ying Ding   Indiana University, USA
Gill Dobbie   University of Auckland, New Zealand
Dejing Dou   University of Oregon, USA
Cedric du Mouza   CNAM, France
Johann Eder   University of Klagenfurt, Austria
Samhaa El-Beltagy   Nile University, Cairo, Egypt
Suzanne Embury   The University of Manchester, UK
Markus Endres   University of Augsburg, Germany
Damiani Ernesto   Universitá degli Studi di Milano, Italy
Noura Faci   Lyon 1 University, France
Bettina Fazzinga   ICAR-CNR, Italy
Leonidas Fegaras   The University of Texas at Arlington, USA
Stefano Ferilli   University of Bari, Italy
Flavio Ferrarotti   Software Competence Center Hagenberg, Austria
Vladimir Fomichov   National Research University, Higher School
     of Economics, Moscow, Russian Federation
Flavius Frasincar   Erasmus University Rotterdam, The Netherlands
Bernhard Freudenthaler   Software Competence Center Hagenberg GmbH,
     Austria
Hiroaki Fukuda   Shibaura Institute of Technology, Japan
Steven Furnell   Plymouth University, UK
Aryya Gangopadhyay   University of Maryland Baltimore County, USA
Yunjun Gao   Zhejiang University, China
Joy Garfield   University of Worcester, UK
Claudio Gennaro   ISTI-CNR, Italy

Chuan-Ming Liu	National Taipei University of Technology, Taiwan
Hong-Cheu Liu	University of South Australia, Australia
Jorge Lloret Gazo	University of Zaragoza, Spain
Jianguo Lu	University of Windsor, Canada
Alessandra Lumini	University of Bologna, Italy
Hui Ma	Victoria University of Wellington, New Zealand
Qiang Ma	Kyoto University, Japan
Stephane Maag	TELECOM SudParis, France
Zakaria Maamar	Zayed University, United Arab Emirates
Elio Masciari	ICAR-CNR, Università della Calabria, Italy
Brahim Medjahed	University of Michigan - Dearborn, USA
Faouzi Mhamdi	ESSTT, University of Tunis, Tunisia
Alok Mishra	Atilim University, Ankara, Turkey
Harekrishna Mishra	Institute of Rural Management Anand, India
Sanjay Misra	University of Technology, Minna, Nigeria
Jose Mocito	Brisa Innovation, Portugal
Lars Moench	University of Hagen, Germany
Riad Mokadem	IRIT, Paul Sabatier University, France
Yang-Sae Moon	Kangwon National University, South Korea
Franck Morvan	IRIT, Paul Sabatier University, France
Dariusz Mrozek	Silesian University of Technology, Poland
Francesc Munoz-Escoi	Universitat Politecnica de Valencia, Spain
Ismael Navas-Delgado	University of Málaga, Spain
Wilfred Ng	Hong Kong University of Science and Technology, Hong Kong, SAR China
Javier Nieves Acedo	University of Deusto, Spain
Mourad Oussalah	University of Nantes, France
Gultekin Ozsoyoglu	Case Western Reserve University, USA
George Pallis	University of Cyprus, Cyprus
Ingrid Pappel	Tallinn University of Technology, Estonia
Marcin Paprzycki	Polish Academy of Sciences, Warsaw Management Academy, Poland
Oscar Pastor Lopez	Universidad Politecnica de Valencia, Spain
Clara Pizzuti	Institute for High Performance Computing and Networking, ICAR, National Research Council, CNR, Italy
Pascal Poncelet	LIRMM, France
Elaheh Pourabbas	National Research Council, Italy
Jianbin Qin	University of New South Wales, Australia
Claudia Raibulet	Università degli Studi di Milano-Bicocca, Italy
Isidro Ramos	Technical University of Valencia, Spain
Praveen Rao	University of Missouri-Kansas City, USA
Manjeet Rege	University of St. Thomas, USA
Rodolfo F. Resende	Federal University of Minas Gerais, Brazil
Claudia Roncancio	Grenoble University/LIG, France
Massimo Ruffolo	ICAR-CNR, Italy

Giovanni Maria Sacco	University of Turin, Italy
Simonas Saltenis	Aalborg University, Denmark
Carlo Sansone	Università di Napoli Federico II, Italy
Igor Santos Grueiro	Deusto University, Spain
N.L. Sarda	I.I.T. Bombay, India
Marinette Savonnet	University of Burgundy, France
Klaus-Dieter Schewe	Software Competence Centre Hagenberg, Austria
Florence Sedes	IRIT, Paul Sabatier University, Toulouse, France
Nazha Selmaoui	University of New Caledonia, New Caledonia
Michael Sheng	Macquarie University, Australia
Patrick Siarry	Université Paris 12, LiSSi, France
Gheorghe Cosmin Silaghi	Babes-Bolyai University of Cluj-Napoca, Romania
Hala Skaf-Molli	Nantes University, France
Leonid Sokolinsky	South Ural State University, Russian Federation
Bala Srinivasan	Monash University, Australia
Umberto Straccia	ISTI, CNR, Italy
Raj Sunderraman	Georgia State University, USA
David Taniar	Monash University, Australia
Maguelonne Teisseire	Irstea, TETIS, France
Sergio Tessaris	Free University of Bozen-Bolzano, Italy
Olivier Teste	IRIT, University of Toulouse, France
Stephanie Teufel	University of Fribourg, Switzerland
Jukka Teuhola	University of Turku, Finland
Jean-Marc Thevenin	University of Toulouse 1 Capitole, France
A Min Tjoa	Vienna University of Technology, Austria
Vicenc Torra	University of Skövde, Sweden
Traian Marius Truta	Northern Kentucky University, USA
Theodoros Tzouramanis	University of the Aegean, Greece
Lucia Vaira	University of Salento, Italy
Ismini Vasileiou	University of Plymouth, UK
Krishnamurthy Vidyasankar	Memorial University of Newfoundland, Canada
Marco Vieira	University of Coimbra, Portugal
Junhu Wang	Griffith University, Brisbane, Australia
Wendy Hui Wang	Stevens Institute of Technology, USA
Piotr Wisniewski	Nicolaus Copernicus University, Poland
Huayu Wu	Institute for Infocomm Research, A*STAR, Singapore
Ming Hour Yang	Chung Yuan Christian University, Taiwan
Xiaochun Yang	Northeastern University, China
Junjie Yao	ECNU, China
Hongzhi Yin	The University of Queensland, Australia
Haruo Yokota	Tokyo Institute of Technology, Japan
Yanchang Zhao	IBM Australia, Australia
Qiang Zhu	The University of Michigan, USA
Yan Zhu	Southwest Jiaotong University, China
Marcin Zimniak	TU Chemnitz, Germany

Additional Reviewers

Mira Abboud	University of Nantes, France
Amine Abdaoui	LIRMM, France
Addi Ait-Mlouk	Cadi Ayyad University, Morocco
Ahmed Bahey	Nile University, Egypt
Cristóbal Barba-González	Universidad de Málaga, Spain
Nagwa M. Baz	Nile University, Egypt
Gema Bello	Autonomous University of Madrid, Spain
Kirill Borodulin	South Ural State University, Russian Federation
Yi Bu	Indiana University, USA
Stephen Carden	Georgia Southern University, USA
Loredana Caruccio	University of Salerno, Italy
Brice Chardin	LIAS/ENSMA, Poitiers, France
Arpita Chatterjee	Georgia Southern University, USA
Hongxu Chen	The University of Queensland, Australia
Weitong Chen	The University of Queensland, Australia
Van-Dat Cung	Grenoble INP, France
Sarah Dahab	Telecom SudParis, France
Matthew Damigos	NTUA, Greece
María del Carmen Rodríguez Hernández	University of Zaragoza, Spain
Yassine Djoudi	USTHB University, Algiers, Algeria
Hai Dong	RMIT University, Australia
Xingzhong Du	The University of Queensland, Australia
Daniel Ernesto Lopez Barron	University of Missouri-Kansas City, USA
Xiu Susie Fang	Macquarie University, Australia
William Ferng	Boeing Research and Technology, USA
Marco Franceschetti	Alpen Adria University Klagenfurt, Austria
Feng George Yu	Youngstown State University, USA
Azadeh Ghari-Neiat	University of Sydney, Australia
Paola Gomez	Université Grenoble-Alpes, France
Antonio Gonzalez	Autonomous University of Madrid, Spain
Senen Gonzalez	Software Competence Center Hagenberg, Austria
Wentian Guo	NUS, Singapore
Rajan Gupta	University of Delhi, India
Hsin-Yu Ha	Florida International University, USA
Ramón Hermoso	University of Zaragoza, Spain
Juan Jose Hernandez Porras	Telecom SudParis, France
Bing Huang	University of Sydney, Australia
Xin Huang	Hong Kong Baptist University, SAR China
Liliana Ibanescu	AgroParisTech and Inria, France
Angelo Impedovo	University of Bari Aldo Moro, Italy
Daniel Kadenbach	University of Applied Sciences and Arts Hannover, Germany

Eleftherios Kalogeros	Ionian University, Greece
Johannes Kastner	University of Augsburg, Germany
Anas Katib	University of Missouri-Kansas City, USA
Shih-Wen George Ke	Chung Yuan Christian University, Taiwan
Julius Köpke	Alpen Adria University Klagenfurt, Austria
Cyril Labbe	Université Grenoble-Alpes, France
Chuan-Chi Lai	National Chiao Tung University, Taiwan
Meriem Laifa	University of Bordj Bou Arreridj, Algeria
Raul Lara	Autonomous University of Madrid, Spain
Hieu Hanh Le	Tokyo Institute of Technology, Japan
Martin Ledvinka	Czech Technical University in Prague, Czech Republic
Jason Jingshi Li	IBM, Australia
Xiao Li	Dalian University of Technology, China
Corrado Loglisci	University of Bari Aldo Moro, Italy
Li Lu	Institute for Infocomm Research, Singapore
Alejandro Martín	Autonomous University of Madrid, Spain
Jorge Martinez-Gil	Software Competence Center Hagenberg, Austria
Riccardo Martoglia	University of Modena and Reggio Emilia, Italy
Amine Mesmoudi	LIAS/Poitiers University, Poitiers, France
Paolo Mignone	University of Bari Aldo Moro, Italy
Sajib Mistry	University of Sydney, Australia
Pascal Molli	University of Nantes, France
Ermelinda Oro	ICAR-CNR, Italy
Alejandra Lorena Paoletti	Software Competence Center Hagenberg, Austria
Horst Pichler	Alpen Adria University Klagenfurt, Austria
Gianvito Pio	University of Bari Aldo Moro, Italy
Valentina Indelli Pisano	University of Salerno, Italy
Samira Pouyanfar	Florida International University, USA
Gang Qian	University of Central Oklahoma, USA
Alfonso Quarati	Italian National Council of Research, Italy
Cristian Ramirez Atencia	Autonomous University of Madrid, Spain
Franck Ravat	IRIT-Université Toulouse1 Capitole, France
Leonard Renners	University of Applied Sciences and Arts Hannover, Germany
Victor Rodriguez	Autonomous University of Madrid, Spain
Lena Rudenko	University of Augsburg, Germany
Lama Saeeda	Czech Technical University in Prague, Czech Republic
Zaidi Sahnoun	Constantine University, Algeria
Arnaud Sallaberry	LIRMM, France
Mo Sha	NUS, Singapore
Xiang Shili	Institute for Infocomm Research, Singapore
Adel Smeda	University of Nantes, France
Bayu Adhi Tama	Pukyong National University, Korea
Loreadana Tec	RISC Software, Austria
Haiman Tian	Florida International University, USA
Raquel Trillo-Lado	University of Zaragoza, Spain

Sponsors of DEXA 2017

Organism	Logo
Université de Lyon	
Claude Bernard University Lyon 1	
Université Lumière Lyon 2	
Jean Moulin Lyon 3 University	
INSA Lyon (National Institute of Applied Science)	
FIL (Fédération Informatique de Lyon)	
CNRS (National Center for Scientific Research)	
AMIES (Agence pour les mathématiques en interaction avec l'entreprise et la société)	
LIRIS Lab	
ERIC Lab	
FAW Company	

Contents – Part I

Preferences and Query Optimization

Data Integration and RDF Matching

Security and Privacy (I)

Web Search

Data Clustering

Top-K and Skyline Queries

Data Mining and Big Data

Contents – Part II

Indexing and Concurrency Control Methods

Data Warehouse and Data Stream Warehouse

Data Mining and Machine Learning

Recommender Systems and Query Recommendation

Graph Algorithms

Semantic Clustering and Data Classification

Semantic Web and Semantics

MULDER: Querying the Linked Data Web by Bridging RDF Molecule Templates

Kemele M. Endris[1], Mikhail Galkin[1,2,3]($\boxtimes$), Ioanna Lytra[1,2],
Mohamed Nadjib Mami[1,2], Maria-Esther Vidal[2], and Sören Auer[1,2]

[1] Enterprise Information Systems (EIS), University of Bonn, Bonn, Germany
{endris,galkin,lytra,mami,auer}@cs.uni-bonn.de
[2] Fraunhofer Institute for Intelligent Analysis and Information Systems (IAIS),
Sankt Augustin, Germany
vidal@cs.uni-bonn.de
[3] ITMO University, Saint Petersburg, Russia

Abstract. The increasing number of RDF data sources that allow for querying Linked Data via Web services form the basis for federated SPARQL query processing. Federated SPARQL query engines provide a unified view of a federation of RDF data sources, and rely on source descriptions for selecting the data sources over which unified queries will be executed. Albeit efficient, existing federated SPARQL query engines usually ignore the meaning of data accessible from a data source, and describe sources only in terms of the vocabularies utilized in the data source. Lack of source description may conduce to the erroneous selection of data sources for a query, thus affecting the performance of query processing over the federation. We tackle the problem of federated SPARQL query processing and devise MULDER, a query engine for federations of RDF data sources. MULDER describes data sources in terms of RDF *molecule templates*, i.e., abstract descriptions of entities belonging to the same RDF class. Moreover, MULDER utilizes RDF *molecule templates* for source selection, and query decomposition and optimization. We empirically study the performance of MULDER on existing benchmarks, and compare MULDER performance with state-of-the-art federated SPARQL query engines. Experimental results suggest that RDF *molecule templates* empower MULDER federated query processing, and allow for the selection of RDF data sources that not only reduce execution time, but also increase answer completeness.

1 Introduction

Linked Open Data initiatives have received support from different communities, e.g., scientific, governmental, and industrial, and as a result, an increasing amount of RDF data sources is publicly available. Many of these data sources are accessible through Web access interfaces, e.g., SPARQL endpoints or other Linked Data Fragment (LDF) clients. The datasets have some value in their own right, but often the usefulness for application scenarios is potentialized

© Springer International Publishing AG 2017
D. Benslimane et al. (Eds.): DEXA 2017, Part I, LNCS 10438, pp. 3–18, 2017.
DOI: 10.1007/978-3-319-64468-4_1

when combining information from several datasets. While it is feasible and beneficial to aggregate and co-locate datasets in certain scenarios, there are also many situations, when this is not possible; for example, due to rapidly evolving data, security or licensing constraints, or simply the data scale, and the resulting resource demands for creating and maintaining an unified, aggregated view.

An approach for dealing with such situations is federated query processing, which has already been investigated by the database community and is currently receiving wide attention within semantic technology research [1,4,9,12]. There are two major challenges in federated query processing: (i) Discovering the structure of the available RDF datasets for facilitating the source selection process; (ii) Optimizing query decomposition and planning for balancing between query answer completeness and query execution time. While the majority of the existing works focus on either *one* of these two challenges, we deem them closely related, since knowledge about the structure of the data can guide the query decomposition and ultimately result in increased performance. Hence, we present with MULDER an integrated approach for federated query processing, which uses *RDF Molecule Templates* (RDF-MTs) for describing the structure of RDF datasets, and for bridging between parts of a query to be executed in a federated manner. Our contributions are in particular: (1) A thorough formalization accompanied by an implementation for federated query processing employing RDF Molecule Templates (RDF-MTs) for selecting relevant Web access interfaces, query decomposition, and execution; (2) An empirical evaluation assessing the performance of MULDER demonstrating a reduction in query execution time and increase in answer completeness with respect to the state-of-the-art.

The remainder of the paper is structured as follows: We describe behavior of existing systems with a motivating example in Fig. 1. We introduce the MULDER approach formally in Sect. 3. In Sects. 4 and 5 we get into the details of the Web interface description model of RDF-MTs, and query decomposition techniques used in MULDER, respectively. We present the results of our experimental study in Sect. 6. We discuss the related work in Sect. 7 and conclude with an outlook on future work in Sect. 8.

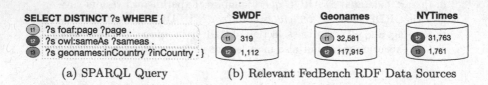

(a) SPARQL Query   (b) Relevant FedBench RDF Data Sources

Fig. 1. Motivating Example. (a) SPARQL query over FedBench RDF data sources; (b) FedBench data sources able to execute the query triple patterns. Each triple pattern can be executed in more than one RDF data source.

2 Motivating Example

We motivate our work by comparing the performance of state-of-the-art federated SPARQL query engines on a federation of RDF data sources from the *FedBench* benchmark. FedBench [11] is a standard benchmark for evaluating federated query processing approaches. It comprises three RDF data collections of interlinked datasets, i.e., the *Cross Domain Collection* (LinkedMDB, DBpedia, GeoNames, NYTimes, SWDF, and Jamendo), the *Life Science Collection* (Drugbank, DBpedia, KEGG, and ChEBI), and the synthetic SP^2Bench *Data Collection*. Although these datasets are from different domains, some RDF vocabularies are utilized in more than one dataset. For instance, `foaf` properties are used in DBpedia, GeoNames, SWDF, LinkedMDB, and NYTimes, while RDF triples with the `owl:sameAs` property are present in all the FedBench datasets.

Federated SPARQL query engines, e.g., *ANAPSID* [1] and *FedX* [12], provide a unified view of the federation of FedBench datasets, and support query processing over this unified view. Figure 1a presents a SPARQL query on the FedBench federation of three data sources. The query comprises three triple patterns: t1 can be answered on SWDF and Geonames; NYTimes can answer t3, while t2 can be executed over SWDF, Geonames, and NYTimes. Figure 1b reports on the number of answers of t1, t2, and t3 over SWDF, Geonames, and NYTimes. Federated query engines rely on source descriptions to select relevant sources for a query. For instance, based on the vocabulary properties utilized in each of the data sources, ANAPSID decides that SWDF, Geonames, and NYTimes are the relevant sources, while FedX contacts each of the federation SPARQL endpoints to determine where t1, t2, and t3 will be executed.

(a) Query Decompositions   (b) Join Cardinality

Fig. 2. Motivating Example. (a) Query Decompositions by FedX and ANAPSID; (b) Cardinality of Joins of triple patterns over relevant RDF data sources. FedX decomposition produces complete answers, but at the cost of execution time. ANAPSID decompositions run faster, but produce incomplete results.

Furthermore, different criteria are followed to decompose the query into the subqueries that will be posed over the relevant sources to collect the data required to answer the query. As presented in Fig. 2a, FedX identifies that t3 composes an exclusive group and can be executed over NYTimes; while t1 is executed over SWDF and Geonames, and t2 on all the three datasets. Thus, FedX produces a complete answer by joining the results obtained from executing these three subqueries. Nevertheless, FedX requires 239.4 s. to execute the query.

ANAPSID offers two query decomposition methods: SSGS and SSGM (Fig. 2a). ANAPSID SSGS only selects one relevant source per triple pattern; execution time is reduced to 0.338 s, but sources are erroneously selected and the query execution produces empty results. Finally, ANAPSID SSGM builds a star-shaped subquery that includes t2 and t3. The star-shaped subquery is executed on NYTimes, while t1 is posed over SWDF and Geonames. Execution time is reduced, but only 19 answers are produced, i.e., results are incomplete.

Based on the values of join cardinality reported in Fig. 2b, the decomposition that produces all the results requires that t2 is executed over NYTimes and Geonames, while t1 and t3 should be only executed in Geonames and NYTimes, respectively. However, because of the lack of source description, neither FedX nor ANAPSID is capable of finding this decomposition. On the one hand, to ensure completeness, FedX selects irrelevant sources for t1 and t2, negatively impacting execution time. On the other hand, ANAPSID SSGS blindly prunes the relevant sources for t1 and t2, and does not collect data from Geonames and NYTimes required to answer the query. Similarly, ANAPSID SSGM prunes Geonames from t2, while it is unable to decide irrelevancy of Geonames in t1.

MULDER describes sources with RDF molecule templates (RDF-MTs), and will be able to select the RDF-MTs of the correct relevant sources. Thus, MULDER will produce complete answers without impacting on execution time.

3 Problem Statement and Proposed Solution

In this section, we formalize the query decomposition and execution problems over a federation of RDF data sources.

Definition 1 (Query Decomposition). *Given a basic graph pattern BGP of triple patterns $\{t_1,\ldots,t_n\}$ and RDF datasets $D = \{D_1,\ldots,D_m\}$, a decomposition P of BGP in D, $\gamma(P|BGP,D)$, is a set of service graph patterns $SGP = (SQ,SD)$, where SQ is a subset of triple patterns in BGP and SD is a subset of D.*

Definition 2 (Query Execution over a Decomposition). *The evaluation of $\gamma(P|BGP,D)$ in D, $[[\gamma(P|BGP,D)]]_D$, is defined as the join of the results of evaluating SQ over RDF datasets D_i in SD:*

$$[[\gamma(P|BGP,D)]]_D = JOIN_{(SQ,SD)\in\gamma(P|BGP,D)} (UNION_{D_i\in SD}[[SQ]]_{D_i}) \quad (1)$$

After we defined what a decomposition of a query is and how such a decomposed query can be evaluated, we can define the problem of finding a suitable decomposition for a query and a given set of data sources.

Definition 3 (Query Decomposition Problem). *Given a SPARQL query Q and RDF datasets $D = \{D_1,\ldots,D_m\}$, the problem of decomposing Q in D is defined as follows. For all BGPs, $BGP = \{t_1,\ldots,t_n\}$ in Q, find a query decomposition $\gamma(P|BGP,D)$ that satisfies the following conditions:*

- *The evaluation of $\gamma(P|BGP, D)$ in D is complete, i.e., if D^* represents the union of RDF datasets in D, then the results of evaluating BGP in D^* and the results of evaluating decomposition $\gamma(P|BGP, D)$ in D are the same, i.e.,*

$$[[BGP]]_{D^*} = [[\gamma(P|BGP, D)]]_D \qquad (2)$$

- *$\gamma(P|BGP, D)$ has the minimal execution cost, i.e., if $cost(\gamma(P'|BGP, D))$ represents the execution time of a decomposition P' of BGP in D, then*

$$\gamma(P|BGP, D) = \underset{\gamma(P'|BGP,D)}{\operatorname{argmin}} \ cost(\gamma(P'|BGP, D)) \qquad (3)$$

To solve the query decomposing problem, we devised MULDER, a federated query engine for RDF datasets accessible through Web access interfaces, e.g., SPARQL endpoints. The MULDER architecture is depicted in Fig. 3. The MULDER *Query Processing Client* receives a SPARQL query, and identifies and executes a bushy plan against RDF datasets. The MULDER *Decomposition & Source Selection* creates a query decomposition with service graph patterns (SGPs) of star-shaped subqueries built according to RDF-MT metadata. RDF-MTs describe the properties of the RDF molecules contained in the RDF datasets, where an RDF molecule is a set of RDF triples that share the same subject. Once the star-shaped subqueries are identified, a bushy plan is built by the MULDER *Query Planning*; the plan leaves correspond to star-shaped subqueries. The MULDER *Query Engine* executes the bushy plan and contacts the MULDER query processing server to evaluate SGPs over the Web access interfaces. Further, the MULDER query processing server receives requests from the MULDER client to retrieve RDF-MT metadata about RDF datasets, e.g., metadata about properties of RDF molecules contained in these RDF datasets.

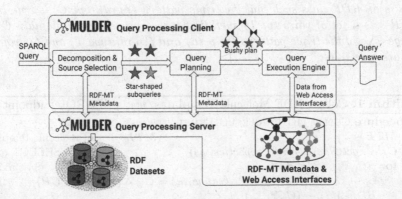

Fig. 3. The MULDER Client-Server Architecture. MULDER query processing client receives SPARQL queries, creates query decompositions with star-shaped subqueries, and identifies and executes bushy plans. MULDER query processing server collects both RDF-MT metadata about RDF datasets and results of executing queries over Web access interfaces, e.g., SPARQL endpoints.

4　The MULDER Web Interface Description Model

Web interfaces provide access to RDF datasets, and can be described in terms of resources and properties in the datasets. MULDER relies on RDF Molecule Templates to describe the set of properties that are associated with subjects of the same type in a given RDF dataset, i.e., the set of properties of the RDF molecules contained in the dataset. Further, an RDF Molecule Template is associated with a Web access interface that allows for accessing the RDF molecules that respect the template. Figure 4 presents the pipeline for RDF-MT creation.

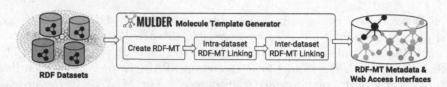

Fig. 4. Pipeline for Creating RDF Molecule Templates (RDF-MTs). `Create RDF-MT`: Queries are executed over RDF datasets for creating RDF-MTs. `Intra-dataset RDF-MT Linking`: RDF-MTs are linked to other RDF-MTs in the same dataset. `Inter-dataset RDF-MT Linking`: RDF-MTs are associated with RDF-MTs in other datasets. Web interfaces are created to access RDF-MTs.

Definition 4 (RDF Molecule Template (RDF-MT)). *An RDF Molecule Template (RDF-MT) is a 5-tuple* $= <WebI, C, DTP, IntraL, InterL>$, *where:*

- *WebI – is a Web service API that provides access to an RDF dataset G via SPARQL protocol;*
- *C – is an RDF class such that the triple pattern (?s rdf:type C) is true in G;*
- *DTP – is a set of pairs (p, T) such that p is a property with domain C and range T, and the triple patterns (?s p ?o) and (?o rdf:type T) and (?s rdf:type C) are true in G;*

Algorithm 1. Collect RDF Molecule Templates: ws: a SPARQL endpoint

---

1: **procedure** COLLECTRDFMOLECULETEMPLATES(ws)
2:　　　$MTL \leftarrow [\,]$　　　　　　　　　　　　　　　$\triangleright$ MTL - list of molecule templates
3:　　　$CP \leftarrow getClassesWithProperties(ws)$　　　　　　　$\triangleright$ SELECT query
4:　　　**for** $(C_i, \mathbb{P}_i) \in CP$ **do**　　　　　　　　　　$\triangleright$ C_i - class axioms
5:　　　　　$\mathbb{L}_i \leftarrow \mathbb{L}_i + (p_i, C_j) | \exists p_i \in \mathbb{P}_i, (range(p_i) = C_j) \wedge ((C_j, \mathbb{P}_j) \in CP)$
6:　　　　　$RDF\text{-}MT_i \leftarrow (C_i, \mathbb{P}_i, \mathbb{L}_i)$
7:　　　　　$MTL \leftarrow MTL + RDF\text{-}MT_i$
8:　　　**end for**
9:　　　**return** MTL
10: **end procedure**

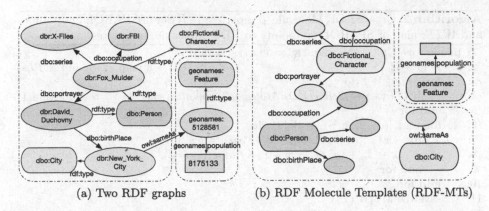

(a) Two RDF graphs   (b) RDF Molecule Templates (RDF-MTs)

Fig. 5. RDF-MT Creation. Two RDF graphs with four RDF molecules of types: dbo:Person, dbo:City, geonames:Feature, and dbo:FictionalCharacter. Four RDF Molecule Templates (RDF-MTs) are created for these RDF classes.

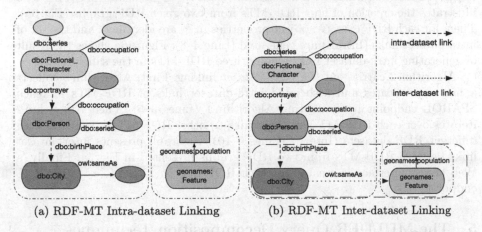

(a) RDF-MT Intra-dataset Linking   (b) RDF-MT Inter-dataset Linking

Fig. 6. RDF-MT Linking. (a) Each RDF-MT is linked to other RDF-MTs in the same RDF dataset: dbo:portrayer and dbo:birthPlace. (b) Each RDF-MT is linked to other RDF-MTs in a different RDF dataset: owl:sameAs.

- *IntraL – is a set of pairs (p, C_j) such that p is an object property with domain C and range C_j, and the triple patterns (?s p ?o) and (?o rdf:type C_j) and (?s rdf:type C) are true in G;*
- *InterL – is a set of triples (p, C_k, SW) such that p is an object property with domain C and range C_k; SW is a Web service API that provides access to an RDF dataset K, and the triple patterns (?s p ?o) and (?s rdf:type C) are true in G, and the triple pattern (?o rdf:type C_k) is true in K.*

Algorithm 1 extracts RDF-MTs by executing SPARQL queries against a Web access interface ws of an RDF dataset, e.g., a SPARQL endpoint. First, RDF classes are collected with their corresponding properties (Line 3), i.e., pairs

Algorithm 2. Create RDF Molecule Templates: WI: Set of SPARQL endpoints, and WIT: map of SPARQL endpoints to RDF Molecule Templates

```
1: procedure CREATEMOLECULETEMPLATES(WI)
2:     WIT ← { }                                    ▷ WIT - a map of ws and MTLs
3:     for wsᵢ ∈ WI do
4:         WIT(wsᵢ) ← CollectRDFMoleculeTemplates(wsᵢ)        ▷ Algorithm 1
5:     end for
6:     for (wsᵢ, [(Cᵢ, ℙᵢ, 𝕃ᵢ)]) ∈ WIT do
7:         for (wsₖ, [(Cₖ, ℙₖ, 𝕃ₖ)]) ∈ WIT|wsₖ ≠ wsᵢ do
8:             𝕃ᵢ ← 𝕃ᵢ + (pᵢ, Cₖ)|∃pᵢ ∈ ℙᵢ ∧ range(pᵢ) = Cₖ
9:         end for
10:    end for
11:    saveRDFMT(WIT)
12: end procedure
```

$(C_i, \mathbb{P}_i)$ are created where C_i is a class and $\mathbb{P}_i$ is a set of predicates of C_i. Figure 5 illustrates the creation of four RDF-MTs from two given RDF graphs (Fig. 5a). Then, for each RDF class C_i, object properties in $\mathbb{P}_i$ are identified, and the set of intra-dataset links (IntraL) are generated (Line 4–8). Figure 6a shows the result of generating intra-dataset links among three RDF-MTs in the same dataset.

Algorithm 2 carries out the *inter-dataset* linking. First, Algorithm 2 collects a list of predicates and associated intra-dataset links of RDF-MTs for each SPARQL endpoint by contacting Algorithm 1 (Lines 3–5). Then Algorithm 2 iterates over each RDF-MT and finds links to other RDF-MTs associated with different RDF datasets (InterL) (Lines 6–10). Figure 6b presents inter-dataset links between RDF-MTs from two RDF graphs presented in Fig. 5a. Finally, a Web access interface is created for each RDF-MT on Line 11, e.g., SPARQL endpoint.

5 The MULDER Query Decomposition Techniques

MULDER identifies query decompositions composed of star-shaped subqueries that match RDF-MTs, and minimize execution time and maximize answer completeness. A star-shaped subquery star(S,?X) on a variable ?X is defined as [14]:

- star(S,?X) is a triple pattern t = {?X p o}, and p and o are different to ?X.
- star(S,?X) is the union of two stars, star(S1,?X) and star(S2,?X), where triple patterns in S1 and S2 only share the variable ?X.

Figure 7 shows an example of query decomposition and source selection. The example query in Fig. 7b, contains eight triple patterns. The first step of query decomposition is to identify the star-shaped subqueries (SSQ). In our example, four subqueries which contain two triples patterns each are identified, i.e., ?drug (t_1, t_2), ?target (t_3, t_4), ?ref (t_5, t_6), and ?Int (t_7, t_8). Each of SSQs are then associated with RDF-MTs that contain predicates in SSQs, as shown in Fig. 7c.

(a) RDF-MTs   (b) SPARQL Query   (c) Star-shaped Subqueries

Fig. 7. Query Decomposition. (a) RDF-MTs about `db:drug_interaction`, `db:drugs`, `db:target`, `d:reference`. (b) SPARQL query composed of eight triple patterns that can be decomposed into four star-shaped subqueries. (c) Four star-shaped subqueries associated with four RDF-MTs in (a).

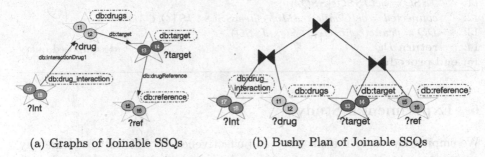

(a) Graphs of Joinable SSQs   (b) Bushy Plan of Joinable SSQs

Fig. 8. Query Planning. (a) Joinable Graph of Star-shaped Subqueries (SSQs) represents joins between SSQs. (b) Bushy plan of joinable Star-shaped Subqueries (SSQs). Graph of Joinable SSQs is utilized by MULDER query planner to create a bushy plan of SSQs where joins between SSQs are maximized.

Figure 8a shows joinable star-shaped subqueries (SSQ) that share at least one variable, i.e., `?Int` is joinable with `?drug` via predicate `db:interactionDrug1`, while `?drug` is joinable with `?target` via `db:target`. Furthermore, `?target` is joinable with `?ref` through `db:drugReference` property. Finally, MULDER query planner generates bushy plans combining SSQs, as the one presented in Fig. 8b. The problem of identifying a bushy plan from conjunctive queries is known to be NP-complete [10]. MULDER planner implements a greedy heuristic based approach to generate a bushy plan, where the leaves correspond to SSQs, and the number of joins between SSQs is maximized while the plan height is minimized.

The MULDER query decomposer is sketched in Algorithm 3. Given a BGP and a set of RDF-MTs (WIT), SSQs are first identified (Line 3). Then, RDF-MTs which contain all predicates in SSQ are determined from WIT as candidate RDF-MTs (Lines 4–10). Furthermore, linked candidate RDF-MTs with respect to Joinable SSQs (Line 11) are identified (Line 12). Finally, candidate RDF-MTs are pruned, i.e., candidate RDF-MTs that contain all predicates in SSQ but not linked to any RDF-MT that matches Joinable SSQ are excluded (Line 13).

Algorithm 3. Molecule template based SPARQL query decomposition: BGP: Basic Graph Pattern, WIT: set of RDF-MTs

---
1: **procedure** DECOMPOSE(BGP, WIT)
2:      $CM \leftarrow \{\}$ ▷ CM - Candidate RDF-MTs
3:      $SSQs = getStarShapedSubqueries(BGP)$ ▷ Subject stars
4:      **for** $s \in SSQs$ **do**
5:          **for** $RDF_{MT} \in WIT$ **do**
6:              **if** $predicatesIn(s) \subseteq predicatesIn(RDF_{MT})$ **then**
7:                  $CM[s].append(RDF_{MT})$
8:              **end if**
9:          **end for**
10:     **end for**
11:     $JSSQ = getJSSQs(SSQs)$
12:     $connected = getConnectedMolecules(SSQs, JSSQ, CM)$
13:     $DQ = prune(connected, SSQs, JSSQ)$
14:     **return** DQ ▷ decomposed query
15: **end procedure**

---

6   Experimental Study

We empirically study the efficiency and effectiveness of MULDER to identify query decompositions and select relevant data sources. In the first experiment, we assess the query performance of MULDER utilizing RDF molecule templates and templates generated using the METIS [5] and SemEP [7] partitioning algorithms. In a second experiment, we compare MULDER with the federated query engines ANAPSID and FedX for two well-established benchmarks – BSBM and FedBench. We address the following research questions: **(RQ1)** Do different MULDER source descriptions impact on query processing in terms of efficiency and effectiveness? **(RQ2)** Are RDF Molecule Template based query processing techniques able to enhance query execution time and completeness?

The experimental configuration is as follows:

Benchmarks and Queries: *(i) BSBM*: We use the synthetic Berlin SPARQL Benchmark to generate 12 queries (with 20 instantiations each) over a generated dataset containing 200 million triples. *(ii) FedBench*: We run 25 FedBench queries including cross-domain queries (CD), linked data queries (LD), and life science queries (LS). Additionally, 10 complex queries (C) proposed by [14] are included. The queries are executed against the FedBench datasets (cf. Fig. 1). A SPARQL endpoint able to access a unified view of all the FedBench datasets (i.e., the RDF dataset $D*$ in Eq. 2) serves as gold standard and baseline.

Metrics: *(i)* Execution Time: Elapsed time between the submission of a query to an engine and the delivery of the answers. Timeout is set to 300 s. *(ii)* Cardinality: Number of answers returned by the query. *(iii)* Completeness: Query result percentage with respect to the answers produced by the unified SPARQL endpoint with the union of all datasets.

Implementation: The MULDER[1] decomposition and source selection, and query planning components are implemented in Python 2.7.10, and integrated into ANAPSID [1], i.e., MULDER plans are executed using ANAPSID physical operators. Experiments are executed on two Dell PowerEdge R805 servers, AMD Opteron 2.4 GHz CPU, 64 cores, 256 GB RAM. FedBench and BSBM datasets are deployed on one machine as SPARQL endpoints using *Virtuoso 6.01.3127*, where each dataset resides in a dedicated Virtuoso docker container.

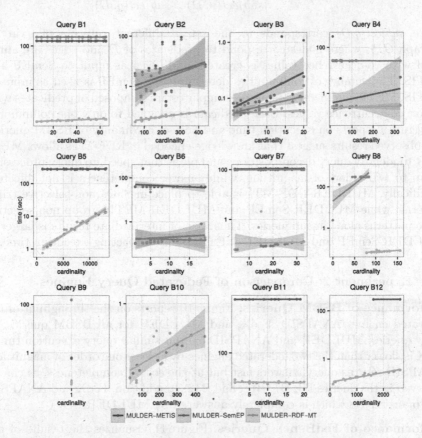

Fig. 9. BSBM: Performance for MULDER source descriptions. RDF molecules are computed using: Algorithm 1, SemEP, and METIS. RDF-MTs allow MULDER to identify query decompositions and plans that speed up query processing by up to two orders of magnitude, without affecting completeness.

6.1   Experiment 1: Comparison of MULDER Source Descriptions

We study the impact of MULDER RDF-MTs on query processing, and compare the effect of computing molecule templates using existing graph partitioning

---
[1] https://github.com/EIS-Bonn/MULDER.

methods: METIS and SemEP. We name MULDER-SemEP and MULDER-METIS, the version of MULDER where molecule templates have been computed using SemEP and METIS, respectively. Co-occurrences of predicates in the RDF triples of a dataset D are computed. Given predicates p and q in D, co-occurrence of p and q $(co(p,q,D))$ is defined as follows:

$$co(p, q, D) = \frac{|subject(p, D) \cap subject(q, D)|}{|subject(p, D) \cup subject(q, D)|} \tag{4}$$

where $subject(q,D)$ corresponds to the set of different subjects of q in D. A graph GP_D where nodes correspond to predicates of D and edges are annotated with co-occurrence values is created, and given as input to SemEP and METIS. The number of communities determined by SemEP is used as input in METIS. METIS and SemEP molecule templates are composed of predicates with similar co-occurrence values. Each predicate is assigned to only one community.

Figure 9 reports on execution time and answer cardinality of BSBM queries. The observed results suggest that knowledge encoded in RDF-MTs allows MULDER to identify query decompositions and plans that speed up query processing by up to two orders of magnitude, while answer completeness is not affected. Specifically, MULDER-RDF-MTs is able to place in SSQs non-selective triple patterns, while MULDER-SemEP and MULDER-METIS group non-selective triple patterns alone in subqueries. Thus, size of intermediate results is larger in MULDER-SemEP and MULDER-METIS plans, impacting execution time.

6.2   Experiment 2: Comparison of Federated Query Engines

Performance of BSBM Queries. Figure 10 reports on the throughput of the federated engines ANAPSID, FedX, and MULDER for all BSBM queries. In many queries, MULDER and ANAPSID exhibit similar query execution times. FedX is slower than the two federated engines by at least one order of magnitude. ANAPSID returns query answers fast but at the cost of completeness, as can be observed in the queries B4, B7, B11, and B12. In addition, FedX and ANAPSID fail to answer B8 which is completely answered by MULDER.

Performance of FedBench Queries. Figure 11 visualizes the results of the four FedBench groups of queries (CD, LD, LS, C) in terms of answer completeness and query execution time. Measurements that are located in Quadrants I and III indicate bad performance and incomplete results, points in Quadrant IV are the best in terms of execution time and completeness, i.e., they correspond to a solution to the query decomposition problem; finally, points in Quadrant II show complete results but slower execution times. MULDER outperforms ANAPSID and FedX with regard to the number of queries it manages to answer: ANAPSID answers 29, FedX 27, and MULDER 31 out of 35 queries (Query C9 could not be answered by any of the engines). In particular, MULDER delivers answers to queries C1, C3, C4, LS4, LS5, and LS6 for which FedX fails and CD6 and LD6 for which ANAPSID fails. FedX returns complete and

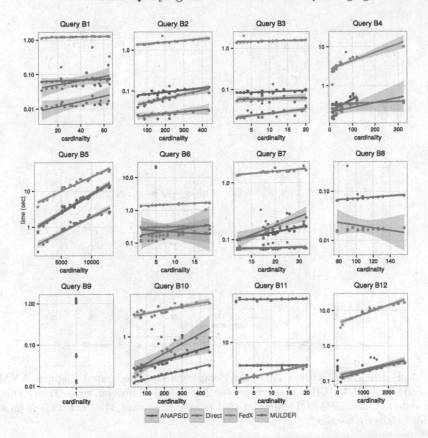

Fig. 10. BSBM: Performance of Federated Engines. MULDER and ANAPSID outperform FedX in terms of query execution time, while MULDER overcomes ANAPSID in terms of completeness. Direct represents a unified SPARQL endpoint over one dataset with all the federation RDF triples.

partially complete results for 20 and 7 queries respectively, exhibiting high execution times though (>1 s). In comparison to ANAPSID, MULDER achieves in general higher completeness of results, but at the cost of query execution time. For instance, C2, C8, and LD1 are answered by ANAPSID faster by almost one order for magnitude.

Results observed in both benchmarks allow us to positively answer **RQ1** and **RQ2**, and conclude that RDF-MTs enable MULDER decomposition and planning methods to identify efficient and effective query plans.

7   Related Work

Saleem et al. [8] study federated query engines with native Web access interfaces. *ANAPSID* [1] is an adaptive federated query engine capable to adjust

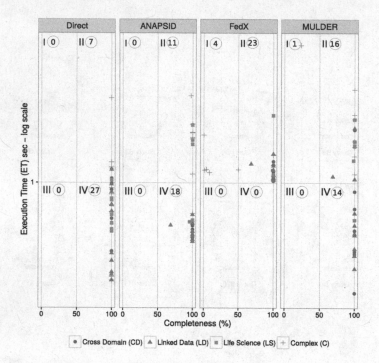

Fig. 11. FedBench: Execution Time and Completeness of Federated Engines. ANAP-SID, FedX and MULDER manage to answer 29, 27, and 31 queries, respectively. Direct represents a unified SPARQL endpoint that is able to answer 34 of the 35 benchmark queries before timing out.

the execution depending on the detected traffic load of a SPARQL endpoint. The engine delivers query results as soon as they arrive from sources, i.e., by implementing non-blocking join operators. *FedX* [12] is a non-adaptive federated query processing approach. FedX optimizes query execution introducing exclusive groups, i.e., triple patterns of a SPARQL query which have only one relevant source. In contrast to ANAPSID, FedX does not require metadata about the sources beforehand, but uses ASK queries. *Avalanche* [3] is a federated query engine that requires no prior knowledge about sources (e.g. SPARQL endpoints), their content, data schema, or accessibility of the endpoints. Instead, Avalanche identifies relevant sources and plans the query based on online statistical information. *Linked Data Fragments* (LDF) [13] is an approach that provides distributed storage and federated querying element. LDF is optimized for processing *triple patterns* while MULDER is able to generalize to *RDF molecule templates* (RDF-MT) which combine several triple patterns in a star-shaped group.

Federated query engines rely on source descriptions for source selection. Görlitz et al. describe SPLENDID, a federated query engine which leverages endpoint metadata available as *Vocabulary of Interlinked Datasets* (VoID) [2]. Although VoID provides dataset statistics, its description is limited and lacks details necessary for efficient query optimization. *HiBISCuS* [9] is a source selection technique based on hypergraphs. HiBISCuS discards irrelevant sources for a particular query by modeling SPARQL queries as hypergraphs. Capability metadata is gathered for each endpoint to label and subsequently prune a query hypergraph. Wylot and Cudré-Mauroux [15] use a notion of RDF molecule templates but provide only an intuitive description without formalization. We formally define RDF-MTs, and devise techniques for utilizing RDF-MTs during source selection, query decomposition, and planning.

As to query decomposition, *FED-DSATUR* presented in [14] applies graph colouring algorithms to perform query decomposition. *LILAC* [6] is a query decomposition technique that tackles data replication across data sources and SPARQL endpoints. MULDER utilizes RDF-MTs in both source selection and query decomposition stages, reducing, therefore, technological granularity and providing the semantics for enhancing federated query processing.

Graph partitioning algorithms may be used to automatically create source descriptions in terms of communities of RDF properties in the sources, i.e., RDF-MTs are defined as communities. *METIS* [5] partitions graph nodes in clusters whereas *semEP* [7] partitions graph edges. Limitations of applying METIS and semEP for RDF reside in the fact that the computed clusters contain unique sets of edges, i.e., it is impossible to have a property in multiple clusters. For instance, rdf:type is allocated to only one cluster which hinders query decomposition and planning. In contrast, MULDER RDF-MTs leverage semantics encoded in RDF sources and create logical partitions without any restrictions on properties.

8   Conclusions and Future Work

We introduced MULDER, a SPARQL query engine for federated access to SPARQL endpoints that uses RDF Molecule Templates (RDF-MTs) to describe these interfaces and guide the query decomposition. MULDER is significantly reducing query execution time and increasing answer completeness by using semantics in the source descriptions. In future work, we plan to integrate additional Web access interfaces, like TPFs (Triple Pattern Fragments), RESTful APIs, and empower RDF-MTs with statistics such cardinality and link selectivity.

Acknowledgements. This work has been partially funded by the EU Horizon 2020 research and innovation programme under the Marie Skłodowska-Curie grant agreement No. 642795 (WDAqua) and the EU H2020 programme for the project BigDataEurope (GA 644564).

References

1. Acosta, M., Vidal, M.-E., Lampo, T., Castillo, J., Ruckhaus, E.: ANAPSID: an adaptive query processing engine for SPARQL endpoints. In: Aroyo, L., Welty, C., Alani, H., Taylor, J., Bernstein, A., Kagal, L., Noy, N., Blomqvist, E. (eds.) ISWC 2011. LNCS, vol. 7031, pp. 18–34. Springer, Heidelberg (2011). doi:10.1007/978-3-642-25073-6_2

2. Alexander, K., Hausenblas, M.: Describing linked datasets-on the design and usage of voiD, the 'vocabulary of interlinked datasets'. In: LDOW (2009)
3. Basca, C., Bernstein, A.: Querying a messy web of data with avalanche. J. Web Semant. **26**, 1–28 (2014)
4. Görlitz, O., Staab, S.: SPLENDID: SPARQL endpoint federation exploiting VOID descriptions. In: COLD (2011)
5. Karypis, G., Kumar, V.: A fast and high quality multilevel scheme for partitioning irregular graphs. SIAM J. Sci. Comput. **20**(1), 359–392 (1998)
6. Montoya, G., Skaf-Molli, H., Molli, P., Vidal, M.: Decomposing federated queries in presence of replicated fragments. J. Web Semant. **42**, 1–18 (2017)
7. Palma, G., Vidal, M.-E., Raschid, L.: Drug-target interaction prediction using semantic similarity and edge partitioning. In: Mika, P., et al. (eds.) ISWC 2014. LNCS, vol. 8796, pp. 131–146. Springer, Cham (2014). doi:10.1007/978-3-319-11964-9_9
8. Saleem, M., Khan, Y., Hasnain, A., Ermilov, I., Ngomo, A.N.: A fine-grained evaluation of SPARQL endpoint federation systems. Semant. Web **7**(5), 493–518 (2015)
9. Saleem, M., Ngonga Ngomo, A.-C.: HiBISCuS: hypergraph-based source selection for SPARQL endpoint federation. In: Presutti, V., d'Amato, C., Gandon, F., d'Aquin, M., Staab, S., Tordai, A. (eds.) ESWC 2014. LNCS, vol. 8465, pp. 176–191. Springer, Cham (2014). doi:10.1007/978-3-319-07443-6_13
10. Scheufele, W., Moerkotte, G.: On the complexity of generating optimal plans with cross products. In: 16th ACM SIGACT-SIGMOD-SIGART Symposium on Principles of Database Systems, pp. 238–248 (1997)
11. Schmidt, M., Görlitz, O., Haase, P., Ladwig, G., Schwarte, A., Tran, T.: FedBench: a benchmark suite for federated semantic data query processing. In: Aroyo, L., Welty, C., Alani, H., Taylor, J., Bernstein, A., Kagal, L., Noy, N., Blomqvist, E. (eds.) ISWC 2011. LNCS, vol. 7031, pp. 585–600. Springer, Heidelberg (2011). doi:10.1007/978-3-642-25073-6_37
12. Schwarte, A., Haase, P., Hose, K., Schenkel, R., Schmidt, M.: FedX: optimization techniques for federated query processing on linked data. In: Aroyo, L., Welty, C., Alani, H., Taylor, J., Bernstein, A., Kagal, L., Noy, N., Blomqvist, E. (eds.) ISWC 2011. LNCS, vol. 7031, pp. 601–616. Springer, Heidelberg (2011). doi:10.1007/978-3-642-25073-6_38
13. Verborgh, R., Sande, M.V., Hartig, O., Herwegen, J.V., Vocht, L.D., Meester, B.D., Haesendonck, G., Colpaert, P.: Triple pattern fragments: a low-cost knowledge graph interface for the web. J. Web Semant. **37**, 184–206 (2016)
14. Vidal, M., Castillo, S., Acosta, M., Montoya, G., Palma, G.: On the selection of SPARQL endpoints to efficiently execute federated SPARQL queries. Trans. Large-Scale Data Knowl.-Cent. Syst. **25**, 109–149 (2016)
15. Wylot, M., Cudré-Mauroux, P.: DiploCloud: efficient and scalable management of RDF data in the cloud. IEEE Trans. Knowl. Data Eng. **28**(3), 659–674 (2016)

QAESTRO – Semantic-Based Composition
of Question Answering Pipelines

Kuldeep Singh[1,2], Ioanna Lytra[1,2(✉)], Maria-Esther Vidal[1],
Dharmen Punjani[3], Harsh Thakkar[2], Christoph Lange[1,2], and Sören Auer[1,2]

[1] Fraunhofer Institute for Intelligent Analysis and Information Systems (IAIS),
Sankt Augustin, Germany
Kuldeep.Singh@iais.fraunhofer.de, {lytra,langec,auer}@cs.uni-bonn.de
[2] Institute for Applied Computer Science, University of Bonn, Bonn, Germany
{vidal,thakkar}@cs.uni-bonn.de
[3] Department of Informatics and Telecommunications,
National and Kapodistrian University of Athens, Athens, Greece
Dharmen.punjani@gmail.com

Abstract. The demand for interfaces that allow users to interact with
computers in an intuitive, effective, and efficient way is increasing. Question Answering (QA) systems address this need by answering questions
posed by humans using knowledge bases. In recent years, many QA systems and related components have been developed both by practitioners
and the research community. Since QA involves a vast number of (partially overlapping) subtasks, existing QA components can be combined in
various ways to build tailored QA systems that perform better in terms
of scalability and accuracy in specific domains and use cases. However, to
the best of our knowledge, no systematic way exists to formally describe
and automatically compose such components. Thus, in this work, we
introduce QAESTRO, a framework for semantically describing both QA
components and developer requirements for QA component composition.
QAESTRO relies on a controlled vocabulary and the Local-as-View (LAV)
approach to model QA tasks and components, respectively. Furthermore,
the problem of QA component composition is mapped to the problem
of LAV query rewriting, and state-of-the-art SAT solvers are utilized to
efficiently enumerate the solutions. We have formalized 51 existing QA
components implemented in 20 QA systems using QAESTRO. Our empirical results suggest that QAESTRO enumerates the combinations of QA
components that effectively implement QA developer requirements.

1 Introduction

The main goal of Question Answering (QA) systems is to allow users to ask
questions in natural language, to find the corresponding answers in knowledge
bases, and to present the answers in an appropriate form. In recent years, QA
systems have received much interest, since they manage to provide intuitive
interfaces to humans for accessing distributed knowledge – structured, semi-
structured, or unstructured – in an efficient and effective way. Since the first

© Springer International Publishing AG 2017
D. Benslimane et al. (Eds.): DEXA 2017, Part I, LNCS 10438, pp. 19–34, 2017.
DOI: 10.1007/978-3-319-64468-4_2

attempts to provide natural language interfaces to databases around 1970 [1], an increasing number of QA systems and QA related components have been developed by both industry and the research community [12,16].

Despite different architectural components and techniques used by the various QA systems, these systems have several high-level functions and tasks in common [22]. For instance, the analysis of a question often includes tasks such as named entity recognition, disambiguation, and relation extraction, to name a few. Recent literature reviews have studied and classified existing QA systems and QA components with regard to the tasks they attempt to solve [22] and the common goals and challenges they tackle [12,16]. In addition, several frameworks have been proposed to address the re-usability of QA components. For instance, openQA [17] suggests a modular QA system consisting of components performing QA tasks that expose well-defined interfaces. The interchangeability of QA components is the main focus of other approaches as well, such as QALL-ME [8] which proposes a service-oriented architecture for the composition of QA components and Qanary [19] which introduces an ontology to tackle interoperability in the information exchange between QA components.

The aforementioned works provide a framework for developing or even integrating QA systems but fail to systematically address how to formally describe and automatically compose existing QA components. Still the composition of new QA systems for a specific domain or use case given the plethora of existing QA components is a rather manual, tedious, and error-prone task.

We introduce QAESTRO, a framework to semantically describe QA components and QA developer requirements and to produce QA component compositions based on these semantic descriptions. In particular, we introduce a controlled vocabulary to model QA tasks and exploit the *Local-As-View* (LAV) approach [15] to express QA components. Furthermore, QA developer requests are represented as conjunctive queries involving the concepts included in the vocabulary. The QA Component Composition problem can be afterwards cast to the LAV *Query Rewriting Problem* (QRP) [11]. Then, state-of-the art SAT solvers [10] can find the solution models in the combinatorial space of all solutions which eventually correspond to QA component compositions. Using QAESTRO, we formalized 51 QA components included in 20 distinct QA systems. In an empirical study, we show that QAESTRO effectively enumerates possible combinations of QA components for different developer requirements. Our main contributions are: (1) a vocabulary for expressing QA tasks and developer requirements; (2) the formalization of existing 51 QA components from 20 QA systems; (3) a mapping of the QA Component Composition problem into QRP; (4) the QAESTRO framework that generates QA component compositions based on developer requirements and (5) an empirical evaluation of QAESTRO behavior on QA developer requirements over the formalized QA components.

The remainder of the paper is structured as follows. We introduce the problem of QA Component Composition in the context of a motivating example in Sect. 2. In Sects. 3 and 4, we introduce the QAESTRO framework and present its details

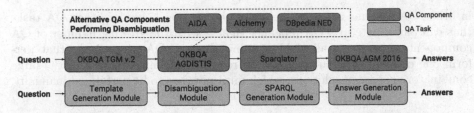

Fig. 1. OKBQA QA Pipeline and Pipeline Instance. OKBQA pipeline consists of four components that implement four core modules: Template Generation Module, Disambiguation Module, Query Generation Module, and Answer Generation Module. In this example, the disambiguation task can be performed by OKBQA AGDISTIS, Alchemy, and DBpedia NED interchangeably.

respectively. The results of our evaluation are reported in Sect. 5. We discuss the related work in Sect. 6 and conclude with an outlook on future work in Sect. 7.

2   Motivating Example

We motivate our work by discussing the problem of QA component composition in the context of the Open Knowledge Base and Question Answering (OKBQA) framework[1]. OKBQA considers QA as a predefined workflow consisting of four core modules providing Web service interfaces: (1) Template Generation Module for analyzing a question in natural language and producing SPARQL query skeletons, (2) Disambiguation Module for mapping words or word sequences to Linked Data resources, (3) Query Generation Module for producing SPARQL queries based on modules (1) and (2), and finally, (4) Answer Generation Module for executing SPARQL queries to get the answers. Figure 1 illustrates an instantiation of a QA pipeline with the components OKBQA TGM v.2, OKBQA AGDISTIS, Sparqlator, and OKBQA AGM 2016 which implement the aforementioned modules (1)–(4), respectively[2]. Although OKBQA provides a public repository comprising several QA components that can be composed in the OKBQA pipeline, still several issues remain open for the QA system developer. First of all, there is no systematic way to identify other existing components – either standalone or parts of other QA systems – that could be part of the OKBQA pipeline. Secondly, there is no way to exploit OKBQA QA components in existing QA systems systematically. Thirdly, it is not clear whether and how other QA-related tasks and/or subtasks can be integrated in the OKBQA framework. For instance, let us consider the disambiguation task. Several components, such as Alchemy API[3], and DBpedia NED [18] may replace OKBQA AGDISTIS in the QA pipeline of Fig. 1 since they perform conceptually the same QA task. Similarly, OKBQA AGDISTIS could serve the purpose of disambiguation

---

[1] http://www.okbqa.org/.
[2] All components can be found at http://repository.okbqa.org.
[3] http://alchemyapi.com.

in other QA systems as well. The same observation holds for other QA tasks that can participate in a QA pipeline. Hence, with the growing number of QA components, identifying all viable combinations of QA components that perform one or more tasks in combination requires a complex search in the large combinatorial space of solutions, which until now has to be performed manually.

3   QAESTRO Framework

QAESTRO is a QA framework that allows for the composition of QA components into QA pipelines. QAESTRO is based on a QAV vocabulary, which encodes the properties of generic QA tasks and is utilized to semantically describe QA components. QAESTRO exploits semantic descriptions of QA components, and enumerates the compositions of the QA components that implement a given QA developer requirement. Thus, QAESTRO provides a semantic framework for QA systems that not only enables a precise description of the properties of generic QA tasks and QA components, but also facilitates composition, integration, and reusability of semantically described QA components.

Formally, QAESTRO is defined as a triple ⟨QAV, QAC, QACM⟩, where: *(i)* QAV is a domain vocabulary composed of predicates describing QA tasks, e.g., disambiguation or entity recognition; *(ii)* QAC is a set of existing QA components that implement QA tasks, e.g., AGDISTIS [23] or Stanford NER [9]; and *(iii)* QACM is a set of mappings that define the QA components in QAC in terms of the QA tasks that they implement. Mappings in QACM correspond to conjunctive rules, where the head is a predicate in QAC and the body is a conjunction of predicates in QAV. QA developer requirements are also represented as conjunctive queries over the predicates in QAC. Moreover, the problem of QA Component Composition corresponds to the enumeration of combinations of QA components that implement a QA developer requirement. In the following sections, QAESTRO and the problem of QA composition are described in detail.

3.1   The Question Answering Tasks

QA systems implement abstract QA tasks to answer questions posed by humans. QA tasks include question analysis, query construction, and the evaluation of the generated query over a knowledge base to answer the input question [22]. Figure 2 depicts an abstract pipeline of the QA tasks [22], which receives a question as input and outputs the answers to this question over a knowledge base.

Question Analysis: Using different techniques, the input question is analyzed linguistically to identify syntactic and semantic features. The following techniques form important subtasks:

Tokenization: A natural language question is fragmented into words, phrases, symbols, or other meaningful units known as tokens.
POS Tagging: The part of speech, such as noun, verb, adjective, and pronoun, of each question word is identified and attached to the word as a tag.

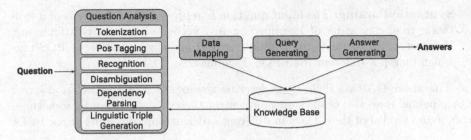

Fig. 2. Pipeline of QA Tasks. A QA pipeline receives a question and outputs the question answers. Question Analysis allows for question linguistic and semantic analysis to identify question features. During Data Mapping, question features are mapped into concepts in a Knowledge Base. A SPARQL query is constructed and executed during Question and Answer Generation.

Dependency Parsing: An alternative form of syntactic representation of the question to form a tree-like structure is created where arcs (edges in the tree) indicate that there is a grammatical relation between two words, whereas the nodes in the tree are the words (or tokens) in the question.

Recognition: An input question is parsed to identify the sequence of words that represent a person, a thing, or any other entity.

Disambiguation: The identity of the entity in the text is retrieved and then linked to its mentions in knowledge bases. Input for this may be one or more of the following: question, entity, and template. The output is a list of disambiguated entities.

Linguistic Triple Generation: Based on the input natural language question, triple patterns of the form ⟨*query term, relation, term*⟩ are generated [22].

Data Mapping: Information generated by Query Analyzer such as entities and tokens is mapped to its mentions in online knowledge bases such as DBpedia.

Query Generating: SPARQL queries are constructed; generated queries represent input questions over entities and predicates in online knowledge bases.

Answer Generating: The SPARQL queries are executed on the end points of knowledge bases to obtain the final answer.

Other QA Tasks include:

- **Question Type Identification:** This task identifies the type of the question. The input is the natural language question; the output is the type of the question, e.g., "yes-no", "location", "person", "time", or "reason".
- **Answer Type Identification:** This task identifies the desired type of the final answer. This task is sometimes performed as a part of the Question Analysis task or as a subtask of Answer Generation.
- **Query Ranking:** In some of the QA systems, the task Query Generation generates multiple candidate queries. This task ranks the generated SPARQL queries using a ranking function and it helps to select the best ranked query.

– **Syntactic Parsing:** The input question is represented in the form of a syntactic tree, consisting of identified nouns, verbs, adjectives, relations, etc. However, this task may use as input natural language question or POS tags which makes it different from POS Tagging [21].

The above QA task definitions describe the logical structure of an abstract QA pipeline. However, QA systems implement these tasks differently, sometimes combining several of these tasks in different order or skipping some of the tasks.

3.2   Controlled Vocabulary for Abstract Question Answering Tasks

A vocabulary QAV of the domain of QA tasks is described as a pair $\langle \delta, A \rangle$, where δ is a signature of a logical language and A is the set of axioms describing the relationships among vocabulary concepts. A signature δ is a set of predicate and constant symbols, from which logical formulas can be constructed, whereas the axioms A describe the vocabulary by illustrating the relationships between concepts. For instance, the term *disambiguation* is a predicate of arity four in δ; $disambig(x, y, z, t)$ denotes that the QA task disambiguation relates an entity x, a question y, a disambiguated entity z, and a template t. Furthermore, the binary predicate $questionAnalysis(x, y)$ models the question analysis task and relates an entity x to a question y. The following axiom states that the disambiguation task is a subtask of the question analysis task:

$$disambig(x, y, z, t) \rightarrow questionAnalysis(x, y)$$

3.3   Semantic Descriptions of Question Answering Components

QAC is a set of predicate signatures $\{QAC_1, \ldots, QAC_n\}$ that model QA components. For example, AGDISTIS [23] is represented with predicate $Agdistis(x, y, z)$ where x, y, and z denote an entity, a question, and a disambiguated entity, respectively. Further, the QA component Stanford NER [9] is modeled with the predicate $StanfordNER(y, x)$, which relates a question y to an entity x.

QAESTRO follows the Local-As-View (LAV) approach to define QA components in QAC based on predicates in QAV. LAV is commonly used by data integration systems to define semantic mappings between local schemas and views that describe integrated data sources and a unified ontology [20]. The LAV formulation allows QAESTRO to scale up to a large number of QA components, as well as to easily be adjusted to new QA components or modifications of existing ones. This property of the LAV approach is particularly important in the area of Question Answering, where new QA systems and components are constantly being proposed by practitioners and the research community. Following the LAV approach, a QA component C is defined using a conjunctive rule R. The head of R corresponds to the predicate in QAC that models C, while the body of R is a conjunction of predicates in QAV that represent the tasks performed by C. LAV rules are safe, i.e., all the variables in the head of a rule are also variables in the predicates in the body of the rule. Additionally, input and output restrictions of the QA components can be represented in LAV rules. The following LAV rules illustrate the semantic description of the QA components AGDISTIS and Stanford NER in terms of predicates in QAV. The symbol "$" denotes an input attribute of the corresponding QA component.

$$Agdistis(\$x, \$y, z) :- disambig(x, y, z, t), entity(x), question(y), disEntity(z)$$
$$StanfordNER(\$y, x) :- recognition(y, x), question(y), entity(x)$$

These rules state the following properties of AGDISTIS and Stanford NER: (i) AGDISTIS implements the QA task of disambiguation; an entity and a question are received as input, and a disambiguated entity is produced as output; (ii) Stanford NER implements the QA task of recognition; it receives a question as input and outputs a recognized entity.

3.4   Question Answering Developer Requirements

A QA developer requirement expresses the QA tasks that are required to be implemented by compositions of existing QA components. QA developer requirements are represented as conjunctive rules, where the body of a rule is composed of a conjunction of QA tasks. Similarly as for LAV mapping rules, input and output conditions can be represented; the symbol "$" denotes attributes assumed as input in the QA developer requirement. For instance, consider a developer who is interested in determining those compositions of QA components that, given a question q, perform entity recognition and disambiguation, and produce as output an entity e; the question q will be given as input to the pipeline.

$$QADevReq(\$q, e) :- recognition(q, e), disambig(e, q, de, t)$$

Now, suppose another developer requires also to know the compositions of QA components able to perform the pipeline of entity recognition and disambiguation. However, given the question as input, she requires to check all the intermediate results produced during the execution of the two tasks. In this case, the body of the rule remains the same, while the head of the rule ($QADevReq$) includes *all* variables corresponding to the arguments of the disambiguation task.

$$QADevReq(\$q, e, de, t) :- recognition(q, e), disambig(e, q, de, t)$$

4   Composition of QA Components in Pipelines

In this section, we describe the QAESTRO solution to the problem of QA Component Composition. We then describe the QAESTRO architecture, and the main features of the QAESTRO components.

4.1   The Problem of QA Component Composition

As previously presented, QAESTRO provides a vocabulary QAV that formalizes QA tasks, and allows for the definition of QA components using LAV rules and QA developer requirements using conjunctive queries based on QAV. In this subsection, we will show how QAESTRO solves the problem of QA Component Composition, i.e., how different QA components are automatically composed for a given developer requirement based on LAV mappings that semantically describe existing QA components. Next, we illustrate the problem of QA Component Composition. Besides the descriptions of AGDISTIS and Stanford NER (Subsect. 3.3), we consider further semantic descriptions for DBpedia NER [18], Alchemy API, and the answer type generator component of the QAKiS QA system [6], which we call *Qakisatype*.

$$DBpediaNER(\$y, x) :- recognition(y, x), question(y), entity(x)$$
$$Alchemy(\$y, z) :- disambig(x, y, z, t), question(y), disEntity(z)$$
$$Qakisatype(\$y, a) :- answertype(y, a, o), question(y), atype(a)$$

These rules state the following properties of the described QA components:

– The predicates $DBpediaNER(\$y, x)$, $Alchemy(\$y, z)$, and $Qakisatype(\$y, a)$ represent the QA components DBpedia NER, the Alchemy API, and Qakisatype, respectively. The symbol $\$$ denotes the input restriction of these QA components, i.e., the three QA components receive a question as input. These predicates belong to QAC.
– The predicates $recognition(y, x)$, $disambig(x, y, z, t)$, and $answertype(y, a, o)$ model the QA tasks: entity recognition, disambiguation, and answer type identification, respectively. These predicates belong to the QAV.
– An input natural language question is modeled by the predicate $question(y)$, while $entity(x)$ represents a named entity identified in a question.
– The QAV predicates $disEntity(z)$ and $atype(a)$ model the QA tasks of generating disambiguated entities and answer type identification, respectively.
– The variables x, y, z, and a correspond to instances of predicates *entity*, *question*, *disEntity*, and *atype*, respectively. The variable o is not bound to any predicate because *Qakisatype* does not produce ontology concepts.

Additionally, consider the following QA developer requirement for QA component compositions in a pipeline of entity recognition, disambiguation, and answer type identification, which receives a question q and outputs an entity e.

$$QADevReq(\$q, e) :- recognition(q, e), disambig(e, q, de, t), answertype(q, a, o)$$

QAESTRO generates two QA compositions as solutions to the problem of QA Component Composition. These compositions correspond to the enumeration of those combinations of QA components that implement the pipeline of the QA tasks of recognition, disambiguation, and answer type identification. Further, each composition satisfies the input restrictions of each QA component.

$QADevReq(\$q, e) :- StanfordNER(\$q, e), Agdistis(\$e, \$q, de), Qakisatype(\$q, a)$ (1)
$QADevReq(\$q, e) :- DBpediaNER(\$q, e), Agdistis(\$e, \$q, de), Qakisatype(\$q, a)$ (2)

Composition (1) indicates that the combination of the QA components Stanford NER, AGDISTIS, and Qakisatype implements the pipeline of recognition, disambiguation, and answer type identification. The input restriction of $StanfordNER(\$q, e)$ is satisfied by the question that is given as input in the pipeline. The QA component $Agdistis(\$e, \$q, de)$ is next in the composition; both the entity e produced by Stanford NER and the question q given by input to the pipeline, satisfy the input restrictions of this QA component. Similarly, input restriction of $Qakisatype(\$q, a)$ is satisfied by the question q. Additionally, Composition (2) implements the pipeline, but the QA component DBpedia NER is utilized for the QA task of entity recognition. The input restriction of DBpedia NER is also satisfied by the question received as input of the pipeline.

Consider the following compositions for the same QA developer requirement:

$QADevReq(\$q, e) :- StanfordNER(\$q, e), Alchemy(\$q, de), Qakisatype(\$q, a)$ (3)
$QADevReq(\$q, e) :- DBpediaNER(\$q, e), Alchemy(\$q, de), Qakisatype(\$q, a)$ (4)

Both compositions implement the pipeline of recognition, disambiguation, and answer type identification; also the input restrictions of the QA components are satisfied. However, these compositions are not valid because the argument e that represents an entity is not generated by Alchemy. This argument is required to be joined with the entity produced by the QA component that implements the entity recognition task and to be output by the compositions.

Formally, the problem of QA Component Composition is cast to the problem of Query Rewriting using LAV views [2]. An instance of QRP receives a set of LAV rules on a set P of predicates that define sources in V, and a conjunctive query Q over predicates in P. The output of Q is the set of valid re-writings of Q on V. Valid rewritings QR of Q on V are composed of sources in V that meet the following conditions:

– Every source in QR implements at least one subgoal of Q.
– If S is a source in QR and implements the set of subgoals SG of Q, then
  • The variables in both the head Q and SG are also in the head of S.
  • The head of the LAV rule where S is defined, includes the variables in SG that are in other subgoals of Q.

Note that the QA component $Alchemy(q, de)$ violates these conditions in Composition (3) and (4), i.e., $Alchemy(q, de)$ does not produce an entity e for a question q. Thus, compositions that implement the QA task of disambiguation with $Alchemy(q, de)$ are not valid solutions for this QA developer requirement.

QAESTRO casts the problem of QA Component Composition into the Query Rewriting Problem (QRP). Deciding if a query rewriting is a solution of QRP is NP-complete in the worst case [20]. However, given the importance of QRP in data integration systems and query optimization, QRP has received a lot of

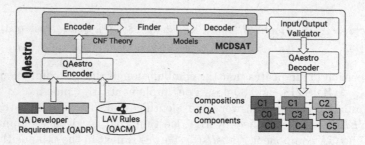

Fig. 3. QAESTRO Architecture. QAESTRO receives as input a QA developer requirement QADR and a set QACM of LAV rules describing QA components, and produces all the valid compositions that implement QADR.

attention in the Database area, and several approaches are able to provide effective and efficient solutions to the problem, e.g., MCDSAT [2,13] or GQR [14]. Thus, building on existing solutions for QRP, we devise a solution to the problem of QA Component Composition that is able to efficiently and effectively enumerate valid compositions of a QA developer requirement. QAESTRO implements a two-fold approach, where first, solutions to the cast instance of QRP are enumerated. Then, input and output restrictions of QA components are validated. Valid compositions of QA components that both implement a QA developer requirement and respect the input and output restrictions, are produced as solutions of an instance of the problem of QA Component Composition.

4.2   The QAESTRO Architecture

QAESTRO relies on MCDSAT, a state-of-the-art solver of QRP to efficiently enumerate the compositions of QA components that correspond to implementations of a QA developer requirement. Figure 3 depicts the QAESTRO architecture. QAESTRO receives as input a QA developer requirement QADR expressed as a conjunctive query over QA tasks in a vocabulary QAV. Furthermore, a set QACM of LAV rules describing QA components in terms of QAV is given as input to QAESTRO. QACM and QADR correspond to an instance of the QA Component Composition which is *cast* into an instance of QRP and passed to MCDSAT, a solver of QRP. MCDSAT encodes the instance of QRP into a CNF theory in a way that *models* of this theory correspond to solutions of QRP. MCD-SAT utilizes an off-the-shelf SAT solver to enumerate all *valid* query rewritings that correspond to *models* of the CNF theory. The output of the SAT solver is decoded, and input and output restrictions are validated in each query rewriting. Finally, QAESTRO decodes valid query rewritings where input and output restrictions are satisfied, and generates the compositions of QA components that implement the pipeline of QA tasks represented by QADR.

5   Empirical Study

We empirically study the behavior of QAESTRO in generating possible QA component compositions given QA developer requirements. We assess the following research questions: **(RQ1)** Given the formal descriptions of QA components using QAV and QA developer requirements are we able to produce sound and correct compositions? **(RQ2)** Are we able to produce efficiently solutions to the problem of QA Component Composition? The experimental configuration is as follows:

QA Components and Developer Requirements. To evaluate QAESTRO empirically, we have semantically described 51 QA components implemented by 20 QA systems which have participated in the first five editions of the QA over Linked Data Challenge (QALD1–5)[4]. Additionally, we studied well-known QA systems such as AskNow [7], TBSL [21], and OKBQA to semantically describe their components. After closely examining more than 50 components of these QA systems, we broadly categorized the components based on the QA tasks they perform, as defined in Sect. 3.1. For defining the LAV mappings, we selected only those QA components, for which there is a clear statement about input, output, and the QA tasks they perform in a publication (i.e., scientific paper, white paper, or source repository) about the respective QA system. Furthermore, we constructed manually 30 QA developer requirements for standalone QA tasks and QA pipelines integrating various numbers of QA tasks.

Metrics. *(i) Number of QA component compositions*: Number of QA component compositions given the semantic descriptions of QA components in QACM and a QA developer requirement; *(ii) Processing Time*: Elapsed time between the submission of a QA developer requirement and the arrival of all the QA component compositions produced by QAESTRO.

Implementation. QAESTRO is implemented in Python 2.7 on top of MCD-SAT [2], which solves QRP with the use of the off-the-shelf model compilation and enumeration tool c2d[5]. QAESTRO source code can be downloaded from https://github.com/WDAqua/QAestro and the evaluation results can be viewed at https://wdaqua.github.io/QAestro/. Experiments were executed on a laptop with Intel i7-4550U, 4×1.50 GHz and 8 GB RAM, running Fedora Linux 25.

5.1   Evaluation Results

Analysis of QA Components. In Fig. 4, we illustrate all QA components that have been formalized using QAESTRO along with their connections to the QA tasks they implement and the QA systems they belong to as an undirected graph[6]. In total, the resulting graph consists of 82 nodes and 102 edges. From the 82 nodes, 20 correspond to QA systems, 11 represent QA tasks, and 51 refer

---

[4] http://qald.sebastianwalter.org/.
[5] http://reasoning.cs.ucla.edu/c2d/.
[6] The graph visualization was generated with cytoscape - http://www.cytoscape.org.

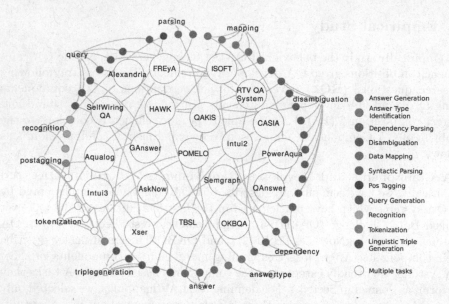

Fig. 4. QA Systems, Components, and Tasks. 51 QA components from 20 QA systems, implementing 11 distinct QA tasks are depicted as a directed graph.

to concrete QA components – 43 are part of the QA systems while 8 are provided also as standalone components (e.g., AGDISTIS, DBpedia NER, etc.). It can be observed in Fig. 5a that the majority of the analyzed QA components implement the Disambiguation task (10 in total) followed by the Query Generation (8), Tokenization (8), and POS Tagging (7) tasks. Many of these components are reused among the different QA systems. In addition, Fig. 5b shows that in almost half of the QA systems, components that implement Tokenization and Query Generation are included, while some less popular QA tasks like Answer Type Identification and Syntactic Parser are part of only two QA systems.

QA Component Compositions. In order to evaluate the efficiency of QAE-STRO, we edited 30 QA developer requirements with different number of QA tasks to be included in the QA pipeline and different expected inputs and outputs. Given these requirements and the semantic descriptions of QA components QAESTRO produced possible QA component compositions. Figure 6a reports on the number of different compositions for all 30 requirements grouped according to the number of QA tasks they include. Figure 6b demonstrates the time needed by QAESTRO to process each of the requirements and generate QA component compositions. We performed the measurements 10 times and calculated the mean values. While for standalone QA components or components that perform two tasks the solution space is relatively limited – from one to 30 combinations – for QA developer requirements that include three or more QA tasks the number of QA compositions may increase significantly. For instance, we notice that for a few requirements with three and four QA tasks the possible compositions are

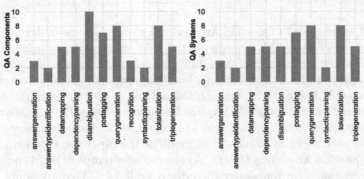

(a) QA Components per QA Task   (b) QA Systems per QA Task

Fig. 5. Frequencies of QA components and Systems per QA Task. Disambiguation, Tokenization, and Query Generation are the most popular tasks.

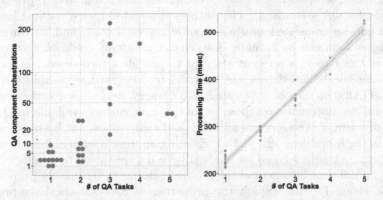

(a) QA Component Compositions (b) Execution Time for Generat-
per # of QA Tasks in the Pipeline ing QA Component Compositions

Fig. 6. Analysis of QA Component Compositions. QAESTRO is able to produce QA component compositions effectively and very fast.

more than 100. In these cases, the requirements do not foresee input or output dependencies between QA components, hence, the number of possible combinations increases significantly. All solutions produced by QAESTRO are sound and complete, since MCDSAT is able to produce every valid solution and all solutions that it provides are valid [2]. Furthermore, the processing time is for all requirements less than half a second and relates linearly to the number of QA tasks, since MCDSAT can perform model enumeration in linear time. Consequently, the experimental results allow us to positively answer **RQ1** and **RQ2**.

6   Related Work

Since 2010, more than 70 Question Answering Systems have been developed [12]. However, most of these QA systems (e.g., [6,7]) are monolithic in their implementation, which restricts their reusability in other QA approaches. Therefore, researchers have shifted their focus to building reusable QA architectures and frameworks. QALL-ME [8] is one such framework; it provides a reusable architecture to build multilingual QA systems. Furthermore, openQA [17] is an extensible framework for building multiple QA pipelines. It includes many external QA systems such as TBSL [21] to build modular QA pipelines. Open Knowledge Base and Question Answering (OKBQA) is a recent attempt to build component-based QA systems. Its repository includes overall 24 QA components solving four core QA tasks. However, there is no formalized way to describe how standalone QA components like Stanford NER [9] and Alchemy API could be used to perform the disambiguation task in OKBQA pipeline besides existing OKBQA disambiguation components. In 1978, researchers first attempted to provide formalization for QA systems [4]. The authors illustrated how a natural language question can be translated into a semantic representation and an underlying knowledge system can be formally described. Qanary [5], co-developed by some authors of this paper, is a recent attempt to provide a formalised methodology for building vocabulary-driven QA systems by integrating reusable QA components. QAESTRO can be integrated into Qanary, and allow for the semantic description and automatic composition of QA components available in Qanary.

The problem of Web services selection and composition has been extensively studied in the literature (e.g., [3,13]). Existing approaches range from heuristic-based [3] to SAT solver-based methods [13], and have shown to be efficient and effective for different instances of the problem. However, techniques proposed by Izquierdo et al. [13] that exploit the properties of SAT solvers, have provided evidence for large-scale composition of Web services. QAESTRO also exploits the benefits of modern SAT solvers and makes available an effective and efficient solution to the problem of QA Component Composition.

7   Conclusions and Future Work

In this work, we have tackled the problem of QA Component Composition by casting it to the Query Rewriting Problem. We introduced QAESTRO, a framework that enables QA developers to semantically describe QA components and developer requirements by exploiting the LAV approach. Moreover, QAESTRO computes compositions of QA components for a given QA developer requirement by taking advantage of SAT solvers. In an empirical evaluation, we tested QAESTRO with various QA developer requirements for QA pipelines of varying complexity, containing from two to five tasks. We observed that QAESTRO can not only produce sound and valid compositions of QA components, but also demonstrates efficient processing times. QAESTRO can successfully deal with the growing number of QA systems and standalone QA components, that is, the appearance of a new QA component only causes the addition of a new mapping describing the QA component in terms of the concepts in the QA vocabulary. Automated composition of QA components will enable subsequent research towards determining and executing best-performing QA pipelines that achieve better performance in terms of accuracy (precision, recall) and execution time. Currently, QAESTRO is not capable of implementing the QA pipeline in an automated way to answer an input question, however, in the future, QAESTRO will be integrated in approaches like Qanary to automatically retrieve all feasible combinations of available QA components and to realize the best-performing QA pipeline in concrete use cases.

Acknowledgements. Parts of this work received funding from the EU Horizon 2020 research and innovation programme under the Marie Skłodowska-Curie grant agreement No. 642795 (WDAqua).

References

1. Androutsopoulos, I., Ritchie, G.D., Thanisch, P.: Natural language interfaces to databases - an introduction. Nat. Lang. Eng. **1**(1), 29–81 (1995)
2. Arvelo, Y., Bonet, B., Vidal, M.: Compilation of query-rewriting problems into tractable fragments of propositional logic. In: Proceedings of the 21st National Conference on Artificial Intelligence and the 18th Innovative Applications of Artificial Intelligence Conference (2006)
3. Berardi, D., Cheikh, F., Giacomo, G.D., Patrizi, F.: Automatic service composition via simulation. Int. J. Found. Comput. Sci. **19**(2), 429–451 (2008)
4. Bolc, L. (ed.): Natural Language Communication with Computers. LNCS, vol. 63. Springer, Heidelberg (1978). doi:10.1007/BFb0031367
5. Both, A., Diefenbach, D., Singh, K., Shekarpour, S., Cherix, D., Lange, C.: Qanary-a methodology for vocabulary-driven open question answering systems. In: ESWC (2016)
6. Cabrio, E., Cojan, J., Aprosio, A.P., Magnini, B., Lavelli, A., Gandon, F.: QAKiS: an open domain QA system based on relational patterns. In: Proceedings of the ISWC 2012 Posters and Demonstrations Track (2012)
7. Dubey, M., Dasgupta, S., Sharma, A., Höffner, K., Lehmann, J.: AskNow: a framework for natural language query formalization in SPARQL. In: Sack, H., Blomqvist, E., d'Aquin, M., Ghidini, C., Ponzetto, S.P., Lange, C. (eds.) ESWC 2016. LNCS, vol. 9678, pp. 300–316. Springer, Cham (2016). doi:10.1007/978-3-319-34129-3_19
8. Ferrández, Ó., Spurk, C., Kouylekov, M., Dornescu, I., Ferrández, S., Negri, M., Izquierdo, R., Tomás, D., Orasan, C., Neumann, G., Magnini, B., González, J.L.V.: The QALL-ME framework: a specifiable-domain multilingual question answering architecture. J. Web Semant. **9**(2), 137–145 (2011)
9. Finkel, J.R., Grenager, T., Manning, C.D.: Incorporating non-local information into information extraction systems by Gibbs sampling. In: 43rd Annual Meeting of the Association for Computational Linguistics ACL (2005)
10. Gomes, C.P., Kautz, H.A., Sabharwal, A., Selman, B.: Satisfiability Solvers (2008)
11. Halevy, A.Y.: Answering queries using views: a survey. VLDB J. **10**(4), 270–294 (2001)

12. Höffner, K., Walter, S., Marx, E., Usbeck, R., Lehmann, J., Ngonga Ngomo, A.-C.: Survey on challenges of question answering in the semantic web. Semant. Web J. (2016). http://www.semantic-web-journal.net/content/survey-challenges-question-answering-semantic-web
13. Izquierdo, D., Vidal, M.-E., Bonet, B.: An expressive and efficient solution to the service selection problem. In: Patel-Schneider, P.F., Pan, Y., Hitzler, P., Mika, P., Zhang, L., Pan, J.Z., Horrocks, I., Glimm, B. (eds.) ISWC 2010. LNCS, vol. 6496, pp. 386–401. Springer, Heidelberg (2010). doi:10.1007/978-3-642-17746-0_25
14. Konstantinidis, G., Ambite, J.L.: Scalable query rewriting: a graph-based approach. In: Proceedings of the ACM SIGMOD International Conference on Management of Data (2011)
15. Levy, A.Y., Rajaraman, A., Ordille, J.J.: Querying heterogeneous information sources using source descriptions. In: Proceedings of 22th International Conference on Very Large Data Bases (1996)
16. López, V., Uren, V.S., Sabou, M., Motta, E.: Is question answering fit for the semantic web?: a survey. Semant. Web $2(2)$, 125–155 (2011)
17. Marx, E., Usbeck, R., Ngomo, A.N., Höffner, K., Lehmann, J., Auer, S.: Towards an open question answering architecture. In: SEMANTICS (2014)
18. Mendes, P.N., Jakob, M., García-Silva, A., Bizer, C.: DBpedia spotlight: shedding light on the web of documents. In: Proceedings of the 7th International Conference on Semantic Systems, I-SEMANTICS (2011)
19. Singh, K., Both, A., Diefenbach, D., Shekarpour, S.: Towards a message-driven vocabulary for promoting the interoperability of question answering systems. In: ICSC (2016)
20. Ullman, J.D.: Information integration using logical views. Theor. Comput. Sci. $239(2)$, 189–210 (2000)
21. Unger, C., Bühmann, L., Lehmann, J., Ngomo, A.N., Gerber, D., Cimiano, P.: Template-based question answering over RDF data. In: WWW (2012)
22. Unger, C., Freitas, A., Cimiano, P.: An introduction to question answering over linked data. In: Koubarakis, M., Stamou, G., Stoilos, G., Horrocks, I., Kolaitis, P., Lausen, G., Weikum, G. (eds.) Reasoning Web 2014. LNCS, vol. 8714, pp. 100–140. Springer, Cham (2014). doi:10.1007/978-3-319-10587-1_2
23. Usbeck, R., Ngonga Ngomo, A.-C., Röder, M., Gerber, D., Coelho, S.A., Auer, S., Both, A.: AGDISTIS - graph-based disambiguation of named entities using linked data. In: Mika, P., et al. (eds.) ISWC 2014. LNCS, vol. 8796, pp. 457–471. Springer, Cham (2014). doi:10.1007/978-3-319-11964-9_29

Nested Forms with Dynamic Suggestions for Quality RDF Authoring

Pierre Maillot[1]([⊠]), Sébastien Ferré[1], Peggy Cellier[2], Mireille Ducassé[2], and Franck Partouche[3]

[1] IRISA/Université de Rennes 1, Rennes, France
{pierre.maillot,sebastien.ferre}@irisa.fr
[2] IRISA/INSA Rennes, Rennes, France
{peggy.cellier,mireille.ducasse}@irisa.fr
[3] IRCGN, Cergy, France
franck.partouche@gendarmerie.interieur.gouv.fr

Abstract. Knowledge acquisition is a central issue of the Semantic Web. Knowledge cannot always be automatically extracted from existing data, thus domain experts are required to manually produce it. On the one hand, learning formal languages such as RDF represents an important obstacle to non-IT experts. On the other hand, well-known data input interfaces do not address well the relational nature and flexibility of RDF. Furthermore, it is difficult to maintain data quality through time, and across contributors. We propose FORMULIS, a form-based interface for guided RDF authoring. It hides RDF notations, addresses the relational aspects with nested forms, and guides users by computing intelligent filling suggestions. Two user experiments show that FORMULIS helps users maintain good data quality, and can be used by users without Semantic Web knowledge.

Keywords: Knowledge acquisition · Forms · Dynamic suggestions · Semantic Web

1 Introduction

The development of the Semantic Web [1] is fueled by the constant creation of data. RDF graphs are created either through automatic extraction from existing sources or from manual acquisition by contributors. The latter is required when there is no digital source or when the extraction cannot easily be automated. There are a number of issues with RDF authoring. A first issue is that contributors have to learn the formal syntax and semantics of RDF. This learning effort is an obstacle to the adoption and growth of the Semantic Web. A second issue is that the graph structure of RDF is difficult to present in the user interface to contributors, who are mostly used to forms and tables. A third issue is that RDF is very flexible, i.e. every entity can be related to any other entity through any property. A too flexible input interface will accept data that is inconsistent

© Springer International Publishing AG 2017
D. Benslimane et al. (Eds.): DEXA 2017, Part I, LNCS 10438, pp. 35–45, 2017.
DOI: 10.1007/978-3-319-64468-4_3

with the intended schema, while a schema-restricted input interface will forbid contributors to extend or amend the schema when needed.

A number of RDF authoring tools based on forms have been proposed, e.g. WebProtégé [10] or semantic wikis [3,8]. They offer a user-friendly interface but require a Semantic Web expert to configure the forms before domain experts can use them. This works for datasets with a stable ontology but not for datasets that are new or that continually evolve. Another limitation is that they do not allow to create graphs composed of several linked entities, working on several entities at the same time. In addition, those forms guide user input in the knowledge base thanks to simple consistency rules (e.g., film actors must be persons). However, in general, those rules are too simple and do not sufficiently reduce the suggestions. This forces the user to scroll through long lists of potentially irrelevant values (e.g., all persons when looking for an actor). That makes data authoring a tedious and error-prone process.

We propose FORMULIS, a system that offers the user-friendliness of forms while producing arbitrary RDF graphs, and guiding users with dynamic suggestions based on existing data. The principle of our system is to leverage the existing data and the already filled elements to make intelligent suggestions during the creation of a new entity. Our contribution to form-based RDF authoring is threefold. First, the forms can be nested as deeply as necessary so as to create several interlinked entities at once. Second, FORMULIS dynamically suggests fields and values that take into account all the fields and values already entered in the base and in the current form, thanks to SEWELIS [7] query relaxation mechanisms. Last but not least, the forms can be extended with new fields and new sub-forms at any time, according to user needs and data evolution. New forms can also be created on the fly for new classes of resources. This enables to leverage the flexibility of the RDF data model, and removes the need for the configuration of the interface by Semantic Web experts. Two user experiments show that FORMULIS helps users maintain good data quality, and can be used by contributors without Semantic Web knowledge. The first one is a controlled experiment involving laymen on the description of cooking recipes. The second is an application in a real setting involving forensic experts from IRCGN (Forensic Science Institute of the French Gendarmerie) on the description of forged Portuguese ID cards seized during a police operation.

Section 2 describes FORMULIS. Section 3 describes the experiments and their results. Section 4 discusses related work and Sect. 5 concludes the article.

2   Proposed Approach: FORMULIS

FORMULIS aims at facilitating RDF authoring without knowledge of RDF by proposing a familiar interface, forms, and by dynamically suggesting values from current and previous inputs. On the one hand, graph-based data representations tend to become quickly illegible with the size of the graph, and text-based notations such as Turtle need prior knowledge to be understood. Forms, on the other hand, are a common interface for data input. FORMULIS generates a data-driven form-based interface for the production of RDF data. Figure 1 summarizes the interaction loop followed by FORMULIS during data input. Starting with an empty form, the user interacts with FORMULIS until it is judged complete. Each interaction trigger an interaction loop to modify the content of the form. At each interaction, the partially-filled form is translated into an initial query that is supposed to retrieve the existing values for the field currently edited, with respect to the already filled fields. From this query, suggestions are computed with the help of the query relaxation and query evaluation mechanisms of a SEWELIS RDF base. Those suggestions have then to be rendered as form elements, i.e. user interface widgets and controls. From there, a new interaction step can start.

In the following, we explain the different parts of the interaction loop.

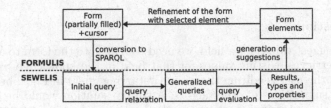

Fig. 1. Interaction loop of FORMULIS.

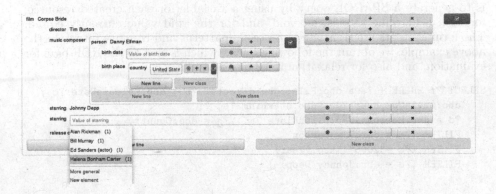

Fig. 2. Screenshot of FORMULIS during the creation of the "Corpse Bride" film, showing the creation of a new person in a nested form as object of the property "music composer" and creating a new country as object of property "birth place".

2.1 Nested Forms and Their RDF Counterpart

A form F in FORMULIS is composed of an RDF class C, a resource URI R, and a set of RDF properties $p_1, \ldots, p_n$, along with their respective sets of values $V_1, \ldots, V_n$. Class C (e.g., dbo:Film) determines the type of the resource being described, and resource R (e.g., dbr:Corpse_Bride) determines its identity. Each property p_i (e.g., dbo:director) defines a field of the form. Each field may have one or several values. Unlike tabular data, the RDF data model allows a resource to have several values for the same property, and it is therefore important to account for this in an RDF editing tool. A good example about films is the property dbo:starring relating films to actors. A field value can be one of: a literal (e.g., a string or a date); void (i.e. the field has not yet been filled); a RDF resource already present in the knowledge base (e.g., dbr:Tim_Burton); or a nested form to create and describe a new RDF resource. Those definitions exhibit the recursive nature of forms in FORMULIS, which allows for nested forms. They allow for cascading descriptions of resources that transcribe the graph structure of the RDF base to the user. Figure 2 shows an example of nested forms, a deeper nested form has been opened to create a new country as the birth place of the new person being created as the music composer.

2.2   Suggestions

In order to suggest values for a field, we need to translate the form to a SPARQL query that retrieves existing values for that field in the knowledge base. Let us suppose that the user is editing the second "starring" field in Fig. 2 because s-he wants to get suggestions of other actors. A simple solution would be to retrieve all values of the "starring" property in the base, using query: SELECT ?v WHERE {?x dbo:starring ?v}. However, this would make suggestions unspecific, here all actors present in the base would be listed. We make suggestions dynamic by taking into account the fields that have already been filled. The principle is to generate a SPARQL query by using a variable for each created resource, for each field value that is not void, and for the void value currently edited. The form contents are translated into triple patterns and equality filters. In the above example, we obtain the following query, which is sent to the RDF base for evaluation, and also for relaxation if it has no results.

```
SELECT ?f WHERE { ?a a dbo:Film; dbo:director ?b ; dbo:releaseDate ?c ;
    dbo:musicComposer ?d ; dbo:starring ?e, ?f .
    ?d a dbo:Person; dbo:birthPlace ?g . ?g a dbo:Country .
    FILTER (?b = dbr:Tim_Burton)
    FILTER (?c ="2005"^^xsd:gYear)
    FILTER (?e = dbr:Johnny_Depp) }
```

2.3   Query Relaxation and Evaluation

The more fields are filled, the more specific the query is, and the more likely it is to have no results, *e.g.* if the film being described is the first in the base to be directed by Tim Burton. The second and essential step is then to relax that initial query in order to have generalized queries, and hence more query results. FORMULIS use SEWELIS [7] as a RDF base. SEWELIS applies relaxation rules that can replace a class by a super-class, a property by a super-property, or remove altogether a triple or an equality. The *relaxation distance* of a generalized query is the number of relaxation rules that have to be applied to generate it. For example, the generalized query SELECT ?f WHERE { ?a a dbo:CreativeWork ; dbo:director ?b ;

`dbo:starring ?f. FILTER (?b = dbr:Tim_Burton) }` is at relaxation distance 9: one rule for replacing `dbo:Film` by the super-class `dbo:CreativeWork`, and 8 times another rule for removing triple patterns and equalities. SEWELIS uses an algorithm that generates the results of generalized queries, by increasing relaxation distance. It is efficient by using dynamic programming principles to improve scalability by avoiding the enumeration of generalized queries. The query results are a set of RDF resources or literals. In the example, it is a set of actor URIs. That set can be made larger by increasing the relaxation distance.

SEWELIS does not only return a set of resources and literals, but also the types and properties that they have in the knowledge base. Those are then used by FORMULIS to provide suggestions to help users further to fill the form.

2.4   Generation of Suggestions and Refinement of the Form

The refinement of a form in FORMULIS is done through interactions with the form elements. Before entering into the interaction loop of a form, users are presented with a list of classes to select the type of the instance to be created. This list is generated from the suggestions returned by SEWELIS for the query `SELECT ?c WHERE { ?c a rdfs:Class }`. Once the user has selected a class, the fields of the form are initialized by the properties obtained from SEWELIS, *e.g.* a film creation form would use the query `SELECT ?p WHERE { ?s a dbo:Film. ?s ?p ?o. }`. In the situation where there is only one proposed class, the form is automatically initialized from it. After that, users enter the interaction loop to create a new resource.

Continuing with the example in previous sections, FORMULIS receives a set of results, here actors, along with their types and properties, because the user is editing the second "starring" field. Suggestions in FORMULIS are derived from those results. Each result, a resource or a literal, is suggested as a value to fill the field. From these suggestions, the user can select an existing value among the suggestions in the displayed drop-down menu, as in Fig. 2 showing suggestions sorted by their number of occurrences in similar films.

The creation of new values is also guided by suggestions. When users choose to create a new value, FORMULIS scans the suggested values to determine the most likely class or datatype of the new value. If all suggestions are literals, the user is given a creation widget enabling the creation of dates, numbers, text or a new resource. The creation widget is then set by default on the datatype appearing most often in the suggestions. If all suggestions are resources, the property field is replaced by a new FORMULIS form, *e.g.* in Fig. 2 a new person is created as music composer and for this new person, a new country is created.

3   User Experiments

We have conducted two user experiments on the collaborative creation of a knowledge base. The first one compares FORMULIS with WebProtégé, a collaborative ontology editor, in a controlled experiment with layman users describ-

Table 1. Annotators grouped by profile.

Semantic web knowledge	Other input interfaces	System used first	
		FORMULIS	WebProtégé
None	No	$User_a$	$User_h$, $User_i$
None	Yes	$User_b$, $User_c$, $User_d$	$User_j$, $User_k$
Basic	No	$User_e$	$User_l$, $User_m$
Basic	Yes	$User_f$, $User_g$	$User_n$

ing cooking recipes. The second one shows the benefits of FORMULIS to create domain knowledge in a real setting (description of forged ID documents by forensic experts) with no defined vocabulary and without starting data. For both experiments, we discuss the quality of the created knowledge as well as the usability of FORMULIS for non-IT experts.

3.1   Cooking Recipes Experiment

Methodology. We selected 42 recipes taken among the top featured recipes of a well-known cooking website[1]. FORMULIS and WebProtégé were each initialized with a small base of 9 recipes extracted from the cooking website.

Each recipe was described by its ingredients (without quantities), preparation time, cooking time, and a qualitative ranking of its difficulty and of its cost (*e.g.*, "Very cheap"). For each qualitative property, only one value was initially present. The annotators (users) were 14 volunteers among students, colleagues, IT workers, and non-IT relatives. One half used WebProtégé before FORMULIS, while the other half used FORMULIS before WebProtégé. Table 1 shows information about the annotators: the first system they used, their level of knowledge of the Semantic Web, and their previous experiences of data input interfaces.

At the beginning of the experiment, each annotator received: (i) a tutorial describing how to create individuals and their properties for each system, and (ii) three recipes to enter with WebProtégé and three other recipes to enter with FORMULIS. Note that each recipe was input once in each system by two different annotators. Some recipes contained nested recipes such as sauces or bases. Each annotator had to fill a System Usability Scale (SUS [2]) survey after using each system.

Results and Interpretation: Data Quality. The 9 recipes present in the knowledge base before the experiment are described as instances of the `Recipe` class with relations to instances of the `Ingredient` class. In the final knowledge base created with WebProtégé, 31 recipes and 34 ingredients do follow the structure of the original data but 14 recipes were created as classes or as instances of `owl:Thing`, 57 ingredients were created as instances of `owl:Thing`, and 62 ingredients were created as classes or properties. The majority of invalid creations in

---

[1] http://www.750g.com/.

WebProtégé were made by users with no knowledge of the Semantic Web. On the contrary, in the final knowledge base created with FORMULIS all recipes and ingredients but one follow the structure of the original data. Only one recipe and one property were created as classes. This better data quality can be attributed to the fact that FORMULIS does not easily let users depart from the original structure or do unexpected operations.

In order to evaluate, for each system, the difficulty for users to extend the model on their own, the quantity of each ingredient was given in recipe descriptions but how to describe them was not explained in the tutorial. In WebProtégé, some annotators used ingredients as properties or even OWL cardinality restrictions on recipe classes to specify quantities. In FORMULIS, $User_c$ created a quantity property attached to ingredients. A positive result is that this new property was then suggested by FORMULIS, and reused by other annotators. However, it inadequately attached quantities to ingredients instead of to relationships between recipes and ingredients.

Globally, better quality data were created with FORMULIS than with WebProtégé. Our interpretation is that the data-driven guidance of FORMULIS helps to maintain homogeneity between new data and old data, while the lack of guidance and feedback in WebProtégé favors a wide range of mistakes. However, when a modelling error is introduced in FORMULIS, it tends to propagate from one annotator to the next. It therefore shows the necessity of a proper initialization of FORMULIS with enough examples, and expert users may be required for extensions of the models.

Results and Interpretation: System Usability. We compare the usability of the two systems as evaluated by annotators through a SUS survey for each system. Table 2 gives the average SUS scores obtained by each system (i) globally, and according to (ii) the order in which the systems were used, (iii) Semantic Web knowledge, and (iv) experience with other input interfaces. (i) FORMULIS was globally evaluated as significantly easier to use than WebProtégé. (ii) Annotators evaluated each system higher if they started with it, FORMULIS is still

Table 2. SUS scores relative to each experiment parameter.

		No. of users	Average SUS score	
			WebProtégé	FORMULIS
(i)	Global	14	30.7	64.1
(ii)	WebProtégé → FORMULIS	7	34.6	56.8
	FORMULIS → WebProtégé	7	26.8	72.1
(iii)	SW knowledge: None	8	25.9	61.9
	SW knowledge: Basic	6	37.1	67.1
(iv)	Other systems: Yes	8	32.2	58.8
	Other systems: No	6	28.8	71.3

rated more than 20 points higher whatever the order of usage. (iii) Annotators with basic Semantic Web knowledge evaluated both WebProtégé and FORMULIS higher than annotators without that knowledge. (iv) The gap between the perception of usability is the highest when users have not previously used another input system. Even for users who had previous experience, FORMULIS is rated more than 25 points higher. In conclusion, FORMULIS is perceived as significantly easier to use by all categories of users.

3.2   Forged ID Experiment

Methodology. In this experiment, the annotators are six forensic experts from the forged ID unit of the document department of IRCGN. We evaluated the data quality and ease of use of FORMULIS when used by national-level experts converting their knowledge into RDF data. All annotators had experience with web forms or spreadsheets. $User_1$ is an IT specialist, $User_4$ is a chemist, $User_2$ and $User_5$ are handwriting experts and $User_3$ and $User_6$ are ID document experts.

Each annotator had to put into a knowledge base the description of a dozen forged Portuguese ID cards seized during a police operation. In order to provide those descriptions each annotator had to carefully examine the documents in order to assess at least eighteen description attributes, such as paper imperfections or ultraviolet reaction. The attributes values vocabulary could not be totally defined prior to the experiment as new elements might be discovered during examination. In addition, annotators have different backgrounds and might use different terminologies.

For the evaluation, the set of annotators was split into two groups of three annotators. Each group shared its own knowledge base. Note that in the sequel, the knowledge base of group 1 (respectively group 2) is denoted by $base_1$ (resp. $base_2$). There were three sessions. During each session two annotators, one of each group, were providing descriptions through FORMULIS for the same documents without exchanging about their observations. At the first session, $User_1$ and $User_4$ were chosen as first annotators as they were the most knowledgeable in forged document detection available at that moment. At the end of each session, annotators were asked to fill a SUS survey in order to provide feedback on the experiment. There was no noticeable differences between the input times of annotators as they spent most of it on the careful examination of each ID document.

Results and Interpretation: Data Quality. We have assessed how helpful FORMULIS is to create quality data in a knowledge base when there are several users. Precisely, we focus on how well the use of FORMULIS is limiting the input of non-standard values inside a base such as spelling variations, synonyms or typos, *i.e.* whether it maintains the homogeneity of the base values. Only the first two documents created by $User_1$ in $base_1$ have non-standard values for three fields, with respect to other documents in the same base. Those non-standard values are either capitalization discrepancies or wording differences, *e.g.* a signature denoted by "written with a pen" instead of "handwritten". We assume that, at the beginning of the experiment, the system was unable to make suggestions, and User$_1$ himself, despite being an expert, was unsure of the appropriate vocabulary. The two knowledge bases use different vocabularies for some attributes, such as "Xerography" in base$_1$ and "Electrophotography" in base$_2$ to describe the same printing technique. Those differences are explained by the different professional specialties of the first annotators. Despite those differences, each base is homogeneous, which is what matters for further forensic analysis. Data quality is thus maintained thanks to the guidance of suggestions. However, the first stumbles of User$_1$ in the first two documents and the vocabulary differences between the two knowledge bases point to the importance of the base initialization.

Results and Interpretation: System Usability. The usability rated by annotators tends to rise with the number of guidance documents in the base, going from an average of 30 during session 1 to an average of 77.5 during session 3. As the number of documents rises, the system makes more accurate suggestions to annotators.

3.3 Discussion

In both experiments, FORMULIS helped annotators to maintain homogeneity in the data structures and values of the knowledge base. This homogeneity ensures that the produced data can be easily used and shared. The first experiment showed that with the same set-up, FORMULIS could be used more easily and with better results than WebProtégé, by users with little to no training. While WebProtégé is not intended for layman annotators, the comparison with FORMULIS shows that our approach of a generic and intelligent data-driven interface is valid.

As illustrated by the errors of the first user in the ID document experiment and the problems with the quantity of ingredients in the recipe experiment, FORMULIS suffer from the problem known as the "cold start problem" in recommendation systems [9].

4 Related Work

In the Semantic Web community, a great deal of effort has been made to produce knowledge. Great quantities of data are extracted from existing sources such as databases, CSV files or web pages. However, to extract new data with no digital sources, it is necessary to use manual acquisition methods, such as FORMULIS. In the following, we discuss other methods of manual acquisition through forms with respect to data quality and usability.

RDF editors aim at making it easier to create RDF data. To that end, they are often coupled with a data browser presenting the current content of the base. Users have to understand the general structure of the base and its vocabulary before editing its content. Wikidata [5] uses a wiki-like interface for crowd-sourced data edition. Users can browse content as they would in a classical wiki, they directly edit RDF-like data by defining properties around a concept. Edition is done through forms allowing the creation of new values and new properties. However, forms cannot be nested, and there are no suggestions for values. Auto-completion helps to select an existing resource but it does not even take into account the field being filled. The quality of Wikidata is maintained by the constant corrections of contributors and by regular inspections from administrators. Semantic Wikis [3,8], another kind of wiki-based editing systems, enable to enhance fulltext wiki pages with semantic annotations, and therefore enable the generation of RDF documents from those pages. The main difficulty is that a special syntax has to be learned for the semantic annotations. Semantic Wikis are most useful when the main data is textual, and when the RDF data is used as metadata. OntoWiki [6] takes the opposite approach to Semantic Wikis by using wiki pages to browse and create RDF instead of extract RDF data from them. OntoWiki aims at easing RDF data authoring by providing the accessibility of wiki pages. It is adapted to cases where there are numerous users for one particular task of knowledge engineering in a distributed environment. Contrary to FORMULIS, OntoWiki does not provide guidance during edition, only basic template and relies on the user-based wiki mechanisms to maintain data quality. WebProtégé [10] is a web-form-based ontology editor. As seen in the cooking recipe experiment, WebProtégé is more adapted to users with Semantic Web training. ActiveRaUL [4] is a web form generator for RDF authoring at instance level. This system generates edition forms from a description made by an expert with the RaUL ontology, describing the various form controls (textboxes, radio buttons, etc.) associated with the elements of the edited ontology. ActiveRaUL has been compared to WebProtégé in an experiment with twelve users with various background regarding the Semantic Web. The experiment showed that a web form-based interface was familiar enough to users to create RDF data correctly, faster and more easily than with WebProtégé. The quality of the edited data is maintained by the constraints defined by experts.

5   Conclusion

In this paper, we have proposed an approach for guided RDF authoring, FORMULIS[2]. This system allows to deal with the flexibility of the RDF data model without requiring supervision by Semantic Web experts. FORMULIS manages at the same time: (i) a user-friendly way to create RDF graphs without Semantic Web training thanks to forms, (ii) a powerful expression tool thanks to the nested forms and the possibility to add new fields at any time, as well as (iii) a way to maintain the homogeneity of values in a knowledge base thanks to refined and dynamic suggestions.

---

[2] Available at http://servolis.irisa.fr:8080/formulis/.

We have conducted two user experiments. One compared FORMULIS to WebProtégé in a controlled experiment with layman users describing cooking recipes on both systems. The other experiment evaluated in a real setting the use of FORMULIS for domain expert users, forged ID experts from IRCGN, to create domain specific knowledge with partially fixed vocabulary. Both experiments showed that even without the supervision of a Semantic Web expert, the system guided users to create good quality data, both well-structured and homogeneous.

Acknowledgement. This work is supported by ANR project IDFRAud (ANR-14-CE28-0012). We thank our partners of project IDFRAud, as well as the experts from IRCGN and the fourteen volunteers who participated in our experiments.

References

1. Berners-Lee, T., Hendler, J., Lassila, O., et al.: The semantic web. Sci. Am. **284**(5), 28–37 (2001)
2. Brooke, J., et al.: SUS-A quick and dirty usability scale. Usability Eval. Ind. **189**(194), 4–7 (1996)
3. Buffa, M., Gandon, F., Ereteo, G., Sander, P., Faron, C.: SweetWiki: a semantic wiki. Web Semant.: Sci. Serv. Agents World Wide Web **6**(1), 84–97 (2008)
4. Butt, A.S., Haller, A., Liu, S., Xie, L.: ActiveRaUL: a web form-based user interface to create and maintain RDF data. In: International Semantic Web Conference Posters & Demos, pp. 117–120 (2013)
5. Erxleben, F., Günther, M., Krötzsch, M., Mendez, J., Vrandečić, D.: Introducing wikidata to the linked data web. In: Mika, P., et al. (eds.) ISWC 2014. LNCS, vol. 8796, pp. 50–65. Springer, Cham (2014). doi:10.1007/978-3-319-11964-9_4
6. Frischmuth, P., Martin, M., Tramp, S., Riechert, T., Auer, S.: OntoWiki - an authoring, publication and visualization interface for the data web. Semant. Web J. **6**(3), 215–240 (2015)
7. Hermann, A., Ferré, S., Ducassé, M.: An interactive guidance process supporting consistent updates of RDFS graphs. In: Teije, A., Völker, J., Handschuh, S., Stuckenschmidt, H., d'Acquin, M., Nikolov, A., Aussenac-Gilles, N., Hernandez, N. (eds.) EKAW 2012. LNCS, vol. 7603, pp. 185–199. Springer, Heidelberg (2012). doi:10.1007/978-3-642-33876-2_18
8. Krötzsch, M., Vrandečić, D., Völkel, M., Haller, H., Studer, R.: Semantic wikipedia. Web Semant.: Sci. Serv. Agents World Wide Web **5**(4), 251–261 (2007)
9. Schein, A.I., Popescul, A., Ungar, L.H., Pennock, D.M.: Methods and metrics for cold-start recommendations. In: Annual International ACM SIGIR Confernce Research and Development in Information Retrieval (2002)
10. Tudorache, T., Nyulas, C., Noy, N.F., Musen, M.A.: WebProtégé: a collaborative ontology editor and knowledge acquisition tool for the web. Semant. Web J. **4**(1), 89–99 (2013)

Graph Matching

A Graph Matching Based Method for Dynamic Passenger-Centered Ridesharing

Jia Shi[1], Yifeng Luo[1], Shuigeng Zhou[1(✉)], and Jihong Guan[2]

[1] School of Computer Science, and Shanghai Key Lab of Intelligent Information
Processing, Fudan University, Shanghai 200433, China
{shij15,luoyf,sgzhou}@fudan.edu.cn
[2] Department of Computer Science and Technology, Tongji University,
Shanghai 201804, China
jhguan@tongji.edu.cn

Abstract. Ridesharing is one transportation service deeply influenced
by the prosperity of Mobile Internet. Existing work focuses on passenger-
vehicle matching, which considers how to optimally dispatch passengers to appropriate vehicles. While dynamic passenger-passenger matching addresses how to optimally handle continually-arriving requests for
ridesharing from passengers, without considering vehicles. It is a kind
of dynamic passenger-centered ridesharing that has not been studied
enough. This paper studies dynamic passenger-centered ridesharing with
both temporal and cost constraints. We first propose a ridesharing
request matching method based on maximum weighted matching on
undirected weighted graphs, aiming to minimize the overall travel distance of targeting passengers. We then devise a distance indexing strategy to prune unnecessary calculations to accelerate ridesharing request
matching and reduce request response time. Experiments on real-life
road networks indicate that our method can successfully match 90% of
ridesharing requests while saving 23% to 35% of travel distance.

Keywords: Intelligent transportation system · Road networks ·
Ridesharing · Graph matching

1 Introduction

Nowadays transportation is deeply influenced by sharing economy, which treats
vehicles as a kind of shared resource and subsequently fosters a variety of sharing-
economy transportation services to facilitate traveling and daily-life of ordinary
people. Ridesharing [4,5] is one typical sharing-economy transportation service,
which rides at least two passengers traveling with roughly similar route together
on the same vehicle to save fare for passengers. What is more, ridesharing is
an environment-friendly service that can reduce fossil energy consumption and
alleviate traffic congestion. Consequently, in recent years ridesharing has been
accepted by more and more car owners and become popular in ordinary people.

© Springer International Publishing AG 2017
D. Benslimane et al. (Eds.): DEXA 2017, Part I, LNCS 10438, pp. 49–64, 2017.
DOI: 10.1007/978-3-319-64468-4_4

Generally, different ridesharing problems are the variants of the Dial a Ride Problem (DARP) [3], which aim to deliver a number of passengers to various destinations economically or time-efficiently, where passengers may join the to-be-delivered queue dynamically. These problems can be divided into two categories: ridesharing between vehicle owners (drivers) and passengers (riders), and ridesharing between passengers only. The former focuses on matching passengers' ad hoc requests with moving vehicles [17,22], where a moving vehicle on its way delivering on-board passengers may be assigned with a new passenger heading for an additional destination, as long as there are empty seats available on the vehicle. While the latter considers only passengers and assumes that there are multiple platforms or services for people to choose vehicles, thus it need only to consider how to match and coordinate people heading with roughly similar route. So we call it passenger-centered rdiesharing in this paper.

Existing studies either focus on designing efficient algorithms and schemes to facilitate ridesharing request matchings in practical applications without global optimization goals, or pay attention to achieving some specific optimization goals without considering the particular requirements of practical applications. In this paper, we address dynamic passenger-centered ridesharing, while considering both global optimization goal and efficient request matching strategies for practical mobile internet applications. On the one hand, We develop a maximum weighted matching algorithm to optimize ridesharing request matchings with specific optimization goal. On the other hand, to boost the processing efficiency, we build indexes over road networks by pre-computing the distances between all the marked locations and a set of selected landmarks, and employ the built indexes to estimate the distance lower bounds of request pairs. With the indexes, we can substantially eliminate unnecessary matching calculations and thus improve matching efficiency.

The rest of this paper is organized as follows: we first introduce some basic concepts and the problem statement in Sect. 2, then present the techniques and algorithms in Sect. 3. Following that, we provide experimental evaluation in Sect. 4, and review the related work in Sect. 5. We finally conclude the paper in Sect. 6.

2   Concepts and Problem Statement

Request: A ridesharing *request* q is a passenger's request for a ridesharing submitted at time $q.t$, with a pickup location $q.o$, a destination location $q.d$, the earliest departure time $q.ot$, and the latest arrival time $q.dt$. The passenger could only depart after the earliest departure time and should arrive at the destination location no later than the latest arrival time. We denote the original shortest distance between the pickup location $q.o$ and destination location $q.d$ as $q.s$.

Paid Distance: The *paid distance ps* is the distance paid by a passenger sharing the ride with other passengers. The cost a passenger should pay for a shared ride depends on the combined distance of traveling alone and ridesharing with other passengers. Assuming two passengers are scheduled to share a ride, with two ridesharing requests q_1 and q_2 respectively, and the shared ride takes the following planned route: $q_1.o \rightarrow q_2.o \rightarrow q_1.d \rightarrow q_2.d$. We denote the distance between $q_1.o$ and $q_2.o$ as d_1, the distance between $q_2.o$ and $q_1.d$ as d_2, and the distance between $q_1.d$ and $q_2.d$ as d_3, the paid distance ps_1 of q_1 would be $d_1 + d_2/2$ and the paid distance ps_2 of q_2 would be $d_2/2 + d_3$, if the two passengers evenly share the cost incurred between $q_2.o$ and $q_1.d$.

Match: A *match* is a possible ridesharing schedule for two ridesharing requests q_1 and q_2, as long as all temporal and cost constraints for the two requests could be satisfied.

Waiting Time: The *waiting time wt* of a request is the interval between the time it is submitted and the time it receives a response. Given a request q submitted at time $q.t$, if it receives a response at time T, the waiting time of the request is $T - q.t$. Of course, it is possible that no matches could be found for the request till it expires at the time T', then the request receives a response with no matches, and the waiting time of the request will be $T' - q.t$.

Reverse Pair: Given two requests q_1 and q_2, if q_1 is submitted earlier than q_2, while q_2 is served earlier than q_1, we call (q_1, q_2) a reverse pair.

Live Request: Suppose the current timestamp is T, if the following inequality holds for a request q:

$$T + (time\ cost\ from\ q.o\ to\ q.d\ without\ ridesharing) \leq q.dt,$$

q is assumed as a live request. Otherwise it's not a live request, and no further endeavor should be devoted to finding matches for it, as the passenger can not arrive at his destination on time even he takes his ride alone instead of sharing the ride with other passengers.

In this paper, we define the ridesharing problem as follows: given a sequence of requests $S = \{q_1, q_2, ..., q_n\}$, to compute matches for these requests, and maximize the overall saved paid distance, namely:

$$Max \sum_{q_i \in S} (q_i.s - ps_i),$$

subjected to the temporal and cost constraints of passengers. In the meanwhile, we aim to reduce the number of reverse pairs and the average waiting time of requests, barely decreasing the saved distance.

3   Techniques and Algorithms

In this section, we first introduce distance indexing and estimation, then present the techniques for checking the match feasibility of two ridesharing requests, finally we give the algorithms for optimizing the ridesharing problem.

3.1 Distance Indexing and Distance Estimation

It is time-consuming to compute the shortest path between two spatial points in a road network, while we need to do multiple shortest path computations each time the match feasibility of two ridesharing requests is checked. Thus it is essential to build necessary indexes over the road network so as to improve the efficiency of match feasibility checks.

We index a road network via pre-computing the shortest distances of each marked locations to some selected landmark locations, and then employ the pre-computed distances to estimate the lower bound of the actual distance between the locations contained in two ridesharing requests to prune obviously unfeasible matches. Specifically, we select k spatial points on the road network as the landmarks, which are denoted as $L = \{l_1, l_2, l_3, ..., l_{k-1}, l_k\}$, and then compute the distances from all marked locations on the road network to the selected k landmarks, all these computed point-landmark distances are recorded as indexes. Then we employ the triangle inequality to estimate the lower bound of the actual distance of the possible routes that we need to compute when checking the match feasibility of two ridesharing requests.

For a possible route $s \rightarrow t$ from the source point s to the destination point t, its distance lower bound could be estimated with the pre-computed point-landmark distances, just as Fig. 1 shows. We can infer from Fig. 1 that:

$$Dist_{s \rightarrow t} \geq Dist_{s \rightarrow l_i} - Dist_{t \rightarrow l_i}. \tag{1}$$

Thus we can get the following inequality:

$$Dist_{s \rightarrow t} \geq \max_{l_i \in L}\{Dist_{s \rightarrow l_i} - Dist_{t \rightarrow l_i}\}, \tag{2}$$

and we can estimate the lower bound of the distance from s to t as follows:

$$MinDist_{s \rightarrow t} = \max_{l_i \in L}\{Dist_{s \rightarrow l_i} - Dist_{t \rightarrow l_i}\} \tag{3}$$

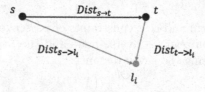

Fig. 1. Triangle inequality.

3.2   Match Feasibility Check

Given two requests q_1 and q_2, there exist four possible ridesharing routes:

- $route_1$: $q_1.o \rightarrow q_2.o \rightarrow q_1.d \rightarrow q_2.d$
- $route_2$: $q_1.o \rightarrow q_2.o \rightarrow q_2.d \rightarrow q_1.d$
- $route_3$: $q_2.o \rightarrow q_1.o \rightarrow q_2.d \rightarrow q_1.d$
- $route_4$: $q_2.o \rightarrow q_1.o \rightarrow q_1.d \rightarrow q_2.d$

In order to check whether two requests q_1 and q_2 can form a match, we need to check whether the time and cost constraints of the two requests can be satisfied in at least one possible ridesharing route: the passengers can arrive at their destinations before the latest arrival time, and each passenger's paid distance with the ridesharing trip is less than his/her original distance. If the time and cost constraints are satisfied from more than one possible ridesharing routes, the two requests form a match.

When checking the match feasibility of two requests q_1 and q_2, the shortest distances of the six sub-routes namely $q_1.o \rightarrow q_2.o$, $q_2.o \rightarrow q_1.o$, $q_1.d \rightarrow q_2.d$, $q_2.d \rightarrow q_1.d$, $q_2.o \rightarrow q_1.d$, $q_1.o \rightarrow q_2.d$ should be computed. From Subsect. 3.1, we can estimate the distance lower bound of the six sub-routes using the distance index, and then check whether two ridesharing requests can match with the estimated overall distance lower bound. If we can judge with the estimated lower bound that either the temporal constraint or the cost constraint could not be satisfied even with the distance lower bound, then the match feasibility of the two ridesharing requests could be ruled out without calculating the accurate distances. Otherwise, we need to continue to calculate the actual distances and further check the match feasibility of the two ridesharing requests using the accurate distances.

Since $Dist_{s \rightarrow l_i}$ and $Dist_{t \rightarrow l_i}$ are pre-computed and indexed, the time-complexity for calculating the distance lower bound for route $s \rightarrow t$ is $O(k)$. Since the number of selected landmarks k is not very large, we spend much less time in estimating the distance lower bounds than calculating the accurate shortest distances for massive number of request pairs, that are clearly not matchable with each other.

3.3   Maximum Weighted Matching Based Ridesharing (MWMR)

Here, we propose an algorithm based on maximum weighted matching to solve the ridesharing problem. For the static ridesharing problem, we employ the maximum weighted matching in an undirected weighted graph to achieve a global optimization result. For the dynamic ridesharing problem, we extend the maximum weighted matching algorithm by dividing the ridesharing service time into continuous equal intervals and treat each interval as an instance of static ridesharing problem.

Static Ridesharing. In the static ridesharing problem, we suppose all requests are known in advance, and we focus on how to maximize the objective function of the static ridesharing problem. Suppose we have a set of ridesharing requests $N = \{q_1, q_2, ..., q_{n-1}, q_n\}$, and a matching graph MG built for these requests, we can prove that the solution of the maximum weighted matching problem in graph MG is the answer of maximizing the objective function of the static ridesharing problem. In graph theory, a *matching* is a subset of edges such that none of the selected edges share a common vertex. With respect to a weighted graph, a *maximum weighted matching* is a matching which has the maximal overall weight of contained edges. In Fig. 2, a maximum weighted matching with weight 15 can be found by pairing vertex b to vertex c and vertex d to vertex e.

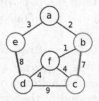

Fig. 2. Maximum weighted matching.

Given a set of requests $N = \{q_1, q_2, ..., q_{n-1}, q_n\}$, we can build an undirected weighted graph based on N, and we call the built graph as the matching graph (MG). The pseudo-code of building the matching graph is described in Algorithm 1. For each request q_i in N, we create a node v_i for it (Line 5). For a request pair q_i and q_j, we determine whether they match with each other using our match feasibility check algorithm (Line 9). If a match is determined, we create an undirected edge e_{ij} between them (Line 10–11), and we use the following equation to calculate the weight w_{ij} of the edge:

$$w_{ij} = q_i.s + q_j.s - \min_{1 \leq k \leq 4} \left\{ \sum_{1 \leq l \leq 3} d_{kl} \right\}, \tag{4}$$

where d_{kl} denotes the distance of the lth part of the path in $route_k$ described in Subsect. 3.2.

We denote the selected edge set from MG as $SE = \{e_{k_1}, e_{k_2}, ..., e_{k_{m-1}}, e_{k_m}\}$, where m is the number of the selected edges. Recall that we want to maximize the objective function:

$$\sum_{q_i \in N} (q_i.s - ps_i),$$

a request q_k has no contribution to the objective function if it does not match with any other request, where we have $q_k.s - ps_k = 0$, and if request q_i matches with q_j in our result, their contribution to the objective function is

$$c_{ij} = q_i.s + q_j.s - ps_i - ps_j \tag{5}$$

Algorithm 1. Matching Graph Construction

Input: a set of requests N
Output: a weighted graph MG generated from N
1: **function** CONSTRUCTGRAPH(N)
2:     $V \leftarrow \emptyset$
3:     $E \leftarrow \emptyset$
4:     **for all** $q_i \in N$ **do**
5:         $v_i \leftarrow$ node created for q_i
6:         $V \leftarrow V \cup \{v_i\}$
7:     **end for**
8:     **for all** $(q_i, q_j) \in N \times N$ **do**
9:         **if** $i < j$ and MATCHCHECK(q_i, q_j) **then**
10:             $w_{ij} \leftarrow q_i.s + q_j.s - \min_{1 \le k \le 4}\{\sum_{1 \le l \le 3} d_{kl}\}$
11:             $E \leftarrow E \cup \{(v_i, v_j, w_{ij})\}$
12:         **end if**
13:     **end for**
14:     **return** $MG = (V, E)$
15: **end function**

We know from the definition of ps that:

$$ps_i + ps_j = \min_{1 \le k \le 4}\{\sum_{1 \le l \le 3} d_{kl}\}, \qquad (6)$$

we can rewrite Eq. (5) to:

$$c_{ij} = q_i.s + q_j.s - \min_{1 \le k \le 4}\{\sum_{1 \le l \le 3} d_{kl}\} \qquad (7)$$

Putting Eqs. (4) and (7) together, we can find that: if requests q_i and q_j form a ridesharing match, their contribution to the objective function is the weight of the edge between the nodes v_i and v_j in MG. We denote $x_{ij} = 1$ if requests q_i and q_j form a ridesharing match in the ridesharing result, $x_{ij} = 0$ otherwise. That is to say, we want to maximize:

$$W = \sum_{q_i \in N, q_j \in N, i < j} x_{ij} c_{ij}. \qquad (8)$$

Since a request can at most match with one request, we have the following constraint:

$$\sum_{q_j \in N, i < j} x_{ij} + \sum_{q_j \in N, j < i} x_{ji} \le 1, \forall q_i \in N \qquad (9)$$

So solving the maximum weighted matching problem for MG is equivalent to solving the ridesharing problem for N, where selecting the edge e_{ij} in MG means that requests q_i and q_j form a ridesharing match in the ridesharing result, and W is the maximal saved distance. The maximum weighted matching problem can be solved in polynomial time [9], and has efficient algorithms on sparse graphs [12].

Algorithm 2. Maximum Weighted Matching Algorithm

Input: a stream of requests S
Output: the ridesharing result
 1: **function** MWMR(S)
 2:    **if** a new time interval $[t_{i-1}, t_i)$ ends **then**
 3:       **for all** q_k which arrived in $[t_{i-1}, t_i)$ **do**
 4:          $N \leftarrow N \cup \{q_k\}$
 5:       **end for**
 6:       **for all** q_k which is live and has not been served **do**
 7:          $N \leftarrow N \cup \{q_k\}$
 8:       **end for**
 9:       $MG \leftarrow$ CONSTRUCTGRAPH(N)
10:       $result \leftarrow$ MAXWEIGHTEDMATCHING(MG)
11:    **end if**
12:    **return** $result$
13: **end function**

Dynamic Ridesharing. In this section, we extend our method to solve the dynamic ridesharing problem. Different from the static ridesharing problem, requests can arrive continually at any time in the dynamic ridesharing problem, and thus the ridesharing requests are not known in advance. Our idea is to divide the ridesharing service time into a sequence of continuous equal time intervals: $[t_0, t_1), [t_1, t_2), [t_2, t_3), ..., [t_{m-1}, t_m)$, and treat each interval as an instance of static ridesharing problem, where newly submitted requests in an interval, together with remaining live requests from preceding intervals, form the request set N. Algorithm 2 presents the pseudo-code for solving the dynamic ridesharing problem.

3.4   Discussion

In the last few subsections, we have theoretically discussed how to solve the dynamic ridesharing problem to maximize the total saved distance. While there are some important reality factors we should take into consideration, including request waiting time, ridesharing fairness and time interval granularity. Reducing request waiting time and improving ridesharing fairness could directly improve users' experience. Request waiting time denotes the time it takes for a request to receive its ridesharing result. For a request q, its waiting time is $T - q.t$ if we can find a match for it at time T, and its waiting time is $T' - q.t$ if we cannot find a match for it until it is no longer a live request at time T'. Intuitively, earlier requests should receive their ridesharing results earlier. If a request with an earlier birth time receives it ridesharing result later than a request with a later birth time, we say that the two requests form a reverse pair, and we use the number of reverse pairs to represent the ridesharing fairness. Thus the fewer reverse pairs we have, the higher the degree of fairness is.

Regarding how to reduce the waiting time and improve ridesharing fairness, we dynamically modify the weight of edges in the matching graph MG, letting edges related to requests with longer waiting time have higher weights, and thus have higher priority to be selected for ridesharing. If the current timestamp is T, we rewrite Eq. (4) as follows:

$$w_{ij} = q_i.s + q_j.s - \min_{1 \leq k \leq 4} \{ \sum_{1 \leq l \leq 3} d_{kl} \} + \alpha[(T - q_i.t) + (T - q_j.t)] \quad (10)$$

We use α to designate the importance of reducing the waiting time and improving ridesharing fairness when modifying the weight of edges, the higher the α value is, the more important we think reducing the waiting time and improving ridesharing fairness are. Our experimental results show that this simple modification could reduce the waiting time and improve ridesharing fairness, while hardly decrease the saved distance.

Obviously, the time interval granularity at which the ridesharing service time is divided into intervals has huge impact on the final ridesharing results. If the time interval is too short, the number of requests falling in an interval would be moderate, and the final saved distance would not be impressive; if the time interval is too long, increased saved distance could be achieved, while requests would acquire extended waiting/response time. Thus it is important to tune the time granularity and to find a good compromise for end users, to balance their demands on economic cost and temporal cost.

4   Performance Evaluation

In this section, we present and analyze the experimental results of evaluating the proposed maximum weighted matching based ridesharing algorithm. Since there are few works on dynamic ridessharing which consider passengers only, for comparison, we implement a simple streaming matching (SM) algorithm for the dynamic ridesharing problem, which does not take the optimization goal into consideration. In SM algorithm, each time a new ridesharing request is submitted, the streaming matching algorithm matches the newly-submitted request with the earliest feasible request within the remaining live requests.

4.1   Datasets

We choose two international metropolis, Beijing and New York, to evaluate our ridesharing algorithms, as they have high population density and are suitable for promoting the ridesharing services. We use the data extracted from the Open Street Map (OSM)[1] to create the road networks of Beijing and New York. The graphs of the road networks of Beijing and New York respectively contain $234,051$ nodes, $252,253$ edges and $949,803$ nodes, $941,396$ edges. One thing worth noting is that the graph of New York contains more nodes and edges than that of Beijing, the main reason is that more people are using the OSM map of New York, and thus they make more contribution to the New York road map. In

---

[1] http://www.openstreetmap.org/.

our experiments, we assume vehicles travel in the road networks at the constant speed of 40 km/h.

For Beijing, we generate 103, 461 ridesharing requests spanning across a whole day as our evaluation ridesharing requests from the trajectory dataset, which contains GPS trajectory records collected from over 33,000 taxis in Beijing [17]; for New York, we generate our evaluation ridesharing requests from the four years' taxi operations in New York [7], where 99, 818 trip announcements were generated per day. In each request q, the earliest pickup time $q.ot$ is set to the submission time of the request $q.t$. As the latest arrival time of a request q is related with the request's earliest pickup time and distance, we generate the request's latest arrival time $q.dt$ according to the following equation:

$$q.dt = q.ot + q.s/speed + \Delta \qquad (11)$$

Here we set $speed = 40$ km/h and $\Delta = 20$ min. Obviously, if the request's submission time is late or its origin distance is long, the latest arrival time of the request will be late accordingly.

4.2   Experimental Results

We evaluate ridesharing algorithms with the performance metrics listed in Table 1.

In Table 1, SR denotes the proportion of satisfied ridesharing requests, $PDSR$ and $PDSRSR$ measure the performance of our algorithm in saving travel distance. AWT implies how long it takes for a user to wait for his/her ridesharing response on average, and RPR shows the fairness of the ridesharing algorithm. $ADCSR$ is proposed to measure the performance of our distance index and pruning strategy, which denotes the proportion of the saved actual distance calculation.

We compare MWMR algorithm's SR, $PDSR$ and $PDSRSR$, AWT and RPR metrics with the SM algorithm. For the dynamic MWMR ridesharing, we set four different time interval durations to 60 s, 120 s, 300 s and 600 s, which are represented as MWMR60, MWMR120, MWMR300 and MWMR600 respectively. The comparison results for various evaluation metrics are shown in Fig. 3. We can see that a relatively higher satisfaction rate is achieved from the SM algorithm, but the achieved distance save rates ($PDSR$ and $PDSRSR$) are much lower than the MWMR algorithm. With different time interval durations, the MWMR algorithm achieves higher satisfaction rates and lower distance save rates as the time interval duration becomes shorter. This is because we run the matching algorithms more frequently as the time interval durations become shorter, leaving the requests having more chances to be matched before it expires. Meanwhile, the MWMR algorithm has fewer requests for ridesharing matching during each interval if the interval duration is short, leaving lower paid distance save rates achieved. As we mainly want to achieve optimized paid distances for passengers, we take the paid distance save rate as the most significant measurement metric, and consider that the MWMR algorithm overtakes the SM

Table 1. Performance metrics.

Performance metric	Definition	Description
Satisfaction rate (SR)	$SR = NSR/NR$	NSR denotes the number of ridesharing requests successfully matched, NR denotes the overall number of requests
Paid distance save rate ($PDSR$)	$PDSR = 1 - D_{ps}/D_s$	D_s denotes the overall original distance of all the requests without ridesharing, D_{ps} denotes the total paid distance of all the requests
Paid distance save rate for satisfied requests ($PDSRSR$)	$PDSRSR = 1 - D'_{ps}/D'_s$	D'_s denotes the overall original distance of the successfully matched requests, D'_{ps} denotes the total paid distance of the successfully matched requests
Average waiting time (AWT)	$AWT = WT/NR$	WT denotes the total waiting time of all the requests
Reverse pair rate (RPR)	$RPR = \frac{NRP}{NR(NR-1)/2}$	NRP denoting the number of reverse pairs
Actual distance calculation save rate ($ADCSR$)	$ADCSR = 1 - ADC/EDC$	EDC denotes the times of estimated distance calculation, ADC denotes the times of accurate distance calculation

algorithm by a large margin. We can also see that the SM algorithm achieves shorter average waiting time and a lower reverse pair rate, as it runs ridesharing matchings more frequently and matches the new request with the remaining live requests according to their submission time.

To evaluate the effectiveness of calculation pruning via the distance indexing technique, we benchmark the actual distance calculation save rate via measuring the times of estimation distance calculation and the times of actual distance calculation. It is evident that landmark selection affects the effectiveness of calculation pruning. Selecting too many landmarks would make computing the lower bound of shortest distances too expensive, while selecting too few landmarks would make estimation of the lower bound inaccurate. We randomly select nodes with high degrees as our landmarks, where 168 and 301 landmarks are selected for the road network of Beijing and New York respectively, Fig. 4 shows the benchmark results. We can see that $ADCSR$ is higher than 0.95 with various time interval durations, which means more than 95% of shortest distance calculation can be reduced for our MWMR algorithm via the distance indexing technique. As the percentage of the landmarks compared to the road network size of Beijing is higher than that of New York, slightly better calculation pruning effect is achieved on the Beijing road network.

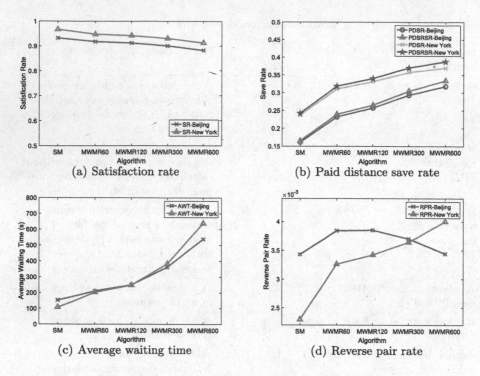

(a) Satisfaction rate

(b) Paid distance save rate

(c) Average waiting time

(d) Reverse pair rate

Fig. 3. Performance results in Beijing and New York.

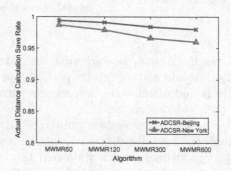

Fig. 4. The actual distance calculation save rate of MWMR with $\alpha = 0$.

We also evaluate the impact of α introduced in Eq. 10 on various evaluation metrics for the MWMR algorithm, where the time interval duration is set to 300 s, the results are shown in Fig. 5. We can see that changing the value of α has little impact on the satisfaction rate and the paid distance save rate, while increasing α can effectively reduce the waiting time and improve the ridesharing fairness.

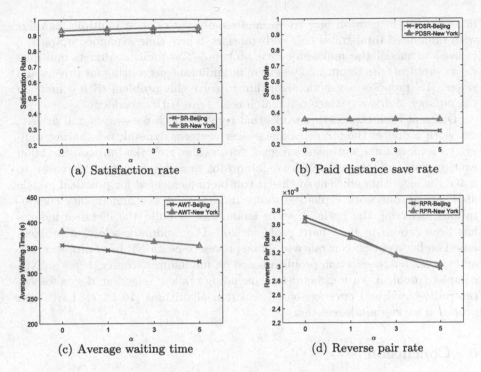

Fig. 5. Performance results for different α values in Beijing and New York.

5   Related Work

As ridesharing is becoming more and more popular, a number of ridesharing frameworks and algorithms have been proposed [1,11]. The ridesharing problems can be categorized into four types: single-driver-single-rider, single-driver-multiple-riders, multiple-drivers-single-rider and multiple-drivers-multiple-riders.

The single-driver-single-rider problem focuses on determining the best driver (rider) among a set of drivers (riders) to share his trip for a given rider (driver), with various optimization objectives, such as minimizing overall detour [13] or overall vehicle kilometers [2]. The single-driver-multiple-riders problem considers how to optimally deliver a number of riders by the same car to their destinations, aiming to achieve minimized overall travel time, travel distance and maximized number of ridesharing matches. Various algorithms and methods including exact and heuristic methods based on integer programming [4], local search based heuristic method [5,6], and genetic algorithm based models [16], are proposed to solve the ridesharing optimization problems. Besides, [6] leverages social network to extract home/work locations and studies the possibility of rider-passenger ridesharing using social relationships. [8] considers each passenger's satisfaction degree in payment, travel time and waiting time. In the multiple-drivers-single-rider problem, riders may be transferred from one vehicle to another on the road network. [14] focuses on how to optimally deliver a rider in multiple transfers with minimized total travel cost and transfers, where time-expanded graph [19] is used to model the multi-objective problem. The multiple-drivers-multiple-riders problem tries to optimally determine simultaneous routing for drivers and riders. [15] proposes a genetic algorithm to solve this problem with a limit on the number of drivers, which can match their trips with two riders.

Different from the existing works that consider both passengers and drivers, our work addresses the problem of passenger-centered dynamic ridesharing, aiming to achieve some optimization goal. Meanwhile, we take into consideration waiting time and fairness, which are important to user experience. Moreover, to make the algorithm efficient so that it can be implemented for practical mobile applications, our work exploits distance indexes to prune unnecessary computations. Selecting the optimal set of landmarks to build the distance indexes has been proven to be a hard problem [20]. The landmark selection problem based on betweenness centrality could be proven to be an NP-hard problem [18], and the landmark selection problem based on minimum K-center [10] is an NP-complete problem. Various heuristics, including random selection, degree-based, centrality-based and coverage-based selection algorithms [10,18,21] have been proposed for landmark selection.

6   Conclusion

In this paper, we define the passenger-centered dynamic ridesharing problem and propose a method based on maximum weighted graph matching to compute the optimal ridesharing matching. In order to improve processing efficiency, we build distance indexes over the road network by pre-computing the distances between all the marked locations and selected landmarks, and then employ the built indexes to estimate the distance lower bounds of ridesharing request pairs, so that unnecessary matching computations are substantially eliminated. Experiments on real-world datasets validate the proposed method.

Acknowledgement. This work was supported by the Key Projects of Fundamental Research Program of Science and Technology Commission of Shanghai Municipality (STCSM) (No. 14JC1400300), Program of Science and Technology Innovation Action of STCSM (No. 17511105204), and China Postdoctoral Science Foundation. Jihong Guan was supported by NSFC (No. 61373036) and the Program of Shanghai Subject Chief Scientist (No. 15XD1503600).

References

1. Agatz, N., Erera, A., Savelsbergh, M., Wang, X.: Optimization for dynamic ridesharing: a review. Eur. J. Oper. Res. **223**(2), 295–303 (2012)
2. Amey, A.: Proposed methodology for estimating rideshare viability within an organization: application to the mit community. In: Proceedings of Transportation Research Board Annual Meeting, pp. 1–16 (2011)

3. Attanasio, A., Cordeau, J.F., Ghiani, G., Laporte, G.: Parallel tabu search heuristics for the dynamic multi-vehicle dial-a-ride problem. Parallel Comput. **30**(3), 377–387 (2004)
4. Baldacci, R., Maniezzo, V., Mingozzi, A.: An exact method for the car pooling problem based on lagrangean column generation. Oper. Res. **52**(3), 422–439 (2004)
5. Calvo, R.W., De Luigi, F., Haastrup, P., Maniezzo, V.: A distributed geographic information system for the daily car pooling problem. Comput. Oper. Res. **31**(13), 2263–2278 (2004)
6. Cici, B., Markopoulou, A., Frias-Martinez, E., Laoutaris, N.: Assessing the potential of ride-sharing using mobile and social data: a tale of four cities. In: Proceedings of the ACM International Joint Conference on Pervasive and Ubiquitous Computing, pp. 201–211 (2014)
7. Donovan, B., Work, D.B.: Using coarse GPS data to quantify city-scale transportation system resilience to extreme events. Comput. Sci. (2015)
8. Duan, X., Jin, C., Wang, X., Zhou, A., Yue, K.: Real-time personalized taxi-sharing. In: Navathe, S.B., Wu, W., Shekhar, S., Du, X., Wang, X.S., Xiong, H. (eds.) DASFAA 2016. LNCS, vol. 9643, pp. 451–465. Springer, Cham (2016). doi:10.1007/978-3-319-32049-6_28
9. Edmonds, J.: Paths, trees, and flowers. Can. J. Math. **17**(3), 361–379 (2009)
10. Francis, P., Jamin, S., Jin, C., Jin, Y., Raz, D., Shavitt, Y., Zhang, L.: IDMaps: a global internet host distance estimation service. IEEE/ACM Trans. Netw. **9**(5), 525–540 (2001)
11. Furuhata, M., Dessouky, M., Ordez, F., Brunet, M.E., Wang, X., Koenig, S.: Ridesharing: the state-of-the-art and future directions. Transp. Res. Part B Methodol. **57**(57), 28–46 (2013)
12. Galil, Z.: Efficient algorithms for finding maximum matching in graphs. ACM Comput. Surv. (CSUR) **18**(1), 23–38 (1986)
13. Geisberger, R., Luxen, D., Neubauer, S., Sanders, P., Volker, L.: Fast detour computation for ride sharing. Comput. Sci. (2009)
14. Herbawi, W., Weber, M.: Evolutionary multiobjective route planning in dynamic multi-hop ridesharing. In: Merz, P., Hao, J.-K. (eds.) EvoCOP 2011. LNCS, vol. 6622, pp. 84–95. Springer, Heidelberg (2011). doi:10.1007/978-3-642-20364-0_8
15. Herbawi, W., Weber, M.: Modeling the multihop ridematching problem with time windows and solving it using genetic algorithms. In: Proceedings of the IEEE International Conference on Tools with Artificial Intelligence, pp. 89–96 (2012)
16. Herbawi, W., Weber, M.: The ridematching problem with time windows in dynamic ridesharing: a model and a genetic algorithm. In: Proceedings of the ACM Genetic and Evolutionary Computation Conference, pp. 1–8 (2012)
17. Ma, S., Zheng, Y., Wolfson, O.: T-share: a large-scale dynamic taxi ridesharing service. In: Proceedings of the IEEE International Conference on Data Engineering, pp. 410–421 (2013)
18. Potamias, M., Bonchi, F., Castillo, C., Gionis, A.: Fast shortest path distance estimation in large networks. In: Proceedings of the ACM Conference on Information and Knowledge Management, pp. 867–876 (2009)
19. Pyrga, E., Schulz, F., Wagner, D., Zaroliagis, C.: Efficient models for timetable information in public transportation systems. J. Exp. Algorithmics **12**(1), 2–4 (2008)
20. Qiao, M., Cheng, H., Chang, L., Yu, J.X.: Approximate shortest distance computing: a query-dependent local landmark scheme. IEEE Trans. Knowl. Data Eng. **26**(1), 55–68 (2013)

21. Qiao, M., Cheng, H., Yu, J.X.: Querying shortest path distance with bounded errors in large graphs. In: Bayard Cushing, J., French, J., Bowers, S. (eds.) SSDBM 2011. LNCS, vol. 6809, pp. 255–273. Springer, Heidelberg (2011). doi:10.1007/978-3-642-22351-8_16
22. Santos, D.O., Xavier, E.C.: Dynamic taxi and ridesharing: a framework and heuristics for the optimization problem. In: Proceedings of the International Joint Conference on Artificial Intelligence, pp. 2885–2891 (2013)

Answering Graph Pattern Matching Using Views: A Revisit

Xin Wang(✉)

Southwest Jiaotong University, Chengdu, China
xinwang@swjtu.cn

Abstract. This paper studies how to answer graph pattern matching defined in terms of subgraph isomorphism by using a set of materialized views. We first propose a notion of pattern containment to characterize graph pattern matching using graph pattern views, and show that graph pattern matching can be answered using a set of views if and only if the pattern query is contained by the views, and develop efficient algorithm to determine pattern containment. Based on this characterization, an efficient algorithm is developed to evaluate graph pattern matching using views. In addition, when a pattern query is not contained in a set of views, we study the problem of approximately answering graph pattern matching using views. We first study maximally contained (resp. containing) rewriting problems, develop algorithms to find such rewritings. We then propose techniques to find approximate answers using maximally contained (resp. containing) rewriting. Using real-life and synthetic data, we experimentally verify that these methods are able to efficiently conduct graph pattern matching on large social graphs.

1 Introduction

Answering graph pattern matching using views has already been studied in [10, 11], where pattern queries are defined in terms of graph simulation [16] and bounded simulation [9]. However, little efforts have been made on evaluating graph pattern matching by using views, for pattern queries defined in terms of subgraph isomorphism. While the need for studying the problem is evident, graph pattern matching via subgraph isomorphism has wide applications in, *e.g.,* social marketing [12], link predication [22], pattern discovery [19] and social group identification [13], and view based technique is shown to be an effective way for evaluating costly graph pattern matching on big social data.

Example 1. A fraction of a recommendation network is depicted as a graph G in Fig. 1 (a), where each node denotes a person with name and job title (*e.g.,* project manager (PM), database administrator (DBA), programmer (PRG), business analyst (BA) and software tester (ST)); and each edge indicates collaboration, *e.g.,* (Bob, Mat) indicates that Mat worked well with Bob on a project led by Bob.

© Springer International Publishing AG 2017
D. Benslimane et al. (Eds.): DEXA 2017, Part I, LNCS 10438, pp. 65–80, 2017.
DOI: 10.1007/978-3-319-64468-4_5

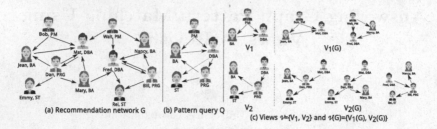

(a) Recommendation network G   (b) Pattern query Q   (c) Views $\mathcal{V}=\{V_1, V_2\}$ and $\mathcal{V}(G)=\{V_1(G), V_2(G)\}$

Fig. 1. Data graph, pattern query and views

To build a team for software development, one issues a pattern query Q depicted in Fig. 1(b) (without dotted line). The team members need to satisfy the following requirements: (1) with expertises: PM, DBA, PRG, BA and ST; (2) meeting the following collaborative experience: (i) DBA and BA worked well under the project manager PM; (ii) BA, DBA and PRG form a collaboration cycle [17], where DBA had supervised a PRG, and had been supervised by a BA, who collaborated well with PRG; and (iii) ST also collaborated well with PRG. When pattern query is defined in subgraph isomorphism [8], the matches Q(G) of Q in G includes a set of subgraphs {{(PM, Bob), (BA, Jean), (DBA, Mat), (PRG, Dan), (ST, Emmy)}, {(PM, Walt), (BA, Nancy), (DBA, Fred), (PRG, Bill), (ST, Rei)}}.

It is a daunting task to find matches of Q in G via subgraph isomorphism, since it takes $O(|G|!|G|)$ time to compute Q(G) [8], where $|G|$ is the size of G. While one can do better by leveraging a set of *views*. Suppose that a set of views $\mathcal{V} = \{V_1, V_2\}$ is defined, materialized and cached ($\mathcal{V}(G) = \{V_1(G), V_2(G)\}$), as shown in Fig. 1(c). As will be shown later, to compute Q(G), (1) we only need to visit $\mathcal{V}(G)$, *without* accessing the original big G; and (2) Q(G) can be efficiently computed by "merging" views in $\mathcal{V}(G)$. Indeed, $\mathcal{V}(G)$ already contains partial answers to Q in G. For example, as one subgraph of Q, the matches of V_1 (resp. V_2) are contained in $V_1(G)$ (resp. $V_2(G)$). These partial answers can be used to construct Q(G). Hence, the cost of computing Q(G) is dominated by $|Q|$ and $|\mathcal{V}(G)|$, where $|\mathcal{V}(G)|$ is the total size of matches, and often *much smaller than* $|G|$.   □

This example suggests that we find matches by capitalizing on views. To do this, several questions have to be settled. (1) How to decide whether a pattern query Q can be answered by a set $\mathcal{V}$ of views? (2) If so, how to compute Q(G) from $\mathcal{V}(G)$? (3) If not, how to find approximate answers to Q(G) with $\mathcal{V}(G)$?

Contributions. This paper investigates these questions for answering *graph pattern queries* using *graph pattern views*. We focus on pattern matching defined in terms of *subgraph isomorphism* [8], since it is (1) commonly used in social network analysis [26], and (2) computational expensive (NP-complete problem [8]).

(1) We propose a notion of *pattern containment* (Sect. 3) to characterize when graph pattern matching via subgraph isomorphism can be answered using views, by revising the notion defined for graph simulation and bounded simulation [10, 11]. Based on the characterization, we provide efficient algorithms

to determine pattern containment, and answer graph pattern matching by using views.

(2) When exact matches of a pattern query Q cannot be found by using available views $\mathcal{V}$, one can find another pattern query Q', which is close to Q and contained in $\mathcal{V}$, and use Q' to approximately answer Q. To this end, we study the problems of *maximally contained rewriting* and *maximally containing rewriting* (Sect. 4), develop one algorithm to approximately find the *aximally contained rewriting* Q_d with performance guarantee, and another algorithm to efficiently compute the *maximally containing rewriting* Q_g. Using Q_d and Q_g, we introduce techniques to find approximate matches of Q using available views.

(3) Using real-life graphs, we experimentally verify the effectiveness, efficiency and accuracy of our view-based matching method (Sect. 5). We find that our method is 11.5 times faster than conventional methods for pattern matching on *WebGraph* [4], a Web graph with 12.1 million nodes (web pages) and 103.6 million edges (hyperlinks). In addition, our matching algorithm scales well with the data size. We further find that our algorithm can compute maximally contained (resp. containing) rewriting Q_d (resp. Q_g) efficiently, and the query results of Q_d (resp. Q_g) on $\mathcal{V}(G)$ has accuracy of 0.77 (resp. 0.81) on average on *WebGraph*.

The work gives a full treatment for answering graph pattern matching using views, for pattern queries defined in terms of subgraph isomorphism. It provides techniques to efficiently compute matches using cached views, or find approximate matches of Q from views, by using a maximally contained (resp. containing) rewriting of Q. In contrast with earlier works, this work fills one critical void for view based answering for graph pattern matching via subgraph isomorphism, and yields a promising approach to querying "big" social data. All the proofs, algorithms and complexity analyses can be found in [2].

2   Preliminary

In this section, we review data graphs, pattern queries and graph pattern matching, and introduce the problem of answering graph pattern matching using views.

2.1   Graphs, Patterns and Graph Pattern Matching

We start with basic notations of graphs.

Data Graphs. A *data graph* is a node-labeled, directed graph $G = (V, E, L)$, where (1) V is a set of nodes; (2) $E \subseteq V \times V$, where $(v, v') \in E$ denotes a *directed* edge from node v to v'; and (3) $L(\cdot)$ is a function such that for each node v in V, $L(v)$ is a label from an alphabet Σ. Intuitively, $L(\cdot)$ specifies *e.g.*, social roles.

To simplify the discussion, we do not explicitly mention edge labels. Nonetheless, our techniques can be readily adapted for edge labels: for each labeled edge e, we can insert a "dummy" node to represent e, carrying e's label.

Pattern Queries. A *pattern query* is a directed graph $Q = (V_q, E_q, f_v)$, where (1) V_q is the set of *query nodes*, (2) E_q is the set of *query edges*, and (3) $f_v(\cdot)$ is a function defined on V_q such that for each node $u \in V_q$, $f_v(u)$ is a label in Σ.

Graph Pattern Matching [8]. A *match* of Q in G via *subgraph isomorphism* is a subgraph G_s of G that is isomorphic to Q, *i.e.*, there exists a *bijective function* h from V_q to the node set of G_s such that (1) for each node $u \in V_q$, $f_v(u) = L(v)$; (2) (u, u') is an edge in Q if and only if $(h(u), h(u'))$ is an edge in G_s.

We use the following notations. The match result of Q in G, denoted as $Q(G)$ is the set including all the matches G_s of Q in G. For a pattern edge $e = (u, u')$, we derive the set $S(e)$ from $Q(G)$ by letting $S(e) = \{(v, v') | v, v'$ can be mapped by h to u, u', respectively, and $(v, v') \in E\}$. A *maximum common subgraph* [7] of G_1 and G_2 is denoted by $\mathsf{mcs}(G_1, G_2)$. A pattern query Q' is a *subquery* of Q, denoted as $Q' \subseteq Q$, if it is a node induced subgraph of Q. We denote $|V_q| + |E_q|$ as $|Q|$, and $|V| + |E|$ as $|G|$, then $Q(G)$ can be computed in $O(|G|!|G|)$ time [8].

Views. A *view* (*a.k.a.* view definition) V is a pattern query, and $V(G)$ in G is denoted as *view extension*, or *extension* when it is clear in the context [15].

Graph Pattern Matching Using Views. Given a pattern query Q and a set $\mathcal{V} = \{V_1, \cdots, V_n\}$ of view definitions, answering *graph pattern matching using views* is to find another query A such that (1) A is equivalent to Q, *i.e.*, $A(G) = Q(G)$ for *all* data graphs G; and (2) A only refers to views $V_i \in \mathcal{V}$ and their extensions $\mathcal{V}(G) = \{V_1(G), \cdots, V_n(G)\}$ in G, without accessing G. If such a query A exists, we say that Q *can be answered by using* $\mathcal{V}$.

We define the size $|Q|$ (resp. $|V|$) of pattern query Q (resp. view V) to be the total number of nodes and edges in Q (resp. V). We also define the size $|\mathcal{V}|$ of $\mathcal{V}$ to be the total size of V's in $\mathcal{V}$, the cardinality $||\mathcal{V}||$ of $\mathcal{V}$ to be the number of view definitions in $\mathcal{V}$, and $|\mathcal{V}(G)|$ to be the total size of matches in $\mathcal{V}(G)$.

3   Evaluating Graph Pattern Matching Using Views

In this section, we first introduce pattern containment problem, followed by an algorithm to determine pattern containment. We then introduce techniques to compute graph pattern matching using views.

3.1   Pattern Containment Problem

Along the same line as the *pattern containment* defined for pattern matching via simulation [11], we say that a pattern query $Q = (V_q, E_q)$, defined in terms of subgraph isomorphism, is contained in a set of views $\mathcal{V} = \{V_1, \cdots, V_n\}$, denoted by $Q \sqsubseteq \mathcal{V}$, if there exists a mapping λ from E_q to powerset $\mathcal{P}(\bigcup_{i \in [1,n]} E_i)$, such that for any data graph G, $S(e) \subseteq \bigcup_{e' \in \lambda(e)} S(e')$ for any edge $e \in E_q$.

Theorem 1. *For any graph* G, *a pattern query* Q *can be answered by using* $\mathcal{V}$ *and* $\mathcal{V}(G)$ *if and only if* $Q \sqsubseteq \mathcal{V}$.

Theorem 1 tells us that pattern containment determines whether graph pattern matching can be answered by using views, and motivates us study the pattern containment problem: that's, given pattern query Q and a set of view definitions $\mathcal{V}$, how to determine $Q \sqsubseteq \mathcal{V}$.

3.2 Pattern Containment Checking

To determine pattern containment, we first introduce a notion of *shadow* as a characterization of pattern containment.

Given a pattern query Q and a view definition V, one can compute V(Q) by treating Q as data graph, and V as pattern query. The *shadow* from V to Q, denoted by S_V^Q, is defined to be the union of edge set of all the matches of V in Q.

Below result tells us how *shadow* is used to determine pattern containment.

Proposition 2: *For a set of view definitions $\mathcal{V}$ and a pattern query Q, $Q \sqsubseteq \mathcal{V}$ if and only if $E_q = \bigcup_{V_i \in \mathcal{V}} S_{V_i}^Q$.*   □

Following Proposition 2, we show that $Q \sqsubseteq \mathcal{V}$ can be efficiently determined.

Theorem 3. *Given a pattern query Q and a set of view definitions $\mathcal{V}$, it is in $O(\|\mathcal{V}\|\|Q\|!|Q|)$ time to decide whether $Q \sqsubseteq \mathcal{V}$, and if so, to compute an associated mapping λ from Q to $\mathcal{V}$.*

We prove Theorem 3 by presenting such an algorithm.

Algorithm. The algorithm, denoted as contain (not shown), takes Q and $\mathcal{V}$ as inputs, and returns true if and only if $Q \sqsubseteq \mathcal{V}$. The algorithm first initializes an empty set E_o to record the *shadow* from V_i to Q. It then checks the condition of Proposition 2 as follows: (1) compute *shadow* $S_{V_i}^Q$ for each V_i in $\mathcal{V}$, by invoking the revised subgraph isomorphism algorithm, which first finds all the matches of V_i in Q with algorithm in [8], and then merges all the matches; (2) extend E_o with $S_{V_i}^Q$, since $S_{V_i}^Q$ is a subset of E_q. When all the *shadows* are merged, contain then checks whether $E_o = E_q$. It returns true if so, and false otherwise.

Correctness and Complexity. To see the correctness, observe that (1) contain correctly computes the "*shadows*" for each V_i in $\mathcal{V}$, by using the revised algorithm of [8]; and (2) when contain halts, it determines whether $Q \sqsubseteq \mathcal{V}$ by checking if the union of all the "*shadows*" covers edge set of Q, following Proposition 2. The correctness of contain then follows from the proof for Proposition 2. Algorithm contain iteratively computes *shadow* $S_{V_i}^Q$ for each $V_i \in \mathcal{V}$. It takes $O(|Q|!|Q|)$ time to compute *shadow* from V_i to Q for a single iteration, and the For loop repeats $\|\mathcal{V}\|$ times, thus, contain is in $O(\|\mathcal{V}\|\|Q\|!|Q|)$ time.

Example 2. Recall pattern query Q and a set of view definitions $\mathcal{V} = \{V_1, V_2\}$ shown in Fig. 1(b) and (c). Algorithm contain first computes shadows $S_{V_1}^Q$ and $S_{V_2}^Q$, which include a set of edges $\{(PM, BA), (PM, DBA), (BA, DBA)\}$ (marked in red in Q) and $\{(DBA, PRG), (PRG, BA), (PRG, ST)\}$ (marked in blue in Q), respectively. It then returns true indicating $Q \sqsubseteq \mathcal{V}$, since $S_{V_1}^Q \bigcup S_{V_2}^Q = E_q$.   □

3.3   An Matching Algorithm

Given Q, $\mathcal{V}$ and $\mathcal{V}(G)$, one can compute matches by using views as following: (1) determine whether $Q \sqsubseteq \mathcal{V}$ and compute the mapping λ with algorithm contain; and (2) compute $Q(G)$ with λ, $\mathcal{V}$ and $\mathcal{V}(G)$, if $Q \sqsubseteq \mathcal{V}$.

Theorem 4. *For any graph G, a pattern query Q can be evaluated by using $\mathcal{V}$ and $\mathcal{V}(G)$ in $O(2^{|Q|}|\mathcal{V}(G)|^3)$ time, if $Q \sqsubseteq \mathcal{V}$.*

We next show Theorem 4 by providing an algorithm as a constructive proof.

Algorithm. The algorithm, denoted as Match (not shown), takes Q, $\mathcal{V}$, $\mathcal{V}(G)$, and the mapping λ as inputs. It first initializes an empty pattern query Q_o, and an empty set M as intermediate results. It then iteratively invokes Merge to "merge" Q_o with V_i, and matches of Q_o with matches of V_i. When all the V_i are merged together, M is returned as the final result.

Given the mapping λ, a pattern query Q_o, its match set M, a view definition V_i and its extension $V_i(G)$, Merge "merges" matches M of Q_o with matches $V_i(G)$ of V_i as following. It iteratively (1) updates Q_o by merging it with G_s, which is mapped via λ^{-1} from V_i to Q; (b) checks whether matches m_1 of Q_o can be "merged" with matches m_2 of V_i in the similar way as the merging process of Q_o and G_s, and updates matches m_1 of Q_o by merging it with m_2. After all the G_s in $\lambda^{-1}(V_i)$ is merged with Q_o, *i.e.*, Q_o can not be expanded, Merge returns updated Q_o and its match set M as results.

Example 3. Consider Q, $\mathcal{V}$ and $\mathcal{V}(G)$ shown in Fig. 1. As Q is contained in $\mathcal{V}$, and a mapping λ, which maps Q to V_1 and V_2, is in place (see Example 2), algorithm Match then invokes Merge to "merge" V_1 with V_2, and $V_1(G)$ with $V_2(G)$. Specifically, in the first round iteration, Merge merges G_s (resp. $V_1(G)$) with the empty pattern query Q_o (resp. set M), as there is only one subgraph G_s of Q in $\lambda^{-1}(V_1)$, with edge set $\{(\text{PM}, \text{BA}), (\text{PM}, \text{DBA}), (\text{BA}, \text{DBA})\}$; in the second round iteration, Merge merges Q_o with another subgraph (with edge set $\{(\text{DBA}, \text{PRG}), (\text{PRG}, \text{BA}), (\text{PRG}, \text{ST})\}$) of Q in $\lambda^{-1}(V_2)$, and matches G_{s_1}, G_{s_2} of Q_o with matches $G_{s_3}, G_{s_4}, G_{s_5}$ of V_2, respectively. As the match G_{s_4} of V_2 can not be merged with G_{s_1} or G_{s_2}, then two matches $\{\text{Bob}, \text{Mat}, \text{Jean}, \text{Dan}, \text{Emmy}\}$, $\{\text{Walt}, \text{Fred}, \text{Nancy}, \text{Bill}, \text{Rei}\}$ of Q are returned.   □

Correctness and Complexity. The correctness is guaranteed by (1) Merge correctly merges Q_o (resp. M) with V_i (resp. $V_i(G)$); and (2) when Match terminates, Q_o (resp. M) is equivalent to Q (resp. $Q(G)$). Algorithm Match iteratively invokes Merge to generate new pattern query and its matches. The merge process repeats at most $||\mathcal{V}||$ times. For a single process, it takes Merge $|\lambda^{-1}(V_i)||M||V_i(G)|$ time to merge Q_o with V_i, and M with $V_i(G)$. As in the worst case, $|\lambda^{-1}(V_i)|$ is bounded by $2^{|Q|}$, and $|M|$ is bounded by $(\frac{|G|(|G|-1)}{2})^{|G|}$ (at the end of iteration), hence Merge is in $O(2^{|Q|} \cdot |V_i(G)|^3)$ time, and Match is in $O(2^{|Q|}|\mathcal{V}(G)|^3)$ time.

These complete the proof of Theorem 4.   □

4 Approximate Answering

In this section, we first propose notions of *maximally contained rewritings*, denoted as MDR and *maximally containing rewritings*, denoted as MGR by revising the notions, introduced in [6,20]. We then answer how to find and use MDR and MGR to approximately answer graph pattern matching via available views.

4.1 Maximally Contained Rewritings Problem

Maximally Contained Rewritings. A pattern query Q_d is a contained rewriting of Q using $\mathcal{V}$ if (a) $Q \sqsubseteq Q_d$, and (b) $Q_d \sqsubseteq \mathcal{V}$. The rewriting Q_d is a *maximally contained rewriting* of Q using $\mathcal{V}$ if no contained rewriting $Q_d{}'$ with $|E'_d| < |E_d|$ exists, where E_d and E'_d are the edge set of Q_d and $Q_d{}'$, respectively.

Given Q and $\mathcal{V}$, it is nontrivial to find the MDR. While we show below that there exists an algorithm that can find a contained rewriting Q_d with performance guarantee: $|E_d| \leq |E_q| + \log(|E_q|)(|E_{\mathsf{OPT}}| - |E_q|)$, where E_{OPT} is the edge set of the contained rewriting of Q with least edges, and E_d and E_q are the edge sets of Q_d and Q, respectively.

Theorem 5. *The* MDR *problem is* NP-*hard (decision problem); (2) there is an algorithm that finds a contained rewriting Q_d of Q using $\mathcal{V}$ with edges no more than $|E_q| + \log(|E_q|)(|E_{\mathsf{OPT}}| - |E_q|)$ in $O(|Q|^{|\mathcal{V}|} + (||\mathcal{V}|||Q|)^{3/2})$ time.*

Proof: We next prove Theorems 5(1), and (2), respectively.

(1) The decision problem of MDR is to decide, given Q, $\mathcal{V}$ and an integer B, whether there exists a pattern query Q_d such that $Q_d \sqsubseteq \mathcal{V}$, $Q \sqsubseteq Q_d$ and $|E_d \setminus E_q| = B$. We show that this problem is NP-hard by reduction from the NP-complete *set cover problem* (SCP).

Given a set U, a collection of subsets of U, $\mathcal{S} = \{S_1, \cdots, S_k\}$, where each subset S_i takes integer weight $\omega(S_i)$, and an integer k, SCP is to decide whether there exists a subset $\mathcal{S}'$ of $\mathcal{S}$ that covers U and with total weight k. Given such an instance of SCP, we construct an instance of MDR problem as follows: (a) for each $x_i \in U$, we create a unique edge e_{x_i} with two distinct nodes u_{x_i} and v_{x_i}; (b) we define a pattern query Q as a graph consisting of all edges e_{x_i} defined in (a); (c) for each subset $S_j \in \mathcal{S}$ and $x_i \in S_j$, we define a corresponding view V_j that consists of all edges e_{x_i} from S_j and $\omega(S_j)$ distinct edges; and (d) we set $k = B$. The construction is in PTIME. We next verify that there exists $\mathcal{S}'$ with total weight no more than k if and only if there exists Q_d with $|E_d \setminus E_q| \leq B$.

Assume that there exists a subset $\mathcal{S}'$ of $\mathcal{S}$ that covers U with total weight at most k. Let $\mathcal{V}'$ be the set of views V_j corresponding to $S_j \in \mathcal{S}'$. One can verify that $Q_d \sqsubseteq \mathcal{V}'$, since the union of all the edges from these $S_{V_j}^{Q_d}$ is E_d. Moreover $|E_d \setminus E_q| = \Sigma_{S_i \in \mathcal{S}'}\omega(S_i) \leq k$. Conversely, if there exists Q_d such that $Q \subseteq Q_d$ and $|E_d \setminus E_q \leq B|$, it is easy to see that the corresponding set $\mathcal{S}'$ is a set cover with weight no more than k.

Input: A pattern query $Q = (V_q, E_q)$, and a set of view definitions $\mathcal{V}$.
Output: A maximally contained rewriting Q_d of Q.

1. initialize set $\mathcal{F} := \emptyset$, an empty pattern query $Q_d = (V_d, E_d)$;
2. **for each** view definition $V_i \in \mathcal{V}$ **do**
3.   compute $\mathsf{IS}_{V_i}^Q$, $\mathsf{CS}_{V_i}^Q$; $\mathcal{F} := \mathcal{F} \cup \{\langle \mathsf{IS}_{V_i}^Q, \mathsf{CS}_{V_i}^Q \rangle\}$;
4.   **if** $(\mathsf{CS}_{V_i}^Q \setminus Q) = \emptyset$ **then** Q_d merges $\mathsf{IS}_{V_i}^Q$;
5. **while** $\mathcal{F} \neq \emptyset$ **do**
6.   find V_i with least $\alpha(V_i)$;
7.   **if** $(\mathsf{IS}_{V_i}^Q \setminus Q_d) = \emptyset$ **then break** ;
8. $\mathcal{F} := \mathcal{F} \setminus \{\langle \mathsf{IS}_{V_i}^Q, \mathsf{CS}_{V_i}^Q \rangle\}$; Q_d merges $\mathsf{CS}_{V_i}^Q$;
9. **if** $(E_q \not\subseteq E_d)$ **then return** $\emptyset$;
10.**return** Q_d;

Fig. 2. Algorithm MCD

(2) We show Theorem 5(2) by an algorithm with detailed analysis.

To illustrate the algorithm, we first introduce two notions: *inner shadow* and *complete shadow* by extending the notion of *shadow* introduced in Sect. 3.

Given a pattern query Q and a view definitions V_i, we define the *inner shadow* of V_i in Q, denoted by $\mathsf{IS}_{V_i}^Q$, to be the union of all the maximum common subgraphs $\mathrm{mcs}(V_i, Q)$ of V_i and Q; similarly, the *complete shadow* of V_i in Q, denoted as $\mathsf{CS}_{V_i}^Q$, is defined to be the union of view defitions V_i, where each V_i is mapped to $\mathsf{CS}_{V_i}^Q$ via the *bijective function* h that maps each subgraph of V_i to the subgraph of Q as maximum common subgraph. With these, we are now ready to present the algorithm.

Algorithm. Following Theorem 5, we present an algorithm denoted as MCD and shown in Fig. 2. The algorithm follows one greedy strategy: it iteratively identifies the "best" view which not only "covers" more edges of Q, but also introduces less "errors", *i.e.*, edges that Q does not have. To measure the goodness of a view V_i, a metric $\alpha(V_i) = \frac{|\mathsf{CS}_{V_i}^Q \setminus Q_d|}{|\mathsf{IS}_{V_i}^Q \setminus Q_d|}$ is defined. Intuitively, $\alpha(V_i)$ indicates the average error rate to cover each uncovered edge of Q when using V_i.

Algorithm MCD first initializes an empty set $\mathcal{F}$ and an empty pattern query Q_d (line 1). It then applies the algorithm [7] to compute maximum common subgraph $\mathrm{mcs}(V_i, Q)$ of V_i and Q, generates $\mathsf{IS}_{V_i}^Q$ (resp. $\mathsf{CS}_{V_i}^Q$) by merging $\mathrm{mcs}(V_i, Q)$ (resp. V_i) following the *bijective functioin* h, maintains a pair $\langle \mathsf{IS}_{V_i}^Q, \mathsf{CS}_{V_i}^Q \rangle$ in set $\mathcal{F}$, and merges Q_d with $\mathsf{IS}_{V_i}^Q$ if $\mathsf{CS}_{V_i}^Q$ does not have any edge that is not in Q, for each V_i in $\mathcal{V}$ (lines 2–4). It next selects V_i with least $\alpha(V_i)$ (line 6). When there is no V_i that can contribute to form a larger pattern query, *i.e.*, $\mathsf{IS}_{V_i}^Q \setminus Q_d = \emptyset$, MCD terminates the **while**loop (line 7). Otherwise, it removes $\langle \mathsf{IS}_{V_i}^Q, \mathsf{CS}_{V_i}^Q \rangle$ from $\mathcal{F}$, and merges Q_d with $\mathsf{CS}_{V_i}^Q$ as the updated pattern query (line 8). After loop terminates, MCD checks whether $E_q \subseteq E_d$, returns an empty pattern query if $E_q \not\subseteq E_d$ (line 9), or Q_d as the maximally contained rewriting of Q (line 10).

Correctness and Complexity. To see the correctness, observe that MCD always terminates since it visits each view definition in a finite set $\mathcal{V}$ only once to compute $\mathsf{IS}_{V_i}^Q$ and $\mathsf{CS}_{V_i}^Q$ (lines 2–3), and runs at most $\|\mathcal{V}\|$ times to find the Q_d (lines 4–7), and when MCD terminates, either an empty pattern query is returned if no MDR exists, or a contained rewriting Q_d with $|E_d| \leq |E_q| + \log(|E_q|)(|E_{\mathsf{OPT}}| - |E_q|)$ is returned, where E_{OPT} is the edge set of a contained rewriting Q_{OPT} of Q using $\mathcal{V}$ with minimum size. For the complexity, MCD firstly iteratively computes $\mathsf{IS}_{V_i}^Q$ and $\mathsf{CS}_{V_i}^Q$ for each view definition V_i in $\mathcal{V}$. Since it takes MCD $O(\frac{(|V_q|+1)!}{(|V_q|-|V_i|+1)!})$ time to compute $\mathsf{mcs}(V_i, Q)$ of V_i and Q [7], and generate $\mathsf{IS}_{V_i}^Q$, $\mathsf{CS}_{V_i}^Q$, for a single iteration (line 3), and the **for** loop repeats $\|\mathcal{V}\|$ times, hence, it is in $O(|Q|^{|\mathcal{V}|})$ time to prepare for the set $\mathcal{F}$. The process for computation of the maximally-contained rewriting Q_d (lines 4–7) takes $O((\|\mathcal{V}\| \|Q\|)^{3/2})$ time. Since the **while** loop is executed in $O((\|\mathcal{V}\| \|Q\|)^{1/2})$ time, and each iteration takes $O(\|\mathcal{V}\| \|Q\|)$ time to find a view definition with least α [25]. Thus, MCD is in $O(|Q|^{|\mathcal{V}|} + (\|\mathcal{V}\| \|Q\|)^{3/2})$ time.

This completes the proof of Theorem 5. □

Example 4. Consider Q with an extra edge (DBA, ST) in Fig. 1(b), and $\mathcal{V} = \{V_1, V_2, V_3, V_4, V_5, V_6\}$ in both Figs. 1(c) and 3. Since Q can not be contained by $\mathcal{V}$, algorithm MCD is applied to find the maximally-contained rewriting of Q using $\mathcal{V}$. It first computes $\mathsf{IS}_{V_i}^Q$ and $\mathsf{CS}_{V_i}^Q$ for each $V_i \in \mathcal{V}$. Take V_4 as example, the *inner shadow* $\mathsf{IS}_{V_4}^Q$ of V_4 in Q consists of a set of edges $\{(\mathsf{DBA}, \mathsf{PRG}), (\mathsf{DBA}, \mathsf{ST}), (\mathsf{PRG}, \mathsf{ST})\}$, while the *complete shadow* $\mathsf{CS}_{V_4}^Q$ of V_4 in Q contains one more edge (PRG, DBA) than $\mathsf{IS}_{V_4}^Q$. After all the $\mathsf{IS}_{V_i}^Q$ and $\mathsf{CS}_{V_i}^Q$ are computed, Q_d as the union of $\mathsf{IS}_{V_1}^Q$, $\mathsf{IS}_{V_2}^Q$, and $\mathsf{IS}_{V_3}^Q$ is generated. Using $\mathcal{F}$, MCD iteratively selects "best" views as following: it first chooses V_4, since $\alpha(V_1) = 1$ is minimum, and merges Q_d with $\mathsf{CS}_{V_4}^Q$; it then finds that Q_d can not be enlarged by any $\mathsf{IS}_{V_i}^Q$, since $(\mathsf{IS}_{V_i}^Q \setminus Q_d) = \emptyset$; MCD hence returns Q_d, which has one more edge (PRG, DBA) than Q as the maximally contained rewriting of Q using $\mathcal{V}$. □

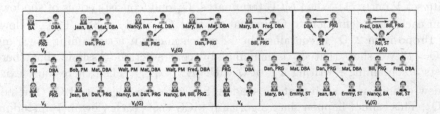

Fig. 3. Views and their extensions

4.2   Maximally Containing Rewritings Problem

Maximally Containing Rewritings. A pattern query Q_g is a containing rewriting of Q using $\mathcal{V}$ if (a) $Q_g \subseteq Q$, and (b) $Q_g \sqsubseteq \mathcal{V}$. Such a rewriting Q_g is a *maximally containing rewriting* of Q using $\mathcal{V}$ if there exists no containing rewriting Q_g' with $|E_g| < |E_g'|$, where E_g and E_g' are the edge set of Q_g and Q_g', respectively.

Though it is nontrivial to compute MGR: its decision problem is NP-hard, we provide efficient algorithm to find *maximally containing rewriting* of Q using $\mathcal{V}$.

Theorem 6. *The* MGR *problem is (1)* NP-*hard (decision problem); and (2) it is in* $O(\|\mathcal{V}\|\|Q\|!|Q|)$ *time to find a maximally-containing rewriting of Q using $\mathcal{V}$.*

Proof: We now prove Theorem 6(1) and (2), respectively.

(1) The decision problem of MGR is to decide, given Q, $\mathcal{V}$ and an integer B, whether there exists a pattern query Q_g such that $Q_g \sqsubseteq \mathcal{V}$, $Q_g \subseteq Q$ and $|E_q \setminus E_g| = B$. We show that this problem is NP-hard by reduction from the NP-complete *subgraph isomorphism problem* (ISO). Given a graph G and a pattern query Q_1 as an instance of ISO, we construct an instance of MGR problem by setting $Q = G$, $\mathcal{V} = \{Q_1\}$, an integer B. The construction is obviously in PTIME. One can easily verify that there exists a subgraph of G that is isomorphic to Q if and only if there exists Q_g such that $Q_g \sqsubseteq \mathcal{V}$, $Q_g \subseteq Q$ and $|E_q \setminus E_g| = B$.

(2) We present an algorithm and its analysis as the proof of Theorem 6(2).

Algorithm. The algorithm, denoted as MCG (not shown), works similarly as the algorithm contain: it takes pattern query Q and a set of views $\mathcal{V}$ as input, maintains a set E_g (initially empty), iteratively computes $S_{V_i}^Q$ and merges it with E_g, for each $V_i \in \mathcal{V}$. In contrast to contain, which determines whether edge set of Q equals to the union of all the *shadows*, MCG directly returns a pattern query Q_g with edge set E_g, as a *maximally containing rewriting* of Q using $\mathcal{V}$.

Correctness and Complexity. The correctness is ensured by that when MCG terminates, Q_g is a containing rewriting, and a *maximally containing rewriting*. Obviously MCG always terminates since it visits each view in a finite set $\mathcal{V}$ once. (1) When MCG terminates, Q_g only consists edges of shadows from each view definition to Q, indicating that $Q_g = \bigcup_{V_i \in \mathcal{V}} S_{V_i}^Q$. Thus, following Proposition 2, Q_g is contained by $\mathcal{V}$, which makes it a containing rewriting of Q. (2) We show that Q_g is maximal by contradiction. Assume that there exists a containing rewriting Q_g' of Q with $|E_g| < |E_g'|$. Then there must exist a view definition V_i such that its shadow $S_{V_i}^Q$ is in the edge set of Q_g' but not in Q_g. This cannot happen since algorithm MCG visits each view in $\mathcal{V}$ including V_i, and hence E_g must include $S_{V_i}^Q$ when algorithm MCG visits V. To see the complexity, note that the only difference of MCG from contain is that MCG generates an induced subgraph of Q with the edge set E_g in $O(|E_q|)$ time, rather than checking equivalence between union of all shadows with edge set of Q. It takes MCG $O(|||\mathcal{V}|||Q|!|Q|)$ time to compute all the shadows. Thus, it is in total $O(|||\mathcal{V}|||Q|!|Q|)$ time to compute Q_g, with the same time complexity as contain.

This completes the proof of Theorem 6.   □

Example 5. Recall Q with edge (DBA, ST) and view definitions V_1–V_6 in Example 4. Algorithm MCG iteratively computes shadows $S_{V_i}^Q$ ($i \in [1, 6]$), and merges them with E_g. After all the shadows are merged together, one may verify that E_g includes a set of edges {(PM, BA), (PM, DBA), (BA, DBA), (DBA, PRG), (PRG, BA), (PRG, ST)}, and the pattern query Q_g induced by E_g, is returned as the maximally containing rewriting of Q.   □

4.3 Approximate Answering Using Views

Given a pattern query Q, a set of view definitions $\mathcal{V}$ and their extensions $\mathcal{V}(G)$, when Q is not contained in $\mathcal{V}$, we can find approximate matches by using views as following: (1) find a maximally contained (resp. containing) rewriting Q_d (resp. Q_g) of Q with algorithm MCD (resp. MCG), and (2) compute $Q_d(G)$ (resp. $Q_g(G)$) as approximate answers to Q(G) by using algorithm Match.

<u>MDR *vs.* MGR</u>. When an MDR Q_d of Q is used to find approximate matches, one may verify that for any graph G, each match of Q_d embeds "true" matches of Q since $Q \subseteq Q_d$; one can even derive a subset of "true" matches from $Q_d(G)$. In contrast to MDR, an MGR Q_g is a subquery of Q, hence for any graph G, each match of Q_g is possibly a subgraph of a "true" match of Q, and can be viewed as a possible match. This said, computing matches using Q_g and views offers a way to find "possible" matches of Q.

Accuracy. To measure the quality of the approximate answers $Q_d(G)$ (resp. $\overline{Q_g(G)}$) versus the true matches in the exact answers Q(G), we used the F-measure [27]: $ACC = \frac{2 \cdot recall \cdot precision}{recall + precision}$, where $recall = \frac{|S|}{|S_t|}$, $precision = \frac{|S|}{|S_m|}$, S_t is the set of matches in Q(G), S_m is the set of matches in $Q_d(G)$ (resp. $Q_g(G)$), and S consists of "true" matches that can be identified from $Q_d(G)$ (resp. $Q_g(G)$).

As will be seen in our experimental studies, approximately answering graph pattern queries using MDR and MGR is quite effective: the average accuracy of matching answer on real-life graphs can reach 70% and 77%, respectively.

5   Experimental Evaluation

Using real-life and synthetic data, we conducted two sets of experiments to evaluate (1) the efficiency of algorithm contain, and the efficiency and scalability of algorithm Match for answering graph pattern; (2) the efficiency of algorithms MCD and MCG; and the efficiency, accuracy and scalability of algorithms Match$_{MDR}$, Match$_{MGR}$ for approximately answering graph pattern matching using views.

Experimental Setting. We used the following data.

(1) Real-life graphs. We used four real-life graphs: (a) *Amazon* [3], a product co-purchasing network with 548K nodes and 1.78M edges. (b) *Citation* [1], a collaboration network with 1.4M nodes and 3M edges. (c) *YouTube* [5], a recommendation network with 1.6M nodes and 4.5M edges. (d) *WebGraph* [4], a web graph including 12.1M nodes and 103.6M edges.

(2) Synthetic graphs. We designed a generator to produce random graphs, controlled by the number $|V|$ of nodes, the number $|E|$ of edges, and an alphabet Σ for node labels.

(3) Pattern queries. We implemented a generator for pattern queries controlled by: the number $|V_q|$ (resp. E_Q) of pattern nodes (resp. edges), and node label f_v from an alphabet Σ of labels drawn from corresponding real-life graphs. We denote $(|V_q|, |E_q|)$ as the size of pattern queries, and generated a set of 30 pattern queries with $(|V_q|, |E_q|)$ ranging from $(3, 2)$ to $(8, 16)$, for each data graph.

(4) Views. We generated views for *Amazon* following [19], designed views to search for papers and authors in computer science for *Citation*, generated views for *Youtube* following [11], and designed views to search Web pages on *WebGraph*. For each of the real-life graph, a set $\mathcal{V}$ of 50 view definitions with different sizes *e.g.*, $(2, 1)$, $(3, 2)$, $(4, 3)$, $(4, 4)$ and structures are generated. For synthetic graphs, we randomly generated a set of 50 views whose node labels are drawn from a set Σ of 10 labels and with sizes of $(2, 1)$, $(3, 2)$, $(4, 3)$ and $(4, 4)$.

(5) Implementation. We implemented the following algorithms, all in Java: (1) contain for pattern containment checking; (2) VF2 [8] and Match for graph pattern matching; (3) MCD and MCG for finding Q_d and Q_g, respectively; and (4) Match$_{MDR}$ and Match$_{MGR}$ for computing approximate matches, which invokes Match to compute matches of Q_d and Q_g using views, respectively.

All the experiments were run on a machine powered by an Intel Core(TM)2 Duo 3.00 GHz CPU with 4 GB of memory, using Ubuntu. Each experiment was run 10 times and the average is reported here.

Experimental Results. We next present our findings.

Exp-1: Graph Pattern Matching Using Views. We evaluated the performance of contain, and compared performance of Match, with VF2 [8].

Efficiency of contain. Fixing a set of synthetic views $\mathcal{V}$, we varied the pattern size from $(4, 4)$ to $(8, 16)$, where each size corresponds to *a set of patterns* with different structures and node labels. As shown in Fig. 4(a), (1) contain is efficient, *e.g.*, it takes in average 143 ms to decide whether a pattern with $|V_q| = 8$ and $|E_q| = 16$ is contained in $\mathcal{V}$; and (2) contain takes more time over larger patterns, which is consistent with its computational complexity.

Efficiency of Match. Figures 4(b), (C), (d) and (e) show the results on *Amazon*, *Citation*, *YouTube* and *WebGraph*, respectively. The x-axis represents pattern size $(|V_q|, |E_q|)$. The results tell us the following. (1) Match substantially outperform VF2, taking only 9.5% of its running time on average over all real-life

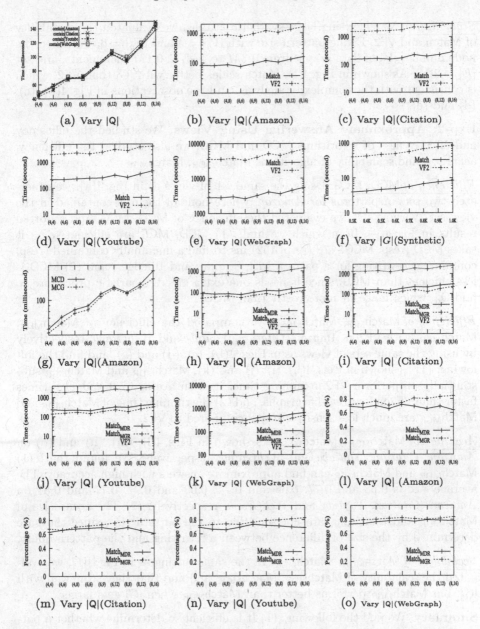

Fig. 4. Performance evaluation

graphs. (2) Two algorithms spend more time on larger patterns. Nonetheless, Match is less sensitive than VF2, since it reuses previous computation cached in view extensions hence saves computational cost.

Scalability of Match. Using large synthetic graphs, we evaluated the scalability of Match and VF2. Fixing pattern size with $|V_q| = 4$ and $|E_q| = 6$, we varied the node number $|V|$ of data graphs from $0.3M$ to $1M$, in $0.1M$ increments, and set $|E| = 2|V|$. As shown in Fig. 4(f), Match scales better with $|G|$ than VF2, which is consistent with the complexity analysis, and the observations in Fig. 4(b), (c), (d) and (e).

Exp-2: Approximate Answering Using Views. We studied the efficiency and effectiveness of algorithms MCD and MCG; we also studied the efficiency, accuracy, and scalability of algorithms Match$_{MDR}$, Match$_{MGR}$.

Efficiency of MCD, MCG. Using the same set of views as in Fig. 4(a), we generated two sets of patterns for *Amazon*, where none of them is contained in the view set. Varying $|Q|$, we evaluated the efficiency of MCD, MCG, and reported results in Fig. 4(g), from which we find: (1) MCD, MCG are efficient, *e.g.*, it takes MCD (resp. MCG) 178 (resp. 152) ms to find a maximally contained (resp. containing) rewriting for a pattern with 8 nodes and 16 edges; and (2) MCD is less efficient than MCG, since it needs one extra greedy heuristic procedure to find Q_d, which makes it more costly than MCG.

Efficiency of Match$_{MDR}$, Match$_{MGR}$. We compared the efficiency of Match$_{MDR}$, Match$_{MGR}$, with VF2on *Amazon*, *Citation*, *Youtube* and *WebGraph*, respectively by using the same sets of views as in Figs. 4(b), (c), (d) and (e), and find the following: (1) as shown in Figs. 4(h), (i), (j) and (k), Match$_{MDR}$ and Match$_{MGR}$ substantially outperform VF2in running time: it is on average 7.9 and 11.8 times faster than VF2on four real-life graphs, and (2) the running time of Match$_{MDR}$ and Match$_{MGR}$ are much less sensitive to $|Q|$ compared to VF2.

Accuracy of Match$_{MDR}$, Match$_{MGR}$. As shown in Figs. 4(l), (m), (n) and (o) on *Amazon*, *Citation*, *Youtube* and *WebGraph*, respectively. We found that: (1) Match$_{MDR}$ and Match$_{MGR}$ can find approximate answers with high accuracy. The average Acc is 0.68 and 0.77, 0.66 and 0.74, 0.69 and 0.77, 0.77 and 0.81 on *Amazon*, *Citation*, *Youtube* and *WebGraph*, respectively. (2) The accuracies of Match$_{MDR}$ and Match$_{MGR}$ are not sensitive to the pattern size, since they are determined by the size of difference between a rewriting and the pattern query.

Scalability of Match$_{MDR}$, Match$_{MGR}$. In the same setting as in Fig. 4(f), we evaluated the scalability of Match$_{MDR}$, Match$_{MGR}$, and find that they scale well with $|G|$, but Match$_{MGR}$ performs better than Match$_{MDR}$ when $|G|$ gets larger.

Summary. We find the following. (1) It is efficient to determine whether a pattern query can be answered using views. (2) Answering graph pattern matching using views is effective in querying large social graphs. For example, by using views, pattern matching via subgraph isomorphism takes only 9.3% of the time needed for computing matches directly in *YouTube*, and 10.6% on synthetic graphs; and our view-based matching algorithms scale well with data size. (3) When pattern queries are not contained by views, our techniques can find approximate matches efficiently with reasonable accuracy. For example, Match$_{MDR}$ is 10.4 times faster than VF2, with accuracy 0.77 over *Webgraph*.

6   Related Work

We categorize previous work as follows.

Query Answering Using Views. Answering queries using views has been well
studied for relational data (see [14,18] for surveys) and XML data [23,24]. For
graph data, this topic has also been investigated by [10,11]. This work differs
from theirs in the following: (1) We adopt subgraph isomorphism as the semantic
of pattern matching, instead of graph simulation [16] and bounded simulation [9].
that are applied as matching semantics by [10,11]. (2) We study *minimum containment problem*, whose objective function is defined on the total size of materialized views, rather than the number of views of a subset of view definitions.

Approximate Graph Pattern Matching Using Views. [11] introduced techniques
to compute approximate matches by using maximally contained rewritings of
pattern queries via graph simulation. [21] studied the problem of computing
upper and lower approximations from views for pattern queries based on graph
simulation and subgraph isomorphism. Unlike [11,21], we studied the problem
of maximally contained rewritings (MDR) and maximally containing rewritings
(MGR) for pattern matching with subgraph isomorphism, and developed techniques to find approximate answers with MDR and MGR.

7   Conclusion

We have studied graph pattern matching using views, for pattern queries defined
via subgraph isomorphism, from theory to algorithms. We have proposed a
notion of pattern containment as a *characterization* to determine whether pattern queries can be answered by using views, and provided an efficient containment checking algorithm. Based on the characterization, we have developed an
algorithm to answer pattern matching using views. We have also studied approximately answering pattern matching using views when pattern queries are not
contained in views, and developed algorithms with high efficiency and accuracy.
These results extend the earlier study of graph pattern matching using views.

The study of graph pattern matching using views is still in its infancy. One
issue is to decide what views to cache such that a set of frequently used pattern
queries can be answered by using the views. Techniques such as adaptive and
incremental query expansion may apply. Moreover, one may need to integrate
techniques such as view-based, distributed, and incremental together to form a
practical method for querying "big" social data.

Acknowledgments. This work is supported by NSFC 61402383 and 71490722,
Sichuan Provincial Science and Technology Project 2014JY0207, and Fundamental
Research Funds for the Central Universities, China.

References

1. Citation. http://www.arnetminer.org/citation/

2. Full version. http://emlc.swjtu.edu.cn/download/WangXin/paper.pdf
3. Stanford large network dataset collection. http://snap.stanford.edu/data/index.html
4. Webgraph data. http://law.dsi.unimi.it/datasets.php
5. Youtube dataset. http://netsg.cs.sfu.ca/youtubedata/
6. Afrati, F., Chandrachud, M., Chirkova, R., Mitra, P.: Approximate rewriting of queries using views. In: Grundspenkis, J., Morzy, T., Vossen, G. (eds.) ADBIS 2009. LNCS, vol. 5739, pp. 164–178. Springer, Heidelberg (2009). doi:10.1007/978-3-642-03973-7_13
7. Conte, D., Foggia, P., Vento, M.: Challenging complexity of maximum common subgraph detection algorithms: a performance analysis of three algorithms on a wide database of graphs. J. Graph Algorithms Appl. 11(1), 99–143 (2007)
8. Cordella, L., Foggia, P., Sansone, C., Vento, M.: A (sub)graph isomorphism algorithm for matching large graphs. TPAMI 26, 1367–1372 (2004)
9. Fan, W., Li, J., Ma, S., Tang, N., Wu, Y., Wu, Y.: Graph pattern matching: from intractability to polynomial time. PVLDB 3, 264–275 (2010)
10. Fan, W., Wang, X., Wu, Y.: Answering graph pattern queries using views. In: ICDE, pp. 184–195 (2014)
11. Fan, W., Wang, X., Wu, Y.: Answering pattern queries using views. IEEE Trans. Knowl. Data Eng. 28(2), 326–341 (2016)
12. Fan, W., Wang, X., Wu, Y., Xu, J.: Association rules with graph patterns. PVLDB 8(12), 1502–1513 (2015)
13. Faris, R., Ennett, S.: Adolescent aggression: the role of peer group status motives, peer aggression, and group characteristics. Soc. Netw. 34(4), 371–378 (2012)
14. Halevy, A.: Answering queries using views: a survey. VLDB J. 10(4), 270–294 (2001)
15. Halevy, A.: Theory of answering queries using views. SIGMOD Rec. 29, 40–47 (2001)
16. Henzinger, M.R., Henzinger, T., Kopke, P.: Computing simulations on finite and infinite graphs. In: FOCS (1995)
17. Lappas, T., Liu, K., Terzi, E.: A survey of algorithms and systems for expert location in social networks. Soc. Netw. Data Anal. 215–241 (2011)
18. Lenzerini, M.: Data integration: a theoretical perspective. In: PODS (2002)
19. Leskovec, J., Singh, A., Kleinberg, J.: Patterns of influence in a recommendation network. In: Ng, W.-K., Kitsuregawa, M., Li, J., Chang, K. (eds.) PAKDD 2006. LNCS, vol. 3918, pp. 380–389. Springer, Heidelberg (2006). doi:10.1007/11731139_44
20. Levy, A.Y., Mendelzon, A.O., Sagiv, Y., Srivastava, D.: Answering queries using views. In: PODS (1995)
21. Li, J., Cao, Y., Liu, X.: Approximating graph pattern queries using views. CIKM 2016, 449–458 (2016)
22. Lu, L., Zhou, T.: Link prediction in complex networks: A survey. CoRR, abs/1010.0725 (2010)
23. Miklau, G., Suciu, D.: Containment and equivalence for an xpath fragment. In: Proceedings of ACM Symposium on Principles of Database Systems (PODS) (2002)
24. Neven, F., Schwentick, T.: XPath containment in the presence of disjunction, DTDs, and variables. In: Calvanese, D., Lenzerini, M., Motwani, R. (eds.) ICDT 2003. LNCS, vol. 2572, pp. 315–329. Springer, Heidelberg (2003). doi:10.1007/3-540-36285-1_21
25. Papadimitriou, C.H.: Computational Complexity. Addison-Wesley, Boston (1994)
26. Terveen, L.G., McDonald, D.W.: Social matching: a framework and research agenda. ACM Trans. Comput.-Hum. Interact. 12, 401–434 (2005)
27. Wikipedia. F-measure. http://en.wikipedia.org/wiki/F-measure

Towards an Integrated Graph Algebra for Graph Pattern Matching with Gremlin

Harsh Thakkar[1]([✉]), Dharmen Punjani[2], Sören Auer[1,3],
and Maria-Esther Vidal[1]

[1] University of Bonn, Bonn, Germany
{Thakkar,Auer,Vidal}@cs.uni-bonn.de
[2] National and Kapodistrian University of Athens, Athens, Greece
dpunjani@di.uoa.gr
[3] Fraunhofer Institute for Intelligent Analysis and Information Systems (IAIS),
Sankt Augustin, Germany

Abstract. Graph data management has revealed beneficial characteristics in terms of flexibility and scalability by differently balancing between query expressivity and schema flexibility. This has resulted into an rapid developing new task specific graph systems, query languages and data models, such as property graphs, key-value, wide column, resource description framework (RDF), etc. Present day graph query languages are focused towards flexible graph pattern matching (aka sub-graph matching), where as graph computing frameworks aim towards providing fast parallel (distributed) execution of instructions. The consequence of this rapid growth in the variety of graph based data management systems has resulted in a lack of standardization. Gremlin, a graph traversal language and machine, provides a common platform for supporting any graph computing system (such as an OLTP graph database or OLAP graph processors). We present a formalization of graph pattern matching for Gremlin queries. We also study, discuss and consolidate various existing graph algebra operators into an integrated graph algebra.

Keywords: Graph pattern matching · Graph traversal · Graph query algebra

1 Introduction

Observing the evolution of information technology, we see a trend from data models and knowledge representation techniques being tightly bound to the capabilities of the underlying hardware towards more intuitive methods resembling human-style information processing. This evolution started with machine assembly languages, went over procedural programming, object-oriented methods and resulted in an ever more loosely coupling of data and code with relational data bases and object-relational mapping (ORM). Recently, we can observe an even further step in this evolution – graph based data models, which organize information in conceptual networks. Graphs are valued distinctly, when it comes

© Springer International Publishing AG 2017
D. Benslimane et al. (Eds.): DEXA 2017, Part I, LNCS 10438, pp. 81–91, 2017.
DOI: 10.1007/978-3-319-64468-4_6

to choosing formalisms for modelling real-world scenarios such as biological, transport and social networks due to their intuitive data model. A property graph data model is capable of representing complex domain networks [11].

Graph analysis tools have turned out to be one of pioneering applications in understanding these natural and man-made networks. Graph analysis is carried out using graph processing techniques which ultimately boil down to efficient graph query processing. Graph Pattern Matching (GPM), also referred to as the sub-graph matching is the foundational problem of graph query processing. Many vendors have proposed a variety of (proprietary) graph query languages to demonstrate the solvability of graph pattern matching problem.

These modern graph query languages focus either on *traversal*, or on *pattern-matching*. *Gremlin* [10] is one such modern graph query language, with a distinctive advantage over others that it offers both of these perspectives. This implies that a user can reap benefits of both declarative and imperative matching style within the same framework. Furthermore, conducting GPM in Gremlin can be of crucial importance in cases such as: (i) Querying very large graphs, where a user is not completely aware of certain dataset-specific statistics of the graph (e.g., the number of `created` vs `knows` edges existing in the graph (Ref. Fig. 1)); (ii) Creating optimal query plans, without the user having to dive deep into traversal optimization strategies; (iii) In application-specific settings such as a question answering, users express information needs (e.g., natural language questions) which can be better represented as graph pattern matching problems than path traversals.

In this work, we contribute to establishing a formal base for a graph query algebra, by surveying and integrating existing graph query operators. The contributions of this work are in particular: (i) We consolidate existing graph algebra operators from the literature and propose two new traversal operators into an integrated graph algebra; (ii) We formalize graph pattern matching construct of Gremlin query language; (iii) We provide a formal specification of pattern matching traversals for the Gremlin language, which can serve as a foundation for implementing a Gremlin based query compilation engine. As a result, the formalization of graph query algebra supports the integration and interoperability of different graph data models (e.g., executing SPARQL queries on top of Gremlin), helps to prevent vendor lock in scenarios and boosts data management benchmarking efforts such as LITMUS [13].

The remainder of this article starts with Sect. 2 that describes the preliminaries of our work, followed by Sect. 3 presents Gremlin as a traversal language and machine, and the notion of GPM. In Sect. 4 we present the proposed mapping algorithm. Section 5 discusses the related work. Finally Sect. 6 concludes the paper.

2   Preliminaries

We now introduce the mathematical concepts used in this article. Our notation closely follows [4] and extends [12] by adding the notion of vertex labels.

2.1 Property Graphs

Property graphs, also referred to as directed, edge-labeled, attributed multi-graphs, have been formally defined in a wide variety of texts, such as [1,5,9,11].

Definition 1 (Property Graph). *A property graph is defined as* $G = \{V, E, \lambda, \mu\}$; *where:*

- *V is the set of vertices,*
- *E is the set of directed edges such that $E \subseteq (V \times Lab \times V)$ where Lab is the set of Labels,*
- *λ is a function that assigns labels to the edges and vertices (i.e. $\lambda : V \cup E \to \Sigma^*$)[1], and*
- *μ is a partial function that maps elements and keys to values (i.e. $\mu : (V \cup E) \times R \to S$) i.e. properties (key $r \in R$, value $s \in S$).*   □

For simplicity, we define disparate sets (of μ and λ) for the labels and properties of vertices and edges respectively, adapting the terminology used in [4]. We define:

- L_v: Set of vertex labels $(L_v \subset \Sigma^*)$, $\lambda_l : V \to L_v$ assigns label to each vertex
- L_e: Set of edge labels $(L_e \subset \Sigma^*)$, $\lambda_e : E \to L_e$ assigns label to each edge. $(L_v \cap L_e = \phi, L_v \cup L_e = Lab$; wrt Definition 1, $\lambda = \lambda_v \cup \lambda_e)$
- P_v: Set of vertex properties $(P_v \subset R)$, $\mu_v : V \times P_v \to S$ assigns a value to each vertex property
- P_e: Set of edge properties $(P_e \subset R)$, $\mu_e : E \times P_e \to S$ assigns a value to each edge property $(P_v \neq P_e$; in the context of Definition 1, $\mu = \mu_v \cup \mu_e)$

Figure 1, presents a visualization of the *Apache TinkerPop* modern crew graph[2], which is used as a running example throughout this paper.

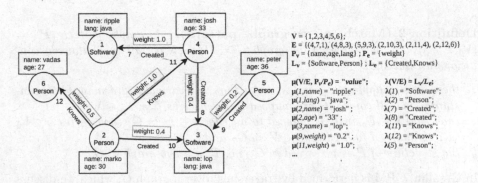

Fig. 1. Running example. This figure represents a collaboration network of employees in a typical software company.

---

[1] set of strings (Σ^*).
[2] http://tinkerpop.apache.org/docs/3.2.3/reference/#intro.

2.2   Graph Pattern Matching

Graph Pattern Matching (GPM) is a computational task consisting of matching graph patterns (P) against a graph (G, ref. Definition 1). Graph databases perform GPM for querying a variety of data models such as RDF, Property Graphs, edge-labelled graphs, etc., and many works address and analyze its solvability, such as [1,5,7,14]. Various graph query languages have been implemented for querying these data models, such as – The SPARQL[3] query language for RDF triple stores (declarative), Neo4J's native query language CYPHER[4] (declarative), Apache TinkerPop's graph traversal language *Gremlin*[5] (a functional language offering both imperative and declarative constructs). GPM queries typically constitute of basic graph patterns (BGP) and complex graph patterns (CGP) [1] (which add operations such as projections, unions, etc. over BGPs) (cf. Fig. 2). We present a formal definition of GPM in our context which closely follows [1].

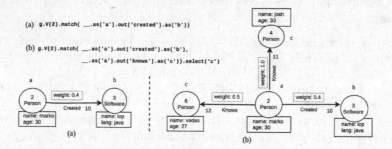

Fig. 2. We illustrate the notion of a sample, (a) BGP and (b) CGP, wrt a gremlin traversal over the graph G as shown in Fig. 1.

Definition 2 (Match of a graph pattern). *A graph pattern $P = (V_p, E_p, \lambda_p, \mu_p)$; is matching the graph $G = (V, E, \lambda, \mu)$, iff the following conditions are satisfied:*

1. *there exist mappings μ_p and λ_p such that, all variables are mapped to constants, and all constants are mapped to themselves (i.e. $\lambda_p \in \lambda, \mu_p \in \mu$),*
2. *each edge $\acute{e} \in E_p$ in P is mapped to an edge $e \in E$ in G, each vertex $\acute{v} \in V_p$ in P is mapped to a vertex $v \in V$ in G, and*
3. *the structure of P is preserved in G (i.e. P is a sub-graph of G)*   □

In Gremlin, GPM is performed by traversing[6] over a graph G, which we discuss wrt the context of Gremlin traversal language in Sect. 3.2.

---

[3] https://www.w3.org/TR/rdf-sparql-query/.
[4] https://neo4j.com/developer/cypher-query-language/.
[5] https://tinkerpop.apache.org/.
[6] The act of visiting of vertices ($v \in V$) and edges ($e \in E$) in a graph in an alternating manner (in some algorithmic fashion) [11].

Example 1. We illustrate the evaluation of graph patterns (a) and (b) from Fig. 2, over the graph G from Fig. 1.

(a) v[3]   (b) c:v[4]

c:v[6]

2.3 Graph Algebra Foundations

We present a consolidated summary of various graph query operators defined in literature [2,4,7,8,11,12]. For brevity, we abstain from dwelling into rigorous formal definitions rather refer interested reader to respective articles and for illustrations to the online resource at https://goo.gl/pC5JGM.

Unary Operators. *Projection* $(\pi_{a,b,..})$: $R \cup S \to \Sigma^*$: operator selects values of specific variables $a, b, .., n$, from the solution set of a matched input graph pattern P, against the graph G. The results returned by $(\pi_{a,b})$ are not deduplicated be default, as it operators on bag semantics. This operator is present in all standard graph query languages (e.g. SELECT in SPARQL, and MATCH in CYPHER).

Selection $(\exists(p))$, filters/restricts the match of a certain graph pattern P against a graph G, by imposing conditional expressions (p) e.g., inequalities and/or other traversal-specific predicates (where predicate is a proposition formula).

Binary Operators. *Concatenation* [12] (∘): $E^* \times E^* \to E^*$: concatenates two paths. For instance, if (i, α, j) and (j, β, k) are two edges in a graph G, then their concatenation is the new path (i, α, j, β, k); where $i, j, k \in V$ and $\alpha, \beta \in L_e$.

Union operator (⊎): $P(E^*) \times P(E^*) \to P(E^*)$: is the multiset union of two path traversals or graph patterns. For instance, {(1,2), (3,4), (3,4), (4,5)} ⊎ {(1,2), (3,4)} = {(1,2), (1,2), (3,4), (3,4), (3,4), (4,5)}. The results of this operator, like projection, are not deduplicated by default.

Join [12] (⋈∘): $P(E^*) \times P(E^*) \to P(E^*)$: produces the concatenative join of two sets of paths (path traversals) such that if $P, R \in P(E^*)$, then[7]

$$P \bowtie_\circ R = \{p \circ r \mid p \in P \land r \in R \land (p = \epsilon \lor r = \epsilon \lor \gamma^+(p) = \gamma^-(r))\}$$

e.g., if $P = \{(v_1, e_1, v_2), (v_2, e_2, v_3)\}$ and $R = \{(v_2, e_2, v_3), (v_2, e_2, v_1)\}$, then,

$$P \bowtie_\circ R = \{(v_1, e_1, v_2, v_2, e_2, v_3), (v_1, e_1, v_2, v_2, e_2, v_1)\},$$

where $v_1, v_2, v_3 \in V; e_1, e_2 \in L_e$.

General Extensions. We borrow the extended relational operators *Grouping* $(\dagger_a(p))$, *Sorting* $(\Re_{\Uparrow a, \Downarrow b}(p))$ and *Deduplication* $(\delta_{a,b,..}(p))$ from [7].

Graph-Specific Extensions. Various graph-specific operators have been defined in works such as [4,12]. We present graph-specific operators, some of

---

[7] Here, (γ^-, γ^+) denote the first and last elements of a path respectively.

which have been adapted from [4,7] and propose two additional operators based on the algebra defined by [11,12].

The *Get-vertices*/*Get-edges* (new) nullary operators (V_g/E_g): return the list of vertices/edges, respectively. These operators, w.r.t. Gremlin query construct, denote the start of a traversal. It is also possible to traverse from a specific vertex/edge in a graph, given their id's.

The *Traverse operator* $(\updownarrow_{v1}^{v2}[e](p))$: $P(V \cup E) \times \Sigma^* \to P(V \cup E)$: is an adapted version, analogues to the *expand-both* operator defined by [4]. The traverse operator represents the traversing over the graph operation (traversing *in* $\downarrow$ or *out* $\uparrow$ from a vertex ($v1$) to an adjacent vertex ($v2$) given the edge label [e], where $(v1, v2) \in V, e \in L_e$).

The *Property filter* operator $(\sigma_{condition}^{v/e}(p))$: $P(V \cup E) \times S \to \Sigma^*$: is a new binary operator which: *(i)* filters the values of selected element (vertex/edge), given a *condition*, *(ii)* otherwise, simply returns the value of the element (Table 1).

The *Restriction* unary operator $(\lambda_l^s(p))$ is an adaptation of [6]. It takes a list as input and returns the top s values, skipping specified l values, analogous to the LIMIT and OFFSET modifiers in SPARQL.

Table 1. A consolidated list of relational algebra graph operators with their corresponding instruction steps in Gremlin traversal language.

Ops	Operation	Operator	Gremlin Step	Step type
0	Get vertices	V_g	`g.V()`	-
	Get edges	E_g	`g.E()`	-
1	Selection	$\exists(p)$	`.where()`	Filter
	Property filter	$\sigma_{condition}^a(p)$	`.has()/.values()`	Filter
	Projection	$\Pi_{a,b,...}(p)$	`.select()`	Map
	De-duplication	$\delta_{a,b,..}(p)$	`.dedup()`	Filter
	Restriction	$\lambda_l^s(p)$	`.limit()`	Filter
	Sorting	$\Re_{\Uparrow a, \Downarrow b}(p)$	`.order().by()`	Map
	Grouping	$\dagger_a(p)$	`.group().by()`	Map/SideEffect
	Traverse (out/in)	$\updownarrow_{v1}^{v2}[e](p)$	`.out()/.in()`	FlatMap
2	Join	$p \bowtie_\circ r$	`.and()`	Filter
	Union	$p \uplus r$	`.union()`	Branch

3   The Gremlin Graph Traversal Language and Machine

Gremlin is the query language of Apache TinkerPop graph computing framework. Gremlin is system agnostic, and enables both – pattern matching (declarative) and graph traversal (imperative) style of querying over property graphs.

The Machine. Theoretically, a set of traversers in T move (traverse) over a graph G (property graph, cf. Sect. 2.1) according to the instruction in (Ψ), and this computation is said to have completed when there are either: (i) no more existing traversers (t), or (ii) no more existing instructions (ψ) that are referenced by the traversers (i.e. program has halted).

The Traversal. A Gremlin traversal can be represented in any host language that supports function composition and function nesting. These steps are either of: (i) *Linear motif* - $f \circ g \circ h$, where the traversal is a linear chain of steps; or (ii) *Nested motif* - $f \circ (g \circ h)$ where, the nested traversal $g \circ h$ is passed as an argument to step f [10]. A step $(f \in \Psi)$ can be, defined as $f : A^* \to B^{*8}$. Where, f maps a set of traversers of type A to a set of traversers of type B. Given that Gremlin is a language and virtual machine, it is possible to design another traversal language that compiles to the Gremlin traversal machine.

3.1 Graph Pattern Matching Queries in Gremlin

Gremlin provides the GPM construct, analogous to SPARQL [2,8] using the `Match()`-step. This enables the user to represent the query using multiple individual connected or disconnected graph patterns. Each of these graph patterns can be perceived as a simple path traversal, starting at a particular source (A) and terminating at a destination (B) by visiting vertices and edges, referred to as traversing), over the graph. Each path query is composed of one or more single-step traversals. Through function composition and currying, it is possible to define a query of arbitrary length [11].

Example 2. For instance, consider a simple path traversal to the oldest person that marko knows over the graph G as show in Fig. 1 from [11]. Listing 1.1 represents the gremlin query for the described traversal.

```
1  g.V().has("name","marko").out("knows").values("age").max()
```

Listing 1.1. Return the age of the oldest person marko knows

Functionally, this query be written using function currying as:

$$max(values_{age}(out_{knows}(has_{name=marko}(V_g))))) \qquad (1)$$

Here out_{knows}, $values_{age}$ and has_{name} are the single-step traversals. In [11], Rodriguez presents the itemization of such single-step traversals which can be used to represent a complex path traversal. Thus, through functional composition and currying one can represent a graph traversal of random length. If i be the starting vertex in G, then the traversal shown in Listing 1.1 can be represented as following function:

$$f(i) = max(\epsilon_{age} \circ v_{in} \circ e_{lab+}^{knows} \circ e_{out} \circ \epsilon_{name+}^{marko}) \ (i) \qquad (2)$$

---

[8] The Kleene star notation (A^*, B^*) denotes presence of multiple traversers in (A,B).

3.2   Evaluation of `match()`-step in Gremlin

The `match()`-step[9] matches/evaluates (cf. Definition 2) the input graph traversals/patterns (t) over a graph G, which we denote by $[\![t]\!]_G$. The start (and the end) of these graph patterns are represented using `as()`-step (step-modulators[10] i.e. naming variables) such as a, b, c, etc.). The order of execution of each graph pattern is up to the `match()`-step implementation, where the variables and path labels are local only to the current `match()`-step. Due to this uniqueness of the Gremlin `match()`-step it is possible to: *(i)* treat each graph pattern individually as a single step traversal and thus, construct composite graph patterns by joining each of these; and *(ii)* combine multiple `match()`-steps for constructing complex navigational traversals using the concatenative join (ref. Sect. 2.3). For instance, consider the GPM gremlin query as shown in Listing 1.2 from [11].

```
1  g.V().match(__.as('a').out('created').as('b'),
2            __.as('b').has('name','lop'),
3        /   __.as('b').in('created').as('c'),
4            __.as('c').has('age', 30)).select('a','c').by('name')
```

Listing 1.2. This traversal returns the names of people who created a project named 'lop' that was also created by someone who is 30 years old.

Each of the comprising four graph patterns (traversals) of the query (Listing 1.2), can be individually represented using the curried functional notation as described in Eq. 1, as:

$$f(i) = (e_{lab+}^{created}) \circ e_{out}\ (i); \quad g(i) = (\epsilon_{name+}^{lop} \circ v_{in})\ (i); \tag{3}$$

$$h(i) = (e_{lab+}^{created}) \circ e_{in}\ (i); \quad j(i) = (\epsilon_{age+}^{30} \circ v_{in})\ (i) \tag{4}$$

The input arguments of the `match()`-step are the set of graph patterns defined above in Eqs. 3 and 4, which form a composite graph pattern (the final traversal (Ψ)). At run-time, when a traverser enters `match()`-step, it propagates through each of these patterns guaranteeing that, for each graph pattern, all the prefix and postfix variables (i.e. "a","b", etc.) are *binded* with their labelled path values. In simple words, though each of these graph patterns is evaluated individually, it is ensured at run-time, that the overall structure of the composite graph pattern is preserved by mapping the path labels to declared variables. This is achieved in Gremlin by `match()` and `bind()` functions respectively.

The recursively defined `match()` function, evaluates each constituting graph pattern and keeps a track of the traversers location in the graph (i.e. path history); and the `bind()` function, maps the declared variables (elements and keys) to their respective values. Putting the recursive definition of `match()` by [10] in the context of Eqs. (3 and 4), we have:

---

[9] http://tinkerpop.apache.org/docs/3.2.3/reference/#match-step.
[10] Rodriguez and Neubauer [11] refer to step modulators as 'syntactic sugar'.

$$[\![t]\!]_g = \begin{cases} [\![bind_b(\mathbf{f}(t_{\Delta_a(t)} \wedge \Delta_{m1}))]\!]_g & : \Delta_a \neq \phi = \Delta_{m1} \\ [\![\mathbf{g}(t_{\Delta_b(t)} \wedge \Delta_{m2})]\!]_g & : \Delta_b \neq \phi = \Delta_{m2} \\ [\![bind_c(\mathbf{h}(t_{\Delta_b(t)} \wedge \Delta_{m3}))]\!]_g & : \Delta_b \neq \phi = \Delta_{m3} \quad [10] \\ [\![\mathbf{j}(t_{\Delta_a(t)} \wedge \Delta_{m4})]\!]_g & : \Delta_c \neq \phi = \Delta_{m4} \\ t & : \text{otherwise,} \end{cases} \qquad (5)$$

where, $t_{\Delta_a}(t)$ is the labelled path of traverser t. A path is labelled "a" via the step-modulator `.as()`, of the traverser in the current traversal (Ψ); Δ_{m1}, Δ_{m2}, Δ_{m3} are hidden path labels which are appended to the traversers labelled path for ensuring that each pattern is executed only once; and $bind_x(t)$ is defined as:

$$bind_x(t) = \begin{cases} t_{\Delta_x}(t) = \mu(t) & : \Delta_x(t) = \phi \\ t & : \Delta_x(t) = \mu'(t) \\ \phi & : \text{otherwise.} \end{cases} \qquad (6)$$

where $\mu'\colon \mathrm{T} \to \mathrm{U}$, is a function that maps a traverser to its current location in the graph G (e.g., $v \in V$, $V \in U$) [10]. It (μ') is analogous to μ (as in Definition 1), except that here its value is the location of a traverser in G.

4   Mapping Gremlin GPM Traversals to Graph Algebra

We now present a mapping algorithm for representing a given Gremlin traversal, in relational graph algebra, following a bottom-up approach.

1. The input query is parsed extracting its constituent individual graph patterns from the `match()` and `where()` steps.
2. For each graph pattern in the query, we first construct the curried functional form (ref. Eq. 1).
3. Map the get-vertices/edges operators for the g.V()/g.E() steps respectively.
4. Append a *traverse-operator* to all the respective `in()` and `out()` traversal steps, declared within a `match()`-step.
5. Append a *property-filter* operator to all the respective `has()` and `values()` steps declared within a `match()`-step.
6. Multiple `match()` steps can be connected by concatenative join operator.
7. Append a *selection* operator, if a `where()` step is declared.
8. Append a *projection* operator, if `select()`-step is declared.
9. Append a *deduplication* operator, if a `dedup()` step is declared.
10. Append a *sorting* operator, if the `order()` step is declared.
11. Append a *grouping* operator, if the `group()` step is declared.
12. Map the *union* operator if the query contains a `union()`-step. however, a union of multiple patterns can be constructed using a left deep join tree representation.

The sample query as shown in Listing 1.2, can be formalized as:

$$\dagger_{name}\left(\Pi_{a,c}\left([\![\underbrace{\sigma^c_{age=30} \downarrow^c_b [created]\sigma^b_{name=lop} \uparrow^b_a [created](V_g)}_{t}]\!]_g \right) \right) \qquad (7)$$

5   Related Work

Property Graphs. The Property Graph data model is one of the popular graph data models that provides a rich set of features for the user to model domain-specific real world data. The developers of Neo4J & Cypher strive at standardizing Cypher by providing open formal specification via the OpenCypher[11] project [7]. One of the limitations of Cypher is that it misses certain graph querying functionality such as the support for regular path queries and graph construction. *PGQL* [14], is an SQL-like syntax based graph query language for the PG data model. Albeit being able to overcome the limitations of Cypher, and lure the SQL community with its SQL-like syntax support, PGQL lacks standardization and support by database technology vendors.

RDF. The Resource Description Framework (RDF), is another graph data model, popular in the semantic web domain. In RDF the data (i.e., entity descriptions) are stored as triples, similar to the node-edge formalism in PGs. SPARQL [9], the query language for RDF triple stores, is a Cypher-like declarative GPM query language for querying RDF graphs. SPARQL is a W3C standard and its query algebra has been formally described in works such as [3,8], whereas its multiset semantics have been well established by works such as [2].

SQL. Relational databases cater rather limited support for executing graph queries. However, certain databases such as PostgreSQL allow the execution of recursive queries. The SAP HANA GSE prototype, using a SQL-type language proposed by [5], supports modelling high level graph queries. -

6   Conclusion and Future Work

We presented the first efforts on formalizing GPM traversals, which is a subset of the Gremlin which is both a query language and a machine. Our current work provides a theoretical foundation to leverage this advantage of Gremlin for querying graphs on various graph engines.

In near future we like to support the translation of SPARQL queries to Gremlin GPM traversals, using the proposed mapping, to leverage the advantage of using graph traversals for querying property graphs. This will enable execution of SPARQL queries over various OLAP-based graph processors and OLTP-based graph engines [13]. Allowing researchers to access a plethora of graph data management systems without the need of adapting a new graph query language.

Acknowledgments. This work is supported by the EU H2020 WDAqua ITN (GA: 642795).

---

[11] http://www.opencypher.org/.

References

1. Angles, R., Arenas, M., et al.: Foundations of modern graph query languages. arXiv preprint arXiv:1610.06264 (2016)
2. Angles, R., Gutierrez, C.: The multiset semantics of SPARQL patterns. In: Groth, P., Simperl, E., Gray, A., Sabou, M., Krötzsch, M., Lecue, F., Flöck, F., Gil, Y. (eds.) ISWC 2016. LNCS, vol. 9981, pp. 20–36. Springer, Cham (2016). doi:10.1007/978-3-319-46523-4_2
3. Cyganiak, R.: A relational algebra for SPARQL. Digital Media Systems Laboratory HP Laboratories Bristol. HPL-2005-170, vol. 35 (2005)
4. Hölsch, J., Grossniklaus, M.: An algebra and equivalences to transform graph patterns in neo4j. In: EDBT/ICDT 2016 Workshops: EDBT Workshop on Querying Graph Structured Data (GraphQ) (2016)
5. Krause, C., Johannsen, D., Deeb, R., Sattler, K.-U., Knacker, D., Niadzelka, A.: An SQL-based query language and engine for graph pattern matching. In: Echahed, R., Minas, M. (eds.) ICGT 2016. LNCS, vol. 9761, pp. 153–169. Springer, Cham (2016). doi:10.1007/978-3-319-40530-8_10
6. Li, C., Chang, K.C.-C., et al.: RankSQL: query algebra and optimization for relational top-k queries. In: Proceedings of the 2005 ACM SIGMOD International Conference on Management of Data. ACM (2005)
7. Marton, G.S.J.: Formalizing opencypher graph queries in relational algebra. Published online on FTSRG archive (2017)
8. Pérez, J., Arenas, M., Gutierrez, C.: Semantics and complexity of SPARQL. In: Cruz, I., Decker, S., Allemang, D., Preist, C., Schwabe, D., Mika, P., Uschold, M., Aroyo, L.M. (eds.) ISWC 2006. LNCS, vol. 4273, pp. 30–43. Springer, Heidelberg (2006). doi:10.1007/11926078_3
9. Prud, E., et al.: SPARQL query language for RDF. Citeulike Online Archive (2006)
10. Rodriguez, M.A.: The gremlin graph traversal machine and language (invited talk). In: Proceedings of the 15th Symposium on Database Programming Languages, Pittsburgh, PA, USA (2015)
11. Rodriguez, M.A., Neubauer, P.: The graph traversal pattern. In: Graph Data Management Techniques and Applications. IGI Global (2011)
12. Rodriguez, M.A., Neubauer, P.: A path algebra for multi-relational graphs. In: Workshops Proceedings of the 27th International Conference on Data Engineering, ICDE 2011 (2011)
13. Thakkar, H.: Towards an open extensible framework for empirical benchmarking of data management solutions: LITMUS. In: Blomqvist, E., Maynard, D., Gangemi, A., Hoekstra, R., Hitzler, P., Hartig, O. (eds.) ESWC 2017. LNCS, vol. 10250, pp. 256–266. Springer, Cham (2017). doi:10.1007/978-3-319-58451-5_20
14. van Rest, O., Hong, S., et al.: PGQL: a property graph query language. In: Proceedings of the Fourth International Workshop on Graph Data Management Experiences and Systems. ACM (2016)

Data Modeling, Data Abstraction, and Uncertainty

Online Lattice-Based Abstraction
of User Groups

Behrooz Omidvar-Tehrani[1(✉)] and Sihem Amer-Yahia[2]

[1] The Ohio State University, Columbus, USA
omidvar-tehrani.1@osu.edu
[2] CNRS, Paris, France
sihem.amer-yahia@imag.fr

Abstract. User data is becoming increasingly available in various domains from the social Web to location check-ins and smartphone usage traces. Due to the sparsity and impurity of user data, we propose to analyze labeled groups of users instead of individuals, e.g., "countryside teachers who watch Woody Allen movies." When chosen appropriately, labeled groups provide quick and useful insights on user data. Analysis of user groups is often non-trivial due its huge volume. In this paper, we introduce AUGMAN, a framework for the efficient summarization of user groups via abstraction. Our framework performs a dynamic, data-driven and lossless abstraction which helps analysts obtain high quality insights on user data without being overwhelmed. Our experiments show that AUGMANoffers representative and informative abstractions in a scalable fashion.

1 Introduction

Nowadays, huge amounts of user data are generated everywhere thanks to various sensors and unprecedented storage capacity. This data typically contains user demographics (e.g., age, gender, location) and activities (interests, eating habits, movie ratings). The analysis of user data has various applications from population studies in the Social Sciences to market segmentation for online recommendation. However, individual-based analysis of user data suffers from two drawbacks: firstly, user data is *sparse*, i.e., many attribute values are missing, and secondly, it is *noisy*, i.e., it may contain incorrect information. To tackle those inefficiencies, group-based analysis of user data has been proposed [1,2]. User groups can be formed by combining any user attributes or activities which describe a set of users. Users in a group may know each other (e.g., researchers working in the same laboratory) or not (e.g., researchers who work on similar topics but in different laboratories).

We focus on the analysis of *labeled user groups* instead of individuals. Those groups exhibit aggregated user behavior and are used in decision making. Examples are "female students in New York who have checked in Daniel restaurant", "male users in Twitter who follow Emma Stone" and "a group of people who suffer from diabetes". The literature contains various applications of user group

© Springer International Publishing AG 2017
D. Benslimane et al. (Eds.): DEXA 2017, Part I, LNCS 10438, pp. 95–110, 2017.
DOI: 10.1007/978-3-319-64468-4_7

analysis in different domains [1, 3–7]. Group-based analysis enables new insights which help data consumers make better decisions [8, 9]. Example 1 depicts an application in practice.

Example 1. Anderson, a data analyst, wants to help a telecommunication company develop a list of "jobs to be done" for new products and services. He has already made some interviews with customers regarding the way they use their mobile devices throughout the day. After analyzing customer stories, he found two interesting groups of users: *i. multi-usage customers* requiring 2 SIM cards for the same smartphone; *ii. customers who are mothers* requiring parental control and other relevant applications which facilitate their lifestyle as mothers. Discovery of these two groups calls for a distinct analysis and can help Anderson purpose.

A critical challenge for user group analysis is Information Overload (IO): for a dataset, there can be millions of groups, hindering their effective analysis. It is a tedious task for an analyst to skim over all groups to find an interesting subset. We consider the challenge of IO in two different stages of group-based user data analysis: *investigation* and *presentation*.

- At the stage of investigation, the analyst explores different options to make initial decisions and pick further analysis directions. A *careful selection of groups* to present to the analyst will provide a big picture that prevents IO and enables effective investigation. The group space should be summarized in such a way that a smaller number of groups faithfully represents the whole group space.
- At the stage of presentation (usually in the form of visualization), the analyst selects a group of interest and asks for details. The analyst will make up her mind based on provided details. A *concise representation* of those details in such a way that they carry informative insights, will address IO.

A variety of approaches is proposed to tackle IO for user groups such as compression [10], interactive navigation [1] and constrained optimization [11]. However most of those approaches are lossy and require a-priori domain knowledge in form of ontologies, rules and constraints. In this paper, we propose an *abstraction* approach to summarize the group space in a lossless fashion. Abstraction provides a concise view of user groups with the ability to ask for details on demand. For instance, the group of "young females who watched the movie Flicka (2006)" and the group of "males who watched the movie Schindler's List (1993)" can be both abstracted to the group of "drama genre lovers". Abstraction addresses both stages of user data analysis: investigation and presentation. Rewriting multiple groups into a single abstracted group will summarize the group space for investigation and reduce visual clutter in presentation.

In the literature, abstraction is often defined based on domain-specific taxonomies which dictate the generalization process [12–15]. However, such taxonomies are often hand-crafted and therefore expensive to produce. Also the abstraction functionality heavily depends on the nature of the taxonomy such as its height and branching factor [13]. In this paper, we exploit the opportunity of using the *user group lattice* for abstraction. The lattice is automatically generated in the process of user group discovery [16,17], hence no input taxonomy is required. An example of a partial user group lattice is illustrated in Fig. 1. For instance, both groups of "Californian students" and "Californian engineers" (in level 2) can be abstracted to the group of "Californians" (in level 1).

At the core of lattice-based abstraction lies the notion of "quality". In user data analysis, "quality" is defined as a function of user behavior: an abstracted group should reflect the behavior of groups it contains. Quality can be captured either as a score (e.g., the rating score in MOVIELENS [18], a movie-rating dataset) or frequency (e.g., number of times a researcher has published a paper in DEXA). A larger score or frequency can be assumed to be equivalent as higher quality. For instance, the group of "females who watched the movie Flicka" can be abstracted to the group of "females who watched drama genre movies" if the average of rating scores (as an indicator of quality) in the latter group is very close to the former group's. However the group of "females who watched drama genre movies" cannot be abstracted to the group of "all drama genre lovers" as their average rating score differs significantly.

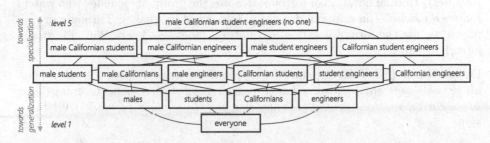

Fig. 1. User group lattice

In this paper, we propose AUGMAN, a framework for user group abstraction. AUGMAN provides an AUGmented MANagement methodology for user groups and makes three contributions: (*i*) a new abstraction semantics based on a lossless notion of quality, (*ii*) a taxonomy-free abstraction approach that relies on the lattice induced by group discovery, and (*iii*) an efficient algorithm for computing abstractions on-the-fly.

The outline of the paper is as follows: in Sect. 2, we discuss our data model. In Sect. 3, we provide necessary definitions and formalize our abstraction problem. Then in Sect. 4, we describe AUGMAN. Section 5 presents detailed experiments. The related work is provided in Sect. 6. We conclude and discuss perspectives in Sect. 7.

2   User Data and Groups

We model user data as a set of items $\mathcal{I}$, users $\mathcal{U}$, and users' activities $\mathcal{R}$. Each record $r \in \mathcal{R}$ is a tuple $\langle u, i, s \rangle$, where $i \in \mathcal{I}$, $u \in \mathcal{U}$ and s is the score that u assigns to i. Note that s is optional and may not be provided in a user dataset. A set of attributes $\mathcal{A}$ are associated to both users and items. For a given item $i \in \mathcal{I}$, $attribs(i)$ returns the set of all attribute-value pairs of i on the schema $\mathcal{A}$. For instance in MOVIELENS, $attribs(Flicka) = \{\langle director, Michael\ Mayer \rangle, \langle genre,$ $drama \rangle\}$. Also for a given user $u \in \mathcal{U}$, $attribs(u)$ returns the set of all attribute-value pairs of u on the schema $\mathcal{A}$. For a given subset of items $\mathcal{I}' \subseteq \mathcal{I}$, $attribs(\mathcal{I}')$ returns the *intersection* of attribute-value pairs for all items $i \in \mathcal{I}'$. Given a subset of users $\mathcal{U}' \subseteq \mathcal{U}$, the same definition holds for $attribs(\mathcal{U}')$.

Definition 1 (User Group). *A group g is a set of records $\langle u, i, s \rangle \in \mathcal{R}$ described by a set of attribute-value pairs shared among the users and the items of those rating records. The description of a group g is defined as $attribs(g_u) \cup attribs(g_i)$ where $g_u \subseteq \mathcal{U}$ and $g_i \subseteq \mathcal{I}$ are the set of all users and items in g's records respectively. By $\mathcal{G}$, we denote the set of all user groups.*

For instance, the user group g described by $\{\langle gender, female \rangle, \langle genre,$ $romance \rangle\}$ (formal notation of a group) denotes the group of "females who watch romance movies" (description of a group in natural language). Throughout the paper, we use the formal notation and natural language description of a group interchangeably.

Definition 2 (User Group Lattice). *Similarly to data cubes [19], the set of all possible user groups form a lattice τ where nodes correspond to groups and edges correspond to ancestor (downwards) and descendant (upwards) relationships.*

Consider the group of "male Californian students" in Fig. 1 with 34 members. An *ancestor* of this group is "male students" with 120 members. We also use the term *parent* to refer to immediate ancestors. The group of "students" with 184 members is another ancestor of that group. Each group in the lattice represents one possible combination of attribute-value pairs on the schema $\mathcal{A}$. Figure 1 illustrates a small part of the lattice. Considering an extremely small case of $|\mathcal{A}| = 4$ where each attribute has only 5 values, the complete lattice will have 1,048,576 nodes (groups), i.e., $2^{4 \times 5}$. In a lattice, the downward direction is *generalization* (towards shorter group descriptions and more records) and the upward direction is *specialization* (towards longer group descriptions and less records).

3   Lattice-Based Abstraction

The size of the user group space is exponential in the number of attribute-value pairs on the schema $\mathcal{A}$. We focus on the challenge of IO which complicates the analysis of user groups and discovery of insights. In this section, we discuss some preliminary concepts and formally define our lattice-based abstraction problem.

3.1 Preliminaries

We define abstraction based on the notion of "quality". In user data, quality can be defined in two different ways.

- **Explicit Quality.** Some user datasets (mostly collaborative rating datasets such as MOVIELENS [18], BOOKCROSSING [20] and LASTFM [21]) provide an explicit notion of quality in form of a *rating score*. For instance, the record ⟨*Tina, Titanic*, 4.5⟩ in MOVIELENS shows that Tina highly appreciates the movie Titanic (1997).
- **Implicit Quality.** In case scores are not available, implicit quality can be approximated using *frequency*. For instance, the record ⟨*Abdelkader Hameurlain, DEXA*⟩ occurs 23 times in DBLP [22] which means that A. Hameurlain has published frequently in DEXA conference.

In this paper, we use the term "quality" to refer to it be it implicit or explicit. For a given group g, we define quality $\vartheta(g)$ as the aggregation of quality scores of g's records. For a given set of n groups, the quality $\vartheta(g_1, g_2 \ldots g_n)$ is defined as the aggregation of quality scores of all records in n groups. The quality function $\vartheta()$ gives an approximate view of the members' tastes in n groups. The aggregation function can be average, minimum, maximum, diameter, etc. We use "average" as the default aggregation function and evaluate the effect of each aggregation function in Sect. 5.

Definition 3 (Lattice-based Quality Waste). *Given a set of groups* $g_1, g_2 \ldots g_n$ *and their lowest common ancestor* $\hat{g}$ *in the lattice* τ, *the lattice-based quality waste denoted* $\omega^\tau(\{g_1, g_2 \cdots g_n\}, \hat{g}) = |\vartheta(g_1, g_2 \ldots g_n) - \vartheta(\hat{g})|$ *is the difference of quality score between groups and their common ancestor in the lattice* τ.

The intuition behind lattice-based quality waste ω is as follows: if the same quality score is preserved by an ancestor group, descendant groups can be safely replaced by their ancestors thereby reducing the size of the group space. In other words, if the same quality is achieved in more general layers of the lattice τ, we can wrap up the lattice up to those layers.

Definition 4 (Valid Abstraction). *Given a sensitivity threshold* ρ, *the lattice* τ, *a set of groups* $g_1, g_2 \ldots g_n$ *and their lowest common ancestor* $\hat{g}$, *we say that* $\hat{g}$ *is a valid abstraction of* $g_1, g_2 \ldots g_n$ *iff* $\omega^\tau(\{g_1, g_2 \ldots g_n\}, \hat{g}) \leq \rho$.

Sensitivity threshold ρ denotes the tolerable amount of quality waste. If ρ is very close to zero, few abstractions will occur with highly preserved quality.

Definition 5 (Maximal Abstraction). *Given a sensitivity threshold* ρ *and the lattice* τ, *we say that* $\hat{g}$ *is a maximal abstraction of* $g_1, g_2 \ldots g_n$ *iff* $\hat{g}$ *is a valid abstraction of the* n *groups (based on* ρ *on the lattice* τ*) and* $\nexists g \in \mathcal{G}$ *s.t.* $\omega^\tau(\{g_1, g_2 \ldots g_n\}, g) \leq \rho$ *and* g *is a descendant of* $\hat{g}$.

Definition 6 (Abstraction Level). *Given a sensitivity threshold ρ and the lattice τ, we say $\hat{g}$ is a h-level abstraction ($h \geq 1$) for the set of groups $g_1, g_2 \ldots g_n$, if $\hat{g}$ is a valid abstraction of the n groups (based on ρ on the lattice τ) and there exists h levels of difference in the lattice τ between $\hat{g}$ and the n groups.*

An example of a one-level abstraction ($h = 1$) is abstracting the group of "Californian engineers" to "engineers" based on the lattice in Fig. 1. We denote the set of all one-level abstractions as Φ with the schema $\langle g, \hat{g}, \omega^\tau(g, \hat{g}) \rangle$.

We consider an example for abstraction in MOVIELENS. Given $\rho = 0.2$, a group g_1 described by "engineers who watch Batman (1966), Jurassic Park (1993) and other Sci-Fi movies" with $\vartheta(g_1) = 3.4$ is abstracted to the group g_2 described by "engineers who watch Sci-Fi movies" with $\vartheta(g_2) = 3.2$ as $\omega^\tau(g_1, g_2) = 0.2 \leq \rho$. Also, the group g_3 described by "engineers who watch Ghostbusters (2016) and other Sci-Fi movies" with $\vartheta(g_3) = 3.1$ is also abstracted to g_2. In case $\rho = 0.1$, g_1 would not be abstracted to g_2. The intuition is that the abstracted group is as "good" as the non-abstracted group(s), hence presenting the abstracted group during the analysis process is adequate for decision making. Indeed, as choices get limited (g_1 and g_3 both become g_2), the analyst has less burden in the decision process by taking care of fewer options.

3.2 Problem Definition

Based on the above definitions, we can now formally define our problem. Given a set of n groups $\{g_1, g_2 \ldots g_n\} \subseteq \mathcal{G}$, the lattice τ induced by $\mathcal{G}$ and a sensitivity threshold ρ, the **abstraction problem** is to provide the set of all valid and maximally abstracted groups for the n given groups.

Abstraction problem is hard due to its exponential search space. All subsets of the lattice should be verified for a potential abstraction. Hence 2^n scans are required where n itself is exponential in the number of attribute-value pairs on the schema $\mathcal{A}$. The ad-hoc nature of input adds up to the complexity of the problem (the analyst may request to abstract any n groups), as the abstraction should be performed on-the-fly within sub-seconds.

4 The AugMan Framework

We present AUGMAN, a framework that enables efficient lattice-based abstraction of user groups. Figure 2 shows the overall framework where gray boxes illustrate different components. A running example is shown below each component. The main focus in the architecture of AUGMAN is on efficiency in order to enable on-the-fly abstractions. AUGMAN first produces user groups (C1) and generates some meta-data for them (C2). Then it exploits meta-data to obtain valid maximal abstracted groups efficiently (C3). We describe each component of AUGMAN as follows.

C1. User Group Discovery. The set of user groups $\mathcal{G}$ is generated from input user data in an offline process. This one-time process is independent from

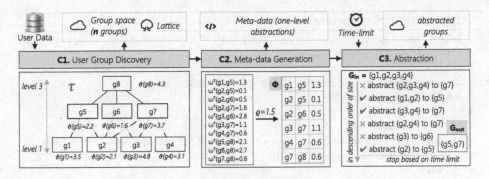

Fig. 2. AugMan framework

Algorithm 1. Meta-data Generation Algorithm

Input: ρ, τ

1   $\Phi \leftarrow \emptyset$
2   **for** *each g generated by α-MOMRI and for each of g's parents $\hat{g}$ in τ* **do**
3 | **if** $\omega^\tau(g, \hat{g}) \leq \rho$ **then** $\Phi.append(g, \hat{g}, \omega^\tau(g, \hat{g}))$
4   **end**
5   **return** Φ

AugMan and can be done with different algorithms such as clustering [23,24], community detection [25], pattern mining [26], and team formation [4]. The choice of the algorithm depends on the nature of the dataset and the analysis tasks. We stress that the focus of abstraction differs from discovery algorithms (such as clustering), where the former is a foundation for the latter. For AugMan, we employ α-MOMRI (an algorithm we developed in [17]) as our choice for group discovery. The algorithm solves a multi-objective group discovery problem efficiently and returns a group-set of size n.

C2. Meta-data Generation. α-MOMRI builds the lattice of user groups while generating $\mathcal{G}$. We extend the functionality of α-MOMRI for AugMan so that it stores some *meta-data* about each generated group. Algorithm 1 illustrates this process. We store meta-data in form of valid one-level abstractions in Φ. As only certain groups constitute a valid one-level abstraction, the size of Φ is much smaller than the group space. For instance in Fig. 2, the one-level abstraction of the group g_3 to g_6 will not be materialized in Φ as $\omega^\tau(g_3, g_6) = 2.8 > \rho = 1.5$ (i.e., not a valid abstraction). The meta-data collection enables AugMan to be independent from the set $\mathcal{G}$ in the online execution and functions only on the smaller set of generated meta-data in Φ. It is shown in Sect. 5.3 that at least 50% of the user space is pruned in Φ.

C3. Abstraction. AugMan takes a set of n groups and exploits abstraction opportunities. Any subset of n groups should be verified for abstraction. Algorithm 2 shows this process. Single abstractions are already computed in Φ and

---

Algorithm 2. Abstraction Algorithm

Input: input groups $G_{in} = \{g_1, g_2 \ldots g_n\} \subseteq \mathcal{G}$, Φ, time limit Γ
Output: output groups G_{out}

1 $G \leftarrow \infty$
2 $\Delta \leftarrow$ all subsets of G_{in} in descending order of cardinality
3 **while** Γ *not exceeded* **do**
4 | $\quad S \leftarrow \Delta.next()$
5 | $\quad aggr\_abstract \leftarrow \emptyset$
6 | $\quad$ **for** $g \in S$ // *aggregate quality wastes using Φ for multiple abstractions* **do**
7 | $\quad$ | $\quad aggr\_abstract.append(\Phi(g))$
8 | $\quad$ **end**
9 | $\quad S' \leftarrow maximal\_abstraction(S, aggr\_abstract)$
10 | $\quad G.append(S')$
11 **end**
12 $G_{out} \leftarrow G$ in descending order of their $aggr\_abstract$ score
13 **return** G_{out}

---

multiple abstractions (for more than one group) are computed using the meta-data in Φ dynamically. We exploit native hash indexes for fast retrieval of meta-data for a specific group in Φ.

We describe an example of the dynamic retrieval of meta-data from Φ to perform multiple abstractions. Consider groups g_1 and g_2 and their parent g_5 in the lattice τ illustrated in Fig. 2. The quality score of g_1's records are $\{4, 2, 3, 5\}$ ($\vartheta(g_1) = 3.5$), for g_2's records are $\{2, 1, 2, 2.5, 3\}$ ($\vartheta(g_2) = 2.1$) and for g_5's records are $\{2, 1, 2, 1, 5\}$ ($\vartheta(g_5) = 2.2$). During the online process, AUGMAN has access only to the content of Φ, i.e., $\langle g_1, g_5, 1.3 \rangle$ and $\langle g_2, g_5, 0.1 \rangle$. Based on Φ, $\vartheta(\{g_1, g_2\}, g_5) = avg(1.3, 0.1) = 0.7$, i.e., the exact same score if computed on non-abstracted groups g_1 and g_2. Hence exact multiple abstractions can be obtained using the data in Φ without scanning $\mathcal{G}$.

Given an input set of groups G_{in}, Algorithm 2 iterates on all its subsets. We ignore subsets whose groups do not exist in Φ (e.g., abstracting $\{g2, g4\}$ to $g7$ in Fig. 2) or do not share a common parent (e.g., $\{g1, g4\}$ in Fig. 2). As the number of possible abstractions is large, we consider a *time limit* parameter Γ to bound user waiting time (line 3). Hence the abstraction algorithm employs a greedy *best-effort* strategy to return the set of abstractions with the least quality waste (Definition 3) within the time limit. The algorithm favors abstractions of subsets with larger cardinality (line 2), hence even with a restrictive Γ, abstractions which contribute to group space reduction will be executed first. In practice, we often set Γ to 400 ms in accordance with productivity degradation limit [27]. Once the time limit expires, abstractions will be returned in decreasing order of their quality waste.

Algorithm 2 returns maximally abstracted groups (line 9). Note that despite the optimization-based and iterative model of maximal abstraction, we do not necessarily reach the root of the lattice (which makes a useless abstraction). This is because the quality function ϑ is not (anti)-monotonic on the lattice.

5   Experiments

In this section, we evaluate the effectiveness and usability of AUGMAN. The main goal is to validate whether AUGMAN is helpful for user group management. We discuss our results for each evaluation. All experiments are implemented in Python 2.7.10 on a 2.4 GHz Intel Core i5 machine with an 8 GB main memory, running OS X 10.12.

We use two user datasets MOVIELENS 1M and BOOKCROSSING. Both datasets have explicit quality scores ranging from 1 to 5. However, the functionality of AUGMAN does not depend on the quality type (being explicit or implicit), hence other datasets can also be employed. MOVIELENS [18] contains 1,000,209 anonymous rating records of 3,952 movies by 6,040 users. It has four user attributes: *gender*, *age*, *occupation* and *zipcode*. We enrich MOVIELENS with IMDb [28] to get movie attributes, i.e., *genres*, *director*, *writer* and *release year*.[1] BOOKCROSSING [20] contains 278,858 users providing 1,149,780 ratings about 271,379 books. The number of users and items are one order of magnitude larger than MOVIELENS. There are only two attributes for each user: *age* and *location*. BOOKCROSSING also offers information on each book (item), i.e., *writer*, *release year* and *publisher*. For the sake of comparability of results, we randomly pick 10M groups from both datasets and focus on those two sets throughout the experiments.

We evaluate qualitative aspects of AUGMAN by performing a user study (Sect. 5.1). We also evaluate quantitative aspects by measuring abstraction quality, space reduction and response time (Sects. 5.2, 5.3 and 5.4). In summary, we observe that users prefer to fulfill a decision-making process over abstracted groups generated by AUGMAN. We also observe that AUGMAN performs on average within one second. Both the sensitivity threshold and group space size influence the execution time. We also show that the number of abstracted groups is a function of the dataset distribution, the sensitivity and aggregation function.

5.1   User Study

We validate qualitative aspects of AUGMAN in a user study. We recruited 25 participants who are all students in Computer Science and evenly distributed in gender. For a given subset of records in MOVIELENS, i.e., ratings of Woody Allen movies, participants were given a set of abstracted and non-abstracted groups side-by-side. They expressed their preference on a 5-star scale ("5" being the most preferred) on the informativeness and representativeness of user groups for describing Woody Allen movies. Figure 3 illustrates results. We only report on MOVIELENS as results on both datasets are closely identical.

Participants received various sets of abstracted and non-abstracted groups obtained from different group space sizes (left chart) and sensitivity values (right chart). In general, abstracted groups are preferred to non-abstracted ones in most cases specifically in larger group spaces.

---

[1] *The dataset is publicly available at* https://goo.gl/ZQ6doV.

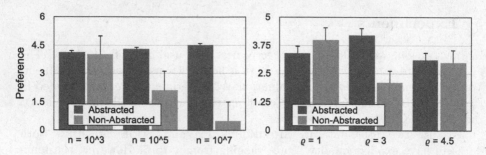

Fig. 3. User study on different group space sizes (left) and sensitivity thresholds (right)

By increasing group space size, we observe that the preference for non-abstracted groups decreases drastically (from 3.9 to 0.5) while for abstracted groups, it stays almost constant (increasing from 4.1 to 4.5). The reason is that participants can potentially make sense out of non-abstracted groups in smaller datasets. But when the space becomes larger (hence less manageable), the need for an abstracted view increases. Note that even in the smallest space size, abstraction is preferred as much as non-abstraction.

Sensitivity threshold manages the tolerance of quality waste. The lower its value, the fewer the abstraction opportunities. We observe a sweet spot, i.e., $\rho = 3$ where the preference for abstraction is highly contrasted in comparison to non-abstraction, i.e., 42% of difference in preference. When $\rho = 1$, non-abstraction is the winner due to lack of abstraction opportunities. However, few abstractions have a high quality and capture a preference of 3.4 on average. When $\rho = 4.5$, preferences are not distinguishable due to many meaningless abstractions. Larger values of sensitivity obstruct the semantics of abstraction, hence participants can hardly promote abstracted groups to non-abstracted ones.

We conclude from our user study that AUGMAN is the method of choice for managing large spaces of groups with a preference margin of 4.0. However, the best functionality of abstraction comes with careful tuning of sensitivity threshold.

5.2   Abstraction Quality

The quality function $\vartheta()$ dictates the functionality of abstraction. It is formulated using the aggregation function and the sensitivity threshold. We evaluate the effect of these two parameters on the quality function. We consider the following aggregation functions in our experiment: average (avg), minimum (min), maximum (max) and diameter (difference of maximum and minimum). We vary ρ from zero (no quality waste permitted) to one (no quality constraint).

To compare different aggregation functions, we measure the ratio of applied abstractions over the total number of possible abstractions. We report the ratio in percentage for a clearer comparison between different cases and datasets. Figure 4 illustrates the results. Naturally, increasing ρ leads a larger number of

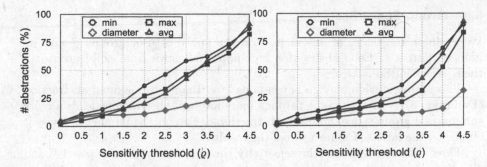

Fig. 4. Effect of sensitivity threshold and aggregation functions on number of abstractions (MOVIELENS on left, BOOKCROSSING on right)

abstractions. This is because a more relaxed sensitivity value (larger ρ) opens room for more abstractions though with more quality waste.

The increase in abstractions is linear in MOVIELENS while in BOOKCROSSING, it is close to exponential. The reason is that MOVIELENS is a denser dataset whose rating score distribution is close to normal: in MOVIELENS, the average number of rating records per user is 4.14 times larger than BOOKCROSSING. Hence increasing ρ leads a linear increase. However, the skewed distribution of BOOKCROSSING leads many more abstractions for larger values of ρ.

The largest number of abstractions is achieved using min and avg functions. The aggregation function $diameter$ is the least successful function as it is very sensitive to outliers: presence of one extreme rating score "1" or "5" can easily return a large value for $diameter$. The function avg is a balance between other extreme functions. Hence, we use average as the default aggregation function throughout the paper.

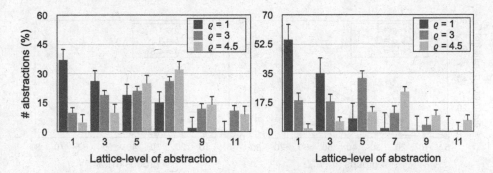

Fig. 5. Lattice-level analysis (MOVIELENS on left, BOOKCROSSING on right)

5.3 Group Space Size

We evaluate the effect of abstraction on the space of user groups $\mathcal{G}$ and its induced lattice τ. Each abstraction is a potential move from child groups to their parent in the lattice.

This representation may go beyond one level thanks to maximal abstraction (Definition 5). Note that going through more levels of abstractions leads a more summarized group space. For each abstraction, we count the number of levels it climbs on the lattice. Figure 5 shows the results.

Three different values of the sensitivity threshold are reported: $\rho = 1, 3$ and 4.5. Note that the largest value of ρ cannot be 5 as it leads an unconditional abstraction which ends up at the τ's root. We observe a peak for each case: for $\rho = 1$, there is a peak on one-level abstractions (37% of abstractions), for $\rho = 3$, on 5- to 7-level (21 to 26% of abstractions), and for $\rho = 4.5$, it is on 7-level (32% of abstractions). This implies that in case of smaller values of sensitivity, most abstractions stop at the first level. Abstractions progress to more levels when relaxing this constraint. In MOVIELENS, the peaks are smoother due to its normal distribution.

Note that in case of BOOKCROSSING, around 27% of abstractions progress even above level 11 (not shown in Fig. 5), while for MOVIELENS, it is only 3%. This is because the height of the lattice for the former dataset is one order of magnitude larger, hence there are higher chances of iterative abstractions.

Beside the vertical analysis (levels of lattice), we also evaluate the horizontal aspect of the lattice: abstraction reduces the number of choices. Human brains get easily distracted by the plethora of options [29]. Given n groups, we evaluate how many abstracted groups will be returned. Figure 6 shows the result. We consider different portions of the group space $\mathcal{G}$ (from 5% to 80%) to observe its corresponding number of abstracted groups. We consider different values of the sensitivity threshold for this experiment.

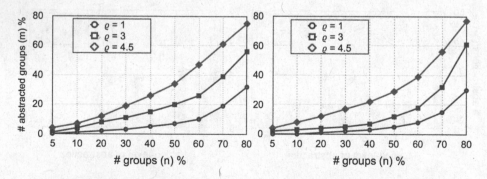

Fig. 6. Choice reduction analysis (MOVIELENS on left, BOOKCROSSING on right)

The general observation is that for larger values of sensitivity, there exists a linear relation between the size of the group space and abstractions. However, by restricting the threshold, an exponential behavior appears. The potential reason is that in a bigger space and a restricted ρ, there are higher chances and a larger number of combinations to make abstractions, hence the number of abstractions increases drastically. Another observation is that the slope of exponentiation for BOOKCROSSING is sharper than the other dataset. It is due to the normal distribution of MOVIELENS and more availability of abstractions in middle stages.

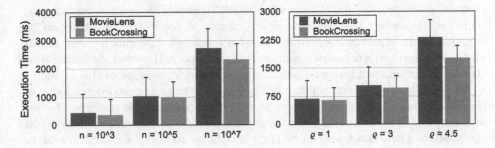

Fig. 7. Performance by varying group space size (left) and sensitivity (right)

5.4 Performance

We evaluate the performance of AUGMAN by considering various sensitivity thresholds and group space sizes. We measure the execution time after 20 runs. In this experiment, we consider $\Gamma = \infty$ (i.e., no time limit is considered.) Fig. 7 illustrates the results. In the left chart, we vary the group space size from 10^3 to 10^7 records by fixing $\rho = 3$. In the right chart, we vary the sensitivity threshold from 1 to 4.5 by fixing $n = 10^5$. MOVIELENS as the denser dataset, requires more time for abstractions by increasing n or relaxing the sensitivity threshold. Note that although the abstraction operates on the huge space of user groups, it is executed in average in 1302 ms. This efficiency is obtained thanks to meta-data table Φ and proper indexing.

6 Related Work

In this paper, we addressed the problem of information overload in the space of user groups. To the best of our knowledge, no prior work has formalized an efficient abstraction framework for managing user groups in all stages of user data analysis. However, our work does relate to some others in its functionality.

User Group Management. To make the space of user groups manageable, different approaches have been proposed in the literature. In [10], Vreeken et al. propose an approach to return a subset of user groups that compresses the whole set of groups in the best possible way. This idea is based on Minimal

Description Length (MDL) principle which leads to pruning many user groups. The approach is a lossy analysis and there is a high probability that one or more interesting groups are pruned. Instead of pruning, another approach is to interactively navigate in the group space while optimizing one or more desiderata at each step to reach a subset of interesting user groups [1, 30]. This methodology is complementary to ours and can be employed (e.g., as an interactive engine) to enrich the framework.

Abstraction. The notion of abstraction has been discussed in various domains such as query processing [31], visualization [32], pervasive computing [33] and NLP [34]. However, there has been few efforts of recognizing abstraction as a management strategy. Constraint-based approaches [35] can be seen as an abstraction method where constraints restrict the space. However, designing constraints is not an easy task and requires an appriori knowledge of the dataset.

In [36, 37], an approach is proposed to learn the model of prior knowledge of the analyst based on her preferences so that the space can be restricted to her interests. In [36], the analyst has to order her preferences which puts burden on the analyst. These methods can be complementary to ours to make the system feedback-based.

The closest work to ours is [13], where a taxonomy-based abstraction methodology is proposed. The idea of abstraction is based on the notion of "usage" (using an application on smartphones, navigating a website, etc.). However, usage is only one specific type of user activity and is not available for all datasets. Moreover, the abstraction requires a hand-crafted taxonomy as input which is expensive and time-consuming to provide. In AUGMAN, we consider two different types of activity quality, i.e., rating score (explicit) and frequency (implicit). Also the framework does not need any apriori knowledge as it builds up on the pre-computed lattice.

7 Conclusion

We addressed the problem of information overload in the context of user groups in different stages of user data analysis. We proposed AUGMAN, a framework for an efficient abstraction of user groups based on the lattice. We validated different aspects of our framework in quantitative and qualitative experiments on MOVIELENS and BOOKCROSSING datasets.

Our immediate course of action is to build a visualization layer on top of AUGMAN to visually present abstracted groups to analysts. The visualization should be able to present group abstractions and provide details on demand. In the future, we would like to consider feedback as a way to capture an analyst's interest and incorporate AUGMAN in an interactive context. Analyst feedback will then influence the way groups get abstracted in consecutive analysis steps. We also aim to address parallelism to enable more efficiency. All aggregation functions in AUGMAN are totally-algebraic hence their functionality can be easily extended to shared-nothing parallel data processing frameworks like Hadoop and Spark [38].

References

1. Omidvar-Tehrani, B., Amer-Yahia, S., Termier, A.: Interactive user group analysis. In: CIKM 2015 (2015)
2. Parida, L.: Redescription mining: structure theory and algorithms. In: AAAI 2005 (2005)
3. Amer-Yahia, S., Tehrani, B.O., Roy, S.B., Shabib, N.: Group recommendation with temporal affinities. In: EDBT (2015)
4. Kargar, M., An, A., Zihayat, M.: Efficient bi-objective team formation in social networks. In: Flach, P.A., Bie, T., Cristianini, N. (eds.) ECML PKDD 2012. LNCS, vol. 7524, pp. 483–498. Springer, Heidelberg (2012). doi:10.1007/978-3-642-33486-3_31
5. Cao, C.C., She, J., Tong, Y., Chen, L.: Whom to ask?: jury selection for decision making tasks on micro-blog services. VLDB (2012)
6. Newman, M.E.J., Girvan, M.: Finding and evaluating community structure in networks. Phys. Rev. E **69**(2), 026113 (2004)
7. Van Leeuwen, M., Ukkonen, A.: Discovering skylines of subgroup sets. In: Blockeel, H., Kersting, K., Nijssen, S., Železný, F. (eds.) ECML PKDD 2013. LNCS, vol. 8190, pp. 272–287. Springer, Heidelberg (2013). doi:10.1007/978-3-642-40994-3_18
8. Jordan, M., Pfarr, N.: Forget the quantified-self, we need to build the quantified-us (2014)
9. Bayer, J., Taillard, M.: Story-driven data analysis (2013)
10. Vreeken, J., Van Leeuwen, M., Siebes, A.: Krimp: mining itemsets that compress. Data Mining Knowl. Discov. **23**(1), 169–214 (2011)
11. Das, M., Amer-Yahia, S., Das, G., Yu, C.: Mri: meaningful interpretations of collaborative ratings. PVLDB 4(11), 1063–1074 (2011)
12. Fopa, L., Jouanot, F., Termier, A., Tchuente, M., Iegorov, O.: Benchmarking of triple stores scalability for MPSoC trace analysis. In: 2nd International workshop on Benchmarking RDF Systems (BeRSys 2014) (2014)
13. Omidvar-Tehrani, B., Amer-Yahia, S., Termier, A., Bertaux, A., Gaussier, É., Rousset, M.-C.: Towards a framework for semantic exploration of frequent patterns. In: IMMoA (2013)
14. Srikant, R., Agrawal, R.: Mining Generalized Association Rules. IBM Research Division, New York (1995)
15. Marinica, C., Guillet, F., Briand, H.: Post-processing of discovered association rules using ontologies. In: ICDMW. IEEE (2008)
16. Uno, T., Kiyomi, M., Arimura, H.: LCM ver. 2: efficient mining algorithms for frequent/closed/maximal itemsets. In: Workshop on Frequent Itemset Mining Implementations (2004)
17. Omidvar-Tehrani, B., Amer-Yahia, S., Dutot, P.-F., Trystram, D.: Multi-objective group discovery on the social web. In: Frasconi, P., Landwehr, N., Manco, G., Vreeken, J. (eds.) ECML PKDD 2016. LNCS, vol. 9851, pp. 296–312. Springer, Cham (2016). doi:10.1007/978-3-319-46128-1_19
18. Grouplens. Movielens dataset: Grouplens research group. http://grouplens.org/datasets/movielens/
19. Gray, J., Chaudhuri, S., Bosworth, A., Layman, A., Reichart, D., Venkatrao, M., Pellow, F., Pirahesh, H.: Data cube: a relational aggregation operator generalizing group-by, cross-tab, and sub-totals. Data Mining Knowl. Discov. **1**(1) (1997)
20. Ziegler, C.-N.: Book-crossing dataset. http://www2.informatik.uni-freiburg.de/~cziegler/BX/

21. LastFM. Million song dataset. https://labrosa.ee.columbia.edu/millionsong/lastfm
22. DBLP. Bibliographic database for computer sciences. https://hpi.de/naumann/projects/repeatability/datasets/dblp-dataset.html
23. Agrawal, R., Gehrke, J., Gunopulos, D., Raghavan, P.: Automatic subspace clustering of high dimensional data for data mining applications, vol. 27. ACM (1998)
24. Amiri, B., Hossain, L., Crowford, J.: A multiobjective hybrid evolutionary algorithm for clustering in social networks. In: Proceedings of the 14th Annual Conference Companion on Genetic and Evolutionary Computation. ACM (2012)
25. Cruz, J.D., Bothorel, C., Poulet, F.: Entropy based community detection in augmented social networks. In: CASoN. IEEE (2011)
26. Agrawal, R., Imieliński, T., Swami, A.: Mining association rules between sets of items in large databases. In: SIGMOD (1993)
27. Liu, Z., Heer, J.: The effects of interactive latency on exploratory visual analysis. IEEE TVCG **20**(12) (2014)
28. IMDb. Internet movie database. http://www.imdb.com
29. Miller, G.: Human memory and the storage of information. IRE Trans. Inf. Theory **2**(3), 129–137 (1956)
30. Kamat, N., Jayachandran, P., Tunga, K., Nandi, A.: Distributed and interactive cube exploration. In: 2014 IEEE 30th International Conference on Data Engineering (ICDE). IEEE (2014)
31. Huh, S.-Y., Moon, K.-H., Lee, H.: A data abstraction approach for query relaxation. Inf. Softw. Technol. **42**(6), 407–418 (2000)
32. Bertini, E., Santucci, G.: Quality metrics for 2D scatterplot graphics: automatically reducing visual clutter. In: Butz, A., Krüger, A., Olivier, P. (eds.) SG 2004. LNCS, vol. 3031, pp. 77–89. Springer, Heidelberg (2004). doi:10.1007/978-3-540-24678-7_8
33. Kabadayi, S., Julien, C.: A local data abstraction and communication paradigm for pervasive computing. In: PerCom. IEEE (2007)
34. Sankar, K., Sobha, L.: An approach to text summarization. In: PCLIAWS3. Association for Computational Linguistics (2009)
35. Bonchi, F., Giannotti, F., Mazzanti, A., Pedreschi, D.: ExAnte: anticipated data reduction in constrained pattern mining. In: Lavrač, N., Gamberger, D., Todorovski, L., Blockeel, H. (eds.) PKDD 2003. LNCS (LNAI), vol. 2838, pp. 59–70. Springer, Heidelberg (2003). doi:10.1007/978-3-540-39804-2_8
36. Xin, D., Shen, X., Mei, Q., Han, J.: Discovering interesting patterns through user's interactive feedback. In: KDD (2006)
37. De Bie, T., Kontonasios, K.-N., Spyropoulou, E.: A framework for mining interesting pattern sets. SIGKDD Explor. **12**, 92–100 (2011)
38. Nandi, A., Yu, C., Bohannon, P., Ramakrishnan, R.: Distributed cube materialization on holistic measures. In: 2011 IEEE 27th International Conference on Data Engineering (ICDE), pp. 183–194. IEEE (2011)

Probabilistic MaxRS Queries on Uncertain Data

Yuki Nakayama[✉], Daichi Amagata, and Takahiro Hara

Department of Multimedia Engineering,
Graduate School of Information Science and Technology, Osaka University,
Yamadaoka 1-5, Suita, Osaka, Japan
nakayama.yuki@ist.osaka-u.ac.jp

Abstract. Given a set of spatial objects with scores and a size of a
rectangle, MaxRS (Maximizing Range Sum) queries retrieve the location
of the rectangle which maximizes the sum of the scores of all objects
covered by the rectangle. MaxRS queries can be employed in many useful
applications such as finding an attractive area for tourists. So far, some
literatures proposed efficient algorithms for MaxRS query processing in
traditional databases. In real environments, however, values of data are
inherently uncertain. Therefore, for the first time, we address the problem
of processing probabilistic MaxRS (P-MaxRS) queries. P-MaxRS queries
retrieve a set of tuples $\langle l, p \rangle$, where p is the probability that l is a MaxRS
location. Our algorithm prunes locations with zero probability to be
MaxRS, and then efficiently calculates p for each location where can be
a MaxRS location. Our experiments demonstrate the efficiency of our
algorithm.

Keywords: Spatial data · Uncertain data · MaxRS query

1 Introduction

Recently, due to the emergence of various location-based services and the prevalence of mobile devices, spatial databases have been receiving much research
attention [2,9]. Since the size of a spatial dataset is often large [4,7], processing
of queries which retrieve only meaningful results is an important task. In this
paper, we consider MaxRS (Maximizing Range Sum) queries [3], which bring
efficient location finding applications. Given a set of spatial objects with scores
and a size of a rectangle, a MaxRS query retrieves the location of the rectangle
which maximizes the sum of the scores of all objects covered by the rectangle.
We illustrate an example of a MaxRS query.

EXAMPLE 1.1. *In Fig. 1, the dashed rectangles and black points respectively
denote rectangles of a user-specified size and spatial objects. For simplicity, let
us assume that the scores of all spatial objects are 1. In this example, the MaxRS
query identifies the shaded rectangle as one of the results, because the rectangle
covers the maximum number of spatial objects among all rectangles.*

© Springer International Publishing AG 2017
D. Benslimane et al. (Eds.): DEXA 2017, Part I, LNCS 10438, pp. 111–119, 2017.
DOI: 10.1007/978-3-319-64468-4_8

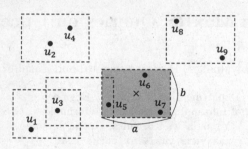

Fig. 1. An examples of a MaxRS query

Example 1.1 shows that a MaxRS query provides an important location, so many real-life applications can employ this query. One of such practical applications is introduced below.

EXAMPLE 1.2. *We consider a recommendation system of the most attractive place for a tourist. The system has spatial objects consisting of locations and scores of sightseeing spots. When a tourist specifies the size of rectangle, which is his/her moving range, a MaxRS query finds the location which maximizes the sum of the scores of all spots covered by the rectangle. The system can recommend the location for the tourist.*

Although spatial objects in Example 1.2 are considered as deterministic data, they are usually *uncertain*. For example, since each sightseeing spot is normally reviewed by many people, different reviewers provide different scores. Therefore, such objects consist of tuples of score and the probability of being the score. This brings the fact that a given location can be MaxRS *probabilistically*. So far, many works addressed the problem of MaxRS query processing [2–4,7,10], but they do not consider data uncertainty. On the other hand, a lot of works addressed the problems of query processing in uncertain databases, e.g., top-k query [5] and TkIS query [9]. However, our problem is totally different from the above ones.

Motivated by this, we tackle the problem of MaxRS query processing on uncertain data. To the best of our knowledge, this problem has not been addressed yet. We first propose probabilistic MaxRS (P-MaxRS) query which retrieves all locations where can be MaxRS with non-zero probability. Specifically, given a set of spatial objects whose scores are uncertain and a size of a rectangle, this query provides a set of tuples $\langle l, p \rangle$, where p is the probability that l is a MaxRS location. To provide probability p, we use the well-known *possible world* semantic. A naive algorithm which calculates the MaxRS location *for all* possible worlds is prohibitive since the number of possible worlds exponentially increases as the number of objects increases. We therefore propose an efficient algorithm for P-MaxRS query processing. Our algorithm prunes locations with zero probability to be MaxRS without probability calculation. Then it efficiently calculates the probabilities that non-pruned locations are MaxRS locations, by utilizing a dynamic programming approach [1].

We summarize our contributions below. (i) For the first time, we address the problem of MaxRS query processing on uncertain data. We define a probabilistic MaxRS (P-MaxRS) query for uncertain data. (ii) We propose an efficient algorithm for processing P-MaxRS queries. First, our algorithm identifies the locations with non-zero probabilities of being MaxRS. Then, it efficiently calculates the exact probabilities that the locations are MaxRS locations. (iii) Through experiments on real data, we show that our algorithm is efficient.

2 Preliminaries

2.1 Data Model

Uncertain Database. An uncertain database U consists of n uncertain objects ($U = \{u_1, u_2, ..., u_n\}$). An uncertain object u_i consists of m instances ($u_i = \{u_{i,1}, u_{i,2}, ..., u_{i,m}\}$), and $u_{i,j} = \langle loc, s, p \rangle$, where loc is a 2-dimensional location, s is a score ($s \in \mathbb{N}$), and p is the probability to be $u_{i,j}$. Note that loc is certain and each uncertain object has no instances whose scores are the same. Also, $\sum_{u_{i,j} \in u_i} u_{i,j}.p = 1$ and uncertain objects are independent of other objects. This model is generally used, e.g., in literatures [1,8].

Possible World. A possible world $\mathcal{W}$ is a set of instances sampled from each uncertain object, i.e., $\mathcal{W} = \{u_{1,\alpha}, u_{2,\beta}, ..., u_{n,\gamma}\}$. The occurrence probability of $\mathcal{W}$ is $\Pr(\mathcal{W}) = \prod_{u_{i,j} \in \mathcal{W}} u_{i,j}.p$. Let Ω be a set of all possible worlds generated by U. Now $|\Omega| = m^n$ and $\sum_{\mathcal{W} \in \Omega} \Pr(\mathcal{W}) = 1$ hold.

2.2 Problem Definition

MaxRS Probability. Given an uncertain database U and a size of a rectangle q, we consider an object set $R^q \subset U$ such that all its elements are covered by a rectangle whose size is q. Let $\mathcal{R}^q$ be a set of all possible R^q, and we define *MaxRS probability* of $R^q \in \mathcal{R}^q$, $\Pr[MaxRS_q = R^q]$.

$$\Pr[MaxRS_q = R^q] = \sum_{\mathcal{W} \in \Omega} \Pr(\mathcal{W}) \cdot \delta(MaxRS_{q^*}^{\mathcal{W}} = R^q)$$

$$\delta(MaxRS_q^{\mathcal{W}} = R^q) = \begin{cases} 1 & (\text{if } R^q = \operatorname*{argmax}_{R \in \mathcal{R}^q} f^{\mathcal{W}}(R)) \\ 0 & (\text{otherwise}) \end{cases} \qquad (1)$$

$$f^{\mathcal{W}}(R) = \sum_{u_i \in R, u_{i,j} \in \mathcal{W}} u_{i,j}.s$$

(Although R^q is not location, the central location of the minimum bounding rectangle consisting of the objects $\in R^q$, l_R, can be easily obtained and is used to consider MaxRS location.) As can be seen from Eq. (1), $f^{\mathcal{W}}(R)$ is the sum of the scores of all objects $\in R$ in a possible world $\mathcal{W}$. Thus, $\delta(MaxRS_q^{\mathcal{W}} = R^q)$ becomes 1 if R^q is MaxRS in $\mathcal{W}$. Otherwise, it is 0. In this paper, we assume

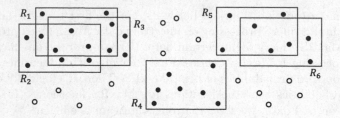

Fig. 2. An example of $\mathcal{AR}$

that there are no object sets with the same scores as that of a MaxRS object set, thereby $\sum_{R^q \in \mathcal{R}^q} \Pr[MaxRS_q = R^q] = 1$. Even if such object sets exist, $\sum_{R^q \in \mathcal{R}^q} \Pr[MaxRS_q = R^q] = 1$ holds by choosing only a single object set as a MaxRS object set. We next define a probabilistic MaxRS query.

DEFINITION 2.1 (PROBABILISTIC MAXRS QUERY (P-MAXRS QUERY)). *Given an uncertain database U and a size of a rectangle q, a P-MaxRS query retrieves a set of tuples $\langle l_R, Pr[MaxRS_q = R] \rangle$ such that $Pr[MaxRS_q = R] > 0$.*

2.3   Existing Algorithms

We here introduce some techniques used in our proposed algorithm. [6] proposed a plane-sweep algorithm, PS, for RI (Rectangle Intersection) query processing. Given a set of rectangles which are the same size, RI queries retrieve the area which intersects with the maximum number of rectangles. We can retrieve an answer of a traditional MaxRS query by processing a RI query when a given object set is transformed into a rectangle set [3]. [1] proposed a dynamic programing based algorithm, DP_PSUM2, for ALL_SUM query processing. Given a set of objects whose scores are uncertain, ALL_SUM queries retrieve a set of tuples which consist of a possible sum of scores and its probability. We omit the detail of the above algorithms (see [1,6]).

3   Proposed Algorithm

Overview. To process P-MaxRS queries efficiently, it is important to quickly remove object sets with zero probabilities to be MaxRS. Our algorithm achieves this and obtains a set of object sets with non-zero MaxRS probabilities by PS. Let $\mathcal{AR}$ be the obtained set, and Fig. 2 illustrates an example of $\mathcal{AR}$. In Fig. 2, rectangles, black points, and white points respectively denote $R_i \in \mathcal{AR}$, objects in R_i, and objects which are not included in any R_i.

We next calculate MaxRS probabilities of object sets in $\mathcal{AR}$. To this end, we first divide $\mathcal{AR}$ into some independent groups of object sets where object sets have common objects. We then calculate local MaxRS, defined later, for each independent group. The MaxRS probability of each object set is calculated while comparing with the local MaxRS scores of the other groups. In a naive

---

Algorithm 1. CalculateAR(U, q)

Input: a set of uncertain data U, a size of a rectangle $q = (a, b)$

1 $U_{min} \leftarrow \emptyset$, $U_{max} \leftarrow \emptyset$, $I \leftarrow \emptyset$
2 **for** $\forall u_i \in U$ **do**
3 $\quad \lfloor \; U_{min} \leftarrow U_{min} \cup \langle u_i.loc, u_i.MinScore \rangle$
4 $I \leftarrow$ PS(U_{min})
5 **for** $\forall u_i \in U$ **do**
6 $\quad \lfloor \; U_{max} \leftarrow U_{max} \cup \langle u_i.loc, u_i.MaxScore \rangle$
7 $\mathcal{AR} \leftarrow$ PS$_{extend}(U_{max}, I.score)$
8 **return** $\mathcal{AR}$

---

manner, we can obtain local MaxRS by calculating a MaxRS location for each possible world generated from objects in a group. The number of possible worlds, however, increases exponentially as the number of objects in a group increases. We therefore decrease the number of possible worlds which need to be calculated, by utilizing DP_PSUM2.

Retrieval of Object Sets with Non-zero MaxRS Probabilities. Algotithm 1 illustrates an algorithm to obtain object sets with non-zero MaxRS probabilities. First, we calculate lb_{score} which is a lower bound score of a MaxRS location among all possible worlds. To this end, the score of each object is set as the minimum score of the object ($MinScore$) and PS is executed (lines 1–4). Note that lb_{score} is the answer of PS, $I.score$. Since lb_{score} is the minimum score of MaxRS among all possible worlds, object sets whose maximum sum is less than lb_{score} have no chance to become MaxRS. That is, the MaxRS probabilities of such object sets are 0. Therefore, we obtain only object sets whose maximum sums are not less than lb_{score}. The score of each object is set as the maximum score of the object ($MaxScore$) and PS$_{extend}$ is executed (lines 5–7). PS$_{extend}$ is an extended version of PS and retrieves all object sets whose scores are not less than lb_{score}.

Calculating MaxRS Probabilities. A straightforward approach to calculating MaxRS probabilities of object sets in $\mathcal{AR}$ is to execute PS for all possible worlds generated from $U' = \bigcup_{R \in \mathcal{AR}} R$. This approach is prohibitive, since the number of possible worlds is $m^{|U'|}$. Instead, DP_SUM2 is employed to efficiently calculate MaxRS probability. However, we can not simply apply DP_PSUM2 to each object set $R \in \mathcal{AR}$ and calculate scores of R and their probabilities, if there is a common object among object sets. We therefore divide object sets into independent groups to effectively employ DP_PSUM2.

DEFINITION 3.1 (INDEPENDENT GROUP). *Given $\mathcal{AR}$, object sets in $\mathcal{AR}$ are divided into groups $G_1, G_2, \ldots, G_{n'}$. If a group G satisfies the following conditions, G is an independent group.*

Algorithm 2. CalculateLocalMaxRS(G)

Input: an independent group G

1 $U_\alpha \leftarrow \{u \mid \exists R_i, R_j \in G, u \in (R_i \cap R_j)\}$
2 **for** $\forall R \in G$ **do**
3      $\Psi_{\hat{R}} \leftarrow$ DP_PSUM2 $(R \backslash U_\alpha)$
4 $U_\beta \leftarrow \{u \mid u \in \cap_{R \in G} R\}$
5 $\Psi_\beta \leftarrow$ DP_PSUM2 (U_β)
6 $U_\gamma \leftarrow U_\alpha \backslash U_\beta$
7 **for** $\forall \mathcal{W} \in \Omega_{U_\gamma}$ **do** $// \ \Omega_{U_\gamma}$ is a set of possible worlds generated from U_γ
8   Calculate $\Pr[LMaxRS_q = R_i, S(R_i) = s]$ for each object set R_i in G by
       utilizing all $\Psi_{\hat{R}}$ and Ψ_β

$$\forall R \in G, \exists R' \in G \backslash R, R \cap R' \neq \emptyset$$
$$\forall R \in G, \forall G \in \mathcal{G} \backslash G, \forall R' \in G', R \cap R' = \emptyset,$$
where $\mathcal{G}$ is the set of groups.

For example, given object sets described in Fig. 2, we can divide them into 3 independent groups, $G_1 = \{R_1, R_2, R_3\}$, $G_2 = \{R_4\}$, and $G_3 = \{R_5, R_6\}$.
We next consider local MaxRS probability.

DEFINITION 3.2 (LOCAL MAxRS PROBABILITY). *Given an independent group G and an object set R in G, the local MaxRS probability of R is the probability that R is MaxRS among the object sets in G.*

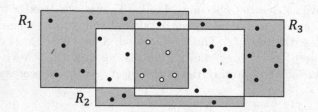

Fig. 3. An example of an independent group

Our algorithm to calculate local MaxRS probabilities reduces the number of objects which generate possible worlds by partially applying DP_PSUM2. We can apply DP_PSUM2 to the following objects: (i) objects included only in a single object set (black points in each shaded area in Fig. 3) and (ii) common objects included in all object sets (white points in the shaded area in Fig. 3).

Algorithm 2 illustrates an algorithm which calculates local MaxRS probabilities by partially applying DP_PSUM2. Line 1 calculates a set of common objects U_α. Lines 2–3 calculate $\Psi_{\hat{R}}$ which consists of tuples of possible sums and the probabilities of $U_{\hat{R}}$. Note that the objects in $U_{\hat{R}}$ are included only in object set R.

Then, line 4 calculates U_β which is a set of common objects of all the object sets in G, and line 5 calculates Ψ_β in the same manner as $\Psi_{\hat{R}}$. After the above calculation, we consider each possible world $\mathcal{W}$ generated from U_γ ($= U_\alpha \backslash U_\beta$). In each possible world, line 8 efficiently calculates $\Pr[LMaxRS_q = R_i, S(R_i) = s]$ for each R_i in G by utilizing all $\Psi_{\hat{R}}$ and Ψ_β. Note that $\Pr[LMaxRS_q = R_i, S(R_i) = s]$ is a probability that a score of R_i is s and R_i is local MaxRS in G. How to utilize all $\Psi_{\hat{R}}$ and Ψ_β is omitted due to space limitation.

MaxRS Probability of Each Object Set. Now we can calculate MaxRS probability of an object set $R_i \in G_a$ based on local MaxRS probabilities. Specifically,

$$
\Pr[MaxRS_q = R_i] =
$$
$$
\sum_{s \geq lb_{score}} \{\Pr[LMaxRS_q = R_i, S(R_i) = s]
$$
$$
\prod_{G_b \in \mathcal{G} \backslash G_a} (\sum_{R_j \in G_b} \sum_{s' < s} \Pr[LMaxRS_q = R_j, S(R_j) = s'])\}.
$$

4   Experiments

This section provides our experimental results. We compared our algorithm with BASELINE and BASELINE+ to verify the efficiency of our algorithm. BASELINE and BASELINE+ are variants of our algorithm, and they generate possible worlds from U_G and U_α respectively. All algorithms were implemented in C++, and all experiments were conducted on a PC with 3.00 GHz Xeon E5-2687W processor and 512 GB RAM.

Setting. In our experiments, we used two real datasets (CAR[1] and RR (See footnote 1)). CAR was generated from road networks in California, and RR was generated from rivers and railways in Los Angeles. CAR and RR respectively have 2,249,727 and 128,971 data points. We normalized the range of each coordinate to [0, 1,000,000,000] as with [4], and provided scores and probabilities of instances for each object by the following manner. Each object was given a normal distribution whose variance and mean are 20 and integer value generated by a uniform distribution of the range [1, 200]. Then, we gave two instances whose scores follow its own normal distribution to each object. Probabilities of instances were randomly set so that $\sum_{u_{i,j} \in u_i} u_{i,j}.p = 1$ holds.

In our experiments, a given rectangle is square and its size is $L \times L$ [4]. For each experiment, scores and probabilities of instances were randomly generated by the above approach. We ran the algorithms 50 times, and measured the average computation time to process a P-MaxRS query.

---

[1] http://chorochronos.datastories.org/?q=node/59.

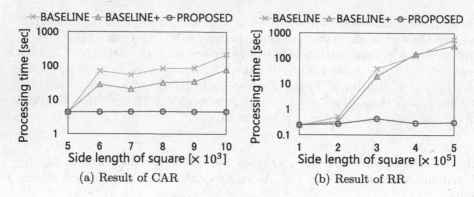

(a) Result of CAR

(b) Result of RR

Fig. 4. Impact of L

Result. Figure 4 shows the result w.r.t. impact of L. We can see that as L increases, the processing time of all algorithms normally increases. It is intuitively known that there are many objects in a rectangle, i.e., an independent group has many objects, when L is large. In this case, the number of possible worlds needed to be considered for calculating local MaxRS probability tends to be large. Compared with BASELINE and BASELINE+, our algorithm significantly reduces processing time, because of its efficient MaxRS probability calculation. Particularly, its processing time is at most 100 times faster than the other algorithms in RR dataset case.

5   Conclusion

In this paper, we tackled the problem of MaxRS query processing on uncertain data. We defined P-MaxRS queries for uncertain data, and proposed an efficient algorithm to process P-MaxRS queries. In our algorithm, the plane sweep based algorithm prunes object sets with zero probability to be MaxRS, and we reduce the number of possible worlds which need to be calculated, by partially utilizing DP_PSUM2. The experimental results demonstrated the efficiency of our algorithm.

Our algorithm provides the exact result, but some applications do not require the exact answer but are satisfied by approximate answers [2,7]. As a part of future work, we plan to design an approximate algorithm to accelerate query processing performance.

Acknowledgement. This research is partially supported by the Grant-in-Aid for Scientific Research (A)(JP2620013) of the Ministry of Education, Culture, Sports, Science and Technology, Japan, and JST, Strategic International Collaborative Research Program, SICORP.

References

1. Akbarinia, R., Valduriez, P., Verger, G.: Efficient evaluation of SUM queries over probabilistic data. TKDE **25**(4), 764–775 (2013)
2. Amagata, D., Hara, T.: Monitoring MaxRS in spatial data streams. In: EDBT, pp. 317–328 (2016)
3. Choi, D.W., Chung, C.W., Tao, Y.: A scalable algorithm for maximizing range sum in spatial databases. PVLDB **5**(11), 1088–1099 (2012)
4. Choi, D.W., Chung, C.W., Tao, Y.: Maximizing range sum in external memory. TODS **39**(3), 21 (2014)
5. Hua, M., Pei, J., Zhang, W., Lin, X.: Ranking queries on uncertain data: a probabilistic threshold approach. In: SIGMOD, pp. 673–686 (2008)
6. Nandy, S.C., Bhattacharya, B.B.: A unified algorithm for finding maximum and minimum object enclosing rectangles and cuboids. Comput. Math. Appl. **29**(8), 45–61 (1995)
7. Tao, Y., Hu, X., Choi, D.W., Chung, C.W.: Approximate MaxRS in spatial databases. PVLDB **6**(13), 1546–1557 (2013)
8. Yuen, S.M., Tao, Y., Xiao, X., Pei, J., Zhang, D.: Superseding nearest neighbor search on uncertain spatial databases. TKDE **22**(7), 1041–1055 (2010)
9. Zhan, L., Zhang, Y., Zhang, W., Lin, X.: Finding top k most influential spatial facilities over uncertain objects. TKDE **27**(12), 3289–3303 (2015)
10. Zhou, X., Wang, W., Xu, J.: General purpose index-based method for efficient MaxRS query. In: Hartmann, S., Ma, H. (eds.) DEXA 2016. LNCS, vol. 9827, pp. 20–36. Springer, Cham (2016). doi:10.1007/978-3-319-44403-1_2

On Addressing the Empty Answer Problem in Uncertain Knowledge Bases

Ibrahim Dellal, Stéphane Jean, Allel Hadjali$^{(\boxtimes)}$, Brice Chardin, and Mickaël Baron

LIAS/ISAE-ENSMA - University of Poitiers, 1, Avenue Clement Ader, 86960 Futuroscope Cedex, France
{ibrahim.dellal,stephane.jean,allel.hadjali, brice.chardin,mickael.baron}@ensma.fr

Abstract. Recently, several large *Knowledge Bases* (KBs) have been constructed by mining the Web for information. As an increasing amount of inconsistent and non-reliable data are available, KBs facts may be *uncertain* and are then associated with an explicit certainty degree. When querying these uncertain KBs, users seek high quality results i.e., results that have a certainty degree greater than a given threshold α. However, as they usually have only a partial knowledge of the KBs contents, their queries may be failing i.e., they return no result for the desired certainty. To prevent this frustrating situation, instead of returning an empty set of answers, our approach explains the reasons of the failure with a set of α*Minimal Failing Subqueries* (αMFSs), and computes alternative relaxed queries, called α *Ma$\underline{X}$imal Succeeding Subqueries* (αXSSs), that are as close as possible to the initial failing query. Moreover, as the user may not always be able to provide an appropriate threshold α, we propose two algorithms to compute the αMFSs and αXSSs for other thresholds. Our experiments on the WatDiv benchmark show the relevance of our algorithms compared to a baseline method.

1 Introduction

A *Knowledge Base* (KB) is a collection of entities and facts about them. Well-known examples of KBs include Knowledge Vault [1] and YAGO [2]. These KBs contain billions of facts captured as RDF triples *(subject, predicate, object)* and are queried with the SPARQL language. As these KBs have been constructed by mining the Web for information, their facts are *uncertain* (i.e., potentially inconsistent). Therefore, an explicit degree of certainty is assigned to KB facts. When querying uncertain KBs, users expect to obtain high quality results i.e., results that have a certainty degree greater than a given threshold α. However, as they rarely know the underlying structure and contents of a KB, they may be faced with the empty answer problem i.e., they obtain no result or results with a degree of certainty lower than α. This is not an uncommon problem. Indeed, the study conducted by Saleem et al. on SPARQL endpoints shows that ten percent of queries submitted to DBpedia between May and July 2010 returned

© Springer International Publishing AG 2017
D. Benslimane et al. (Eds.): DEXA 2017, Part I, LNCS 10438, pp. 120–129, 2017.
DOI: 10.1007/978-3-319-64468-4_9

empty results [3]. Instead of solely returning an empty set as the answer of a query, the system might help the user understand the reasons of this failure by providing him/her with a set of *Minimal Failing Subqueries* (MFSs). Moreover, interesting non-failing relaxed queries, called *MaXimal Succeeding Subqueries* (XSSs), might be suggested to the user by the system as well.

The problem of computing MFSs and XSSs of SPARQL queries expressed on KBs has already been addressed by Fokou et al. [4]. In this paper, we consider a generalization of MFSs and XSSs in the context of uncertain KBs. We call αMFSs and αXSSs the failure causes and maximal relaxed subqueries of a query that filters results according to their certainty degree and a given threshold α. We first show under which conditions the computation of MFSs and XSSs can be directly adapted to αMFSs and αXSSs. In this setting, the user has to define the threshold α. However, as she/he may not have an idea of the certainty degrees assigned to the KB RDF triples, we also investigate the idea of suggesting relaxed queries with lower α thresholds. This kind of relaxation requires the computation of αMFSs and αXSSs for multiple thresholds α. To save computation time, some properties between αMFSs and αXSSs of different thresholds are established and exploited. Thus, depending on which order the α values are considered, two approaches *Top-Down* and *Bottom-Up* are discussed. We run the experiments on the WatDiv benchmark with the Jena TDB quadstore to show the impact of our approaches.

The paper is structured as follows. Section 2 formalizes the problem. Section 3 defines the conditions under which a previous work algorithm can be directly adapted to find the αMFSs and αXSSs for a given α. Section 4 describes the two proposed approaches to compute αMFSs and αXSSs for a set of thresholds. Section 5 discusses the experimental evaluation performed. Section 6 details related work and Sect. 7 concludes.

2   Problem Statement

An *RDF triple* is a triple (subject, predicate, object) $\in (U \cup B) \times U \times (U \cup B \cup L)$ where U is a set of URIs, B is a set of blank nodes and L is a set of literals. We denote by T the union $U \cup B \cup L$. An *RDF database* (or *triplestore*) is a set of *RDF* triples (denoted by T_{RDF}). Each RDF triple has a *trust score* representing the trustworthiness of the triple. This score is assigned with the function $tv : T_{RDF} \rightarrow [0, 1]$.

An *RDF triple pattern* t is a triple (subject, predicate, object) $\in (U \cup V) \times (U \cup V) \times (U \cup V \cup L)$, where V is a set of variables disjoint from the sets U, B and L. We denote by $var(t) \subseteq V$ the set of variables occurring in t. We consider *RDF queries* defined as a conjunction of triple patterns: $Q = t_1 \wedge \cdots \wedge t_n$. The number of triple patterns of a query Q is denoted by $|Q|$ and its variables $var(Q) = \bigcup var(t_i)$.

A *mapping* μ from V to T is a partial function $\mu : V \rightarrow T$. For a triple pattern t, we denote by $\mu(t)$ the triple obtained by replacing in t its variables $var(t)$ by their mapping $\mu(var(t))$. Let D be an *RDF* database, t a triple pattern. The evaluation of the triple pattern t over D denoted by $[[t]]_D$ is defined by: $[[t]]_D = \{\mu \mid dom(\mu) = var(t) \wedge \mu(t) \in D\}$. Let Q be a query, the evaluation of Q over D is defined by: $[[Q]]_D = [[t_1]]_D \bowtie \cdots \bowtie [[t_n]]_D$. Let μ be a solution of the query $Q = t_1 \wedge \cdots \wedge t_n$ and *aggreg* be an aggregation function (e.g., the minimum), the trust value of μ is defined by $tv(\mu, Q) = aggreg(tv(\mu(t_1)), \cdots, tv(\mu(t_n)))$. The evaluation of Q over D that returns trust weighted results with a threshold α is defined by: $[[Q]]_D^\alpha = \{\mu \in [[Q]]_D \mid tv(\mu) \geq \alpha\}$.

Given a query $Q = t_1 \wedge \cdots \wedge t_n$, a query $Q' = t_i \wedge \cdots \wedge t_j$ is a *subquery* of Q, $Q' \subseteq Q$, iff $\{i, \cdots, j\} \subseteq \{1, \cdots, n\}$. If $\{i, \cdots, j\} \subset \{1, \cdots, n\}$, we say that Q' is a *proper subquery* of Q $(Q' \subset Q)$. An α*Minimal Failing Subquery* $(\alpha$MFS$)$ Q^* of a query Q for a given α is defined by: $[[Q^*]]_D^\alpha = \emptyset \wedge \nexists Q' \subset Q^*$ such that $[[Q']]_D^\alpha = \emptyset$. The set of all αMFSs of a query Q for a given α is denoted by $mfs^\alpha(Q)$. An α*Maximal Succeeding Subquery* $(\alpha$XSS$)$ Q^* of a query Q for a given α is defined by: $[[Q^*]]_D^\alpha \neq \emptyset \wedge \nexists Q'$ such that $Q^* \subset Q' \wedge [[Q']]_D^\alpha \neq \emptyset$. The set of all αXSSs of a query Q for a given α is denoted by $xss^\alpha(Q)$.

Problem Statement. We are concerned with computing $mfs^{\alpha_i}(Q)$ and $xss^{\alpha_i}(Q)$ of a failing *RDF* query Q for a set of thresholds $\{\alpha_1, \cdots, \alpha_n\}$.

3 αMFSs and αXSSs Computation for a Single α

In this section, we first give a direct adaptation of the *Lattice-Based Approach* (LBA) proposed in [4] to compute the αMFSs and αXSSs of a query for a given α. It has the same algorithmic complexity as LBA (detailed in [4]). αLBA explores the lattice of subqueries by following a three-steps procedure.

1. Find an αMFS Q^* of Q. Following Algorithm 1, αLBA removes iteratively each triple pattern t_i from Q, resulting in the proper subquery Q'. If Q' fails for α, then Q' contains an αMFS. Conversely, if Q' succeeds, then each αMFS of Q contains t_i. The proof of this property relies on the fact that a successful query cannot contain a failing query [4].

Algorithm 1. Find an αMFS of a failing RDF query Q

FindAnαMFS(Q, D, α)
    inputs : A failing query $Q = t_1 \wedge ... \wedge t_n$; an RDF database D;
            a threshold α
    output: An αMFS of Q denoted by Q^*

1    $Q^* \leftarrow \emptyset$; $Q' \leftarrow Q$;
2   **foreach** *triple pattern* $t_i \in Q$ **do**
3        $Q' \leftarrow Q' - t_i$;
4   **if** $[[Q' \wedge Q^*]]_D^\alpha \neq \emptyset$ **then**
5            $Q^* \leftarrow Q^* \wedge t_i$;

6   **return** Q^*;

---

2. Compute *potential αXSSs* i.e., the maximal queries that do not include the αMFS previously found. The set of *potential αXSSs* is denoted by $pxss(Q, Q^*)$. This set can be computed as follows:

$$pxss(Q, Q^*) = \begin{cases} \emptyset, & \text{if } |Q| = 1. \\ \{Q - t_i \mid t_i \in Q^*\}, & \text{otherwise.} \end{cases}$$

3. Test potential αXSSs. If a subquery found during step 2 succeeds, it is then an αXSS. If it fails, we apply the two previous steps on this particular subquery to find a new αMFS and its associated potential αXSSs. This is illustrated by Algorithm 2. It is worth noting that this algorithm includes mechanisms to avoid discovering the same αMFSs multiple times (lines 11–13).

Algorithm 2. Find the αMFSs and αXSSs of a query Q

α**LBA**(Q, D, α)

    inputs : A failing query $Q = t_1 \wedge ... \wedge t_n$; an RDF database D;
               a threshold α
    outputs: The αMFSs and αXSSs of Q

1      $Q^* \leftarrow$ FindAnαMFS(Q, D, α);
2      $pxss \leftarrow pxss(Q, Q^*)$;
3      $mfs^\alpha(Q) \leftarrow \{Q^*\}$; $xss^\alpha(Q) \leftarrow \emptyset$;
4   **while** $pxss \neq \emptyset$ **do**
5          $Q' \leftarrow pxss.element()$; // **choose an element of** $pxss$
6   **if** $[[Q']]_D^\alpha \neq \emptyset$ **then** // Q' **is an** αXSS
7              $xss^\alpha(Q) \leftarrow xss^\alpha(Q) \cup \{Q'\}$; $pxss \leftarrow pxss - \{Q'\}$;
8   **else** // Q' **contains an** αMFS
9              $Q^{**} \leftarrow$ FindAnαMFS(Q', D, α);
10             $mfs^\alpha(Q) \leftarrow mfs^\alpha(Q) \cup \{Q^{**}\}$;
11   **foreach** $Q'' \in pxss$ such that $Q^{**} \subseteq Q''$ **do**
12                 $pxss \leftarrow pxss - \{Q''\}$;
13                 $pxss \leftarrow pxss \cup \{Q_j \in pxss(Q'', Q^{**}) \mid \nexists Q_k \in$
                   $pxss \cup xss^\alpha(Q)$ such that $Q_j \subseteq Q_k\}$;

14   **return** $\{mfs^\alpha(Q), xss^\alpha(Q)\}$;

The αLBA algorithm relies on the fact that a successful query cannot contain a failing query. In the context of uncertain KBs, this property does not always hold depending on the chosen trust value aggregate function (*aggreg*). For example, with the maximum aggregate function, the degree of certainty of results potentially increases with additional triple patterns. Thus, a query may be failing but not its subqueries. The algorithm αLBA can only be used if the aggregate function *aggreg* is monotonic decreasing with respect to the subset partial order. We omit the proof of this property.

Definition 1. *Let aggreg* : $[0,1]^n \to [0,1]$ *be an aggregate function, aggreg is monotonic decreasing with respect to set[1] inclusion if for all sets A and B $\in [0,1]^n$, $A \subseteq B \Rightarrow aggreg(A) \geq aggreg(B)$.*

As examples of monotonic decreasing aggregate functions, we can cite the *minimum* or the *product* restricted to values $\in [0,1]$.

Proposition 1. *Let aggreg be monotonic decreasing.* $[[Q]]_D^\alpha = \emptyset \wedge Q' \subset Q \Rightarrow$ $[[Q']]_D^\alpha = \emptyset$. *That is to say, if a proper subquery Q' of Q fails for a given α (using the aggreg function) then Q also fails for α.*

4 αMFSs and αXSSs Computation for a Set of α

To find αMFSs and αXSSs for a set of α: $\{\alpha_1, \cdots \alpha_n\}$, the αLBA algorithm can be applied for each α_i. This baseline method is named *NLBA*. In this section, we discuss various improvements of this approach. The idea is that the αMFSs and αXSSs for a given threshold provide a set of hints to deduce some αMFSs and αXSSs with higher (or lower) thresholds.

4.1   Bottom-Up Approach

In this section, we consider two thresholds α_i and α_j such that $\alpha_i < \alpha_j$. If Q^* is an α_iMFS of the query Q, then Q^* also fails for α_j. However, this subquery is not necessarily minimal for α_j and therefore might not be an α_jMFS. The following proposition provides a condition under which an α_iMFS is also an α_jMFS. Due to space constraints, proofs are omitted.

Proposition 2. *Let α_i and α_j be two thresholds such that $\alpha_i < \alpha_j$ and Q^* be an α_iMFS of Q on a dataset D. If $|Q^*| = 1$, then Q^* is also an α_jMFS of Q.*

As pointed out previously, for a subquery Q^* to be an α_jMFS of a query Q, all its proper subqueries have to succeed. As stated in Proposition 2, this property is always true if the query contains a single triple pattern. Checking if a query has a single triple pattern does not require any database access. Thus, this case is checked first and all discovered α_jMFS of Q are put in a set of *discovered αMFSs* denoted by $dmfs^{\alpha_j}(Q)$. Otherwise, proving that Q^* is an α_jMFS requires checking that all its subqueries succeed, by executing those $|Q^*|$ queries. In the worst case where Q^* is not an α_jMFS, $|Q^*|$ queries are executed without finding any α_jMFS. Conversely, the algorithm FindAnαMFS of αLBA (Algorithm 1) also requires $|Q^*|$ queries but guarantees that an αMFS will be found. Thus, our approach favors FindAnαMFS over executing the subqueries of the α_iMFS to discover new α_jMFSs, as shown in Algorithm 3.

As for the αXSSs, an α_iXSS of Q may fail for α_j. The following proposition shows that if it succeeds, it is then an α_jXSS of Q.

---

[1] For simplicity, this definition is restricted to sets but could be extended to multisets.

Proposition 3. *Let α_i and α_j be two thresholds such that $\alpha_i < \alpha_j$ and Q^* be an $\alpha_i XSS$ of Q on a dataset D. If $[[Q^*]]_D^{\alpha_j} \neq \emptyset$, then Q^* is an $\alpha_j XSS$ of Q.*

Thus, discovering if an $\alpha_i XSS$ is also an $\alpha_j XSS$ only requires the execution of a single query ($\alpha_i XSS$ with the new threshold α_j). This enables us to find a set of discovered $\alpha_j XSSs$, denoted $dxss^{\alpha_j}(Q)$.

Algorithm 3 presents our complete approach to find some $\alpha_j MFSs$ and $\alpha_j XSSs$ from the set of $\alpha_i MFSs$ and $\alpha_i XSSs$. All $\alpha_i MFSs$ that have one triple pattern ($oneAtom$) are inserted in $dmfs^{\alpha_j}(Q)$ (line 1). Then, the algorithm iterates over the $\alpha_i MFSs$ with at least two triple patterns (the set FQ). It searches an $\alpha_j MFS$ Q^* in a query Q' of FQ with the FindAnαMFS algorithm (line 5). Then, it removes all the failing queries of FQ that contain Q^* since they cannot be minimal. This process stops when all the queries in FQ have been processed (they have either been used to find an $\alpha_j MFS$ or removed as they contain a found $\alpha_j MFS$). Some $\alpha_j XSSs$ are then identified simply by executing each $\alpha_i XSS$ and keeping those that are succeeding (lines 9–10).

Algorithm 3. Find some $\alpha_j MFSs$ and $\alpha_j XSSs$ for Bottom-Up

---

    DiscoverαMFSXSS($mfs^{\alpha_i}(Q)$, $xss^{\alpha_i}(Q)$ D, α_j)

        inputs : The $\alpha_i MFSs$ $mfs^{\alpha_i}(Q)$ of a query Q for a threshold α_i;

                 The $\alpha_i XSSs$ $xss^{\alpha_i}(Q)$ of a query Q for a threshold α_i;

                 an RDF database D; a threshold $\alpha_j > \alpha_i$

        outputs: A set of $\alpha_j MFSs$ of Q denoted by $dmfs^{\alpha_j}(Q)$;

                 A set of $\alpha_j XSSs$ of Q denoted by $dxss^{\alpha_j}(Q)$;

1       $oneAtom \leftarrow \{Q_a \in mfs^{\alpha_i}(Q) \mid |Q_a| = 1\}$;

2       $dmfs^{\alpha_j}(Q) \leftarrow oneAtom$; $FQ \leftarrow mfs^{\alpha_i}(Q) - oneAtom$;

3   **while** $FQ \neq \emptyset$ **do**

4           $Q' \leftarrow fQ.dequeue()$;

5           $Q^* \leftarrow FindAn\alpha MFS(Q', D, \alpha_j)$;

6           $dmfs^{\alpha_j}(Q) \leftarrow dmfs^{\alpha_j}(Q) \cup \{Q^*\}$;

7   **foreach** $Q'' \in FQ$ such that $Q^{**} \subseteq Q''$ **do**

8               $FQ \leftarrow FQ - \{Q''\}$;

9   **foreach** $Q^* \in xss^{\alpha_i}(Q)$ such that $[[Q^*]]_D^{\alpha_j} \neq \emptyset$ **do**

10      $dxss^{\alpha_j}(Q) \leftarrow dxss^{\alpha_j}(Q) \cup \{Q^*\}$;

11   **return** $\{dmfs^{\alpha_j}(Q), dxss^{\alpha_j}(Q)\}$;

Once some $\alpha_j MFSs$ and $\alpha_j XSSs$ have been discovered, an optimized version of αLBA is executed that takes these discovered $\alpha_j MFSs$ and $\alpha_j XSSs$ as inputs, then computes the remaining $\alpha_j MFSs$ and $\alpha_j XSSs$.

4.2 Top-Down Approach

We also consider a Top-Down approach that computes the $\alpha MFSs$ and $\alpha XSSs$ using threshold values in descending order. Thanks to the duality relation that holds between αMFS and αXSS, the properties used in this approach are dual to the ones used in the bottom-up approach.

5   Experimental Evaluation

Here we investigate the scalability of our approaches and compare them with the baseline method $NLBA$ (executing αLBA for each of the N thresholds).

Experimental Setup. We have implemented the proposed algorithms in JAVA 1.8 64 bits. In our current implementation, these algorithms are run on top of Jena TDB. We chose Jena TDB because it is a quadstore that allows us to store the degree of certainty for each triple. Moreover, Jena TDB provides a low level quad filter hook[2] that we use to retrieve results satisfying the provided threshold. Our implementation is available at https://forge.lias-lab.fr/projects/qars4ukb with a tutorial to reproduce our experiments.

Our experiments were conducted on a Ubuntu Server 16.04 LTS system with Intel XEON CPU E5-2630 v3 @2.4 Ghz CPU and 16 GB RAM. For our experiments, we chose the *min* aggregate function.

Dataset and Queries. We used a dataset of 20M triples generated with the WatDiv benchmark [5]. The certainty degree of each RDF triple were generated randomly. We consider 7 failing queries[3]. These queries range between 1 and 15 triple patterns and cover the main query patterns: star (characterized by *subject-subject* joins between triple patterns), chain (composed of *object-subject* joins) and composite (made of other join patterns).

Experiment. Figure 1 shows the execution times of each algorithm for each query on Jena TDB with the 20M triples dataset. Figure 2 gives the number of executed queries by each algorithm. This experiment has been run with the thresholds arbitrarily set to $\{0.2, 0.4, 0.6, 0.8\}$. In comparison with $NLBA$, our algorithms execute fewer queries for finding the αMFSs and αXSSs of each workload query. Overall, Bottom-Up and Top-Down execute respectively 39% and 44% fewer queries than $NLBA$. As a consequence, these algorithms have shorter execution times (a decrease of respectively 30% and 42% execution times for Bottom-Up and Top-Down). For some queries, this improvement is important. For example, $NLBA$ needs 7 s to find the αMFSs and αXSSs of the query Q2, whereas our algorithms need around 1 s. Execution times depend heavily on the queries that our algorithms avoid executing. For example, our algorithms execute around 30 queries for Q4 whereas $NLBA$ needs 120 queries. For Top-Down, this result is an important performance gain 94%. This is not the case for Bottom-Up that has nearly the same execution time than $NLBA$. By analyzing

---

[2] http://jena.apache.org/documentation/tdb/quadfilter.html.
[3] Available at https://forge.lias-lab.fr/projects/qars4ukb/wiki/Doc#Queries.

the executed queries, we find that Bottom-Up prevents the execution of queries that have short execution times but keeps the most expensive ones. Thus, the overall execution time is almost unchanged.

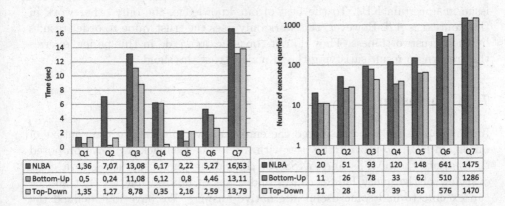

	Q1	Q2	Q3	Q4	Q5	Q6	Q7
■ NLBA	1,36	7,07	13,08	6,17	2,22	5,27	16,63
▨ Bottom-Up	0,5	0,24	11,08	6,12	0,8	4,46	13,11
□ Top-Down	1,35	1,27	8,78	0,35	2,16	2,59	13,79

	Q1	Q2	Q3	Q4	Q5	Q6	Q7
■ NLBA	20	51	93	120	148	641	1475
▨ Bottom-Up	11	26	78	33	62	510	1286
□ Top-Down	11	28	43	39	65	576	1470

Fig. 1. Execution time (20M triples)   **Fig. 2.** # Executed queries (log scale)

This experiment also shows that none of our algorithms provides the best result for every query. Bottom-Up offers the best execution times for Q1, Q2 and Q5 whereas Top-Down is the most efficient for Q3, Q4 and Q6. Despite executing the least number of queries, Bottom-Up does not offer the best total execution time for this workload. Conversely, Top-Down executes the greatest number of queries but has the best execution time. This is due to the fact that our algorithm executes different queries that have different execution times. In particular, Top-Down starts by searching the αMFSs and αXSSs for the highest thresholds. The executed queries tend to be selective as the threshold is high and thus, they have short execution times. Once the αMFSs and αXSSs for the highest thresholds are found, they avoid the execution of queries with a lower threshold that are likely to be more expensive. As Bottom-Up follows the dual approach, it tends to execute non-selective queries and has the overall worst performance.

6 Related Work

Several approaches proposed relaxation operators in the RDF context. These operators are mainly based on RDFS semantics [6–8], similarity measures [9,10] and user preferences [11]. They generate a set of relaxed queries, ordered by similarity with the original query and executed in this order [6,7,12]. Relaxation operators are directly used by the user in her/his query [8] or combined with query rewriting rules to perform relaxation [11]. In these approaches, the failure causes of the query are unknown, which may lead to executing unnecessary relaxed queries. Fokou et al. [4] tackled this problem by defining the LBA and MBA approaches to compute the MFSs and XSSs of the query. Our approach is based on the LBA algorithm. We have extended this work by identifying the condition under which LBA can be used in the context of uncertain KBs and by defining two algorithms to compute αMFSs and αXSSs for several thresholds. Our work is among the pioneering works aiming at exploring the query relaxation issue in uncertain KBs. To the best of our knowledge, the only other work in this context is [12]. However, this work only uses the trust value to order results by their trustworthiness. They do not consider, as we do in this paper, queries that return no result satisfying the provided trust threshold.

7   Conclusion

In this paper, we have considered the empty answer problem in the context of uncertain KBs. To provide the user with a relevant feedback, we have proposed to compute the αMFSs and αXSSs of the failing query as they give a clear overview of the query failure causes and a set of relaxed queries that she/he can execute to find some useful alternative answers. We have first defined the condition under which a previous work algorithm can be directly adapted to the context of uncertain KBs. Then, we have studied the problem of computing the αMFSs and αXSSs for multiple thresholds by defining two approaches that consider α thresholds in different orders. We have done a complete implementation of these algorithms and shown experimentally on WatDiv benchmark that our approaches outperform the baseline method.

In our experiments, none of our algorithms has the best performance for all queries. As a future work, we plan to study the conditions under which an algorithm may provide the best results. An analysis of the queries executed by our algorithms shows that they share some triple patterns. Thus, we will investigate multiple-query optimization techniques to further improve their execution times.

References

1. Dong, X., Gabrilovich, E., Heitz, G., Horn, W., Lao, N., Murphy, K., Strohmann,T., Sun, S., Zhang, W.: Knowledge vault: a web-scale approach to probabilistic knowledge fusion. In: KDD 2014, pp. 601–610 (2014)
2. Hoffart, J., Suchanek, F.M., Berberich, K., Weikum, G.: YAGO2: a spatially and temporally enhanced knowledge base from wikipedia. Artif. Intell. **194**, 28–61 (2013)
3. Saleem, M., Ali, M.I., Hogan, A., Mehmood, Q., Ngomo, A.-C.N.: LSQ: the linked SPARQL queries dataset. In: Arenas, M., et al. (eds.) ISWC 2015. LNCS, vol. 9367, pp. 261–269. Springer, Cham (2015). doi:10.1007/978-3-319-25010-6_15
4. Fokou, G., Jean, S., Hadjali, A., Baron, M.: Handling failing RDF queries: from diagnosis to relaxation. Knowl. Inf. Syst. (KAIS) **50**(1), 167–195 (2017)
5. Aluç, G., Hartig, O., Özsu, M.T., Daudjee, K.: Diversified stress testing of RDF data management systems. In: Mika, P., Tudorache, T., Bernstein, A., Welty, C., Knoblock, C., Vrandečić, D., Groth, P., Noy, N., Janowicz, K., Goble, C. (eds.) ISWC 2014. LNCS, vol. 8796, pp. 197–212. Springer, Cham (2014). doi:10.1007/978-3-319-11964-9_13

6. Hurtado, C.A., Poulovassilis, A., Wood, P.T.: Ranking approximate answers to semantic web queries. In: Aroyo, L., Traverso, P., Ciravegna, F., Cimiano, P., Heath, T., Hyvönen, E., Mizoguchi, R., Oren, E., Sabou, M., Simperl, E. (eds.) ESWC 2009. LNCS, vol. 5554, pp. 263–277. Springer, Heidelberg (2009). doi:10.1007/978-3-642-02121-3_22

7. Huang, H., Liu, C., Zhou, X.: Approximating query answering on RDF databases. J. World Wide Web: Internet Web Inf. Syst. (WWW) 15(1), 89–114 (2012)

8. Calì, A., Frosini, R., Poulovassilis, A., Wood, P.T.: Flexible querying for SPARQL. In: Meersman, R., Panetto, H., Dillon, T., Missikoff, M., Liu, L., Pastor, O., Cuzzocrea, A., Sellis, T. (eds.) OTM 2014. LNCS, vol. 8841, pp. 473–490. Springer, Heidelberg (2014). doi:10.1007/978-3-662-45563-0_28

9. Hogan, A., Mellotte, M., Powell, G., Stampouli, D.: Towards fuzzy query-relaxation for RDF. In: Simperl, E., Cimiano, P., Polleres, A., Corcho, O., Presutti, V. (eds.) ESWC 2012. LNCS, vol. 7295, pp. 687–702. Springer, Heidelberg (2012). doi:10.1007/978-3-642-30284-8_53

10. Elbassuoni, S., Ramanath, M., Weikum, G.: Query relaxation for entity-relationship search. In: Antoniou, G., Grobelnik, M., Simperl, E., Parsia, B., Plexousakis, D., Leenheer, P., Pan, J. (eds.) ESWC 2011. LNCS, vol. 6644, pp. 62–76. Springer, Heidelberg (2011). doi:10.1007/978-3-642-21064-8_5

11. Dolog, P., Stuckenschmidt, H., Wache, H., Diederich, J.: Relaxing RDF queries based on user and domain preferences. J. Intell. Inf. Syst. (JIIS) 33(3), 239–260 (2009)

12. Reddy, K.B.R., Sreenivasa Kumar, P.: Efficient trust-based approximate SPARQL querying of the web of linked data. In: Bobillo, F., Costa, P.C.G., d'Amato, C., Fanizzi, N., Laskey, K.B., Laskey, K.J., Lukasiewicz, T., Nickles, M., Pool, M. (eds.) UniDL/URSW 2008-2010. LNCS, vol. 7123, pp. 315–330. Springer, Heidelberg (2013). doi:10.1007/978-3-642-35975-0_17

Language Constructs for a Datalog Compiler

Stefan Brass[✉]

Institut für Informatik, Martin-Luther-Universität Halle-Wittenberg,
Von-Seckendorff-Platz 1, 06099 Halle (Saale), Germany
brass@informatik.uni-halle.de

Abstract. Deductive databases promise that an important part of the
application program can be developed in a declarative language, seam-
lessly integrated with the query language. The author is currently developing a Datalog-to-C++ compiler that implements the "Push" method for
bottom-up evaluation. While such a compiler is certainly not a typical
database application, our goal is to write a significant portion of it in an
extended Datalog language. In this paper, we propose some extensions of
the Datalog language that are needed for our application: (1) templates
for declarative output, (2) arrays/tuples as a restricted form of structured terms, (3) a way to assign integer levels based on orders (e.g. for
computing stratifications). All is done in a way such that termination is
guaranteed. Termination is also considered for arithmetic computations.

1 Introduction

Database application programs usually consist of queries, written in a declarative language (SQL), and surrounding program code, written in a procedural
language. The interface between the two languages is not smooth, and frameworks or libraries for improving the situation often reduce the declarative part.

The goal of deductive databases is to write a larger part of the application,
ideally the entire program, in a declarative language based on Prolog/Datalog.
There is currently a revival of deductive database technology, mainly caused
by the simplicity and integrative power of Datalog. For instance, there is the
successful commercial deductive database system LogicBlox [1]. There were two
"Datalog 2.0" workshops. The programming of multicore processors and cloud
computing can be significantly simplified with declarative languages based on
Datalog [13,18]. The renewed interest in deductive database technology is also
triggered by semantic web applications, see, e.g. [9,16]. Furthermore, Datalog is
used as a domain-specific language for program analysis [17].

Heike Stephan and the author have developed the "Push" method for bottom-
up evaluation of Datalog [3,5,7]. It applies the rules from body to head (right to
left) as any form of bottom-up evaluation, but it immediately "pushes" a derived
fact to other rules with matching body literals. In this way, the derived facts often
do not have to be materialized, and temporary storage can be saved. First implementations of some test cases of the OpenRuleBench benchmark suite [14] give

© Springer International Publishing AG 2017
D. Benslimane et al. (Eds.): DEXA 2017, Part I, LNCS 10438, pp. 130–140, 2017.
DOI: 10.1007/978-3-319-64468-4_10

very promising performance results [6]. Our implementation works by translating Datalog to C++, and then using a standard compiler to get an executable program. This program can be applied to different input data sets.

We are trying to write a large part of the system in Datalog or an extended version of Datalog. While it is common that Prolog compilers are written in Prolog, deductive databases so far have been mostly written in C++. It might look like system implementators do not like the language they implement, or do not trust their own system. One could answer that standard relational databases are not written in SQL, too. This is even impossible because SQL is not a computationally complete language (otherwise termination of query evaluation could not be guaranteed). However, if one claims that deductive databases permit more programming than standard relational databases, one can demonstrate this by writing at least some part of the system in Datalog. This is our goal (while keeping the termination guarantee). Note that compiling to C++ (on the way to machine code) does not contradict our goal that human-written code should be in a declarative language as far as possible.

Pure Horn-clause Datalog is a quite restricted language. It is the purpose of the current paper to study some extensions which are used in the new implementation of our Datalog-to-C++ translation.

First, realistic programs need computed values, such as the successor of a given number. In Sect. 3, we give a simple sufficient condition for ensuring termination of bottom-up evaluation for such programs. We also need to compute level mappings, such as stratifications. This is the subject of Sect. 4.

Datalog differs from Prolog by not allowing arbitrary terms, usually only constants and variables. This restriction is too severe, because sometimes one needs a tuple (e.g., a binding pattern) as a single data value. The input to our Datalog compiler is a program with predicates of unknown (arbirarily large) arity. The number of such tuples can be exponential in the arity of the input predicates, thus no fixed Datalog program could compute them all. But arbitrary terms as in Prolog can lead to termination problems. Our implementation uses array data values (briefly sketched in Sect. 5). A Datalog extension with arrays has been studied in [12], but our version is different.

Finally, one obviously needs to specify the output of the transformation (the generated C++ code). We investigated declarative output for Datalog already in [4]. However, in our current implementation the use of "output templates" seemed natural. This is the subject of Sect. 6. Note that output formatting is a large part of many database application programs. Therefore, if one wants that at least simple database applications can be written completely in a declarative language, language constructs for declarative output are certainly important. For Prolog, the PiLLOW library for writing web interfaces is well known [8]. However, it is based on using deeply nested Prolog terms for representing the tree structure of, e.g., web-pages. A different solution for declarative output (with an input state and output state) is used in the Mercury system [2].

Since many of the features we need are available in Prolog, one must ask, why not simply use Prolog? But it is well-known that Prolog has impure features such as the cut, and the order of rules and body literals within a rule is important for most applications. And although there are tools for helping to prove termination, it is at least not trivial. Furthermore, thinking "bottom-up" instead of "top-down" seems simpler and leads sometimes to different formulations. In the "top-down" view, one must describe how queries are transformed into simpler queries. In the "bottom-up" view, one defines how new facts can be derived from known facts. Parallelization might also be simpler if one starts with a large number of facts instead of a single query. Actually, many Prolog predicates working with lists or structured terms can only be understood "top-down". With the language constructs we propose in this paper, we want to increase the power of the simple and natural "bottom-up" view of logical rules.

2   Representation of Input Programs (Datalog Rules)

We assume that the input program is basic Datalog with stratified negation, i.e. the rules are logical formulae of the form $A \leftarrow B_1 \wedge \cdots \wedge B_m$, where A is an atomic formula $p(t_1, \ldots, t_n)$, the head of the rule, and the B_i, $i = 1, \ldots, m$, are atomic formulae or negated atomic formulae, the body literals of the rule. Argument terms t_j of the atomic formulae are variables or constants. Facts are atomic formulae without variables, i.e. of the form $p(c_1, \ldots, c_n)$. We apply the syntax conventions of Prolog, e.g. variables start with an uppercase letter, "_" is the anonymous variable, $\leftarrow$ is represented as :-, $\wedge$ as ",", and $\neg$ as \+.

As usual in deductive databases, we distinguish different types of predicates:

- IDB predicates ("intensional database") are defined by rules of the given logic program, i.e. they appear in rule heads.
- EDB predicates ("extensional database") are defined by given facts in the database. The database is considered the input of the program.
- Built-in predicates are defined by program code in the system. E.g. one can use X \= Y to check that $X \neq Y$. Also the sum of two numbers can be computed with N is I + J (subject to the restrictions explained in Sect. 3 for ensuring termination).

Rules must be range-restricted (allowed), i.e. every variable that is used in a rule must appear at least once in a positive body literal, furthermore variable occurrences in "input" argument positions of built-in predicates cannot be the first occurrence of that variable. These restrictions ensure that rules can be evaluated "bottom-up" to derive facts for the head literal from given facts for the body literals.

The input program is represented as Datalog-facts in the following form:

- prog(NumRules)
  This is a single fact that contains the total number of rules in the input program.
- rule(RuleNo, LineNo, NumBodyLit)
  Rules are numbered from 1 to NumRules. The line number is intended for error messages.

- lit(RuleNo, LitNo, Pred, NumArgs),
  There is one such fact for every literal occurring in a rule. LitNo = 0 means
  the head literal.
- arg(RuleNo, LitNo, ArgNo, var, Var)
  arg(RuleNo, LitNo, ArgNo, const, Const)
  There is one such fact for very argument of a literal.
- negated(RuleNo, LitNo)
  This marks the literal as a negative literal.

Note that facts for the EDB-predicates are known only at runtime, they are
not part of the program to be compiled. However, EDB predicates must be
declared in the input program with argument types (which are needed for the
translation to C++) and optional argument names (which make attributes of
generated classes look nicer). This information is stored in the following facts:

- db(RuleNo, LineNo, Pred, NumArgs)
  Declarations are assigned a "Rule Number", too (in this way, the sequence of
  declarations and rules is encoded in the data).
- db_arg(RuleNo, ArgNo, ArgName, Type).

If one wants to avoid non-ground structured terms, which are normally used in
Prolog for meta-programming, this seems a reasonable representation as facts
in Datalog. Note that also in Prolog, the structured terms will often be split
into lists of arguments, so that one can loop over them. We have these lists
represented as facts with an explicit array index.

As an example, we can compute the set of IDB predicates from this representation (remember that literal number 0 means the head literal).

$$idb\_pred(Pred) :- lit(\_, 0, Pred, \_).$$

3   Computed Values

Realistic programs need to compute data values. As in Prolog, we permit body
literals that compute e.g. the sum of two numbers:

$$NextLitNo\ is\ LitNo + 1.$$

For instance, we might want to write a loop over all body literals in this way
(although such loops can often be avoided by expressing a "for all"-condition
with negation). We have promised that termination is easily guaranteed for
bottom-up evaluation of all programs allowed in our system, therefore programs
such as the following must be excluded:

```
p(0).
p(Y) :- p(X), Y is X + 1.
```

The idea is quite similar to the well-known stratification: We allow the computation of new values, and we allow recursion, but not computation of new values during recursion. We give a simple sufficient condition for termination by assigning levels to variables in rules and to predicate arguments.

If a variable or predicate argument is assigned level 0, it can only contain values that appear in the database or in the rules of the program. Of course, one could assign any value to a variable, but the condition means that the rule body will certainly not be satisfied in the minimal Herbrand model if the variable value is not contained in the given database or the program rules. Obviously, the set of level 0 values is finite (for a given database).

If a predicate argument or variable has level 1, it might contain values that are computed from level 0 values by a single application of an is condition that appears in the program. E.g. if somewhere in the program there is a body literal Z is X + Y and X and Y have level 0 in that rule, then Z can be assigned level 1. If Z happens to appear also in a positive literal within a predicate argument that is assigned level 0, Z can be assigned level 0, too. Since all literals in the rule body must be satisfied at the same time, a variable can be assigned the minimum level that results from all its occurrences in the rule. Again, the set of level 1 values is finite: If one inserts the finitely many level 0 values into the finitely many arithmetic expressions in is-conditions, one can compute only finitely many new values.

Definition 1 (Level-Mapping for Computed Values). *Let a Datalog program P be given (rules for the IDB-predicates, numbered from 1 to R).*

- *Let ℓ_A be a mapping that assigns to each IDB predicate p and argument number i, $1 \leq i \leq \mathsf{arity}(p)$, a natural number $\ell_A(p,i)$.*
- *Let ℓ_V be a mapping that assigns to each rule r, $1 \leq r \leq R$, and each variable X appearing in that rule a natural number $\ell_V(r,X)$.*

The combination of ℓ_A and ℓ_V is valid iff

- *For each predicate p of arity n and each rule $p(t_1,\ldots,t_n) \leftarrow B_1,\ldots,B_m$ about p: Let this rule be rule r. Then for $i = 1,\ldots,n$, the argument term t_i is either a constant or a variable X with $\ell_V(r,X) \leq \ell_A(p,i)$.*
- *For each rule $A \leftarrow B_1,\ldots,B_m$ and each variable X appearing in this rule: Let this be rule r. Then there is a positive body literal B_j, $1 \leq j \leq m$, in which X appears such that one of the following conditions is satisfied:*
  - *The predicate of the literal is a database (EDB) predicate.*
  - *The literal has the form $p(t_1,\ldots,t_n)$ with an IDB predicate p and there is i, $1 \leq i \leq n$, such that X is t_i and $\ell_A(p,i) \leq \ell_V(r,X)$.*
  - *The literal has the form X is t and $\ell_V(r,X) > \ell_V(r,Y)$ for every variable Y appearing in t.*

Corollary 1. *If there is a valid level-mapping (ℓ_A,ℓ_V) for a given Datalog program, the extensions of all IDB predicates are finite in the minimal Herbrand model for each given finite database.*

There is quite a lot of recent research on termination of Datalog with function symbols, see, e.g. [11]. However, computed values are not quite the same. The above definition is somewhat similar to λ-restricted programs [10], but ensures that every standard Datalog program (without is) satisfies it.

4   Level Mappings

A Datalog compiler must be able to compute level mappings such as the one defined in Sect. 3. We use the standard stratification here for demonstration, since it is technically simpler. With standard Datalog it is easily possible to check for cycles in the "depends on" graph between predicates that contain an edge marked as a negative dependency:

```
uses_pos(Pred1, Pred2)    :- lit(RuleNo, 0, Pred1, _),
                             lit(RuleNo, LitNo, Pred2, _),
                             LitNo > 0,
                             \+ negated(RuleNo, LitNo).
uses_neg(Pred1, Pred2)    :- lit(RuleNo, 0, Pred1, _),
                             lit(RuleNo, LitNo, Pred2, _),
                             LitNo > 0,
                             negated(RuleNo, LitNo).
uses(Pred1, Pred2)        :- uses_pos(Pred1, Pred2).
uses(Pred1, Pred2)        :- uses_neg(Pred1, Pred2).
depends_on(Pred1, Pred2)  :- uses(Pred1, Pred2).
depends_on(Pred1, Pred3)  :- uses(Pred1, Pred2),
                             depends_on(Pred2, Pred3).
not_stratified            :- uses_neg(Pred1, Pred2),
                             depends_on(Pred2, Pred1).
stratified                :- \+ not_stratified.
```

However, if the program is stratified, and one wants to generate code to compute predicate extensions in the order of the stratification levels, one should be able to compute these levels. So basically the question is how to compute a topological order of the dependency graph in Datalog (the following solution is inspired by [15]). We propose a special aggregation function "level" for that purpose:

```
strat_level(Pred,  level(0))        :- idb_pred(Pred).
strat_level(Pred1, level(Level+1)) :- uses_neg(Pred1, Pred2),
                                      strat_level(Pred2, Level).
strat_level(Pred1, level(Level))    :- depends_on(Pred1, Pred2),
                                      strat_level(Pred2, Level).
```

These rules encode the well-known algorithm for computing a stratification that first assigns level 0 to each predicate, and then increases levels as needed.

We require that if a predicate uses level(0), level(X) or level(X + 1) in argument i, then all rules about that predicate must have one of these expressions in argument i. Furthermore, the predicate may not be mutually recursive with any other predicate (it may use itself, however). In this way the result of the aggregation function can be computed before it is used.

The `level` aggregation function basically computes a maximum: If a fact `strat_level(P, N)` is derived, any other fact `strat_level(P, M)` with M < N is subsumed and can be deleted. However, we must handle the case that the given program is not stratified, then the computation would not terminate. Therefore, we ensure that the assigned level numbers are dense (without holes): When we delete `strat_level(P, M)`, we check whether there is some other fact `strat_level(Q, M)` with the same level M, but a different predicate. If there is not, we store `strat_level(P, -1)`. This subsumes all facts `strat_level(P, N)` with the same predicate P, i.e. -1 encodes ∞.

5   Arrays

As explained in the introduction, we do not want to allow arbitrary terms as in Prolog, but we do need arrays/tuples as data values. For space reasons, we cannot give the details here. The basic idea is again a stratification: arrays that are arguments of predicates at level n must have index and data values that are contained in the extensions of lower level predicates. Syntactically, an array can be constructed by using an aggregation rule with aggregation function `arr(Index,Value)`.

6   Output Templates

Of course, an important part of the translation from Datalog to C++ is the specification of the output. Also for many standard database applications, the generation of output (e.g., web pages) is a large fraction of the work.

Our program first computes Datalog facts that specify what needs to be generated, and then we execute a set of "templates" that use these data to actually generate output. The templates do not use Datalog syntax, but specify a sequence of output elements, parameterized by values from the Datalog facts. Templates can call other templates, in this way a tree structure is created which is typically used for defining document structure. Recursive calls of the templates are excluded, therefore termination is guaranteed. However, templates have the possibility to create a list of calls corresponding to a set of Datalog facts, ordered by data values.

A template definition starts with the name of the template and an optional parameter list. There can be only one template with a given name. All template names start with a lowercase letter like a Prolog predicate. All template parameters start with an uppercase letter like a Prolog variable. Thus, this "template head" looks like a Prolog/Datalog literal with only variables as arguments. It is followed by the "template body", which is written in "[...]". This is a list consisting of

– Text pieces written in '...' (string constants),

- parameters of the template,
- calls to other templates possibly with values for the parameters in (...), and possibly followed by an iterator query (see below),
- one of a few predefined templates, such as `tab_in` and `tab_out` for controlling indentation, and `nl` for forcing a line break.

From this "code mode", one can switch to "verbatim mode" with "|", then all characters except "|" and "[" are copied literally to output. With "|" one ends verbatim mode, and gets back to code mode. With "[" one can nest the interpreted code mode within the verbatim mode, e.g. "[Param]" can be used to include a parameter value.

What makes templates powerful is the possibility to specify "loops" by writing a kind of Datalog rule instead of a simple template call:

$$\text{template(Args)} <\text{Sort}> \text{ :- Lit1, ..., Litn.}$$

This instantiates `template(Args)` once for each answer to `Lit1, ..., Litn`, ordered by the values for the variables listed in the `Sort` part. Furthermore, one can optionally specify a separator string to be inserted between each two template invocations (after the `Sort` part, marked with "+"). For instance, a constructor for a class to store tuples of a predicate is generated by the following template:

```
rel_class_constructor(Pred): [
    '// Constructor:' nl
    rel_class(Pred) '('
        constructor_arg(ArgNo,Type)<ArgNo>+', ' :-
            constructor_arg(Pred, ArgNo, Type)
    ') {' nl
        constructor_assign(ArgNo)<ArgNo> :-
            arg_type(Pred, ArgNo, Type).
    '}' nl
].
```

This makes use of three further simple templates (defined with verbatim mode):

```
rel_class_(Pred): [
    |[Pred]_row|
].
constructor_arg(ArgNo,Type): [
    |[Type] arg_[ArgNo]|
].
constructor_assign(ArgNo): [
    |    this->col_[ArgNo] = arg_[ArgNo];
    |].
```

The last template includes the indentation and the line break.

There is a "main" template where the generation of output starts. It forms the root of the call tree. It has a sequence of child nodes for text pieces and for calls to other templates. In [4], we defined output in Datalog by deriving facts of the form output(Sort, Text). The output consists of the Text pieces ordered by the Sort argument (which is a list/tuple of values). The same method can be used to translate the templates to Datalog (with arrays/tuples). However, the actual implementation in our prototype is much more direct and avoids the explicit construction of arrays as node IDs.

7   Conclusions

Datalog (in all its variants) is a powerful language, suitable for many applications. The author is currently developing an efficient implementation by translating Datalog to C++. It seems natural that the analysis of the input program and the core of the translation should be written in Datalog itself. In this paper, we presented Datalog extensions that turned out to be important for this task. For each of these constructs, as well as for standard arithmetic computations, we defined simple criteria to ensure termination. The current state of the project is reported at

http://www.informatik.uni-halle.de/~brass/push/

The constructs defined in this paper are useful also for standard database applications. We believe that deductive databases will be successful if they are powerful enough for many programming tasks, while staying declarative and guaranteeing termination.

Acknowledgements. I would like to thank Heike Stephan and Andreas Behrend for very valuable discussions.

References

1. Aref, M., ten Cate, B., Green, T.J., Kimelfeld, B., Olteanu, D., Pasalic, E., Veldhuizen, T.L., Washburn, G.: Design and implementation of the LogicBlox system. In: Proceedings of the 2015 ACM SIGMOD International Conference on Management of Data, pp. 1371–1382. ACM (2015). https://developer.logicblox.com/wp-content/uploads/2016/01/logicblox-sigmod15.pdf
2. Becket, R.: Mercury tutorial. Technical report, University of Melbourne, Department of Computer Science (2010). http://mercurylang.org/documentation/papers/book.pdf
3. Brass, S.: Implementation alternatives for bottom-up evaluation. In: Hermenegildo, M., Schaub, T. (eds.) Technical Communications of the 26th International Conference on Logic Programming (ICLP 2010). Leibniz International Proceedings in Informatics (LIPIcs), vol. 7, pp. 44–53. Schloss Dagstuhl (2010). http://drops.dagstuhl.de/opus/volltexte/2010/2582

4. Brass, S.: Order in Datalog with applications to declarative output. In: Barceló, P., Pichler, R. (eds.) Datalog 2.0 2012. LNCS, vol. 7494, pp. 56–67. Springer, Heidelberg (2012). doi:10.1007/978-3-642-32925-8_7. http://users.informatik.uni-halle.de/~brass/order/

5. Brass, S., Stephan, H.: Bottom-up evaluation of Datalog: preliminary report. In: Schwarz, S., Voigtländer, J. (eds.) Proceedings 29th and 30th Workshops on (Constraint) Logic Programming and 24th International Workshop on Functional and (Constraint) Logic Programming (WLP 2015/2016/WFLP 2016). Electronic Proceedings in Theoretical Computer Science, no. 234, pp. 13–26. Open Publishing Association (2017). https://arxiv.org/abs/1701.00623

6. Brass, S., Stephan, H.: Experiences with some benchmarks for deductive databases and implementations of bottom-up evaluation. In: Schwarz, S., Voigtländer, J. (eds.) Proceedings 29th and 30th Workshops on (Constraint) Logic Programming and 24th International Workshop on Functional and (Constraint) Logic Programming (WLP 2015/2016/WFLP 2016). Electronic Proceedings in Theoretical Computer Science, no. 234, pp. 57–72. Open Publishing Association (2017). https://arxiv.org/abs/1701.00627

7. Brass, S., Stephan, H.: Pipelined bottom-up evaluation of Datalog: The push method. In: Petrenko, A.K., Voronkov, A. (eds.) 11th A.P. Ershov Informatics Conference (PSI 2017) (2017). http://www.informatik.uni-halle.de/brass/push/publ/psi17.pdf

8. Cabeza, D., Hermenegildo, M.: Distributed WWW programming using (Ciao-) Prolog and the PiLLoW library. Theory Pract. Logic Programm. 1(3), 251–282 (2001). https://arxiv.org/abs/cs/0312031

9. Calì, A., Gottlob, G., Lukasiewicz, T.: A general Datalog-based framework for tractable query answering over ontologies. In: Proceedings of the 28th ACM SIGMOD-SIGACT-SIGART Symposium on Principles of Database Systems (PODS 2009), pp. 77–86. ACM (2009)

10. Gebser, M., Schaub, T., Thiele, S.: GrinGo: A new grounder for answer set programming. In: Baral, C., Brewka, G., Schlipf, J. (eds.) LPNMR 2007. LNCS (LNAI), vol. 4483, pp. 266–271. Springer, Heidelberg (2007). doi:10.1007/978-3-540-72200-7_24

11. Greco, S., Molinaro, C.: Datalog and Logic Databases. Morgan and Claypool Publishers, Burlington (2015)

12. Greco, S., Palopoli, L., Spadafora, E.: Datalog^A: Array manipulations in a deductive database language. In: Ling, T.W., Masunaga, Y. (eds.) Proceedings of the Fourth International Conference on Database Systems for Advanced Applications (DASFAA 1995), pp. 180–188. World Scientific (1995). http://www.comp.nus.edu.sg/lingtw/dasfaa_proceedings/DASFAA95/P180.pdf

13. Hellerstein, J.M.: The declarative imperative. SIGMOD Rec. 39(1), 5–19 (2010). http://db.cs.berkeley.edu/papers/sigrec10-declimperative.pdf

14. Liang, S., Fodor, P., Wan, H., Kifer, M.: OpenRuleBench: an analysis of the performance of rule engines. In: Proceedings of the 18th International Conference on World Wide Web (WWW 2009), pp. 601–610. ACM (2009). http://rulebench.projects.semwebcentral.org/

15. Meyerovich, L.: Topological sort in Datalog (2011). http://lmeyerov.blogspot.de/2011/04/topological-sort-in-datalog.html

16. Polleres, A.: How (well) do Datalog, SPARQL and RIF interplay? In: Barceló, P., Pichler, R. (eds.) Datalog 2.0 2012. LNCS, vol. 7494, pp. 27–30. Springer, Heidelberg (2012). doi:10.1007/978-3-642-32925-8_4

17. Scholz, B., Jordan, H., Subotić, P., Westmann, T.: On fast large-scale program analysis in Datalog. In: Proceedings of the 25th International Conference on Compiler Construction (CC 2016), pp. 196–206. ACM (2016)
18. Shkapsky, A., Yang, M., Interlandi, M., Chiu, H., Condie, T., Zaniolo, C.: Big data analytics with Datalog queries on Spark. In: Proceedings of the 2016 International Conference on Management of Data (SIGMOD 2016), pp. 1135–1149. ACM (2016). http://yellowstone.cs.ucla.edu/~yang/paper/sigmod2016-p958.pdf

Preferences and Query Optimization

Temporal Conditional Preference
Queries on Streams

Marcos Roberto Ribeiro[1,2]([✉]), Maria Camila N. Barioni[2], Sandra de Amo[2],
Claudia Roncancio[3], and Cyril Labbé[3]

[1] Instituto Federal de Minas Gerais, Bambuí, Brazil
`marcos.ribeiro@ifmg.edu.br`
[2] Universidade Federal de Uberlândia, Uberlândia, Brazil
`{camila.barioni,deamo}@ufu.br`
[3] Univ. Grenoble Alpes, CNRS, Grenoble INP, LIG, 38000 Grenoble, France
`{claudia.roncancio,cyril.labbe}@imag.fr`

Abstract. Preference queries on data streams have been proved very useful for many application areas. Despite of the existence of research studies dedicated to this issue, they lack to support the use of an important implicit information of data streams, the temporal preferences. In this paper we define new operators and an algorithm for the efficient evaluation of temporal conditional preference queries on data streams. We also demonstrate how the proposed operators can be translated to the Continuous Query Language (CQL). The experiments performed show that our proposed operators have considerably superior performance when compared to the equivalent operations in CQL.

Keywords: Data streams · Preference queries · Temporal preferences

1 Introduction

There is a variety of application domains which data naturally occur in the form of a sequence of values, such as financial applications, sport players monitoring, telecommunications, web applications, sensor networks, among others. An especially useful model explored by many research works to deal with this type of data is the data stream model [2,5,6,10]. Great part of these research works has focused on the development of new techniques to answer continuous queries efficiently [3,9]. Other research works have been concerned with the evaluation of preferences in continuous queries to monitor for information that best fit the users wishes when processing data streams [7].

The research literature regarding this later issue is rich in works dealing with processing of continuous skyline queries where the preferences are simple independent preferences for minimum or maximum values over attributes [4,8]. However, these works do not meet the needs of many domain applications that require the users to express conditional preferences. That is, those applications where the preferences over a data attribute can be affected by values of another

© Springer International Publishing AG 2017
D. Benslimane et al. (Eds.): DEXA 2017, Part I, LNCS 10438, pp. 143–158, 2017.
DOI: 10.1007/978-3-319-64468-4_11

data attribute. Moreover, they do not take advantage of the implicit temporal information of data streams to deal with temporal preferences.

Temporal preferences may allow users to express how an instant of time may influence his preferences at another time moment. Therefore, it allows an application to employ continuous queries to find sequences of patterns in data according to user preferences. For example, considering a soccer game where players are monitored, it is possible for a coach to check if some player behavior matches certain preferences before making an intervention in the game. Thus, it is possible to evaluate queries such as "Which are the best players considering that if a player was at defensive intermediary, then I prefer that this player go to the middle-field instead of staying in the same place?".

The evaluation of continuous queries with *conditional preference* has already been explored in previous works [1,12]. Nevertheless, to the best of our knowledge, the support for temporal conditional preference queries on data streams began to be exploited recently in our previous paper [14]. The main goal herein is to present an extension for the *Continuous Query Language (CQL)* that is specially tailored to efficiently process temporal conditional preference queries on data streams. Although CQL is an expressive SQL-based declarative language [2,3], it was not designed to deal with temporal preferences. In order to cope with this issue, we define appropriate data structures for keeping the temporal order of tuples and new specific operators for selecting the best sequences according to users preferences. These new features allow our approach to achieve a considerable better performance.

Main Contributions. The main contributions of this paper can be summarized as follows: **(1)** The definition of new operators for the evaluation of continuous queries containing temporal conditional preferences; **(2)** A new and efficient incremental method for the evaluation of the proposed preference operator; **(3)** A detailed demonstration of the CQL equivalences for the proposed operators; **(4)** An extensive set of experiments showing that our proposed operators have better performance than their equivalent operations in CQL.

In the following sections, we introduce a motivating example. In addition, we present the fundamental concepts regarding the temporal conditional preferences and describe the operators and algorithms proposed for the evaluation of temporal conditional preference queries on data streams. We also discuss the equivalent operations in CQL and the experimental results. Finally, at the end we give the conclusions of this paper.

2   A Motivating Example

Philip is a soccer coach who uses technology to make decisions. He has access to an information system that provides real-time data about players during a match. The information available is the stream Event(PID, PC, PE) containing the match events along with an identification of the involved players. The attributes of the stream Event are: player identification (PID), current place (PC) and the match event (PE). The values for PC are the regions showed in Fig. 1. The match events (PE) are: carrying the ball (*ca*), completed pass (*cp*), dribble (*dr*), losing the ball (*lb*), non-completed pass (*ncp*) and pass reception (*re*).

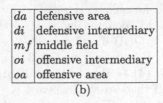

da	defensive area
di	defensive intermediary
mf	middle field
oi	offensive intermediary
oa	offensive area

(a)   (b)

Fig. 1. Values for attribute PC: (a) field division; (b) values description.

Based on his experience, Philip has the following preferences: [**P1**] If the previous in-game event was a pass reception then I prefer a dribble than a completed pass, independent of the place; [**P2**] Completed passes are better than non-completed passes; [**P3**] If all previous in-game events were in the middle-field then I prefer events in the middle-field than events in the defensive intermediary.

These preferences can be used by Philip to submit the following continuous query to the information system: [**Q1**] Every instant, give me the in-game event sequences that best fit my preferences in the last six seconds. When the coach says "best fit my preferences" it means that if a data item X is in the query result then it is not possible to find another response better than X according to his preferences. The query answers could help the coach to give special attention to a particular player having behavior fitting the coach preferences.

3   Background: Temporal Conditional Preferences

Our proposed language, called *StreamPref*, uses the formalism introduced in our previous work [14] to compare sequences of tuples. Let $\textbf{Dom}(A)$ be the domain of the attribute A. Let $R(A_1, ..., A_l)$ be a relational schema. The set of all tuples over R is denoted by $\textbf{Tup}(R) = \textbf{Dom}(A_1) \times ... \times \textbf{Dom}(A_l)$. A sequence $s = \langle t_1, ..., t_n \rangle$ over R is an ordered set of tuples, such that $t_i \in \textbf{Tup}(R)$ for all $i \in \{1, ..., n\}$. The length of a sequence s is denoted by $|s|$. A tuple in the position i of a sequence s is denoted by $s[i]$ and $s[i].A$ represents the attribute A in the position i of s. We use $s[i, j]$ to denote the subsequence $s' = \langle t_i, ..., t_j \rangle$ of $s = \langle t_1, ..., t_n \rangle$ such that $1 \le i \le n$ and $i \le j \le n$. The concatenation s'' of two sequences $s = \langle t_1, ..., t_n \rangle$ and $s' = \langle t'_1, ..., t'_{n'} \rangle$, denoted by $s + s'$, is $s'' = \langle t_1, ..., t_n, t'_1, ..., t'_{n'} \rangle$. We denote by $\textbf{Seq}(R)$ the set of all possible sequences over R.

Our preference model uses StreamPref Temporal Logic (STL) formulas composed by propositions in the form $A\theta a$, where $a \in \textbf{Dom}(A)$ and $\theta \in \{<, \le, =, \ne, \ge, >\}$ (see Definition 1). Let $Q(A)$ be a proposition, $S_{Q(A)} = \{a \in \textbf{Dom}(A) \mid a \models Q(A)\}$ denotes the set of values satisfying $Q(A)$.

Definition 1 (STL Formulas). *The STL formulas are defined as follows: (1)* **true** *and* **false** *are STL formulas; (2) If F is a proposition then F is a STL formula; (3) If F and G are STL formulas then* $(F \wedge G)$, $(F \vee G)$, $(F \text{ \textbf{Since} } G)$, $\neg F$ *and* $\neg G$ *are STL formulas.*

A STL formula F is satisfied by a sequence $s = \langle t_1, ..., t_n \rangle$ at a position $i \in \{1, ..., n\}$, denoted by $(s, i) \models F$, according to the following conditions: **(1)** $(s, i) \models Q(A)$ if and only if $s[i].A \models Q(A)$; **(2)** $(s, i) \models F \wedge G$ if and only if $(s, i) \models F$ and $(s, i) \models G$; **(3)** $(s, i) \models F \vee G$ if and only if $(s, i) \models F$ or $(s, i) \models G$; **(4)** $(s, i) \models \neg F$ if and only if $(s, i) \not\models F$; **(5)** $(s, i) \models (F \text{ \textbf{since} } G)$ if and only if there exists j where $1 \leq j < i$ and $(s, j) \models G$ and $(s, k) \models F$ for all $k \in \{j + 1, ..., i\}$. The **true** formula is always satisfied and the **false** formula is never satisfied. We also define the following derived formulas:

Prev $Q(A)$: Equivalent to (**false since** $Q(A)$), $(s, i) \models$ **Prev** $Q(A)$ if and only if $i > 1$ and $(s, i - 1) \models F$;

SomePrev $Q(A)$: Equivalent to (**true since** $Q(A)$), $(s, i) \models$ **SomePrev** $Q(A)$ if and only if there exists j such that $1 \leq j < i$ and $(s, i) \models Q(A)$;

AllPrev $Q(A)$: Equivalent to $\neg$(**SomePrev**$\neg Q(A)$), $(s, i) \models$ **AllPrev** $Q(A)$ if and only if $(s, j) \models F$ for all $j \in \{1, ..., i - 1\}$;

First: Equivalent to $\neg$(**Prev**(**true**)), $(s, i) \models$ **First** if and only if $i = 1$.

The Definition 2 formalizes the *temporal conditions* used by Definition 3 (*tcp-rules* and *tcp-theories*).

Definition 2 (Temporal Conditions). *A temporal condition is a formula* $F = F_1 \wedge ... \wedge F_n$, *where* $F_1, ..., F_n$ *are propositions or derived formulas. The temporal components of F, denoted by* $F^{\leftarrow}$, *is the conjunction of all derived formulas in F. The non-temporal components of F, denoted by* $F^{\bullet}$, *is the conjunction of all propositions in F and not present in* $F^{\leftarrow}$. *We use* **Att**(F) *to denote the attributes appearing in F.*

Definition 3 (TCP-Rules and TCP-Theories). *Let R be a relational schema. A temporal conditional preference rule, or tcp-rule, is an expression in the format* $\varphi : C_\varphi \rightarrow Q_\varphi^+ \succ Q_\varphi^- [W_\varphi]$, *where:* **(1)** *The propositions* Q_φ^+ *and* Q_φ^- *represent the preferred values and non-preferred values for the preference attribute* A_φ, *respectively, such that* $S_{Q_\varphi^+} \cap S_{Q_\varphi^-} = \emptyset$; **(2)** $W_\varphi \subset R$ *is the set of indifferent attributes such that* $A_\varphi \notin W_\varphi$; **(3)** C_φ *is a temporal condition such that* **Att**$(C_\varphi^{\bullet}) \cap (\{A_\varphi\} \cup W_\varphi) = \emptyset$. *A temporal conditional preference theory, or tcp-theory, is a finite set of tcp-rules.*

Example 1. Consider the coach preferences of Sect. 2. We can express them by the tcp-theory $\Phi = \{\varphi_1, \varphi_2, \varphi_3\}$, where $\varphi_1 :$ **Prev**(PE $= re$) $\rightarrow$ (PE $= dr$) $\succ$ (PE $= cp$)[PC]; $\varphi_2 : \rightarrow$ (PE $= cp$) $\succ$ (PE $= ncp$); $\varphi_3 :$ **AllPrev**(PC $= mf$) $\rightarrow$ (PC $= mf$) $\succ$ (PC $= di$).

Given a tcp-rule φ and two sequences s, s'. We say that s is preferred to s' (or s dominates s') according to φ, denoted by $s \succ_\varphi s'$ if and only if there exists a position i such that: **(1)** $s[j] = s'[j]$ for all $j \in \{1, ..., i-1\}$; **(2)** $(s,i) \models C_\varphi$ and $(s',i) \models C_\varphi$; **(3)** $s[i].A_\varphi \models Q_\varphi^+$ and $s'[i].A_\varphi \models Q_\varphi^-$; **(4)** $s[i].A' = s'[i].A'$ for all $A' \notin (\{A_\varphi\} \cup W_\varphi)$ (*ceteris paribus* semantic).

The notation $\succ_\varphi$ represents the transitive closure of $\bigcup_{\varphi \in \Phi} \succ_\varphi$. The notation $s \succ_\Phi s'$ means that s is preferred to s' according to Φ. When two sequences cannot be compared, we say that they are incomparable. We also must consider consistency issues when dealing with order induced by rules to avoid inferences like "a sequence is preferred to itself". So, we check the consistency of tcp-theories using the test proposed in [14] before the query execution.

4 Proposed Operators

Our StreamPref language introduces the operators **SEQ** and **BESTSEQ**. The **SEQ** operator extracts sequences from data streams preserving the temporal order of tuples and the **BESTSEQ** operator selects the best extracted sequences according to the defined temporal preferences. If it is necessary, our operators can be combined with the existing CQL operators to create more sophisticated queries. As we will see in the next section, our operators can be processed by equivalent CQL operations. However, the definition of these equivalences is not trivial and our operators have better performance than their CQL equivalent operations.

*The **SEQ** Operator.* The **SEQ** operator retrieves identified sequences (Definition 4) over a data stream according to: a set of identifier attributes (X), a temporal range (n) and a slide interval (d). The parameters n and d are used to select a portion of tuples from a data stream analogous to the selection performed by the *sliding window* approach [3,13]. The parameter X is used to group the tuples with the same identifier in a sequence. It is important to note that the values for the identifier attributes must be unique at every instant to keep a relation one-to-one between tuples and sequences.

Definition 4 (Identified Sequences). *Let $S(A_1, ..., A_l)$ be a stream. Let Y and X be two disjoint sets such that $X \cup Y = \{A_1, ..., A_l\}$. An identified sequence $s_x = \langle t_1, ..., t_n \rangle$ from S is a sequence where $t_i \in \textbf{Tup}(Y)$ for all $i \in \{1, ..., n\}$ and $x \in \textbf{Tup}(X)$.*

Example 2. Consider the Event stream of Fig. 2 where TS is the timestamp (instant). The sequence extraction needed by query **Q1** presented in the motivating example is performed by operation $\textbf{SEQ}_{\{\text{PID}\},6,1}$ (Event) as follows:

TS 1: $s_1 = \langle (mf, re) \rangle$; **TS 2:** $s_1 = \langle (mf, re), (oi, dr) \rangle$, **TS 3:** $s_1 = \langle (mf, re), (oi, dr), (oi, cp) \rangle$, $s_2 = \langle (mf, re) \rangle$; **TS 4:** $s_1 = \langle (mf, re), (oi, dr), (oi, cp) \rangle$, $s_2 = \langle (mf, re), (oi, cp) \rangle$; **TS 5:** $s_1 = \langle (mf, re), (oi, dr), (oi, cp) \rangle$, $s_2 = \langle (mf, re), (oi, cp), (oi, lb) \rangle$; **TS 6:** $s_1 = \langle (mf, re), (oi, dr), (oi, cp) \rangle$, $s_2 = \langle (mf, re), (oi, cp), (oi, lb) \rangle$, $s_3 = \langle (mf, ca) \rangle$; **TS 7:** $s_1 = \langle (oi, dr), (oi, cp) \rangle$, $s_3 = \langle (mf, ca), (mf, dr) \rangle$; **TS 8:** $s_1 = \langle (oi, cp) \rangle$, $s_2 = \langle (mf, re), (oi, cp), (oi, lb) \rangle$, $s_3 = \langle (mf, ca), (mf, dr), (di, ncp) \rangle$, $s_4 = \langle (mf, ca) \rangle$; **TS 9:** $s_2 = \langle (oi, cp), (oi, lb) \rangle$, $s_3 = \langle (mf, ca), (mf, dr),$

TS	PID	PC	PE
1	1	mf	re
2	1	oi	dr
3	1	oi	cp

TS	PID	PC	PE
3	2	mf	re
4	2	oi	cp
5	2	oi	lb

TS	PID	PC	PE
6	3	mf	ca
7	3	mf	dr
8	3	di	ncp

TS	PID	PC	PE
8	4	mf	ca
9	4	mf	dr
10	4	mf	cp

Fig. 2. Event stream

$(di, ncp)\rangle$, $s_4 = \langle(mf, ca), (mf, dr)\rangle$; TS **10**: $s_2 = \langle(oi, lb)\rangle$, $s_3 = \langle(mf, ca),$ $(mf, dr), (di, ncp)\rangle$, $s_4 = \langle(mf, ca), (mf, dr), (mf, cp)\rangle$. Note that from TS 7 the **SEQ** operator appends the new tuples and drops the expired positions in the beginning of the sequences.

The **BESTSEQ** *Operator.* Let Z be a set of sequences and Φ be a tcp-theory. The operation **BESTSEQ**$_\Phi(Z)$ returns the *dominant* sequences in Z according to Φ. A sequence $s \in Z$ is dominant according to Φ, if $\nexists s' \in Z$ such that $s' \succ_\Phi s$.

Example 3. Let Z be the extracted sequences of Example 2 and Φ be the tcp-theory of Example 1. The query **Q1** is computed by the operation **BESTSEQ**$_\Phi(\textbf{SEQ}_{\{\text{PID}\},6,1}(\text{Event}))$ as follows: TS **1**: $\{s_1\}$ (the unique input sequence); TS **2**: $\{s_1\}$ (same result of TS 1); TS **3**: $\{s_1, s_2\}$ (incomparable sequences); TS **4**: $\{s_1\}$ ($s_1 \succ_{\varphi_1} s_2$); TS **5**: $\{s_1\}$ (same result of TS 4); TS **6**: $\{s_1, s_3\}$ ($s_1 \succ_{\varphi_1} s_2$ and s_3 is incomparable); TS **7**: $\{s_1, s_2, s_3\}$ (incomparable sequences); TS **8**: $\{s_1, s_2, s_3, s_4\}$ (incomparable sequences); TS **9**: $\{s_2, s_3, s_4\}$ (incomparable sequences); TS **10**: $\{s_2, s_4\}$ ($s_4 \succ_{\varphi_2} ... \succ_{\varphi_3} s_3$ and s_2 is incomparable).

5   CQL Equivalences

This section demonstrates how our StreamPref operators can be translated to CQL equivalent operations. It is worth noting that although this means the StreamPref does not increase the expression power of CQL, the equivalences are not trivial. Moreover, the evaluation of the StreamPref operators are more efficient than their CQL equivalent operations (see Sect. 7). The equivalences consider a stream $S(A_1, ..., A_l)$ and the identifier $X = \{A_1\}$. Although, we can use any subset of $\{A_1, ..., A_l\}$ as identifier without lost of generality. In addition, the CQL equivalences represent the sequences using relations containing the attribute POS to identify the position of the tuples.

We use the symbols π, $-$, $\bowtie$, γ, σ and $\cup$ for the CQL operators that are equivalent to the traditional operations: projection, set difference, join, aggregation function, selection and union, respectively. The **RSTREAM** is a CQL operator to convert a relation to a stream and the symbol $\boxplus$ is the CQL sliding window operator. The notation TS() returns the original timestamp of the tuple. We rename an attribute A to A' by using the notation $A \mapsto A'$ in the projection operator.

Equivalence for the **SEQ** *Operator.* Equation (1) establishes the CQL equivalence for the **SEQ** operator such that $P_0 = \{\}$ and $i \in \{1, ..., n\}$.

$$W_0 = \pi_{\text{POS}, A_1}(\boxplus_{n,d}(\textbf{RSTREAM}(\pi_{\text{TS}()\mapsto\text{POS}, A_1, ..., A_l}(\boxplus_{1,1}(S))))) \tag{1a}$$

$$W_i = W_{i-1} - P_{i-1} \tag{1b}$$

$$P_i = \gamma_{A_1, \min(\text{POS})\mapsto\text{POS}}(W_i) \bowtie_{A_1, \text{POS}} W \tag{1c}$$

$$\textbf{SEQ}_{\{A_1\}, n, d}(S) = \pi_{1\mapsto\text{POS}, A_1, ... A_l}(P_1) \cup ... \cup \pi_{n\mapsto\text{POS}, A_1, ... A_l}(P_n) \tag{1d}$$

Equivalence for the **BESTSEQ** *Operator.* The CQL equivalence for the **BESTSEQ** operator is computed over a relation $Z(\text{POS}, A_1, ..., A_l)$ containing the input sequences where $\{A_1\}$ is the identifier and POS is the position attribute. First, Eq. (2) calculates the position to compare every pair of sequences.

$$Z' = \pi_{\text{POS}, A_1\mapsto B', A_2\mapsto A_2', ..., A_l\mapsto A_l'}(Z) \tag{2a}$$

$$P_{nc} = \sigma_{A_2\neq A_2'\vee...\vee A_l\neq A_l'}(\pi_{\text{POS}, A_1\mapsto B, A_2, ..., A_l}(Z) \bowtie_{\text{POS}} Z') \tag{2b}$$

$$P = \gamma_{B, B', \min(\text{POS})\mapsto\text{POS}}(P_{nc}) \tag{2c}$$

The next step is to identify the sequence positions satisfying the temporal components of the rule conditions. Equation (3) calculates the positions satisfied by every derived formula.

$$P^{\textbf{First}} = \pi_{\text{POS}, B}(\sigma_{\text{POS}=1}(P)) \tag{3a}$$

$$P_{Q(A)}^{\textbf{Prev}} = \pi_{\text{POS}, B}(P \bowtie_{\text{POS}, B} (\pi_{(\text{POS}+1)\mapsto\text{POS}, A_1\mapsto B}(\sigma_{Q(A)}(Z)))) \tag{3b}$$

$$P_{Q(A)}^{\textbf{SomePrev}} = \pi_{\text{POS}, B}(\sigma_{\text{POS}>\text{POS}'}(P \bowtie_B (\gamma_{A_1\mapsto B, \min(\text{POS})\mapsto\text{POS}'}(\sigma_{Q(A)}(Z))))) \tag{3c}$$

$$P^{\max} = \gamma_{A_1\mapsto B, \max(\text{POS})\mapsto\text{POS}}(P) \tag{3d}$$

$$P_{\neg Q(A)}' = \gamma_{B, \min(\text{POS})\mapsto\text{POS}'}(\pi_{\text{POS}, A_1\mapsto B}(\sigma_{\neg Q(A)}(Z)) \cup P^{\max}) \tag{3e}$$

$$P_{Q(A)}^{\textbf{AllPrev}} = \pi_{\text{POS}, B}(\sigma_{\text{POS}\leq\text{POS}'\wedge\text{POS}>1}(P \bowtie_B (P_{\neg Q(A)}'))) \tag{3f}$$

Next, Eq. (4) computes the relation R_i containing the positions satisfied by condition $C_{\varphi_i}^{\leftarrow} = F_1 \wedge ... \wedge F_p$ for every tcp-rule $\varphi_i \in \Phi$.

$$P_j = \begin{cases} P^{\textbf{First}}, \text{if } F_j = \textbf{First} \\ P_{Q(A)}^{\textbf{Prev}}, \text{if } F_j = \textbf{Prev}(Q(A)) \\ P_{Q(A)}^{\textbf{SomePrev}}, \text{if } F_j = \textbf{SomePrev}(Q(A)) \\ P_{Q(A)}^{\textbf{AllPrev}}, \text{if } F_j = \textbf{AllPrev}(Q(A)) \end{cases} \tag{4a}$$

$$R_i = (P_1) \bowtie_{\text{POS}, B} ... \bowtie_{\text{POS}, B} (P_p) \tag{4b}$$

Equation (5) performs the direct comparisons for every $\varphi_i \in \Phi$. This equation also consider the tuples of $\textbf{Tup}(S)$ for posterior computation of the transitive closure. The relations D_i^+ and D_i^- represent the tuples satisfying respectively the preferred values and the non-preferred values of φ_i. These tuples include *original tuples* (from existing positions) and *fake tuples* from $\textbf{Tup}(S)$. We use the attribute A_t to separate these tuples ($A_t = 1$ for original tuples and $A_t = 0$

for fake tuples). Note that the Eq. (5e) applies the filter $E_{\varphi_i} : (A_{i_1} = A'_{i_1}) \wedge ... \wedge$ $(A_{i_j} = A'_{i_j})$ such that $\{A_{i_1}, ..., A_{i_j}\} = (\{A_1, ..., A_l\} - \{A_{\varphi_i}, B, B'\} - W_{\varphi_i})$. This filter is required to follow the *ceteris paribus* semantic.

$$Z_i^+ = \pi_{\mathrm{POS},B}(R_i) \bowtie_{\mathrm{POS},B} (\sigma_{C_{\varphi_i}^\bullet \wedge Q_{\varphi_i}^+} (\pi_{\mathrm{POS},A_1 \mapsto B,...,A_l,1 \mapsto A_t}(Z))) \tag{5a}$$

$$D_i^+ = Z_i^+ \cup (\pi_{\mathrm{POS},B}(R_i) \bowtie_B (\pi_{A_1 \mapsto B,...,A_l,0 \mapsto A_t}(\sigma_{C_{\varphi_i}^\bullet \wedge Q_{\varphi_i}^+}(\mathbf{Tup}(S))))) \tag{5b}$$

$$Z_i^- = \pi_{\mathrm{POS},B'}(R_i) \bowtie_{\mathrm{POS},B'} (\sigma_{C_{\varphi_i}^\bullet \wedge Q_{\varphi_i}^-} (\pi_{\mathrm{POS},A_1 \mapsto B',...,A_l,1 \mapsto A_t}(Z))) \tag{5c}$$

$$D_i^- = Z_i^- \cup (\pi_{\mathrm{POS},B'}(R_i) \bowtie_{B'} (\pi_{A_1 \mapsto B',...,A_l,0 \mapsto A_t}(\sigma_{C_{\varphi_i}^\bullet \wedge Q_{\varphi_i}^-}(\mathbf{Tup}(S))))) \tag{5d}$$

$$D_i = \sigma_{E_{\varphi_i}}(P \bowtie_{\mathrm{POS},B,B'} (D_i^+ \bowtie_{\mathrm{POS}} (\pi_{\mathrm{POS},B',A_2 \mapsto A'_2,...,A_l \mapsto A'_l,A_t \mapsto A'_t}(D_i^-)))) \tag{5e}$$

Equation (6) calculates the transitive closure. The relations T_i', T_i'' and T_i are computed for $i \in \{2, ..., m\}$. In the end, T_m has all comparisons imposed by Φ where $m = |\Phi|$ is the number of tcp-rules.

$$T_1 = D_1 \cup ... \cup D_m \tag{6a}$$

$$T_i' = \pi_{(\mathrm{POS},B,B',A_2,...,A_l,A_t,A_2' \mapsto A_2'',...,A_l' \mapsto A_l'',A_t' \mapsto A_t'')}(T_{i-1}) \tag{6b}$$

$$T_i'' = \pi_{(\mathrm{POS},B,B',A_2 \mapsto A_2'',...,A_l \mapsto A_l'',A_t \mapsto A_t'',A_2',...,A_l',A_t')}(T_{i-1}) \tag{6c}$$

$$T_i = \pi_{\mathrm{POS},B,B',A_2,...,A_l,A_t,A_2',...,A_l',A_t'}(T_i' \bowtie_{\mathrm{POS},B,B',A_2'',...,A_l''} T_i'') \cup T_{i-1} \tag{6d}$$

Equation (7) calculates the dominant sequences. Observe that just comparisons between original tuples are considered ($A_t = 1 \wedge A_t' = 1$).

$$\mathbf{BESTSEQ}_\Phi(Z) = Z \bowtie_{A_1} (\pi'_{A_1}(Z) - \pi'_{B' \mapsto A_1}(\sigma_{A_t=1 \wedge A_t'=1}(T_m))) \tag{7}$$

6   Data Structures and Algorithms

Our previous work [14] proposed the algorithm *ExtractSeq* for extracting sequences and the algorithm *BestSeq* for computing the dominant sequences. The StreamPref operators **SEQ** and **BESTSEQ** can be evaluated by the algorithms *ExtractSeq* and *BestSeq*, respectively. However, only the algorithm *ExtractSeq* uses an incremental method suitable for data streams scenarios. In this paper we propose a new incremental method to evaluate the **BESTSEQ** operator.

Index Structure. The main idea of our incremental method is to keep an index tree built using the sequence tuples. Given a sequence $s = \langle t_1, ..., t_n \rangle$, every tuple t_i is represented by a node in the tree. The tuple t_1 is a child of the root node. For the remaining tuples, every t_i is a father of t_{i+1}. The sequence s is stored in the node t_n.

Example 4. Consider the sequences of the Example 2 at TS 6. Figure 3(a) shows how these sequences are stored in the index tree. The root node, represented by black circle, is an empty node without an associated tuple.

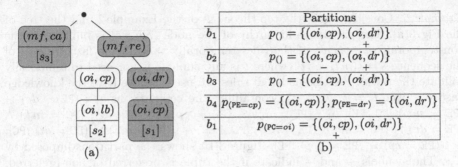

Fig. 3. Preference hierarchy: (a) index tree; (b) partitions imposed by K_Γ of $(oi, 1, la)$.

Starting from the root node, it is possible to find the position where two sequences must be compared. For instance, consider the sequences s_1 and s_2 in the tree of Fig. 3(a). The paths from the root to these sequences are different in the second node. Thus, the comparison of s_1 and s_2 happens in the position 2.

The index is updated only for changed sequences and new sequences. The new sequences are just inserted in the tree. When positions are deleted from a sequence s (and s is still no empty), we reinsert s in the tree. The empty sequences are dropped from the tree. If a sequence s has new tuples (and no expired tuples), we move s to a child branch of its current node.

Given a node nd, $nd.t$ is the tuple associated to nd and $nd.Z$ represents the set of sequences stored in nd. The children of nd are stored in a hash-table $nd.Ch$ mapping the associated tuples to the respective child nodes. In addition, each node nd stores a preference hierarchy $nd.H$ over the tuples of the child nodes. The preference hierarchy allows to determine if a child node is dominant or is dominated. Thus, it is possible to know if a sequence dominates another one.

Preference Hierarchy. Our preference hierarchy structure is based on the preference partition technique originally proposed in our previous work [15]. The main idea is to build a knowledge base for the preferences valid in a node and keep a structure containing the preferred and non-preferred tuples according to such preferences. Given a set of non-temporal preference rules Γ. The knowledge base K_Γ over Γ is a set of comparisons in the format $b : F_b^+ \succ F_b^- [W_b]$. The terms F_b^+ and F_b^- are formulas representing the preferred values and non-preferred values, respectively. The term W_b is the set of indifferent attributes of b. For more details about the construction of the knowledge base, please see [15].

After the construction of K_Γ, the preference hierarchy is built by grouping the child node tuples into subsets called partitions. For every comparison $b \in K_\Gamma$, we group the tuples into partitions according to the attribute values not in W_b. If a partition does not contain *preferred tuples* (those satisfying F_b^+), then all tuples of this partition are dominant. On the other hand, if a partition has at least one preferred tuple, then all *non-preferred tuples* (those satisfying F_b^-) are dominated. Therefore, a tuple t is dominant if t is not dominated in any partition.

Example 5. Consider again the tcp-theory Φ of the Example 2 and the tree of the Fig. 3(a). The preference hierarchy of the node (mf, re) is built using the non-temporal components of the rules temporally valid in the last position of the sequence $s = \langle (mf, re), t \rangle$, where t is any tuple (t is not used to temporally validate the rules). For this node, all rules are used. So, we have the knowledge base K_Γ over $\Gamma = \{\varphi_1^\bullet, \varphi_2^\bullet, \varphi_3^\bullet\}$ containing the comparisons $b_1 : (\text{PE} = dr) \succ (\text{PE} = ncp)$ [PC, PE]; $b_2 : (\text{PE} = dr) \succ (\text{PE} = cp)$ [PC, PE]; $b_3 : (\text{PC} = mf) \wedge (\text{PE} = dr) \succ (\text{PC} = di) \wedge (\text{PE} = ncp)$ [PC, PE]; $b_4 : (\text{PC} = mf) \succ (\text{PC} = di)$ [PC]; $b_5 : (\text{PE} = cp) \succ (\text{PE} = ncp)$ [PE]. Figure 3(b) shows the partitions imposed by K_Γ. The symbols $+$ and $-$ indicate if the tuple is preferred or non-preferred, respectively. We can see that (oi, cp) is dominated because it is a non-preferred tuple in the partition of the comparison b_2 containing a preferred tuple.

Our technique uses just the essential information to update the index efficiently. For every partition p, we keep the mappings $Pref(p)$ and $NonPref(p)$ representing the number of preferred tuples in p and the set on non-preferred tuples in p, respectively. Thus, a node nd is dominated if there exists p such that $Pref(p) > 0$ and $nd.t \in NonPref(p)$.

The construction of the knowledge base is not a trivial task since we must compute comparisons representing the transitive closure imposed by the preferences [15]. Thus, we also use a pruning strategy to avoid the construction of unnecessary preference hierarchies. Nodes having a unique child, do not need preference hierarchy since this child is always dominant. In addition, dominated nodes and their descendants do not require preference hierarchy. For example, in the tree of Fig. 3(a), we know that (oi, dr) dominates (oi, cp) according to the hierarchy of (mf, re). Thus, we do not need preference hierarchies for nodes in the branch starting at (oi, cp).

Algorithms. Our full index structure is composed by the tree nodes represented by its *root* and the mapping *SeqNod*. For every identified sequence s_x, $SeqNod(x)$ stores (s_x, nd) where nd is the node where s_x is stored. In addition, we use the sequence attributes *deleted* and *inserted* to keep the number of deletions and insertions in the last instant.

The algorithm *IndexUpdate* (see Algorithm 1) incrementally updates the index according to sequence changes. The first loop (lines 2–2) processes the changes for every sequence s_x already stored. If s_x has expired positions ($s_x.deleted > 0$), we remove s_x from the index. When s_x is not empty ($|s_x| > 0$), we add s_x into I to be reinserted later since s_x must be repositioned in the index tree. If s_x has no expired positions and has inserted positions, the routine *AddSeq* reallocates s_x from its current node. The second loop (lines 12-12) looks for sequences in Z not stored in the index and adds them into I. At the end, the algorithm stores the sequences of I and calls the routine *Clean* to remove empty nodes.

Algorithm 1. $IndexUpdate(idx, Z)$	Algorithm 2. $AddSeq(nd, s_x)$
1 $I \leftarrow \{\}$;	1 $d \leftarrow Depth(nd)$;
2 **foreach** $x \in idx.SeqNod$ **do**	2 **if** $d = |s_x|$ **then**
3 $(s_x, nd) \leftarrow idx.SeqNod(x)$;	3 $nd.Z.Add(s_x)$;
4 **if** $s_x.deleted > 0$ **then**	4 **return** nd;
5 $nd.Z.Del(s_x)$;	5 $t \leftarrow s_x[d + 1]$;
6 $idx.SeqNod.Del(x)$;	6 **if** $t \in nd.Ch$ **then**
7 **if** $|s_x| > 0$ **then** $I.add(s_x)$;	7 $child \leftarrow nd.Ch(t)$;
8 **else if** $s_x.inserted > 0$ **then**	8 **else**
9 $nd.Z.Del(s_x)$;	9 $child \leftarrow NewChild(nd, t)$;
10 $new \leftarrow AddSeq(nd, s_x)$;	10 **return** $AddSeq(child, s)$;
11 $idx.SeqNod.Put(x \mapsto (s_x, new))$;	
12 **foreach** $s_x \in Z$ **do**	**Algorithm 3.** $IncBestSeq(nd)$
13 **if** $x \notin idx.SeqNod$ **then** $I.add(s_x)$;	1 $Z \leftarrow nd.Z$;
14 **foreach** $s_x \in I$ **do**	2 **foreach** dominant $child$ of nd **do**
15 $nd \leftarrow AddSeq(idx.root, s_x)$;	3 $Z \leftarrow Z \cup IncBestSeq(child)$;
16 $idx.SeqNod.Put(x \mapsto (s_x, nd))$;	4 **return** Z;
17 $Clean(idx.root)$;	

The routine *AddSeq* (see Algorithm 2) performs the insertion of a sequence s_x in the index tree. First, the routine checks if the depth (d) of node nd is equal to the length of s_x ($d = |s_x|$). If true, s_x is stored into nd since the full path containing the tuples of s_x is already created. Otherwise, the routine selects an existing child or creates a new one. At the end, the routine makes a recursion over this *child* node.

The algorithm *IncBestSeq* (see Algorithm 3) employs the index tree to evaluate the **BESTSEQ** operator incrementally. The execution starts at *idx.root*. The algorithm acquires the sequences of input node nd and uses the preference hierarchy to select the dominant children of nd. Thus, for every dominant child, the algorithm makes a recursive call to retrieve all dominant sequences.

Example 6. Consider again the index tree of Fig. 3(a) (the dominant nodes are in gray). The execution of *IncBestSeq* over this index tree works as follows:

(1) The execution starts at the root node. This node has no sequences and the algorithm makes a recursión over the dominant children (mf, ca) and (mf, re);
(2) At (mf, ca) the algorithm reaches the sequence s_3. So, $Z = \{s_3\}$;
(3) At (mf, re), the algorithm performs a recursion over (oi, dr);
(4) At (oi, dr), the algorithm starts a recursion over (oi, cp);
(5) The algorithm reaches the sequence s_1 and returns $Z = \{s_1, s_3\}$.

Complexity. The complexity analysis of the algorithms takes into account the number of input sequences (k), the length of the largest sequence (n) and the number of tcp-rules in Φ (m). We also assume a constant factor for the number of attributes. In the worst case, the insertion or the deletion of a node tuple in the preference hierarchy has the cost $O(m^4)$ (the size of K_Γ [15]). Moreover, in the worst case scenario, the degree of nodes is $O(k)$, the tree depth is $O(n)$ and the number of partitions associated to a child node is $O(|K_\Gamma|)$. Our mapping structures *SeqNod*, *Ch*, *Pref* and *NonPrefSet* are implemented using hash-tables. So, the retrieval and the storage of elements is performed with a cost of $O(1)$.

In the worst case, the routine *AddSeq* reaches a leaf node and the routine *Clean* scans all tree nodes. So, the costs of *AddSeq* and *Clean* are $O(nm^4)$ and $O(k^n m^4)$, respectively. The complexity of the algorithm *IndexUpdate* is $O(k^n m^4)$. Regarding the algorithm *IncBestSeq*, the selection of the dominant children has a cost of $O(km^4)$. Thus, the complexity of *IncBestSeq* is $O(k^n m^4)$.

7   Experimental Results

Our experiments confront our proposed operators against their CQL counterparts to analyze the performance (runtime) and memory usage of both approaches. All experiments were carried out on a machine with a 3.2 GHz twelve-core processor and 32 GB of main memory, running Linux. The algorithms and all CQL operators were implemented in Python language.

Synthetic Datasets. Due to the nonexistence of data generators suitable for the experiment parameters employed herein, we designed our own generator of synthetic datasets[1]. The synthetic datasets are in the format of streams composed by integer attributes. Table 1 shows the parameters (with default values in bold).

Table 1. Parameters for the experiments over synthetic data: (a) dataset generation; (b) sequence extraction; (c) preferences.

(a)		(b)		(c)	
Param.	Variation	Param.	Variation	Param.	Variation
ATT	8, **10**, 12, 14, 16	RAN	10, 20, **40**, 60, 80, 100	RUL	4, **8**, 16, 24, 32
NSQ	4, 8, **16**, 24, 32	SLI	1, **10**, 20, 30, 40	LEV	1, **2**, 3, 4, 5, 6

Table 1(a) presents the parameters related to the dataset generation. The number of attributes (ATT) allows to evaluate the behavior of the algorithms according to data dimensionality. The number of sequences (NSQ) allows to evaluate how the number of tuples per instant (equal to NSQ $\times$ 0.5) affects the algorithms. Table 1(b) displays the parameters used for sequence extraction. These parameters are temporal range (RAN) and slide interval (SLI) and they allow to evaluate how the selection of the stream elements influences the algorithms. Table 1(c) shows the parameters number of rules (RUL) and maximum preference level (LEV) employed for the generation of the preferences. These parameters allow us to evaluate how different preferences affect the cost

---

[1] http://streampref.github.io.

of the sequence comparison done by the algorithms. We use rules in the form φ_i : **First** $\wedge\, Q(A_3) \to Q^+(A_2) \succ Q^-(A_2)[A_4, A_5]$ and φ_{i+1} : **Prev**$Q(A_3) \wedge$ **SomePrev**$Q(A_4) \wedge$ **AllPrev**$Q(A_5) \wedge Q(A_3) \to Q^+(A_2) \succ Q^-(A_2)[A_4, A_5]$ having variations on propositions $Q^+(A_2)$, $Q^-(A_2)$, $Q(A_3)$, $Q(A_4)$, $Q(A_5)$. The number of iterations is RAN plus·maximum slide interval and the sequence identifier is the attribute A_1. The definition of the parameter values was based on the experiments of related works [8,11,13–15]. For each experiment, we varied one parameter and fixed the default value for the others.

Real Datasets. We also used a real dataset containing play-by-play data of the 2014 soccer world cup[2]. This dataset contains 10,282 tuples from the last 4 matches. For this dataset we varied the parameters RAN and SLI. The values for RAN were 6, 12, 18, 24 and 30 s, where the default was 12 s. The values for SLI were 1, 3, 6, 9 and 12 s, where the default was 1 s. The experiments consider the average runtime per match which is the total runtime of all matches divided by the number of matches.

*Experiments with the **SEQ** Operator.* Figure 4(a) and (b) show the experiment results with synthetic data confronting the **SEQ** operator (evaluated by the algorithm *ExtractSeq*) and its CQL equivalence. The first experiment considers the variation on the parameter ATT. Even for few attributes, the **SEQ** operator outperforms the CQL equivalence. Regarding the NSQ parameter, the **SEQ** operator has the best performance again. When there are more sequences, there are more tuples to be processed and the CQL operations are more expensive.

When examining the results obtained with different temporal ranges, it is possible to see that higher temporal ranges had greater impact on the CQL equivalence due to the generation of bigger sequences. Considering the results obtained for the SLI parameter, bigger slides caused more tuples expiration. So, once the sequences are smaller, the **SEQ** operator had the best performance. Finally, considering all experiments shown in Fig. 4(a), it is possible to see that the **SEQ** operator is more efficient than its CQL equivalence. Moreover, when comparing the memory usage displayed on Fig. 4(b), it is possible to verify that the CQL equivalence for the **SEQ** operator had a high memory usage in all experiments due to the extra tuples stored by the intermediary operations.

Figure 4(d) and (c) present the results obtained with the real data. These results are analogous to the ones obtained with the synthetic data. Analyzing these figures, it is possible to see that the **SEQ** operator outperforms the CQL equivalence for all the experiments.

*Experiments with the **BESTSEQ** Operator.* Figuer 5(a) shows the runtime and Fig. 5(b) shows the memory usage for the experiments with the **BESTSEQ** operator with synthetic data. Notice that the runtime graphs are in logarithm scale. We can see that the CQL equivalence is slower than the remaining algorithms due to the processing of the intermediary operations. In addition, the

---

[2] Extracted from data available in http://data.huffingtonpost.com/2014/world-cup.

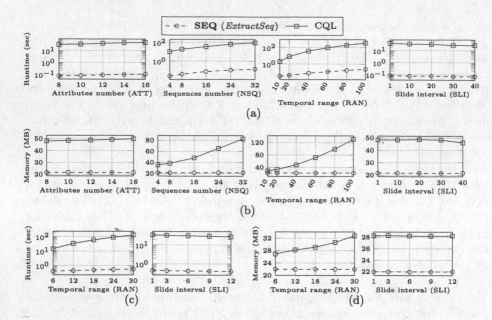

Fig. 4. Experiment results for the **SEQ** operator: (a) synthetic data runtime; (b) synthetic data memory usage; (c) real data runtime; (d) real data memory usage.

incremental algorithm outperforms the naive algorithm due to the index tree and the pruning strategy. The same behavior is observed for the memory usage.

The results obtained for the parameters NSQ, RAN and RUL deserve to be highlighted. Considering the NSQ parameter, the behavior of the **BESTSEQ** algorithms is explained by the fact that when the number of sequences increases, we have more repetition of sequence identifiers and more chances for pruning. So, the updates in the index tree affect fewer branches and the incremental algorithm outperforms the naive algorithm.

Considering the RAN parameter, the incremental algorithm presents an advantage over the naive algorithm since longer sequences have more chances for overlapping. This behavior results in a more compact index tree and in a better performance for the incremental algorithm.

Regarding the results obtained by the naive algorithm considering the RUL parameter, it is worth noting that the number of rules has a great impact in its complexity [14]. Moreover, more rules means more intermediary relations in the CQL equivalence as addressed by Eqs. (4) and (5).

Figure 5(c) and (d) show the results obtained for the **BESTSEQ** operator with the real data. Analyzing this figure, it is possible to see that the naive and the incremental algorithms outperform the CQL equivalence again. In addition, the CQL runtime cannot be applied in a real situation since a regular soccer match has duration of 5400 s. Regarding the memory usage, as expected, the CQL equivalence presented the greatest memory usage due to the storage of

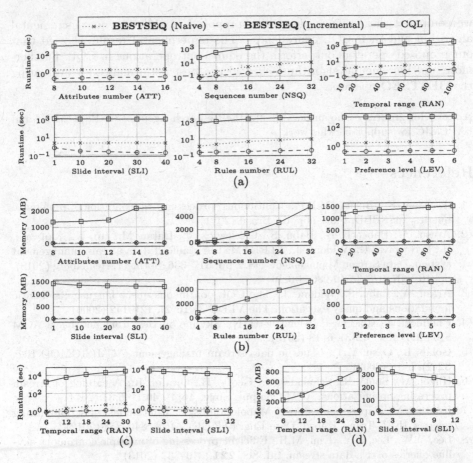

Fig. 5. Experiment results of **BESTSEQ** operator: (a) synthetic data runtime; (b) synthetic data memory usage; (c) real data runtime. (d) real data memory usage.

the intermediary relations. Both versions of **BESTSEQ** have a stable memory usage in all executions (around 20 MB).

8 Conclusion

In this paper we described the StreamPref query language presenting new operators to support temporal conditional preference queries on data streams. Stream-Pref extends the CQL language including the **SEQ** operator for the sequence extraction and the **BESTSEQ** preference operator for the selection of dominant sequences. Regarding the evaluation of this later operator, a new incremental algorithm was proposed. In addition, we also demonstrated the CQL equivalences for the proposed operators. It is worth noting that these equivalences are not trivial since they involve many complex operations. In our experiments, we compared the previous algorithms proposed in [14], the new incremental algorithm and their CQL equivalences. The experimental results showed that our proposed operators outperform the equivalent operations in CQL. Furthermore, our incremental algorithm achieved the best performance for evaluating the **BESTSEQ** operator.

Acknowledgments. The authors thanks the Research Agencies CNPq, CAPES and FAPEMIG for supporting this work.

References

1. de Amo, S., Bueno, M.L.P.: Continuous processing of conditional preference queries. In: SBBD, Florianópolis, Brasil (2011)
2. Arasu, A., Babcock, B., Babu, S., Cieslewicz, J., Datar, M., Ito, K., Motwani, R., Srivastava, U., Widom, J.: STREAM: the stanford data stream management system. Data Stream Management. DSA, pp. 317–336. Springer, Heidelberg (2016). doi:10.1007/978-3-540-28608-0_16
3. Arasu, A., Babu, S., Widom, J.: The CQL continuous query language: semantic foundations and query execution. The VLDB J. **15**(2), 121–142 (2006)
4. Chomicki, J., Ciaccia, P., Meneghetti, N.: Skyline queries, front and back. ACM SIGMOD Rec. **42**(3), 6–18 (2013)
5. Golab, L., Özsu, M.T.: Issues in data stream management. ACM SIGMOD Rec. **32**(2), 5–14 (2003)
6. Hirzel, M., Soulé, R., Schneider, S., Gedik, B., Grimm, R.: A catalog of stream processing optimizations. ACM Comput. Surv. **46**(4), 46:1–46:34 (2014)
7. Kontaki, M., Papadopoulos, A.N., Manolopoulos, Y.: Continuous top-k dominating queries. IEEE Trans. Knowl. Data Eng. (TKDE) **24**(5), 840–853 (2012)
8. Lee, Y.W., Lee, K.Y., Kim, M.H.: Efficient processing of multiple continuous skyline queries over a data stream. Inf. Sci. **221**, 316–337 (2013)
9. Liu, W., Shen, Y.M., Wang, P.: An efficient approach of processing multiple continuous queries. J. Comput. Sci. Technol. **31**(6), 1212–1227 (2016)
10. Margara, A., Urbani, J., van Harmelen, F., Bal, H.: Streaming the web: reasoning over dynamic data. Web Semant.: Sci. Serv. Agents World Wide Web **25**, 24–44 (2014)
11. Pereira, F.S.F., de Amo, S.: Evaluation of conditional preference queries. JIDM **1**(3), 503–518 (2010)
12. Petit, L., Amo, S., Roncancio, C., Labbé, C.: Top-k context-aware queries on streams. In: Liddle, S.W., Schewe, K.-D., Tjoa, A.M., Zhou, X. (eds.) DEXA 2012. LNCS, vol. 7446, pp. 397–411. Springer, Heidelberg (2012). doi:10.1007/978-3-642-32600-4_29
13. Petit, L., Labbé, C., Roncancio, C.: An algebric window model for data stream management. In: ACM MobiDE, Indianapolis, Indiana, USA, pp. 17–24 (2010)
14. Ribeiro, M.R., Barioni, M.C.N., de Amo, S., Roncancio, C., Labbé, C.: Reasoning with temporal preferences over data streams. In: FLAIRS, Marco Island, USA (2017)
15. Ribeiro, M.R., Pereira, F.S.F., Dias, V.V.S.: Efficient algorithms for processing preference queries. In: ACM SAC, Pisa, Italy, pp. 972–979 (2016)

Efficient Processing of Aggregate Reverse Rank Queries

Yuyang Dong[(✉)], Hanxiong Chen, Kazutaka Furuse, and Hiroyuki Kitagawa

Department of Computer Science, University of Tsukuba, Ibaraki, Japan
tou@dblab.is.tsukuba.ac.jp, {chx,furuse,kitagawa}@cs.tsukuba.ac.jp

Abstract. Given two data sets of user preferences and product attributes in addition to a set of query products, the aggregate reverse rank (ARR) query returns top-k users who regard the given query products as the highest aggregate rank than other users.

In this paper, we reveal two limitations of the state-of-the-art solution to ARR query; that is, (a) It has poor efficiency when the distribution of the query set is dispersive. (b) It processes a lot of user data. To address these limitations, we develop a cluster-and-process method and a sophisticated indexing strategy. From the theoretical analysis of the results and experimental comparisons, we conclude that our proposals have superior performance.

Keywords: Aggregate reverse rank queries · Tree-based method

1 Introduction

In the user-product mode, there are two different datasets: user preferences and products. A top-k query retrieves the top-k products for a given user preference in this model. However, manufacturers also want to know the potential customers for their products. Therefore, reverse k-rank query [5] is proposed to obtain the top-k user preferences for a given product. Because most manufacturers offer several products as part of product bundling, the aggregate reverse rank query (ARR) [1] responds to this requirement by retrieving the top-k user preferences for a set of products. Not limited to shopping, the concept of ARR can be extended to a wider range of applications such as team (multiple members) reviewing and area (multiple businesses) reviewing.

An example of ARR query is shown in Fig. 1. There are five different books ($p_1 - p_5$) scored on "price" and "ratings" in Table(b). Three preferences of users Tom, Jerry, and Spike are listed in table(a), which are the weights for each attribute in the book. The score of a book under a user preference is the defined value of the inner product of the book attributes vector and user preference vector $[1, 2, 4, 5]$.[1] Now, assume that the book shop selects two bundles from books, say $\{p_1, p_2\}$ and $\{p_4, p_5\}$. The result of ARR query when $k = 1$ for them

---

[1] Without loss of generality, we assume that the minimum values are preferable.

© Springer International Publishing AG 2017
D. Benslimane et al. (Eds.): DEXA 2017, Part I, LNCS 10438, pp. 159–166, 2017.
DOI: 10.1007/978-3-319-64468-4_12

(a) User preferences data and the ranks

	w[price]	w[rating]	Ranking
Tom	0.8	0.2	p_3,p_2,p_1,p_4,p_5
Jerry	0.3	0.7	p_2,p_5,p_3,p_4,p_1
Spike	0.1	0.9	p_5,p_2,p_4,p_3,p_1

(b) Books data and their ranks on users.

	p[price]	p[rating]	Rank on Tom	Rank on Jerry	Rank on Spike
p_1	6	7	3rd	5th	5th
p_2	2	3	2nd	1st	2nd
p_3	1	6	1st	3rd	4th
p_4	7	5	4th	4th	3rd
p_5	8	2	5th	2nd	1st

(c) Bundled books and their ARR-1Rank

	ARank on Tom	ARank on Jerry	ARank on Spike	AR-1Rank
p_1,p_2	5 (3+2)	6 (5+1)	7 (5+2)	Tom
p_4,p_5	9 (4+5)	6 (4+2)	4 (3+1)	Spike

Fig. 1. The example of aggregate reverse rank query.

are shown in table(c). The ARR query evaluates the aggregate rank (*ARank*) with the sum of each book's rank, e.g., $\{p_1, p_2\}$ ranks as $3 + 2 = 5$ based on Tom's preference. The ARR query returns Tom as the result because Tom thinks the books $\{p_1, p_2\}$ has the highest rank than the rest.

1.1 Aggregate Reverse Rank Query

The assumptions made on the product database, preferences database and the score function between them are the same as those made in the related research [1,4,5]. The querying problems are based on a User-Product model, which has two kinds of database: product data set P and user's preference data set W. Each product $p \in P$ is a d-dimensional vector that contains d non-negative scoring attributes. The product p can be represented as a point $p = (p[1], p[2], ..., p[d])$, where $p[i]$ is the ith attribute value. The preference $w \in W$ is also a d-dimensional vector and $w[i]$ is a non-negative weight that affects the value of $p[i]$, where $\sum_{i=1}^{d} w[i] = 1$. The score of product p w.r.t preference w is defined as the inner product between p and w, denoted by $f(w,p) = \sum_{i=1}^{d} w[i] \cdot p[i]$. All products are ranked with their scores and the minimum score is preferable. Now, let Q denote a query set containing a set of products.

For a specific w, the rank of a single q is defined as the number of products such that $f(w,p)$ is smaller than q's score $f(w,q)$; that is,

Definition 1 $(rank(w,q))$. *Given a product data set P, a preference w and a query q, the rank of q by w is given by $rank(w,q) = |S|$, where $S \subseteq P$ and $\forall p_i \in S, f(w,p_i) < f(w,q) \land \forall p_j \in (P - S), f(w,p_j) \geq f(w,q)$.*

The aggregate reverse rank query [1] retrieves the top-k w's which produce Q better aggregate rank (ARank) than other $w's$. The aggregate evaluation functions ARank is defined as the sum of each $rank(w,q)$, $q \in Q$; that is, $ARank(w,Q) = \sum_{q_i \in Q} rank(w,q_i)$.

Definition 2 *(aggregate reverse rank query, ARR). Given a data set P and data set W, a positive integer k and a query set Q, the ARR query returns the*

set S, $S \subseteq W$ and $|S| = k$ such that $\forall w_i \in S, \forall w_j \in (W - S)$, the identity $ARank(w_i, Q) \leq ARank(w_j, Q)$ holds.

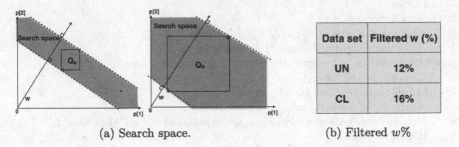

Data set	Filtered w (%)
UN	12%
CL	16%

(a) Search space.   (b) Filtered $w\%$

Fig. 2. (a) The search space gets large when the issued Q is distributed wildly. (b) Only a small portion of w's are filtered.

1.2 Motivation

The purpose of this research is to address the following two limitations of the state-of-the-art method employed to resolve ARR queries:

1. **The efficiency degrades when Q is distributed widely.** The state-of-the-art method for solving ARR queries in [1] is shown in Fig. 2a. In the worst case, when Q is distributed as wide as the whole space, then the efficiency will degrade to a brute-force search.
2. **Only a few of user preferences data $w \in W$ be filtered.** Fig. 2b shows how many $w \in W$ are filtered in our experiments with the synthetic data (UN, CL) (refer to Sect. 4). We can see that the filtered $w \in W$ is less than 16% in all data sets, which is not satisfactory.

2   Clustering Processing Method (CPM)

Regarding the situation where Q distributes widely, we can divide Q into clusters instead of treating all $q \in Q$ entirely. We can then estimate the $ARank$ by counting the rank against each cluster. We propose an efficient algorithm named the Clustering Processing Method (CPM), which is also based on the bound-and-filter framework in [1] but counts rank for all clusters in only one R-tree traversal.

Algorithm 1 shows the CPM. In CPM, a *buffer* keeps the top-k w's for the result of the ARR query which, initially, is simply the first k w's $\in W$ and their aggregate ranks (Line 1). The value $minRank$ is a threshold which changes according as the *buffer*. First, we divide Q into clusters Q_c (Line 2) and, during the bounding phase, we compute the bounds of each cluster (Line 3). In the

---

Algorithm 1. Cluster Processing Method (**CPM**)

Input: P, W, Q

Output: result set *heap*

1: Initialize *buffer* to store the first k w's w.r.t the $ARank(w, Q)$ and sorted in ascending order.
2: $Qc \Leftarrow$ Clustering(Q).
3: Get bounds for each $Q_i \in Qc$ // Bounding phase.
4: $heapW.enqueue(RtreeW.root)$ // Filtering phase start.
5: **while** $heapW$ is not empty **do**
6:    $e_w \Leftarrow heapW.dequeue()$
7:    $heapP.enqueue(RtreeP.root)$
8:    $minRank \Leftarrow$ the last rank in *buffer*.
9:    $flag \Leftarrow$ ARank-Cluster($heapP, e_w, Qc, minRank$) //Call Algorithm 2.
10:   **if** $flag = 0$ **then**
11:      $heapW.enqueue(e_w.children)$
12:   **else**
13:     **if** $flag = 1$ **then**
14:       **if** e_w is a single vector **then**
15:         Update *buffer* with e_w and it's ARank.
16:       **else**
17:         $heapW.enqueue(e_w.children)$
18: **return** *buffer*

---

filtering phase (Line 4–17) Algorithm 2 is called to search with *RtreeP* and count the ARank with clusters Q_c (Line 7–9). According to the *flag* returned by Algorithm 2, we update the *buffer* (Line 15) or add the children of e_w for recursion (Line 11, 17).

The key point of CPM is to recursively count ARank with clusters Q_c but not independently in Algorithm 2. We use an array of flags (*Info*) to mark the state and avoid unnecessary processing; for example, the flag $Info["Q_1'']$ is true means that the current e_p has been filtered by Q_1 and we don't need to consider it again. Parent nodes in the R-tree pass *Info* to their children (Line 15). The child node first confirms *Info* (Line 3) to decide whether to skip processing (Lines 6, 18), and updates *Info* after it determines that it can filter new clusters (Lines 8, 13).

3 $Cone^+$ Tree and Methods

In this section, we address the second limitation in Sect. 1.2. Our solution is to develop the $cone^+$ tree to index the user data (W), as well as the corresponding search methods based on it.

Indexing W in R-tree is not efficient because the R-tree groups data into their MBR and the right-up and left-low points are used to estimate the upper and lower score bounds, respectively. The point $w \in W$ is located on a line L where $\sum_{i=1}^{d} w[i] = 1$, based on the definitions in Sect. 1.1. However, one diagonal of an MBR always makes the largest angle with L; thus, the right-up and left-low points of MBR enlarges (i.e., loosens) the bound of the score unnecessarily for an arbitrary product point p.

Algorithm 2. ARank-Cluster

Input: $heapP, e_w, Qc, minRank$
Output: include: 1; discard: -1; uncertain : 0;
1: $rnk \Leftarrow 0$
2: **while** $heapP$ is not empty **do**
3:     $\{e_p, Info\} \Leftarrow heapP.dequeue()$
4:     **if** e_p is non-leaf node **then**
5:         **for each** $Q_i \in Qc$ **do**
6:             **if** $Info["Q_i"] = false$ **then**
7:                 **if** $e_p \prec Q_i$ **then**
8:                     $Info["Q_i"] \Leftarrow true$
9:                     $rnk \Leftarrow rnk + e_p.size \times |Q_i|$
10:                     **if** $rnk \geq minRank$ **then**
11:                         **return** -1
12:                 **else if** $e_p \succ Q_i$ **then**
13:                     $Info["Q_i"] \Leftarrow true$
14:                 **else**
15:                     $heapP.enqueue(\{e_p.children, Info\})$
16:     **if** e_p is a leaf node **then**
17:         **for each** $Q_i \in Qc$ **do**
18:             **if** $Info["Q_i"] = false$ **then**
19:                 Calculate $f(q_i, w)$ for each $q_i \in Q_i$ and update rnk.
20: **if** $rnk \leq minRank$ **then**
21:     **return** 1
22: **else**
23:     **return** 0

Cone tree [3] is a binary construction tree. Every node in the tree is indexed with a center and encloses all points, which are close to this center up to cosine similarity. The node splits into two, left and right, child nodes if it has more points than a set threshold value M_n. The tree is built hierarchically by splitting itself until the points are fewer than M_n.

The actual bounds of the set of w's are always found from the boundary points. This is because for an arbitrary vector p, the inner product $f(w, p)$ is the length of the projection of w onto p and both the shortest and the longest projection of a set of w's come from the boundary points of the set. We took advantage of this feature and proposed a $cone^+$ tree which keeps the boundary points for each node. Therefore the precise score bounds can be computed directly. Algorithms 3 and 4 shows the construction of $cone^+$ tree. The indexed boundary points are the points containing the maximum value on a single dimension (Algorithm 4, Line 2).

Algorithm 3. $Cone^+$TreeSplit($Data$)

Input: points set, $Data$
Output: two centering points of children, a, b ;
 1: Select a random point $x \in Data$.
 2: $a \Leftarrow \arg \max_{x' \in Data} cosineSim(x, x')$
 3: $b \Leftarrow \arg \max_{x' \in Data} cosineSim(a, x')$
 4: **return** {a,b}

Algorithm 4. Build$Cone^+$Tree($Data$)

Input: points set, $Data$
Output: $cone^+$ tree, $tree$;
 1: $tree.data \Leftarrow Data$
 2: $tree.boundary \Leftarrow \{w \in Data : \arg \max_{i \in [0,d)} w[i]\}$
 3: **if** $|Data| \leq M_n$ **then**
 4:     **return** $tree$
 5: **else**
 6:     $\{a, b\} \Leftarrow Cone^+TreeSplit(Data)$
 7:     $left \Leftarrow \{p \in Data : cosineSim(p, a) > cosineSim(p, b)\}$
 8:     $tree.leftChild \Leftarrow BuildCone^+Tree(left)$
 9:     $tree.rightChild \Leftarrow BuildCone^+Tree(Data - left)$
10:     **return** $tree$

When processing ARR with $cone^+$ tree (W) and R-tree (P) based in the filtering phase, we can compute the bounds between a $cone^+$ and an MBR by the following Theorem.

Theorem 1 (*The bounds with $cone^+$ and MBR*): *Given a set of w's in a $cone^+$ node c_w, a set of points in an MBR e_p. $\forall w \in c_w$, $\forall p \in e_p$, $f(w, p)$ is upper bounded by $Max_{w_b \in c_w.boundar}\{f(w_b, e_p.up)\}$. Similarly, it is lower bounded by $min_{w_b \in c_w.boundar}\{f(w_b, e_p.low)\}$.*

For a Query set Q and an MBR e_p of points, the relationship between them on a w's $cone^+$ can be inferred from the following.

$$\begin{cases} e_p \prec Q : Max_{w_b \in c_w.boundar}\{f(w_b, e_p.up)\} < Min_{w_b \in c_w.boundar}\{f(w_b, Q.low)\} \\ e_p \succ Q : Min_{w_b \in c_w.boundar}\{f(w_b, e_p.low)\} > Max_{w_b \in c_w.boundar}\{f(w_b, Q.up)\} \\ Unknow : otherwise \end{cases}$$

$$(1)$$

The $Cone^+$ Tree Method (C$^+$TM) can be implemented easily by indexing W in the $cone^+$ tree and using Eq. (1) to apply the bound-and-filter framework described in [1]. We can also combine the features of the two methods in this paper, i.e., the (CC$^+$M) method, which can be implemented by using cone$^+$ tree with W and replacing Line 7 and 12 in Algorithm 1 with the above Eq. (1).

4 Experiment

We present the experimental evaluation of the previous DTM [1], the proposed CPM and C⁺TM, and the method (CC⁺M) which combines the features of both CPM and C⁺TM. All algorithms were implemented in C++, and the experiments were run in-memory on a Mac with 2.6 GHz Intel Core i7, 16 GB RAM.

Data Set. The synthetic data sets were uniform (UN) and clustered (CL) with an attribute value range of $[0, 1)$. The details of the generation can be found in [1,2,4,5].

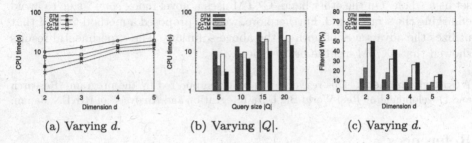

(a) Varying d.   (b) Varying $|Q|$.   (c) Varying d.

Fig. 3. Comparison results on UN data, $|P| = |W| = 100\text{K}$, $k = 10$.

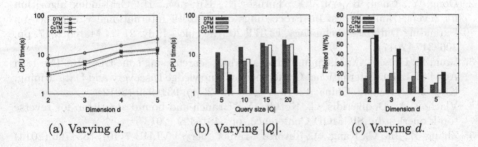

(a) Varying d.   (b) Varying $|Q|$.   (c) Varying d.

Fig. 4. Comparison results on CL data, $|P| = |W| = 100\,\text{K}$, $k = 10$.

Experimental Results. Figures 3 and 4 show the experimental results for the synthetic data sets (UN, CL) with varying dimensions d (2–5), where both data sets P and W contained 100 K tuples. Q had five query points, and we wanted to find the five best preferences ($k = 5$) for this Q. Figures 3a and 4a, show the runtime on varying dimensionality. CPM and C⁺TM boost the performance at least 1.5 times in all cases. The combined method CC⁺M takes advantages of them and be the best. We can also see that all methods have better performance in CL data than others, it is because that the CL data can be indexed (R-tree and Cone⁺ tree) in tighter bounds. We varied the $|Q|$ size to test the performance of clustered processing in CPM, the results are shown in Figs. 3b and 4b. We also observed the percentage of W been filtered. Figures 3c and 4c react the superior filtering of the proposed cone$^+$ tree structure.

5   Conclusion

In this paper, we pointed out two serious limitations of the state-of-the-art method of aggregate reverse rank query and proposed two methods CPM and C$^+$TM to solve them. CPM divides a query set into clusters and processes them to overcome the low efficiency caused by bad filtering when considering the query set as a whole. On the other hand, C$^+$TM uses a novel index $cone^+$ tree to avoid enlarging the score bound. Furthermore, we also proposed a method CC$^+$M that utilizes the advantages of both of the above solutions. The experimental results showed that CC$^+$M has the best efficiency.

Acknowledgement. This research was partly supported by the program "Research and Development on Real World Big Data Integration and Analysis" of RIKEN, Japan.

References

1. Dong, Y., Chen, H., Furuse, K., Kitagawa, H.: Aggregate reverse rank queries. In: Hartmann, S., Ma, H. (eds.) DEXA 2016. LNCS, vol. 9828, pp. 87–101. Springer, Cham (2016). doi:10.1007/978-3-319-44406-2_8
2. Dong, Y., Chen, H., Yu, J.X., Furuse, K., Kitagawa, H.: Grid-index algorithm for reverse rank queries. In: Proceedings of the 20th International Conference on Extending Database Technology, EDBT 2017, Venice, Italy, 21–24 March 2017, pp. 306–317 (2017)
3. Ram, P., Gray, A.G.: Maximum inner-product search using cone trees. In: The 18th ACM SIGKDD International Conference on Knowledge Discovery and Data Mining, KDD 2012, Beijing, China, 12–16 August 2012, pp. 931–939 (2012)
4. Vlachou, A., Doulkeridis, C., Kotidis, Y.: Branch-and-bound algorithm for reverse top-k queries. In: SIGMOD Conference, pp. 481–492 (2013)
5. Zhang, Z., Jin, C., Kang, Q.: Reverse k-ranks query. PVLDB **7**(10), 785–796 (2014)

Storing Join Relationships for Fast Join Query Processing

Mohammed Hamdi[1(✉)], Feng Yu[2], Sarah Alswedani[1],
and Wen-Chi Hou[1]

[1] Department of Computer Science, Southern Illinois University at Carbondale,
Carbondale, IL, USA
{mhamdi, sarah.swy, hou}@cs.siu.edu
[2] Department of Computer Science and Information Systems,
Youngstown State University, Youngstown, OH, USA
fyu@ysu.edu

Abstract. We propose to store equi-join relationships of tuples on inexpensive and space abundant devices, such as disks, to facilitate query processing. The equi-join relationships are captured, grouped, and stored as various tables, which are collectively called the Join Core. Queries involving arbitrary legitimate sequences of equi-joins, semi-joins, outer-joins, anti-joins, unions, differences, and intersections can all be answered quickly by merely merging these tables. The Join Core can also be updated dynamically. Experimental results showed that all test queries began to generate results instantly, and many completed instantly too. The proposed methodology can be very useful for queries with complex joins of large relations as there are fewer or even no relations or intermediate results needed to be retrieved and generated.

Keywords: Query processing · Join queries · Equi-join · Semi-join · Outer-join · Anti-join · Set operations

1 Introduction

As hardware technologies advance, the price of disks drops significantly while the capacity increases drastically. Database researchers now have the luxury of exploring innovative ways to utilize these cheap and abundant spaces to improve query processing.

In relational databases, data are spread among relations. The equi-join operation, which includes the natural join, is the most commonly used operator to combine data spread across relations. Other join operators, such as semi-joins, outer-joins, and anti-joins, are also very useful. Unfortunately, these join operations are generally expensive to execute. Complex queries involving multiple joins of large relations can easily take minutes or even hours to compute. Consequently, much effort in the past few decades has been devoted to developing efficient join algorithms [3, 7]. Even today, improving join operations remains a focus of database research [2, 5].

In this research, we propose to pre-store the equi-join relationships of tuples to facilitate query processing. We have designed a simple method to capture the equi-join

© Springer International Publishing AG 2017
D. Benslimane et al. (Eds.): DEXA 2017, Part I, LNCS 10438, pp. 167–177, 2017.
DOI: 10.1007/978-3-319-64468-4_13

relationships in the form of maximally extended match tuples. A simple and novel naming technique has been designed to group and store the equi-join relationships in tables on disks, which are collectively called the Join Core. The Join Core is an efficient data structure from which not only the results of all possible equi-joins can be obtained, but also the results of all legitimate combinations of equi-joins, outer-joins, anti-joins, unions, differences, and intersections can be derived. Without having to perform joins, memory consumptions are dramatically reduced. In addition, Join Core can be updated dynamically in the face of updates.

In this research, we also discuss heuristics that can effectively cut down the sizes of Join Cores. We believe the benefits of Join Core, namely instant responses, fast query processing, and small memory consumptions, are well worth the additional storage space incurred.

In the literature, materialized views are, to a certain extent, related to our work as both attempt to use precomputed data to facilitate query processing.

Materialized views generally focus on SPJ (Select-Project-Join) queries and, perhaps, with final grouping and aggregate functions. The select and project operations in the views confine and complicate the uses of the views. As a result, much research has focused on how to select the most beneficial views to materialize [1, 6, 11] and how to choose an appropriate set of materialized views to answer a query [4, 9].

Materialized views materialize selected query results while Join Core materializes selected equi-join relationships. Therefore, materialized views may benefit queries that are relevant to the selected queries, while Join Core can benefit queries that are related to the selected equi-join relationships, which include queries with arbitrary sequences of equi-, semi-, outer-, anti-joins and set operators.

A join index [8, 10] for a join stores the (equi-) join result in a concise manner as pairs of identifiers of tuples that would match in the join operation. It has been shown that joins can be performed more efficiently with join indices than the traditional join algorithms. However, it still requires at least one scan of the operand relations, writes and reads of temporary files (as large as the source relations), and generating intermediate result relations (for queries with more than one join). One the other hand, with Join Core, join results are readily available without accessing any source or intermediate relation. Therefore, very little memory and computations are required to process a query. In addition, join indices are not useful for other join operators, such as outer-joins and anti-joins.

The rest of the paper is organized as follows. Section 2 introduces the terminology. Section 3 shows a sample Join Core and how it can be used to answer equi-join queries. Section 4 extends the framework to queries with other types of joins and set operations. Section 5 reports experimental results. Finally, conclusions are presented in Sect. 6.

Due to space limitation, readers are referred to [12] for an extended version of the paper that includes detailed discussions on proofs of theorems, answering cyclic join queries, queries with other joins, the dynamic maintenance of the Join Core in the face of updates, applications to bag semantics, literature survey, and experimental results.

2 Terminology

In this paper, we assume the data model and queries are based on the set semantics. The equi-join operator is the most commonly used operator to combine data spread across relations. Therefore, we shall first lay down the theoretical foundation based on the equi-join, and then extend the framework to other joins in Sect. 6. Hereafter, we shall use a join for an equi-join, unless otherwise stated.

Definition 1 (Join Graph of a Database). Let D be a database with n relations $R_1, R_2, \ldots, R_n$, and $G(V, E)$ be the join graph of D, where V is a set of nodes that represents the set of relations in D, i.e., $V = \{R_1, R_2, R_3, \ldots, R_n\}$, and $E = \{\langle R_i, R_j \rangle \mid R_i, R_j \in V, i \neq j)\}$, is a set of edges, in which each represents a join that has been defined between R_i and R_j, $i \neq j$.

If a join graph is not connected, one can consider each connected component separately. Therefore, we shall assume all join graphs are connected. Each join comes with a predicate, omitted in the graph, specifying the requirements that a result tuple of the join must satisfy, e.g., $R_1.attr1 = R_2.attr2$. For simplicity, we shall use a join, a join edge, and a join predicate interchangeably. We also assume all relations and join edges are numbered.

Example 1 (Join Graph). Figure 1(a) shows the join graph of a database with five relations R_1, R_2, R_3, R_4, and R_5, connected by join edges, numbered from 6 to 9.

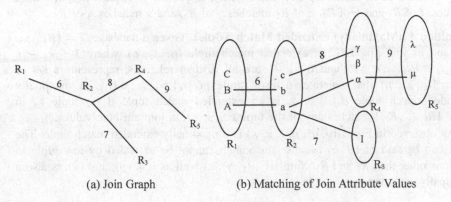

(a) Join Graph   (b) Matching of Join Attribute Values

Fig. 1. A join graph and matching tuples

To round out the theoretical framework, we shall introduce a concept, called the *trivial (equi-)join*. Each tuple in a relation R_i can be considered as a result tuple of a trivial join between R_i and itself with a join predicate $R_i.key = R_i.key$, where *key* is the (set of) key attribute(s) of R_i. Trivial join predicates are not shown explicitly in the join graphs. All join edges in Fig. 1(a), such as 6, 7, 8, and 9, are non-trivial or regular joins.

We have reserved predicate number i, $1 \leq i \leq 5$, for trivial join predicate i, which is automatically satisfied by every tuple in relation R_i. The concept of trivial join predicates will be useful later when we discuss a query that contains outer-joins, anti-joins, or no joins. Hereafter, all joins and join predicates refer to non-trivial ones, unless otherwise stated.

To conserve space, a database and its join graph refer to only the parts of the database and join graphs that are of our interest and for which we intend to build Join Cores.

Definition 2 (Join Queries). Let $\bowtie (\{R_i, \ldots, R_j\}, E')$ be a join query, representing joins of the set of relations $\{R_i, \ldots, R_j\} \subseteq V$, $1 \leq i, \ldots, j \leq n$, with respect to the set of join predicates $E' \subseteq E$.

Definition 3 (Join Graph of a Join Query). The join graph of a join query $\bowtie (\{R_i, \ldots, R_j\}, E')$, denoted by $G'(V', E')$, is a connected subgraph of $G (V, E)$, where $V' = \{R_i, \ldots, R_j\} \subseteq V$, and $E' \subseteq E$ is the set of join predicates specified in the query.

The join graph of a join query is also called *a query graph*. We shall exclude queries that must execute Cartesian products or θ-joins, where $\theta \neq$ "=", from discussion as Join Core cannot facilitate executions of such operators.

Example 2 (Matching of Join Attribute Values). Figure 1(b) shows the matching of join attribute values between tuples. Tuples are represented by their IDs in the figure. That is, R_1 has 3 tuples, A, B, C, i.e., $R_1 = \{A, B, C\}$. $R_2 = \{a, b, c\}$, $R_3 = \{I\}$, $R_4 = \{\alpha, \beta, \gamma\}$, $R_5 = \{\mu, \lambda\}$.

The edges between tuples represent matches of join attribute values. For example, tuples A and B of R_1 match tuples a and b of R_2, respectively. Tuple a has two other matches, I of R_3 and α of R_4. c of R_2 matches γ of R_4, and α matches μ of R_5.

Definition 4 ((Maximally) Extended Match Tuple). Given a database $D = \{R_1, \ldots, R_n\}$ and its join graph G, an extended match tuple $(t_k, \ldots, t_l)$, where $1 \leq k, \ldots, l \leq n$, $t_k \in R_k, \ldots, t_l \in R_l$, and $R_k, \ldots, R_l$ are all distinct relations, represents a set of tuples $\{t_k, \ldots, t_l\}$ that generates a result tuple in $\{t_k\} \bowtie \ldots \bowtie \{t_l\}$. A maximally extended match tuple $(t_k, \ldots, t_l)$, is an extended match tuple if no tuple t_m in $R_m (\notin \{R_k, \ldots, R_l\})$ matches any of the tuples $t_k, \ldots, t_l$ in join attribute values.

As observed in Fig. 1(b), (A, a, I, α, μ) is a maximally extended match tuple. The same can be said of (B, b) because the match cannot be extended by any tuple in relations other than R_1 and R_2. Similarly, (c, γ), as well as (C), (β), and (λ), is also a maximally extended match tuple.

3   Join Core Structure and Construction

In this section, we show an example of a Join Core and explain how it is structured and used to answer equ-join queries.

3.1   Join Core Structure and Naming

Consider Fig. 1 again. The join relationships we wish to store are (A, a, I, α, μ), $(B, b), (c, \gamma), (C), (\beta)$, and (λ), each representing a maximally extended match tuple.

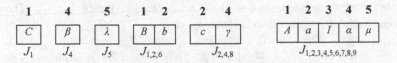

Fig. 2. Join core

We intend to store these maximally extended match tuples in various tables based on the join predicates, both trivial and non-trivial ones, they satisfy. These tables form the *Join Core*.

Example 3 (Sample Join Core). Figure 2 shows the Join Core for the database in Fig. 1. The attributes of the Join Core tables, i.e., 1, 2, 3, 4, and 5, represent the sets of (interested) attributes of $R_1, R_2, R_3, R_4,$ and R_5, respectively, and are called the $R_1, R_2, \ldots, R_5$ components of the tables.

(B, b) is stored in $J_{1,2,6}$ because (B, b) satisfies join predicate 6, and trivial predicates 1 $(B \in R_1)$ and 2 $(b \in R_2)$. Similarly, (c, γ) is stored in $J_{2,4,8}$ and (A, a, I, α, μ) is stored in $J_{1,2,3,4,5,6,7,8,9}$. $C(\in R_1), \beta(\in R_4),$ and $\lambda(\in R_5)$ satisfy only trivial predicates and thus are stored in $J_1, J_4,$ and J_5, respectively.

Assume join predicate numbers 1, ..., n are reserved for trivial joins between R_1, ..., R_n and themselves, respectively, and non-trivial predicates are numbered from $n + 1$ to $n + e$, where e is the number of join edges in the join graph.

Definition 5 (Join Core). A Join core is a set of tables $J_{k, \ldots, l}, 1 \leq k, \ldots, l \leq n + e$, each of which stores a set of maximally extended match tuples that satisfy *all and only* the join predicates $k, \ldots, l$. Each table $J_{k, \ldots, l}$ is called a *Join Core table* (or *relation*). The indices $k, \ldots l$ of the table $J_{k, \ldots, l}$ is called the name of the table for convenience.

For simplicity, we shall call a maximally extended match tuple in a Join Core table a match tuple, to be differentiated from a tuple in a regular relation.

3.2 Join Core Construction

Now, let us discuss how to construct a Join Core for a database. Tuples that find no match in one join may find matches in another join. For example, b finds no match in $R_2 \bowtie R_3$, but finds a match B in $R_1 \bowtie R_2$. Unfortunately, such join relationships can be lost in successive joins, for example, in $(R_1 \bowtie R_2) \bowtie R_3$.

Full outer-joins, or simply outer-joins, retain matching tuples as well as dangling tuples, and thus can capture all the join relationships. Any graph traversal method can be used here as long as it incurs no Cartesian products during the traversal. An outer-join is performed for each join edge. The output of the previous outer-join is used as an input to the next outer-join. The result tuples are distributed to Join Core tables based on the join predicates, both trivial and non-trivial ones, they have satisfied during the traversal.

Example 4 (Join Core Construction). Assume a breadth-first traversal of the join graph (Fig. 1(a)) from R_1 is performed. An outer-join is first performed between R_1 and R_2. It generates intermediate result tuples $(A, a), (B, b), (C, -),$ and $(-, c)$. The next outer-join with R_3 generates $(A,a,I), (B,b,-), (C,-,-)$ and $(-,c,-)$. Then, the outer-join with R_4 generates (A,a,I,α), $(B,b,-,-)$, $(C,-,-,-)$, $(-,c,-,\gamma)$, and $(-,-,-,\beta)$. The final outer-join with R_5 generates (A,a,I,α,μ), $(B,b,-,-,-), (C,-,-,-,-), (-,c,-,\gamma,-)$, $(-,-,-,\beta,-)$, and $(-,-,-,-,\lambda)$, which are written, without nulls, to $J_{1,2,3,4,5,6,7,8,9}$, $J_{1,2,6}$, J_1, $J_{2,4,8}$, J_4, and J_5, respectively, based on the join predicates they satisfy.

3.3 Answering Equi-Queries Using Join Core

The name of a Join Core table specifies the join predicates satisfied by the match tuples stored in it. On the other hand, a join query specifies predicates that must be satisfied by the result tuples. Therefore, to answer a query is to look for Join Core tables whose names contain the predicates of the query.

Consider Figs. 1 and 2 and the query $\bowtie (\{R_1,R_2,R_3,R_4,R_5\}, \{6,7,8,9\})$. The components of the result tuples must satisfy predicates 6, 7, 8, and 9. In addition, the components themselves also satisfy trivial predicates 1, 2, 3, 4, 5. Thus, we look for Join Core tables whose names contain predicates 1, 2, 3, 4, 5, 6, 7, 8, and 9. That is, $\bowtie (\{R_1,R_2,R_3,R_4,R_5\}, \{6,7,8,9\}) = J_{1,2,3,4,5,6,7,8,9}$.

As for $\bowtie (\{R_1,R_2\}, \{6\})$, while $J_{1,2,6}$ certainly contains some result tuples, $J_{1,2,3,4,5,6,7,8,9}$ also contains some result tuples because tuples in $J_{1,2,3,4,5,6,7,8,9}$ also satisfy 1, 2, and 6. That is, $\bowtie (\{R_1,R_2\}, \{6\}) = \pi_{1,2}(J_{1,2,6}) \cup \pi_{1,2}(J_{1,2,3,4,5,6,7,8,9})$. Similarly, $\bowtie (\{R_2,R_4\}, \{8\}) = \pi_{2,4}(J_{2,4,8}) \cup \pi_{2,4}(J_{1,2,3,4,5,6,7,8,9}); \bowtie (\{R_2,R_3\}, \{7\}) = \pi_{2,3}(J_{1,2,3,4,5,6,7,8,9})$.

It even holds for queries containing no non-trivial joins. For example, $R_1 = \pi_1 J_1 \cup \pi_1 (J_{1,2,6}) \cup \pi_1 (J_{1,2,3,4,5,6,7,8,9})$, $R_2 = \pi_2 (J_{1,2,6}) \cup \pi_2 (J_{2,4,8}) \cup \pi_2 (J_{1,2,3,4,5,6,7,8,9})$, $R_3 = \pi_3 (J_{1,2,3,4,5,6,7,8,9})$, $R_4 = \pi_4 J_4 \cup \pi_4 (J_{2,4,8}) \cup \pi_4 (J_{1,2,3,4,5,6,7,8,9})$, and $R_5 = \pi_5 J_5 \cup \pi_5 (J_{1,2,3,4,5,6,7,8,9})$. It is observed that a relation R_i can be reconstructed from the Join Core, which implies that a Join Core can itself be the database, if one wishes to not store the relations in traditional ways.

Theorem 1. Let $\bowtie (\{R_i, \ldots, R_j\}, \{u, \ldots, v\})$ be joins of the set of relations $\{R_i, \ldots, R_j\}$ with respect to a set of join predicates $\{u, \ldots, v\} \neq \emptyset$. Let e be the number of join edges in the join graph,

$$\bowtie (\{R_i, \ldots, R_j\}, \{u, \ldots, v\}) = \cup_{\{k,\ldots,l\} \supseteq \{u,\ldots,v\}} \pi_{i,\ldots,j}(J_k, \ldots, l)$$

where $1 \leq i, \ldots, j \leq n, 1 \leq k, \ldots, l, u, \ldots, v \leq n+e$.
Proof. See [12].

As you may notice that there is no need to match the trivial predicates of a query against the Join Core table names. That is, given a join query with a non-empty set of predicates $\{u, \ldots, v\}$, the result tuples can be found in Join Core tables whose names contain $u, \ldots, v$, without regard to trivial predicates. Trivial predicates cannot be ignored when a query contains no non-trivial joins, or contains outer- or anti-joins.

The database system can begin to generate result tuples once the first block of a relevant Join Core table is read into memory, that is, instantly. The total computation time is also drastically reduced because there are no (or fewer) joins to perform.

4 Queries with Other Joins

Now, a join can be an equi-, semi-, outer- or anti-join. A little deliberation reveals that match tuples that do not satisfy an equi-join predicate must be in Join Core tables whose names do not contain that predicate, recalling that Join Core table names specify *all and only* the equi-join predicates satisfied. An outer-join generates a result tuple no matter whether the equi-join predicate is satisfied or not.

A join query consisting of a sequence of join operators has a *query predicate* that is a logical combination of the individual join predicates of constituent joins. We attempt to obtain query result tuples from Join Core tables whose names satisfy the query predicates. Here, we focus on how to formula the query predicates as (*table name*) *selection criteria* for Join Core tables that contain the query result tuples. For example, satisfying predicate p is rewritten as $p \in \{k, ..., l\}$, where $\{k, ..., l\}$ is the set of indices of a Join Core table name.

Afterward, specific handlings, such as removal of unwanted attributes, and padding null values for "missing" attributes (for outer-joins), are performed.

4.1 Single-Join Queries

We start by deriving the (table name) selection criteria, denoted by S, for queries with only one join operator. Let p be the equi-join predicate between R_i and R_j in the join graph. Consider R_i op R_j, where op is either an equi-join, semi-join, outer-join, or anti-join.

Readers are referred to [12] for discussions on more single-join queries.

Example 6 (Outer-Join). Let us consider Figs. 1 and 2.

$R_1 \bowtie R_2$: $S = 1 \in \{k, ..., l\} \vee 2 \in \{k, ..., l\}$ Only $J_1, J_{1,2,6}, J_{2,4,8}$, and $J_{1,2,3,4,5,6,7,8,9}$ satisfy S. The answer is $\{(C, \text{-}), (B, b), (\text{-}, c) (A, a)\}$. Note that tuples in J_1 and J_8 need to be padded with null values for the set of attributes of the other operand relations, while unwanted components 3, 4, and 5 need to be removed from $J_{1,2,3,4,5,6,7,8,9}$.

$R_1 \bowtie R_2$: $S = 1 \in \{k, ..., l\}$. Only $J_1, J_{1,2,6}, J_{1,2,3,4,5,6,7,8,9}$ satisfy S, and the result is $\{(C, \text{-}), (B, b), (A, a)\}$.

$R_1 \bowtie R_2$: $S = 2 \in \{k, ..., l\}$. Only $J_1, J_{1,2,6}, J_{1,2,3,4,5,6,7,8,9}$ satisfy S, and the result is $\{(B, b), (\text{-}, c) (A, a)\}$.

Example 7 (Anti-Join). Let us consider Figs. 1 and 2.

$R_1 \triangleright R_2 : S = 1 \in \{k, ..., l\} \wedge \neg(6 \in \{k, ..., l\})$. Only J_1 satisfies and the answer is $\{C\}$.

$R_2 \triangleright R_4 : S = 2 \in \{k, ..., l\} \wedge \neg(8 \in \{k, ..., l\})$. Only $J_{1,2,6}$ satisfies and the answer is $\{b\}$.

4.2 Multi-join Queries

Let $E = E_1$ op E_2, where E, E_1, and E_2 are expressions that contain arbitrary legitimate sequences of equi-, semi, outer- and anti-join operators, and op is one of these join operators with a join predicate p. We assume the query graphs for E, E_1, and E_2 are all connected subgraphs of G. Let S_1 and S_2 be the selection criteria on the Join Core tables for E_1 and E_2, respectively, and S the criteria for E.

Readers are referred to [12] for discussions on more multi-join queries.

Example 8 (Multi-anti-join Queries). Let us consider Figs. 1 and 2.

$(R_1 \bowtie R_2) \triangleright R_3 : S = 6 \in \{k, \ldots, l\} \wedge \neg(7 \in \{k, \ldots, l\})$. Only $J_{1,2,6}$ satisfies S and the answer is $\{(B, b)\}$.

$(R_2 \triangleright R_1) \triangleright (R_4 \bowtie R_5) : S = (2 \in \{k, \ldots, l\} \wedge \neg(6 \in \{k, \ldots, l\})) \wedge \neg(9 \in \{k, \ldots, l\} \wedge 8 \in \{k, \ldots, l\})$. Only $J_{2,4,8}$ satisfies S, and the answer is $\{(c)\}$.

Theorem 3. Let $E = E_1$ op E_2, where E, E_1, and E_2 are arbitrary legitimate expressions that contain equi-, semi-, outer- and anti-joins, and op is any of these join operations with a join predicate p. Let S_1 and S_2 be the selection criteria for Join Core tables from which the resulting tuples of E_1 and E_2 can be derived, respectively. Then, the selection criteria S for E is (i) if $op = \bowtie$, $S = S_1 \wedge S_2 \wedge p \in \{k, \ldots, l\}$; (ii) if $op = \ltimes$ or $\rtimes$, $S = S_1 \wedge S_2 \wedge p \in \{k, \ldots, l\}$; (iii) if $op = \bowtie$, $S = S_1 \vee S_2$; if $op = \bowtie$, $S = S_1$; if $op = \bowtie$, $S = S_2$; (iv) if $op = \triangleright$, $S = S_1 \wedge \neg(S_2 \wedge p \in \{k, \ldots, l\})$.

Proof. See [12].

5 Experimental Results

We compare the performance of Join Core with a MySQL database system. In this preliminary study, we use only the simplest set up to see how Join Core alone can improve query processing, leaving other performance improving factors, such as multiple CPUs, disks (magnetic or SSDs), etc., to future work. All experiments are performed on a laptop computer with a 1.60 GHz CPU, 8 GB RAM, and a 1 TB hard drive.

5.1 Datasets

We generate 1, 4, and 10 GB TPC-H datasets for experiments. Figure 3 shows the join graph of the TPC-H datasets with arrows indicating many-one relationships.

5.2 Space Consumptions

As shown in Table 1, the full Join Core sizes, without applying any space reduction methods, are 4, 13.8, and 39.7 GB for the 1, 4, and 10 GB TPC-H datasets, respectively. "Reduced 1" is obtained by removing the smallest relations Region, Nation, and Supplier, which have 5, 25, and 10,000 tuples from the graph, respectively. "Reduced 2", is obtained by further removing the Customer relation from "Reduced 1".

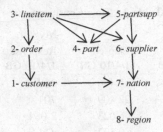

Fig. 3. TPC-H join graph

Table 1. Space consumptions

Join core size	Datasets		
	1 GB	4 GB	10 GB
Full	4 GB	13.8 GB	39.7 GB
Reduced 1	2.3 GB	7.1 GB	20.1 GB
Reduced 2	1.7 GB	5.4 GB	15.8 GB

While removing relations can certainly reduce the space consumption, joins would have to be performed when removed relations are referenced in the queries. Fortunately, removed relations are generally small and joins with them are relatively quick.

5.3   Time Consumptions

Query Processing Time. We measure the response and elapsed time of queries (in seconds) that come with the TPC-H datasets. While keeping the selections and projections of the queries, we remove "group by", "order by", "limit", aggregate functions, etc., to focus on only the join queries. We add "distinct" to the queries as we have assumed the set semantics in the paper.

In Table 2, the response time measures the time up until the first result tuple is written to the disk (or displayed), while the elapsed time measures the time from beginning to end. The ID of the TPC-H query is shown in the first column, followed by the relations involved in the join operations. Relations are referenced by the numbers assigned to them in Fig. 3. Queries were aborted if they took more than 4 h (=14,400 s), as indicated by −'s in the table.

With Join Cores, all queries saw their first result tuples less than 0.01 s, namely instantly. As explained, all it takes is the retrieval of a block of a relevant Join Core table into memory and output it after some simple manipulations. On the other hand, MySQL took minutes to hours to output its first result tuple. As explained, the result size, not the complexity, of the query determines the query processing time because the join result is readily available in the Join Core. Queries 12 and 18 best illustrate this characteristic. Query 12 has only one join but generates large numbers of result tuples.

Table 2. Time Consumptions

Query	Join core		MySQL		Result
	Response 1/4/10 GB	Elapsed 1/4/10 GB	Response 1/4/10 GB	Elapsed 1/4/10 GB	tuples
12: ⋈ {R_2, R_3}	0.008	5.456	360	367	38,928
	0.008	22.409	701	725	155,585
	0.008	56.023	2,084	2,107	388,058
14: ⋈ {R_3, R_4}	0.008	0.502	411	411	1,717
	0.008	1.865	1,307	1,310	6,718
	0.008	3.865	2,014	2,018	16,943
16: ⋈ {R_4, R_5}	0.008	0.812	79	81	3,795
	0.008	3.005	300	306	15,208
	0.008	9.686	856	867	38,195
3: ⋈ {R_1, R_2, R_3}	0.008	1.579	6,782	6,785	11,620
	0.008	8.016	–	–	45,395
	0.008	17.455	–	–	114,003
18: ⋈ {R_1, R_2, R_3}	0.007	0.010	61	61	6
	0.007	0.012	91	91	11
	0.007	0.013	291	291	22
10: ⋈ {R_1, R_2, R_3, R_7}	0.009	1.706	5,060	5,063	3,773
	0.009	5.667	7,562	7,573	14,800
	0.009	14.560	–	–	36,975
2: ⋈ {R_4, R_5, R_6, R_7, R_8}	0.010	1.890	322	325	3,162
	0.010	7.005	838	845	12,723
	0.010	18.609	2,112	2,131	31,871
5: ⋈ {R_1, R_2, R_3, R_6, R_7, R_8}	0.010	1.760	–	–	15,196
	0.010	6.809	–	–	60,798
	0.010	16.355	–	–	152,102

On the other hand, Query 18 has two joins, including the join of Query 12, but generates much smaller numbers of result tuples. Therefore, it took much longer to process Query 12 than Query 18. As shown in Table 2, it took 5.456, 22.409, and 56.023 s to process Query 12 for 1, 4, and 10 GB datasets, respectively, but it took only 0.010, 0.012, and 0.013 s, respectively, to process Query 18. Since there were no joins to perform in the proposed method, many queries completed instantly too. On the other hand, many queries took hours to complete on MySQL. Another advantage of the proposed methodology is that it does not consume much memory. All it needs is to build a hash table for the final duplicate elimination.

We believe the instant responses, fast query processing, and small memory consumption of the Join Core are well worth its required additional storage space.

6   Conclusion

In this paper, we have presented an innovative way to process queries without having to perform expensive joins and set operations. We proposed to store the equi-join relationships in the form of maximally extended match tuples to facilitate query processing. We have designed an innovative way to group the join relationships into tables, called the Join Core, so that queries can be answered quickly, if not instantly, by merely merging subsets of these tables. The Join Core is applicable to queries involving arbitrary sequences of equi-joins, semi-joins, outer-joins, anti-joins, unions, differences, and intersections. Preliminary experimental results have confirmed that with Join Core, join queries can be responded to instantly and the total elapsed time can also be dramatically reduced. We will discuss concurrency control in the face of updates, and perform extensive experiments in different environments in the future.

References

1. Agarawal, S., Chaudhuri, S., and Narasayya, V.: Automated selection of materialized views and indexes for SQL databses. In: VLDB, pp. 496–505 (2000)
2. Chu, S., Balazinska, M., and Suciu, D.: From theory to practice: efficient join query evaluation in a parallel database system. In: ACM SIGMOD Conference, pp. 63–78 (2015)
3. DeWitt, D., Gerber, R.: Multiprocessor hash-based join algorithms. In: VLDB, pp. 151–164 (1985)
4. Goldstein, J., Larson, P.-A.: Optimizing queries using materialized views: a practical, scalable solution. In: ACM SIGMOD, pp. 331–342 (2001)
5. He, B., Yang, K., Fang, R., Lu, M., Govindaraju, N., Luo, Q., Sander, P.: Relational joins on graphics processors. In: ACM SIGMOD Conference, pp. 511–524 (2008)
6. Karloff, H., Mihail, M.: On the complexity of the view-selection problem. In: ACM PODS Conference, 167–173 (1999)
7. Kitsuregawa, M., Tanaka, H., Moto-Oka, T.: Application of hash to data base machine and its architecture. New Gener. Comput. 1(1), 63–74 (1983)
8. Li, Z., Ross, K.A.: Fast joins using join indices. VLDB J.—Int. J. Very Large Data Bases 8 (1), 1–24 (1999)
9. Pottinger, R., Levy, A.: A scalable algorithm for answering queries using views. In: VLDB Conference, pp. 484–495 (2000)
10. Valduriez, P.: Join indices. ACM Trans. Datab. Syst. (TODS) 12(2), 218–246 (1987)
11. Yang, J., Karlapalem, K., Li, Q.: Algorithms for materialized view design in data warehousing environment. In: VLDB, pp. 25–29 (1997)
12. Storing Join Relationships for Fast Join Query Processing. https://goo.gl/7N3JSd

Access Patterns Optimization in Distributed Databases Using Data Reallocation

Adrian Sergiu Darabant[✉], Leon Tambulea, and Viorica Varga

Department of Mathematics and Computer Science, Babes-Bolyai University,
Cluj-Napoca, Romania
dadi@cs.ubbcluj.ro

Abstract. Large distributed databases are split into fragments stored on far distant nodes that communicate through a communication network. Query execution requires data transfers between the processing sites of the system. In this paper we propose a solution for minimizing raw data transfers by re-arranging and replicating existing data within the constraints of the original database architecture. The proposed method gathers incremental knowledge about data access patterns and database statistics to solve the following problem: online re-allocation of the fragments in order to constantly optimize the query response time. We model our solution as a transport network and show in the final section the experimental numerical results we obtain by comparing the improvements obtained between various database configurations, before and after optimization.

1 Introduction

Let us consider a distributed database with a set of fragments/shards F stored on a set of sites S in a communication network. A set Q of *applications/queries* is executed against the database. We start from the assumptions that in order to minimize the query execution time, the data transferred needs to be minimized. Thus the data allocation to the sites of the system needs to be implemented such that data transfer during query execution is minimized. This paper assumes that the fragments have been already determined and (eventually) allocated, and focuses on the problem of allocating (re-allocating) them in order to minimize the total cost of data transmission. In practice, the optimal initial allocation of fragments is not possible without apriori knowledge of the applications running on the database. Our solution relaxes this requirement by allowing an initial allocation unaware of the applications querying data. It then optimizes the allocation by observing, at the database level, the access patterns incurred by the queries and performing redundant re-allocation.

2 Related Work

Many aspects of the data allocation problem have been studied in the literature. Reid and Orlowska [6] studied the communication cost minimization problem

© Springer International Publishing AG 2017
D. Benslimane et al. (Eds.): DEXA 2017, Part I, LNCS 10438, pp. 178–186, 2017.
DOI: 10.1007/978-3-319-64468-4_14

while replica allocation modeled as an integer linear programming with minimizing the execution cost has been approached in [1,3,6].

Menon in [7] considered non-redundant allocation. This paper focuses on redundant allocation. Wiese in [13] use clustering and clustered attributes. Huang and Chen in [5] propose a simple and comprehensive model that reflects transaction behavior in distributed databases.

The fragment allocation problem is NP-complete [9]. A genetic algorithm is proposed in [12], a genetic search-based clustering in [2] and an evolutionary approach in [4].

A reinforcement learning solution for allocating replicated fragments is presented in [8].

3   Query Evaluation and Data Transfer

Let $A(q)$ be the query evaluation tree for query q. We add a root node to this tree and we obtain a sub-tree rooted in the new node that corresponds to the entire query q (the new root represents the overall q query). Leaf nodes in $A(q)$ correspond to fragments, while internal nodes represent relational operators (unary or binary). When evaluating an operator op from q we get a transfer cost for data from the nodes where each operand of op is evaluated/stored to the node where op is evaluated.

In the next paragraphs we will use the following notations: $F = \{f_i | i = \overline{1,n}\}$ - fragment set of the database, $df_i = dim(f_i)$ - size of fragment $f_i, i = 1,n$, $S = \{s_i | i = \overline{1,m}\}$ - the sites of the system where fragments of F are stored, $S(f)$ - sites of the system where a fragment $f, f \in F$ is stored, $F(s)$ - fragments stored on site $s \in S$.

Starting from a predefined (current) state of the database (fragments, sites, current fragment allocation), in [10] we attach two values to each node of the query tree $A(q)$: d and c as follows: d - the size of data associated to the site (fragment size if the current node is a leaf node, or an estimation of the relational operator result size for internal nodes), and the costs vector $c = (c_1, \ldots, c_m)$ (m is the number of sites) of evaluating the query on all sites. For leaf nodes this equates to the size of the fragment or zero. For an internal node corresponding to an operator op, c_i is the minimal cost of the required data transfers when the operator op is evaluated on site s_i. See our previous work [10].

The following paragraphs describe the computation method for the vector c in the case of the two possible cases: an unary and a binary operator.

Let op be an unary/binary operator and its current operand(s) $A(B)$ with its associated values: d_A and $c_A = [c_1^A, \ldots, c_m^A]$. The incurred data transfer in the evaluation of op on site s_i depending on the location of operand(s) is given by the bellow expression (binop=1 for binary operators and 0 otherwise):

$$
\begin{aligned}
c_i = \min\{c_1^A + d_A, \ldots, c_{i-1}^A + d_A, c_i^A, c_{i+1}^A + d_A, \ldots, c_m^A + d_A\} \\
+ binop \times \min\{c_1^B + d_B, \ldots, c_{i-1}^B + d_B, c_i^B, c_{i+1}^B + d_B, \ldots, c_m^B + d_B\}
\end{aligned}
\tag{1}
$$

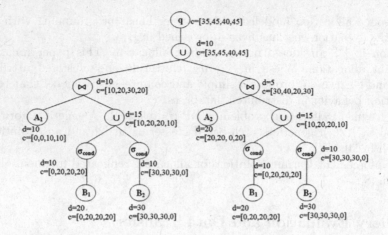

Fig. 1. The evaluation tree and the values associated to an example query.

We show in Fig. 1 the $A(q)$ tree built for some real values of the fragments size and results of the relational operators.

Using (1), we can compute the values for the vector c associated to query q - that is the root of the $A(q)$ tree in Fig. 1. Query q is executed on a specific site of the system. The vector c that labels the root of the $A(q)$ tree provides the minimal cost of the data transfers during the execution of query q. In the following we will analyze the required data transfer for the two possible cases: when f is a sub-tree of q, or f is used in a combination of unary operators. These cases are depicted in Figs. 2 and 3.

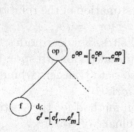

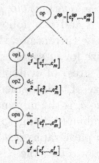

Fig. 2. Fragment used by a binary operator - one operand is always a leaf node (fragment)

Fig. 3. Fragment used by a sequence of unary operators

The two cases described above are valid for a sub-tree of the binary operator op. The same applies for the second sub-tree, corresponding to the second operand. If the fragment used by the second sub-tree is also stored on site s_i, then c_i^{op} will be null. If all fragments used by a query q are stored on the site s_i where the query is evaluated, then $c_i^{op} = 0$ for all operator nodes from $A(q)$.

Using the above analysis we can infer that by storing the fragment f on a site where the query accessing f is executed (let this site be s), we can reduce the data transfer cost by an amount r - equal to the size of fragment f or with the size of the result of the last unary operator applied to f, but before a binary operator applied to f - i.e. the first unary operator applied to f appearing strictly before a binary operator on f, if such exists. If fragment f appears multiple times in the evaluation tree $A(q)$ (as for example is the case for B_1 in Fig. 1), then the data transfer is reduced on all accesses to fragment f.

Given a distributed database and a time interval, we denote by Q the set of observed queries that are executed against the database and their access patterns in the given time period. Information about the access patterns is stored by the database's statistical module in views and can be retrieved for analysis. We should note that apriori knowledge about the database queries is not needed (as for the case of fragmentation). Instead we retrieve statistical observed information about operators, their evaluation and operator to query membership relations from the database statistics.

In order to speed up the evaluation of a query $q \in Q$ on a site $s \in S$, we can infer a set of *replication hints* for the fragments accessed by q, denoted as a triple (f, s, c) and signifying that by storing fragment f on site s we can reduce the data transfer cost by c. Since query q can be observed running a number $r, r \geq 1$, of times on site s, then by using the replicas according to the *replication hints* the data transfer costs are reduced by an amount of $r * c$. Considering the *replication hints* (f, s) proposed by all queries $q \in Q$ and the amount of reduction in data transfer cost for the resulting fragment storage policy we obtain a set of *replication hints* for Q denoted as:

$$P = \{(f_i, s_j, c_{ij}) | f_i \in F, s_j \in S, (f_i, s_j) \neq (f_k, s_l), \forall i \neq k \, and \, j \neq l\} \quad (2)$$

In the following we assume that the database dictionary after an observed running interval contains information about: fragments, fragment allocation, queries and fragments accessed by a query. Suppose that this information is made available throughout computed views like an usual database would.

4   Induced Fragment Replication

Let $s_i \in S$ be a site containing some fragments of the database. Let ds_i be the available memory space on site s_i. We can only store new fragments within the limits of the available memory space. In the trivial case, if the available memory is infinite, the solution to the replication problems is total replication where the data transfer cost is null for any query. When the available memory space is limited our proposed replication model needs to find the optimal set of replicas within the memory space constraint such that data transfer cost is minimal for the overall set of queries in Q. v If the size of a fragment $f_i \in F$ is df_i and the available memory/storage space on site $s_j \in S$ is ds_j, then we propose a solution modeled as a transport network compatible flow problem.

A *replication hint* (f_i, s_j, c_{ij}) as mentioned in (2) has two possible implementation choices: to be retained/applied or dismissed. We introduce a new variable $r_{ij}, r_{ij} \in \{0, 1\}$ that denotes the above possibilities.

4.1   Network Flow Solution

When modeling the induced replication as a network flow problem we have two possible options: a global variant for the whole set S of sites, or individually for each site $s \in S$. We will describe the solution model for the former variant.

Given all the above described elements we propose a transport network denoted as:

$$N = (V, A, lo, up, co, start, fin) \tag{3}$$

where: V - is the vertex set, $V = F \cup S \cup \{start, fin\}$. We add two new vertices: $start$ and fin; A - the set of edges of the graph; lo and up correspond to the lower and upper bound, while co is a set of functions that associates a real non-negative value to each edge;

The set of edges A and the functions lo, up, co are defined as following, where $|S|$ denotes the number of sites:

$$\forall f_i \in F, a_i = (start, f_i) \in A; lo(a_i) = 0; up(a_i) = df_i \times |S|; co(a_i) = 0; \tag{4}$$

$$\forall s_j \in S, a_j = (s_j, fin) \in A; lo(a_j) = 0; up(a_j) = ds_j; co(a_j) = 0; \tag{5}$$

$$\begin{cases} \forall (f_i, s_j, c_{ij}) \in P, a_{ij} = (f_i, s_j) \in A; \\ lo(a_{ij}) = up(a_{ij}) = df_i; co(a_{ij}) = \dfrac{c_{ij}}{df_i}; \end{cases} \tag{6}$$

A flow in the above transport network N is a real function: $fl : A \longrightarrow \Re$ having the following properties:

(1) Capacity Restrictions: $fl(a) = 0$ or $lo(a) \leq fl(a) \leq up(a), \forall a \in A$.
(2) Flow Conservation $\forall v \in V - \{start, fin\}$:

$$\sum_{\substack{u \in V, \\ (u,v) \in A}} fl(u, v) = \sum_{\substack{u \in V, \\ (v,u) \in A}} fl(v, u) \text{ or } \sum_{\substack{a = (u,v) \in A, \\ u \in V}} fl(a) = \sum_{\substack{a = (v,u), \\ u \in V}} fl(a)$$

The *value of the flow* can be computed by: $\sum_{s \in S} fl(s, fin)$. Given a flow in the network N, we can determine a cost given according to the following formula:

$$cost(fl) = \sum_{a \in A} fl(a) \times co(a).$$

where $co(a)$ is the value associated to each edge, introduced above and computed according to (4, 5, 6). A flow has a maximum cost if there is no other flow with a higher cost. Conditions from (6) state that storing a fragment f on site s is done on the entire fragment or not at all. Equation (5) state that storing fragments in a site cannot exceed the available storage space on that site. Equation (4) state that the number of fragment replicas is unbounded. The flow cost is only influenced by the values of the *co* function in (6) and its value represents the amount of cost reduction for data transfers.

Finding the allocation schema (with replication) that maximizes the data transfer cost reduction can be solved in the above conditions by finding the maximum compatible cost flow in the transport network.

This transport problem is a special one due to its capacity restrictions and the maximum cost requirement, but not the maximum flow. We elaborate a backtrack type algorithm which determines for every site s_i a set of fragments that can be replicated on that site. From all the constructed sets we only keep the one with the maximum cost. The main issue of the backtracking algorithm is that it performs an exhaustive search of the solution space. The explored space grows proportionally with the product of the number of fragments and sites and the required time to solution grows and becomes unrealistic for an online system. As a solution to this issue we elaborate an approximate algorithm based on a greedy approach to find the maximum cost flow. The previous algorithm is simplified by considering only one set R in step 2. Candidate replication fragments will be allocated to a site in cost descending order - as long as there is available space. This approach reduces algorithm complexity while still allowing a close approximation of the solution. We thought our proposal as a module in a database system, that runs quasi-continuously and provides replication hints whenever these exist and are possible. As a consequence the algorithm should be as fast as possible and with minimal impact on the database.

```
Algorithm Max Cost Flow Greedy:
INPUT.n (number of fragments);  df (n dimensional array - fragments' dimension);
      m (number of sites);  ds (m dim. array - the available space on sites);
      c (m x n dim. array - transfer cost of fragment F[i] to size S[j];
1.INIT.
   FOR j=1,...,m
      fragm[j] = EMPTY SET  (index of the fragment replicated on site j)
2.FOR every site j=1,...,m construct array fragm[j]
  2.1 SumDimF:=0; R = empty set;
      sort descending fragment transport costs: c[j][1],...,c[j][n]
      LET PF[1],...,PF[n] be the fragments in descending costs order
      FOR i=1,...,n
        --site j has enough space to store fragment PF[i]
        IF (SumDimF+df[PF[i]]<=ds[j] and c[j][PF[i]]>0)
        Add PF[i] to R;
        update SumDimF;
        store R in fragm[j];
OUTPUT: fragm[j], j=1,...,m  --fragments to replicate in site j
```

5 Experimental Results

In order to evaluate the efficiency of the proposed solutions, we run a battery of simulations and tests. For assessing the generality of our model we randomly generate a set of database configurations. To test the proposed Induced Fragment Replication (IFR) we generate different sets of large distributed databases. The

Table 1. Test distributed
DB configurations

DDB	Frag	Sites	Queries	Max reps
DDB1	20	5	10000	100
DDB2	50	5	10000	100
DDB3	100	5	10000	100
DDB4	200	5	10000	100
DDB5	300	5	10000	100
DDB6	20	10	10000	100
DDB7	50	10	10000	100
DDB8	100	10	10000	100
DDB9	200	10	10000	100
DDB10	300	10	10000	100
DDB11	20	20	10000	100
DDB12	50	20	10000	100
DDB13	100	20	10000	100
DDB14	200	20	10000	100
DDB15	300	20	10000	100

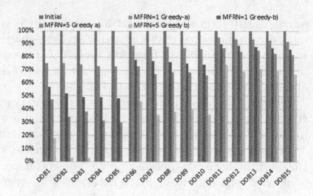

Fig. 4. Cost Improvements Percents for MFRN $= 1$
and MFRN $= 5$

synthetic experiments were preferred due to the lack of large and statistically
complete and consistent real databases. We choose to generate statistically database configurations as we only need to process the meta-information from the
database and not the actual data. We first generated an initial database state
by averaging the evaluation costs over a number of uniformly sampled generated distribution configurations. Then we generate fifteen small to large sample
database configurations drawn from the same distribution (see [11] about the
configuration generator, test data and results) (Fig. 4).

The fifteen distributed database configurations are presented in Table 1.

In the following we present the analysis of the tests' results. In order to asses
the improvements we measure the network transfer before and after applying
the induced fragmentation on a series of test databases. We consider the percent of data transfer cost needed in query processing after applying the IFR
solution compared to the transfer cost in the initial database as the measure
of query optimization. We test the replication problem for the next cases: (a)
the available space is equal with the space occupied by the fragments; and (b)
the available space is 2 * space occupied by the fragments. Table 2 presents the
percents and execution times in seconds for IFR problem when (Maximum Fragment Replication Number - the maximal number of generated fragment replicas)
MFRN=1 and MFRN=5 and the available free space corresponds to above space
constraints (a) and (b).

The proposed network flow problem is a bit uncommon as it needs a flow of
maximum cost (regardless of the value of the flow). There is no solver, to our
knowledge, for this problem formulation and thus we implemented a backtracking solution to solve the flow problem and then we proposed a faster Greedy
approximation algorithm for the same problem.

In Table 2 we show the transport cost expressed as a percentage of the original
database (before applying induced fragmentation). We also present the execution
times in seconds for each algorithm variant. The backtracking variant has an

Table 2. Costs and exec times for MFRN = 1 and MFRN = 5, cases (a) and (b)

	MFRN=1								MFRN=5							
	case (a)				case (b)				case (a)				case (b)			
	Greedy		Backtrack		Greedy		Backtrack		Greedy		Backtrack		Greedy		Backtrack	
DDB	cost (%)	time (s)	cost (%)	time (s)	cost (%)	time (s)	cost (%)	time (s)	cost (%)	time (s)	cost (%)	time (s)	cost (%)	time (s)	cost (%)	time (s)
DDB1	75	0.2	74	0.97	57	0.22	55	2.6	47	0.22	46	1.01	18	0.23	17	4.51
DDB2	75	0.21			52	0.27			34	0.29			3	0.28		
DDB3	74	0.31			49	0.3			38	0.31			3	0.26		
DDB4	73	0.39			49	0.33			31	0.34			0	0.43		
DDB5	73	0.46			48	0.51			30	0.51			0	0.48		
DDB6	89	0.27	88	0.1	78	0.33	77	0.1	73	0.21	72	1.31	46	0.25	45	9.65
DDB7	88	0.51			77	0.31			67	0.33			36	0.33		
DDB8	88	0.43			76	0.43			68	0.33			38	0.36		
DDB9	87	0.53			75	0.63			68	0.56			40	0.59		
DDB10	86	0.7			74	0.74			66	0.79			36	0.7		
DDB11	95	0.31	95	0.1	90	0.32	89	0.1	87	0.27	87	1.1	74	0.29		
DDB12	94	0.39	94	225	89	0.42			84	0.39			69	0.38		
DDB13	94	0.51			88	0.46			85	0.49			71	0.48		
DDB14	93	0.83			87	0.93			83	0.82			69	0.82		
DDB15	92	1.24			86	1.13			82	1.04			67	1.19		

exact solution but the search space explodes exponentially with the product of the number of fragments, sites and queries and thus its execution times explode exponentially with this product (search space). In Table 2 we left empty the cells where the backtracking solution failed to give a solution within the time required for a Mathematical solver to solve the equivalent linear programming problem.

The proposed Greedy solution has an almost constant execution time with a ratio of 6:1 between the fastest and slowest solution. As execution time this is more than appropriate for a system where our module is run quasi-continuously and produces fragmentation hints.

The cost penalty obtained by applying the proposed approximation Greedy solution is around 1.75% higher compared to the exact backtracking solution. However the execution time for the Greedy approach compared to the exact solutions is in the order of 550 times less. The Greedy solution average running time is around 0.5 s with the largest execution time being 1.24 s.

6   Conclusions and Future Work

In this paper we provide a solution for query response time improvement modeled as a maximal compatible flow cost in transport networks. We perform online perpetual data replication within the original space constraints of the database. The major contribution is the Greedy algorithm that solves the transport flow problem in a fraction of the time needed to a classical solver or algorithm with an approximation penalty cost under 2% making this algorithm suitable to online execution within the database and as a replacement to the classical solvers.

We only considered so far the transport cost as the argument driving data replication and allocation in order to improve query response time. While it solves a complex problem this model is in many cases too simplistic. As future work we would like to extend our model to cases where data allocation is driven by more parameters (CPU, storage, network capacities, etc.) or to non-relational cloud databases where principles are different.

References

1. Apers, P.M.G.: Data allocation in distributed database systems. ACM T Datab. Syst. **13**(3), 263–304 (1988). Applied Mathematical Programming. Addison-Wesley (1977)
2. Cheng, C.H., Lee, W.K., Wong, K.F.: A genetic algorithm-based clustering approach for database partitioning. IEEE Trans. Syst. Man Cybern. Part C Appl. Rev. **32**(3), 215–230 (2002)
3. Dokeroglu, T., Bayır, M.A., Cosar, A.: Integer linear programming solution for the multiple query optimization problem. In: Czachórski, T., Gelenbe, E., Lent, R. (eds.) Information Sciences and Systems 2014, pp. 51–60. Springer, Cham (2014). doi:10.1007/978-3-319-09465-6_6
4. Graham, J., Foss, J.A.: Efficient allocation in distributed object oriented databases. In: Proceedings of 16th International Conference on Parallel and Distributed Computing Systems (ISCA), pp. 471–412 (2003)
5. Huang, Y., Chen, J.: Fragment allocation in distributed database design. J. Inf. Sci. Eng. **17**, 491–506 (2001)
6. Lin, X., Orlowska, M.: An integer linear programming approach to data allocation with the minimum total communication cost in distributed database systems. Inf. Sci. **85**, 1–10 (1995)
7. Menon, S.: Allocating fragments in distributed databases. IEEE Trans. Parallel Distrib. **16**(7), 577–585 (2005)
8. Morffi, A.R., et al.: A reinforcement learning solution for allocating replicated fragments in a distributed database. Comput. Sist. **11**(2), 117–128 (2007)
9. Ozsu, M.T., Valduriez, P.: Principles of Distributed Database Systems. Springer, Heidelberg (2011)
10. Tambulea, L., Darabant, A.S., Varga, V.: Data transfer optimization in distributed database query processing. Studia Univ Babes Bolyai, Informatica **LIX**(1), 71–82 (2014)
11. Tambulea, L., Darabant, A. S., Varga, V.: Query Evaluation Optimization in a Distributed Database using Data Reorganization (2015). http://www.cs.ubbcluj.ro/~ivarga/ddbpaper
12. Virk, R.S., Singh, D.G.: Optimizing access strategies for a distributed database using genetic fragmentation. Int. J. Comput. Sci. Netw. Secur. **11**(6), 180–183 (2011)
13. Wiese, L.: Clustering-based fragmentation and data replication for flexible query answering in distributed databases. Int. J. Cloud Comput. **3**(1), 3–18 (2014)

Data Integration and RDF Matching

Semantic Web Datatype Similarity: Towards Better RDF Document Matching

Irvin Dongo[1(✉)], Firas Al Khalil[2], Richard Chbeir[1], and Yudith Cardinale[1,3]

[1] University Pau & Pays Adour, LIUPPA, EA3000, 64600 Anglet, France
{irvin.dongo,richard.chbeir}@univ-pau.fr
[2] University College Cork, CRCTC, 13 South Mall, Cork, Ireland
firas.alkhalil@ucc.ie
[3] Departamento de Computación, Universidad Simón Bolívar, Caracas, Venezuela
ycardinale@usb.ve

Abstract. With the advance of the Semantic Web, the need to integrate and combine data from different sources has increased considerably. Many efforts have focused on RDF document matching. However, they present limited approaches in the context of datatype similarity. This paper addresses the issue of datatype similarity for the Semantic Web as a first step towards a better RDF document matching. We propose a datatype hierarchy, based on W3C's XSD datatype hierarchy, that better captures the *subsumption* relationship among primitive and derived datatypes. We also propose a new datatype similarity measure, that takes into consideration several aspects related to the new hierarchical relations between compared datatypes. Our experiments show that the new similarity measure, along with the new hierarchy, produces better results (closer to what a human expert would think about the similarity of compared datatypes) than the ones described in the literature.

Keywords: Datatype hierarchy · Datatype similarity · XML · XML Schema · Ontology · RDF · Semantic Web

1 Introduction

One of the benefits offered by the Semantic Web initiative is the increased support for data sharing and the description of real resources on the web, by defining standard data representation models such as RDF, the Resource Description Framework. The adoption of RDF increases the need to identify similar information (resources) in order to integrate and combine data from different sources (e.g., Linked Open Data integration, ontology matching). Many recent works have focused on describing the similarity between concepts, properties, and relations in the context of RDF document integration and combination [22].

Indeed, RDF describes resources as triples: ⟨subject, predicate, object⟩, where subjects, predicates, and objects are all resources identified by their IRIs[1].

---

[1] Internationalized Resource Identifier. An extension of URIs that allows characters from the Unicode character set.

© Springer International Publishing AG 2017
D. Benslimane et al. (Eds.): DEXA 2017, Part I, LNCS 10438, pp. 189–205, 2017.
DOI: 10.1007/978-3-319-64468-4_15

Objects can also be literals (e.g., a number, a string), which can be annotated with optional type information, called a datatype; RDF adopts the datatypes from XML Schema. The W3C Recommendation proposed in [1] points out the importance of the existence of datatype annotations to detect entailments between objects that have the same datatype but a different value representation. For example, if we consider two distinct triples containing the objects "20.000" and "20.0", then these objects are considered as different, because of the missing datatype. However, if they were annotated as follows: "20.000"^^xml : decimal and "20.0"^^xml : decimal then we can conclude that both objects are identical. Moreover, works on XML Schema matching proved that the presence of datatype information, constraints, and annotations on an object improves the similarity between two documents (up to 14%) [4].

Another W3C Recommendation [2] proposes a simple method to determine the similarity of two distinct datatypes: the similarity between two primitive datatypes is 0 (disjoint), while the similarity between two datatypes derived from the same primitive datatype is 1 (compatible). Obviously, this method is straightforward and does not capture the degree of similarity of datatypes; for instance, `float` is more similar to `int` than to `date`. This observation lead to the development of *compatibility tables*, that encodes the similarity ($\in [0,1]$) of two datatypes. They were used in several studies [9,23] for XML Schema matching. These compatibility tables were either populated manually by a designated person, as in [9,23] or generated automatically using a similarity measure that relies on a hierarchical classification of datatypes, as in [15,28].

Hence, in the context of RDF document matching, these works present the following limitations:

1. The Disjoint/Compatible similarity method as proposed by the W3C is too restrictive, especially when similar objects can have different, yet related, datatypes (e.g., `float` and `int` *vs* `float` and `double`).
2. The use of a true similarity measure, expressed in a *compatibility table*, is very reasonable; however, we cannot rely on an arbitrary judgment of similarity as done in [9,23]; moreover, for 44 datatypes (primitive and derived ones, according to W3C hierarchy), there are 946 similarity values ($n \times (n-1)/2, n = 44$), which makes the *compatibility table* incomplete as in [9]; a similarity measure that relies on a hierarchical relation of datatypes is needed.
3. The W3C datatype hierarchy, used in other works, does not properly capture any semantically meaningful relationship between datatypes (see, for instance, how datatypes related to `dateTime` and `time` are flattened in Fig. 2).

From these limitations, there is a need to provide a better solution for any RDF document matching approach, where simple datatype similarity is considered. To achieve this, we propose:

1. An extended version of the W3C datatype hierarchy, where a parent-child relationship expresses subsumption (parent subsumes child), which makes it a taxonomy of datatypes.

2. A new similarity measure: extending the one presented in [15], to take into account several aspects related to the new hierarchical relations between compared datatypes (e.g., children, depth of datatypes).

We experimentally compare the effectiveness of our proposal (datatype hierarchy and similarity measure) against existing related works. Our approach produces better results (closer to what a human expert would think about the similarity of compared datatypes) than the ones described in the literature.

The paper is organized as follows. In Sect. 2, we present a motivating scenario. In Sect. 3, we survey the literature on datatype similarity and compare them using our motivating scenario. In Sect. 4, we describe the new datatype hierarchy and the new similarity measure. In Sect. 5, we present the experiments we performed. And finally, we conclude in Sect. 6.

2 Motivating Scenario

In order to illustrate the limitations of existing approaches for datatype similarity, we consider a scenario in which we need to integrate three RDF documents with similar concepts (resources) but based on different vocabularies. Fig. 1 shows three concepts from three different RDF documents to be integrate. Figure 1a describes the concept of a `Light Bulb` with properties (predicates) `Light`, `Efficiency`, and `Manufacturing_Date`, Fig. 1b describes the concept of `Lamp` with properties `Light` and `MFGDT` (manufacturing date), and Fig. 1c shows the concept of `Light Switch` with properties `Light` and `Model_Year`.

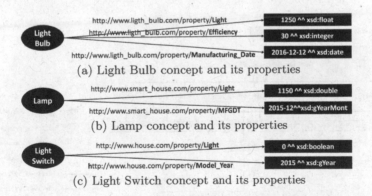

(a) Light Bulb concept and its properties

(b) Lamp concept and its properties

(c) Light Switch concept and its properties

Fig. 1. Three concepts from three different RDF documents

To integrate these RDF documents, it is necessary to determine the similarity of the concepts expressed in them, based on the similarity of their properties. More precisely, we can determine the similarity of two properties by inspecting the datatypes of their *ranges*[2] (i.e., of their objects).

---

[2] A range (*rdfs:range*) defines the object type that is associated to a property.

Intuitively, considering the datatype information, we can say that:

1. `Light Bulb` and `Lamp` are similar, since their properties are similar: the `Light` property is of type `float` for `Light Bulb` and `double` for `Lamp`, we know that both `float` and `double` express floating points, and they differ only by their precisions; the same thing can be said about the properties `Manufacturing_Date` and `MFGDT`.
2. `Light Switch` is different from the other concepts; indeed, the `Light` property is expressed in `binary`, and can hold one of two values, namely 0 and 1, expressing the state of the light switch (i.e., on and off, respectively).

Hence, to support automatic matching of RDF documents based on their concepts similarity, it is necessary to have a datatype hierarchy establishing semantically meaningful relationship among datatypes and a measure able to extract these relations from the hierarchy. In the following section, we survey the literature on datatype similarity and compare them using this motivating scenario.

3   Related Work

To the best of our knowledge, there is no existing work tackling datatype similarity specifically targeting RDF documents. Hence, we review works on datatype similarity described for XML and XSD, since RDF uses the same XML datatypes proposed by the W3C (the datatype hierarchy is shown in Fig. 2), and we also consider works in the context of ontology matching. We evaluate these works in an RDF document matching/integration scenario in the discussion.

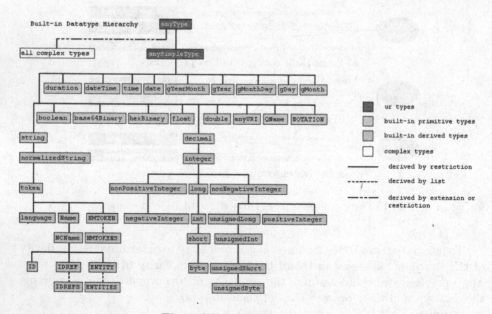

Fig. 2. W3C datatype hierarchy

Most of the existing works in the XML and XSD area are focused on schema matching in contexts of, for example, XML message mapping, web data sources integration, and data warehouse loading. The main approaches taken to establish the datatype similarity are either: 1. based on user-defined compatibility tables [5–7,9,11,23,24,27], or 2. constraining facets[3] [28], or 3. extended W3C hierarchy and measure [3,8,15].

User-defined compatibility tables, as the one presented in Table 1 (taken from [9]), express the judgment and perception of users regarding the similarity between each pair of datatypes. Hence, these tables present similarity values that are not objective, complete, or reliable.

Table 1. Datatype compatibility table of work [9]

Type (s)	Type (t)	Compatibility coefficient (s, t)
string	string	1.0
string	date	0.2
decimal	float	0.8
float	float	1.0
float	integer	0.9
integer	short	0.8

When constraining facets are considered as in [28], the similarity value between two different datatypes is calculated by the number of common facets divided by the union of them. For example, datatypes `date` and `gYearMonth` have the same facets (i.e., pattern, enumeration, whiteSpace, maxInclusive, maxExclusive, minExclusive, and minInclusive), thus, their similarity is equal to 1. This method allows to create an objective, complete, and reliable *compatibility table*; however, suitability is still missing: besides facets, which are only syntactic restrictions, other information should be considered for the Semantic Web (e.g., common datatypes attributes[4] – datatype subsumption).

Other works have proposed a new datatype hierarchy by extending the one proposed by the W3C. In [15], the author proposes five new datatype groups: `Text`, `Calendar`, `Logic`, `Numeric`, and `Other`. They also propose a new datatype similarity function that relies on that hierarchy and takes into account the proximity of nodes to the root and the level of the Least Common Subsumer[5] (LCS) of the two compared datatypes. The works presented in [3,8], combine semantic similarity, structural similarity, and datatype compatibility

---

[3] Constraining facets are sets of aspects that can be used to constrain the values of simple types (https://www.w3.org/TR/2001/REC-xmlschema-2-20010502/#rf-facets).

[4] An attribute is the minimum classification of data, which does not subsume another one. For example, datatype `date` has the attributes year, month, and day.

[5] It is the most specific common ancestor of two concepts/nodes, found in a given taxonomy/hierarchy.

of XML schemas in a function, by using the hierarchy and similarity function proposed by [15]. Even though these works improve the similarity values, we will see their limitations in the context of our motivational scenario, concerning to misdefined datatype relations in the datatype hierarchy.

In the context of ontology matching, most of the works classify datatypes as either Disjoint or Compatible (similarity $\in \{0,1\}$). Some of them are based on the W3C hierarchy, such as [13,17], while others take into account properties of the datatypes (domain, range, etc.) [10,14,16,19–22,25,26]. When domain and range properties are considered, if two datatypes have the same properties, the similarity value is 1, otherwise it is 0. In the context of RDF matching, in which similar objects can have different but related datatypes, this binary similarity is too restrictive. The authors in [12] generate a vector space for each ontology by extracting all distinct concepts, properties and the ranges of datatype properties. To calculate the similarity between the two vectors, they use the cosine similarity measure. However, as the measure proposed in [15], the problem remains in the datatype hierarchy that does not represent more semantically meaningful relationships between datatypes.

Table 2. Related work classification

Group	Work	Datatype similarity	Datatype requirements			
			Simple datatype	Common attributes	SW context	
					XML/XSD	RDF/OWL
1	W3C [2,10,12–14,16,17,19–22,25,26]	Disjoint/compatible (binary values)	✓	X	✓	✓
2	[4–7,9,11,23,24,27]	User-defined compatibility table	✓	X	✓	X
3	[28]	Constraining facets	✓	X	✓	X
4	[3,8,15]	Formula on extended W3C hierarchy	✓	X	✓	X

We classify the existing works into four groups (see Table 2) and we evaluate them in our motivating scenario in the upcoming section.

Resolving Motivating Scenario and Discussion: Now, we evaluate our scenario using the defined groups in Table 2. We have the datatypes float and date from the concept Light Bulb (Fig. 1a), datatypes double and gYearMonth from the concept Lamp (Fig. 1b), and boolean and gYear from concept Light Switch (Fig. 1c).

According to the Disjoint/Compatible similarity, either defined by the W3C or not (Group 1 in Table 2), the similarity between the three pairs of datatypes related to Light property (float–double, float–boolean, and double–boolean) is 0, because the three datatypes are primitives. We have the same similarity result regarding Manufacturing_Date, MFGDT, Model_Year properties, since their datatypes are also primitives. It means that there is no possible integration for these concepts using this similarity method. However, the concepts Light Bulb and Lamp are strongly related according to our scenario.

Based on the user-defined compatibility table shown in Table 1 (as works in Group 2 do), the similarity between float–double is a given constant > 0 (as decimal–float has in the compatibility table), however the similarity values of double–boolean, date–gYearMonth, date–gYear, and gYearMonth–gYear are not present in the compatibility table, therefore leading to a similarity value of 0 as in [15] do. In this case, concepts Light Bulb and Lamp have their respective properties Light considered similar, while Manufacturing_Date and MFGDT are considered disjoint, even though they are clearly related.

According to the methods of Group 3 (based on constraining facets), similarity values for float–double, date–gYearMonth, date–gYear, and gYearMont–gYear are all equal to 1 (because they have the same facets), and for float–boolean and double–boolean, the similarities are equal to 0.29 (2 common facets divided by the union of them, which is 7). Thus, the three concepts can be integrated as similar, which is incorrect. Additionally, datatypes date, gYearMonth, and gYear are related but not equal: besides their facets, other information (such as datatype attributes - year, month, day) should count to decide about their similarities.

Finally, according to the works in Group 4, which are based on similarity measures applied on a datatype hierarchy extended from the W3C hierarchy [15], similarity between float–double is 0.30, similarity between float–boolean and double -boolean is 0.09, for date–gYearMonth, date–gYear, and gYearMonth–gYear the similarity value is 0.296^6. Even though these works manage in a better way the datatype similarity than all other Groups, there is still the issue of considering common datatypes attributes (as for work in Group 3). We can note that date–gYearMonth share year and month as common attributes, while date–gYear only have year as common attribute; thus, similarity between date–gYearMonth should be bigger than the other.

Table 3. Integration results for our motivating scenario

Concept integration	G. 1 (Sim)	G. 2 (Sim)	G. 3 (Sim)	G. 4 (Sim)	Appropriate
Light Bulb and Lamp	NI (0.00)	NI (0.40)	I (1.00)	NI (0.30)	I
Lamp and Light Switch	NI (0.00)	NI (0.00)	I (0.65)	NI (0.19)	**NI**
Light Bulb and Light Switch	NI (0.00)	NI (0.00)	I (0.65)	NI (0.19)	**NI**

Results were obtained by applying a threshold 0.50 for average of properties; NI = Not Integrable, I = Integrable

Table 3 summarizes the integration results of the motivating scenario. Column *Appropriate* shows the correct integration according to our intuition. One can note that existing works cannot properly determine a correct integration.

---

[6] We show the results according the measure proposed on [15], all other works in Group 4 propose similar measures.

With this analysis, we can observe the importance of datatypes for data integration and the limitations of the existing works, from which, the following requirements were identified:

1. The measure should consider at least all simple datatypes (primitive and derived datatypes); complex datatypes are out of the scope in this work.
2. The datatype hierarchy and similarity measure should consider common datatype attributes (subsumption relation) in order to establish a more appropriate similarity.
3. The whole approach should be objective, complete, reliable, and suitable for the Semantic Web.

We can note from Table 2, that all works consider primitive and derived datatypes and are suitable in XML and XSD contexts. Only the works in the context of ontology matching (Group 1) consider RDF data. None of these works consider common datatype attributes. The following section describes our approach, based on a new hierarchy and a new similarity measure, that overcomes the limitations of existing works and addresses these requirements.

4   Our Proposal

In this section, we describe our datatype similarity approach that mainly relies on an extended W3C datatype hierarchy and a new similarity measure.

4.1   New Datatype Hierarchy

As we mentioned before, the W3C datatype hierarchy does not properly capture any semantically meaningful relationship between datatypes and their common attributes. This issue is clearly identified in all datatypes related to date and time (e.g., `dateTime`, `date`, `time`, `gYearMonth`), which are treated as isolated datatypes in the hierarchy (see Fig. 2).

Our proposed datatype hierarchy extends the W3C hierarchy as it is shown in Fig. 3. White squares represent our new datatypes, black squares represent original W3C datatypes, and gray squares represent W3C datatypes that have changed their location in the hierarchy. We propose four new primitive datatypes: `period`, `numeric`, `logic`, and `binary`. Thus, we organize datatypes into eight more coherent groups of primitive datatypes (`string`, `period`, `numeric`, `logic`, `binary`, `anyURI`, `QName`, and `NOTATION`). All other datatypes are considered as derived datatypes (e.g., `duration`, `dateTime`, `time`) because their attributes are part of one particular primitive datatype defined into the eight groups.

We also add two new derived datatypes (`yearMonthDuration` and `dayTimeDuration`), which are recommended by W3C to increase the precision of `duration`, useful for `XPath` and `XQuery`. We classify each derived datatype under one of the eight groups (e.g., `Period` subsumes `duration`, `numeric` subsumes `decimal`) and, in each group, we specify the proximity of datatypes by a sub-hierarchy (e.g., `date` is closer to `gYearMonth` than to `gYear`).

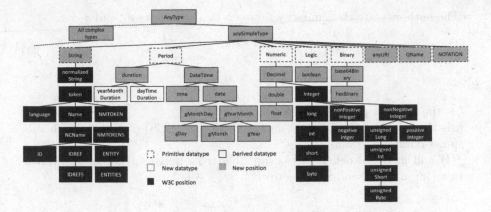

Fig. 3. New datatype hierarchy

The distribution of the hierarchy for derived datatypes is established based on the subsumption relation and stated in the following assumption:

Assumption 1. *If a datatype d_1 contains at least all the attributes of a datatype d_2 and more, d_1 is more general than d_2 (d_1 subsumes d_2).*

As a consequence of Assumption 1, the hierarchy designates datatypes more general to more specific, from the root to the bottom, which in turn defines datatypes more related than others according to their depths in the hierarchy. With regards to this scenario, we have the following assumption:

Assumption 2. *Datatypes in the top of the hierarchy are less related than datatypes in the bottom, because datatypes in the top are more general than the ones in the bottom.*

Thus, according to Assumption 2, the datatype similarity value will depend on their position (depth) in the hierarchy (e.g., gYearMonth–gYear are more similar than period–dateTime), as we show in the next section.

4.2 Similarity Measure

Our proposed similarity measure is inspired by the one presented in [15]. The authors establish the similarity function based on the following intuition:

> *"The similarity between two datatype d_1 and d_2 is related to the distance separating them and their depths in the datatype hierarchy. The bigger the distance separating them, the less similar they are. The deeper they are the more similar they are, since at deeper levels, the difference between nodes is less significant [15]."*

The authors state the similarity between two datatypes d_1 and d_2 as:

$$c(d_1, d_2) = \begin{cases} f(l) \times g(h) & \text{if } d_1 \neq d_2 \\ 1 & \text{otherwise} \end{cases} \tag{1}$$

where:

- l is the shortest path length between d_1 and d_2;
- h is the depth of the Least Common Subsumer (LCS) datatype which subsumes datatype d_1 and d_2.
- $f(l)$ and $g(h)$ are defined based on Shepard's universal law of generalization [18] in Eqs. 2 and 3, respectively.

$$f(l) = e^{-\beta l} \tag{2} \qquad g(h) = \frac{e^{\alpha h} - e^{-\alpha h}}{e^{\alpha h} + e^{-\alpha h}} \tag{3}$$

where α and β are user-defined parameters.

The work in [15] does not analyze the common attributes (children) of compared datatypes. For example, the datatype pair date–gYearMonth (with 2 attributes, namely year and month, in common) involves more attributes than date–gYear (with only 1 attribute, namely year, in common). The authors of [15] consider that the similarity values of both cases are exactly the same.

In order to consider this analysis, we assume that:

Assumption 3. *Two datatypes d_1 and d_2 are more similar if their children in the datatype hierarchy are more similar.*

Furthermore, the depth of the LCS is not enough to calculate the similarity according to Assumption 2. Notice that the difference in levels in the hierarchy is also related to similarity. For example, according to [15], we have c(time, gYearMonth) $= c$(dateTime, gYear), because in both cases the distance between the datatypes is $l = 3$, and the LCS is dateTime, whose $h = 3$ (see Fig. 3). However, the difference between levels of time and gYearMonth is smaller than the one of dateTime and gYear, thus the similarity of time–gYearMonth should be bigger than the second pair (i.e., c(time, gYearMonth) $> c$(dateTime, gYear)). Hence, we assume:

Assumption 4. *The similarity of two datatypes d_1 and d_2 is inversely proportional to the difference between their levels.*

Based on Assumptions 3 and 4, we defined the cross-children similarity measure in the following.

Let $V_{d_1 p, d_2 q}$ be the children similarity vector of a datatype d_1, with respect to datatype d_2 in levels p and q, respectively. In d_1 sub-hierarchy, d_1 has i children in level p and in d_2 sub-hierarchy, d_2 has j children in level q. Thus, $V_{d_1 p, d_2 q}$ is calculated as in Eq. 4.

$$V_{d_1 p, d_2 q} = [c(d_1, d_{1p}^1), \ldots, c(d_1, d_{1p}^i), c(d_1, d_{2q}^1), \ldots, c(d_1, d_{2q}^j)] \tag{4}$$

where d_{1p}^x represents the child x of d_1 (with x from 1 to i) in level p and d_{2q}^y represents the child y (with y from 1 to j) of d_2 in level q.

Similarly, let V_{d_2q,d_1p} be the children similarity vector of a datatype d_2, with respect to datatype d_1 in the levels q and p respectively, defined as in Eq. 5.

$$V_{d_2q,d_1p} = [c(d_2, d_{1p}^1), \ldots, c(d_2, d_{1p}^i), c(d_2, d_{2q}^1), \ldots, c(d_2, d_{2q}^j)] \tag{5}$$

For each pair of vectors V_{d_1p,d_2q} and V_{d_2q,d_1p}, we formally define the cross-children similarity for level p and q, in Definition 1.

Definition 1. *The cross-children similarity of two datatypes d_1 and d_2 for levels p and q, respectively, is the cosine similarity of their children similarity vectors V_{d_1p,d_2q} and V_{d_2q,d_1p}, calculated as:*

$$CCS_{d1p,d2q} = \frac{V_{d_1p,d_2q} \cdot V_{d_2q,d_1p}}{\|V_{d_1p,d_2q}\|\|V_{d_2q,d_1p}\|}$$

Now, considering all pairs of V (i.e., all levels of both sub-hierarchies), we define the total cross-children similarity between d_1 and d_2 in Definition 2.

Definition 2. *The total cross-children similarity of two datatypes d_1 and d_2 is calculated as:*

$$S(d_1, d_2) = \frac{1}{L_1} \times \sum_{p=1}^{L_1} \sum_{q=1}^{L_2} m(d1p, d2q) \times CCS_{d1p,d2q}$$

where $m(d1p, d2q)$ is a Gaussian function based on Assumption 4: L_1 and L_2 are the number of levels of sub-hierarchies of d_1 and d_2, respectively.

The Gaussian function is defined as follows:

$$m(d1p, d2q) = e^{-\pi \times (\frac{(depth(d1p) - depth(d2q))}{H - 1})^2}$$

where $depth(d_{1p})$ and $depth(d_{2q})$ are the depths of the levels p and q respectively. H is the maximum depth of the hierarchy. Note that the depth of the hierarchy starts from 0. We name $S'(d_1, d_2)$ the average between $S(d_1, d_2)$ and $S(d_2, d_1)$.

$$S'(d_1, d_2) = 0.5 \times S(d_1, d_2) + 0.5 \times S(d_2, d_1) \tag{6}$$

Finally, we define similarity between datatypes d_1 and d_2 in Definition 3 as an extension of Eq. 1.

Definition 3. *Similarity between two datatypes d_1 and d_2, denoted as $sim(d_1, d_2)$, is determined as:*

$$sim(d_1, d_2) = \begin{cases} (1 - \omega) \times f(l) \times g(h) + \omega \times S'(d_1, d_2) & \text{if } d_1 \neq d_2 \\ 1 & \text{otherwise} \end{cases}$$

where $\omega \in [0, 1]$ is a user-defined parameter that indicates the weight to be assigned to the cross-children similarity.

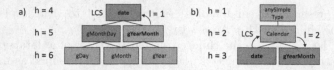

Fig. 4. (a) Sub-hierarchy from our new hierarchy; (b) sub-hierarchy from [15]

With our RDF similarity approach, we satisfy all identified requirements. This measure generates similarity values based on a hierarchy (objective, complete and reliable) for simple datatypes. The whole approach is more suitable for the Semantic Web, because common attributes among datatypes are taking into account both in the hierarchy by Assumption 1 and in the similarity measure by Definition 1.

The following section illustrates how our approach is applied to calculate similarity between the properties of the concepts Light Bulb and Lamp from our motivating scenario and, it is compared with the work in [15].

4.3   Illustrative Example

To better understand our similarity approach, we illustrate step by step the process to obtain the similarity between datatypes date from Light Bulb and gYearMonth from Lamp. We compare it with the one obtained by [15]. To do so, we fix the parameters with the following values: $\alpha = \beta = 0.3057$ (taken from [15]), and $\omega = 0.20$, which means a weigh of 20% for cross-children similarity and 80% for the distance between datatypes and their depths (i.e., $f(l)$ and $g(h)$).

According to our new datatype hierarchy, we have $l = 1$, as the distance between date-gYearMonth, and $h = 4$ the depth of date, which is the LCS. Figure 4(a) shows these values and the sub-hierarchy from the LCS, according to our new hierarchy. For [15], the distance between date-gYearMonth is $l = 2$ and $h = 2$ is the depth of the LCS, which is Calendar. Figure 4(b) shows these values and the sub-hierarchy, according to the hierarchy in [15].

Then, the similarity value for our similarity approach is (see Definition 3):

$$sim(\text{date}, \text{gYearMonth}) = 0.80 \times f(1) \times g(4) + 0.20 \times S'(\text{date}, \text{gYearMonth})$$

and for [15] is (see Eq. 1): $c(\text{date}, \text{gYearMonth}) = f(2) \times g(2)$.

According to Eqs. 2 and 3, $f(1) = 0.74$, $g(4) = 0.84$ (for our similarity approach) and $f(2) = 0.54$, $g(2) = 0.55$ (for [15]). Hence, for [15] the similarity value between date-gYearMonth is: $c(\text{date}, \text{gYearMonth}) = 0.297$.

For our similarity approach, the cross-children similarity is taken into account to finally calculate the similarity between date-gYearMonth (see Eq. 6):

$$S'(\text{date}, \text{gYearMonth}) = 0.5 \times S(\text{date}, \text{gYearMonth}) + 0.5 \times S(\text{gYearMonth}, \text{date})$$

To calculate $S'(\text{date}, \text{gYearMonth})$, we have to calculate before the total cross-children similarities, S(date,gYearMonth) and S(gYearMonth,date). From

Def. 2, we obtain:

$$S(\text{date}, \text{gYearMonth}) = \frac{1}{2} \times \sum_{p=1}^{2} \sum_{q=1}^{1} e^{-\pi \times (\frac{(depth(d1p) - depth(d2q))}{9-1})^2} \times CCS_{\text{date}p, \text{gYearMonth}q}$$

Note that `date` has two levels of children (thus, $p = 1$ to 2 in the sum), while `gYearMonth` has one level of children (thus, $q = 1$ to 1 in its sum). Replacing values, we have $S(\text{date}, \text{gYearMonth}) = 0.945$. An equivalent process is done to calculate $S(\text{gYearMonth}, \text{date}) = 0.978$. Now, we replace the obtained values in the equation: $S'(\text{date}, \text{gYearMonth}) = 0.5 \times 0.945 + 0.5 \times 0.978 = 0.961$.

The $S'(\text{date}, \text{gYearMonth})$ is replaced by the respective value in the similarity equation to finally have: $sim(\text{date}, \text{gYearMonth}) = 0.497 + 0.20 \times 0.961 = 0.688$.

Table 4. Datatypes similarity using the proposal of [15] and our approach

$Datatype_1$	$Datatype_2$	Similarity value [15]	Our similarity value
date	gYearMonth	0.30	0.69
date	gYear	0.30	0.46
dateTime	duration	0.30	0.37
dateTime	time	0.30	0.53
dateTime	gDay	0.30	0.29
decimal	float	0.30	0.39
double	float	0.30	0.62

Using our approach, the similarity value between `date-gYearMonth` has increased from 0.30 (according [15]) to 0.69. Table 4 compares our approach and [15], with other pairs of datatypes and their respective similarity values. Note that datatypes with attributes in common (e.g., `dateTime` and `time` have in common *time*) have greater similarity value than the ones obtained by [15]. Next section evaluates the accuracy of our approach.

5   Experiments

In order to evaluate our approach, we adopted the experimental set of datatypes proposed in [15], since there is not a benchmark available in the literature for datatype similarity. This set has 20 pairs of datatypes taken from the W3C hierarchy. These pairs were chosen according to three criteria: (i) same branch but at different depth levels (e.g., `int-long`); (ii) different branches with different depth levels (e.g., `string-int`); and (iii) identical pairs (e.g., `int-int`).

In [15], the authors used the human perception as reference values for the 20 pairs. The closer their similarity measure is to the human perception, the better the measure performs. We used the *Human Average* similarity values presented by [15] to benchmark our approach and a new *Human Average-2* dataset that we obtained by surveying 80 persons that have under- and pots-graduate degrees in computer science[3]. We also compared our work with the similarity values obtained from the compatibility table found in [9,28], and with the disjoint/compatible similarity from W3C.

Table 5. Experimental results: for the first and second experiments

Datatype 1	Datatype 2	Work [9] (Cupic)	Work [28]	W3C	Work [15]	Measure [15] + our hierarchy	Our Mea. + our hierarchy	H. Avg. from [15]	Our H. Avg-2
string	normalizedString	0.00	1.00	1.00	0.53	0.40	0.47	0.27	0.77
string	NCName	0.00	1.00	1.00	0.21	0.16	0.29	0.11	0.55
string	hexBinary	0.50	1.00	0.00	0.09	0.09	0.09	0.36	0.23
string	int	0.40	0.25	0.00	0.03	0.05	0.08	0.28	0.13
token	boolean	0.00	0.17	0.00	0.05	0.05	0.05	0.37	0.15
dateTime	time	0.90	1.00	0.00	0.30	0.53	0.53	0.70	0.71
boolean	time	0.00	0.58	0.00	0.09	0.06	0.06	0.04	0.13
int	byte	0.00	1.00	1.00	0.52	0.52	0.52	0.71	0.58
int	long	0.00	1.00	1.00	0.67	0.67	0.73	0.79	0.72
int	decimal	0.00	1.00	0.00	0.29	0.29	0.38	0.59	0.55
int	double	0.00	0.83	0.00	0.12	0.21	0.23	0.51	0.50
decimal	double	0.00	0.83	0.00	0.30	0.53	0.55	0.60	0.72
byte	positiveInteger	0.00	1.00	1.00	0.13	0.13	0.13	0.57	0.49
gYear	gYearMonth	0.00	1.00	0.00	0.30	0.67	0.67	0.65	0.65
int	int	1.00	1.00	1.00	1.00	1.00	1.00	1.00	1.00
string	byte	0.00	0.25	0.00	0.02	0.03	0.03	0.34	0.21
token	byte	0.00	0.25	0.00	0.01	0.01	0.01	0.46	0.24
float	double	0.00	1.00	0.00	0.30	0.62	0.62	0.60	0.75
float	int	0.00	0.83	0.00	0.12	0.16	0.16	0.46	0.47
gYear	negativeInteger	0.00	0.83	0.00	0.03	0.01	0.01	0.02	0.10
CC. wrt. H. Avg [15]		40.33%	38.32%	27.45%	69.45%	80.21%	77.15%	100.0%	-
CC. wrt. our H. Avg-2		29.48%	70.69%	51.09%	83.93%	90.23%	92.39%	-	100.0%

To compare how close are the similarity values to the human perception, we calculate the correlation coefficient (CC) of every work (i.e., [9,15,28], and our approach) with respect to *Human Average* and *Human Average-2*. A higher CC shows that the approach is closer to the human perception (*Human Average* and *Human Average-2*), and viceversa. The CC is calculated as follows:

$$CC = \frac{1}{n-1} \sum_{i=1}^{n} \frac{(x_i - \bar{x})}{\sigma_x} \frac{(y_i - \bar{y})}{\sigma_y}$$

where n is the number of datatype pairs to compare ($n = 20$ in this case), x_i is the similarity value between datatype pair i, and y_i is its respective human average value, $\bar{x}$ and $\bar{y}$ are averages, and σ_x and σ_y are standard deviations with respect to all similarity values x and all human average values y. Results are shown in Table 5.

Since the similarity measures for work [15] and our work depend on the values of α and β, we evaluate the results under different assignments of α and β. To that end, we devised four experiments:

---

[3] Results are available: http://cloud.sigappfr.org/index.php/s/yRRbUQUeHs0NJnW.

1. In the first experiment, we fix $\alpha = \beta = 0.3057$ as chosen by [15], which they report to be the optimal value obtained by experimentation. We calculated the similarity values as in Eq. 1 to: (i) the W3C extended hierarchy [15] (column 6 in Table 5); and (ii) our proposed datatype hierarchy (column 7 in Table 5). We calculated the CC for both scenarios with respect to *Human Average* and *Human Average-2*. With this experiment, we evaluated the quality of our proposed datatype hierarchy.
2. In the second experiment, we fix $\alpha = \beta = 0.3057$ as chosen by [15], but instead of using their measure (Eq. 1), we used our cross-children similarity measure (see Def. 3) with our proposed datatype hierarchy (column 8 in Table 5). We fixed the $\omega = 0.20^8$. With this experiment, we compared the quality of our approach against all other works.
3. In the third experiment, we chose values for α and β from the range $(0, 1]$, with a 0.02 step. In this case, 2010 possibilities were taken into account.
4. The fourth experiment is similar to the third one, except that a smaller step of 0.001 is considered. Therefore, there were 999181 possibilities.

As shown in Table 5, for experiments 1 and 2, we obtained a CC of 80.21% and 77.15% respectively, with respect to the *Human Average*. With respect to our *Human Average-2*, we obtained even better CC (90.23% and 92.39%).

In the third experiment, we obtained our best results for $\alpha = 0.20$ and $\beta = 0.02$, $CC = 82.60\%$ with respect to the *Human Average* (see Table 6(a), row 1). For $\alpha = 0.50$ and $\beta = 0.18$, $CC = 95.13\%$ with respect to our *Human Average-2* (see Table 6(a), row 2). In general, the similarity values generated by our work were closer to both human perception values than the other works (99.90% of the 2010 possible cases).

Similarly, for the fourth experiment, we obtained our best results for $\alpha = 0.208$ and $\beta = 0.034$ with a $CC = 82.76\%$ with respect to the *Human Average* of the work [15] (see Table 6(b), row 1). With respect to our *Human Average-2*, we obtained the best results for $\alpha = 0.476$ and $\beta = 0.165$, with a $CC = 95.26\%$ (see Table 6(b), row 2). In general, the similarity values generated by our work were closer to both human perceptions (99.97% of the 999181 possible cases).

Table 6. Results of the third and fourth experiments

	α	β	CC.
Human Average [15]	0.20	0.02	82,60%
Human Average-2	0.50	0.18	95,13%

(a) Third experiment with step = 0.02

	α	β	CC.
Human Average [15]	0.208	0.034	82.76%
Human Average-2	0.476	0.165	95.126%

(b) Forth experiment with step = 0.001

In conclusion, our approach outperforms all other works that we surveyed by considering a new hierarchy that captures a semantically more meaningful relation among datatypes, in addition to a measure based on cross-children similarity. Note that our work is not exclusive to RDF data; it can be also applied to XML data similarity and XSD/ontology matching.

---

[8] By experimentation, we determined this value as the optimal one.

6   Conclusions

In this paper, we investigated the issue of datatype similarity for the application of RDF matching/integration. In this context, we proposed a new simple datatype hierarchy aligned with the W3C hierarchy, containing additional types to cope with XPath and XQuery requirements in order to ensure an easy adoption by the community. Also, a new datatype similarity measure inspired by the work in [15], is proposed to take into account the cross-children similarity. We evaluated the new similarity measure experimentally. Our results show that our proposal presents a significant improvement, over the other works described in the literature.

We are currently working on extending and evaluating this work to include complex datatypes that can be defined for the Semantic Web. Also, we plan to evaluate the improvement of using our datatype similarity approach in existing matching tools [11,12].

Acknowledgments. This work has been partly supported by FINCyT/ INOVATE PERU - Convenio No. 104-FINCyT-BDE-2014.

References

1. RDF 1.1 Semantics, W3C Recommendation 25 February 2014. https://www.w3.org/TR/rdf11-mt/#literals-and-datatypes
2. XML Schema Datatypes in RDF and OWL, W3C Working Group Note 14 March 2006. https://www.w3.org/TR/swbp-xsch-datatypes/#sec-values
3. Al-Bakri, M., Fairbairn, D.: Assessing similarity matching for possible integration of feature classifications of geospatial data from official and informal sources. Int. J. Geogr. Inf. Sci. **26**(8), 1437–1456 (2012)
4. Algergawy, A., Nayak, R., Saake, G.: XML schema element similarity measures: a schema matching context. In: Meersman, R., Dillon, T., Herrero, P. (eds.) OTM 2009. LNCS, vol. 5871, pp. 1246–1253. Springer, Heidelberg (2009). doi:10.1007/978-3-642-05151-7_36
5. Algergawy, A., Nayak, R., Saake, G.: Element similarity measures in xml schema matching. Inf. Sci. **180**(24), 4975–4998 (2010)
6. Algergawy, A., Schallehn, E., Saake, G.: A sequence-based ontology matching approach. In: Proceedings of European Conference on Artificial Intelligence, Workshop on Contexts and Ontologies, pp. 26–30 (2008)
7. Algergawy, A., Schallehn, E., Saake, G.: Improving XML schema matching performance using prufer sequences. Data Knowl. Eng. **68**(8), 728–747 (2009)
8. Amarintrarak, N., Runapongsa, S., Tongsima, S., Wiwatwattana, N.: SAXM: semi-automatic xml schema mapping. In: Proceedings of International Technical Conference on Circuits/Systems, Computers and Communications, pp. 374–377 (2009)
9. Bernstein, P.A., Madhavan, J., Rahm, E.: Generic schema matching with cupid. Technical report MSR-TR-2001-58, pp. 1–14. Microsoft Research(2001)
10. Cruz, I.F., Antonelli, F.P., Stroe, C.: Agreementmaker: efficient matching for large real-world schemas and ontologies. Proc. VLDB **2**(2), 1586–1589 (2009)
11. Do, H.-H., Rahm, E.: Coma: a system for flexible combination of schema matching approaches. In: Proceedings of VLDB, pp. 610–621 (2002)

12. Eidoon, Z., Yazdani, N., Oroumchian, F.: Ontology matching using vector space. In: Macdonald, C., Ounis, I., Plachouras, V., Ruthven, I., White, R.W. (eds.) ECIR 2008. LNCS, vol. 4956, pp. 472–481. Springer, Heidelberg (2008). doi:10.1007/978-3-540-78646-7_45

13. Euzenat, J., Shvaiko, P. (eds.): Ontology Matching, vol. 18. Springer-Verlag New York Inc., New York (2007)

14. Hanif, M.S., Aono, M.: An efficient and scalable algorithm for segmented alignment of ontologies of arbitrary size. J. Web Semant. **7**(4), 344–356 (2009)

15. Hong-Minh, T., Smith, D.: Hierarchical approach for datatype matching in xml schemas. In: 24th British National Conference on Databases, pp. 120–129 (2007)

16. Hu, W., Qu, Y., Cheng, G.: Matching large ontologies: a divide-and-conquer approach. Data Knowl. Eng. **67**(1), 140–160 (2008)

17. Jean-Mary, Y.R., Shironoshita, E.P., Kabuka, M.R.: Ontology matching with semantic verification. Web Semant. **7**(3), 235–251 (2009)

18. Jiang, J.J., Conrath, D.W.: Semantic similarity based on corpus statistics and lexical taxonomy. In: Proceedings of Conference on Research in Computational Linguistics, pp. 1–15 (1997)

19. Jiang, S., Lowd, D., Dou, D.: Ontology matching with knowledge rules. CoRR, abs/1507.03097 (2015)

20. Lambrix, P., Tan, H.: Sambo-a system for aligning and merging biomedical ontologies. Web Semant. **4**(3), 196–206 (2006)

21. Li, J., Tang, J., Li, Y., Luo, Q.: RiMOM: a dynamic multistrategy ontology alignment framework. Trans. Knowl. Data Eng. **21**(8), 1218–1232 (2009)

22. Mukkala, L., Arvo, J., Lehtonen, T., Knuutila, T., et al.: Current state of ontology matching. A survey of ontology and schema matching. Technical report 4, University of Turku, pp. 1–18 (2015)

23. Nayak, R., Tran, T.: A progressive clustering algorithm to group the XML data by structural and semantic similarity. Int. J. Pattern Recogn. Artif. Intell. **21**(04), 723–743 (2007)

24. Nayak, R., Xia, F.B.: Automatic integration of heterogenous XML-schemas. In: Proceedings of Information Integration and Web Based Appslications & Services, pp. 1–10 (2004)

25. Ngo, D., Bellahsene, Z.: Overview of YAM++(not) yet another matcher for ontology alignment task. Web Semant.: Sci. Serv. Agents WWW **41**, 30–49 (2016)

26. Stoilos, G., Stamou, G., Kollias, S.: A string metric for ontology alignment. In: Proceedings of International Conference on the SW, pp. 624–637 (2005)

27. Thang, H.Q., Nam, V.S.: Xml schema automatic matching solution. Comput. Electr. Autom. Control Inf. Eng. **4**(3), 456–462 (2010)

28. Thuy, P.T., Lee, Y.-K., Lee, S.: Semantic and structural similarities between XML schemas for integration of ubiquitous healthcare data. Pers. Ubiquitous Comput. **17**(7), 1331–1339 (2013)

SJoin: A Semantic Join Operator to Integrate Heterogeneous RDF Graphs

Mikhail Galkin[1,2,5(✉)], Diego Collarana[1,2], Ignacio Traverso-Ribón[3],
Maria-Esther Vidal[2,4], and Sören Auer[1,2]

[1] Enterprise Information Systems (EIS), University of Bonn, Bonn, Germany
{galkin,collaran,auer}@cs.uni-bonn.de
[2] Fraunhofer Institute for Intelligent Analysis and Information Systems (IAIS),
Sankt Augustin, Germany
vidal@cs.uni-bonn.de
[3] FZI Research Center for Information Technology, Karlsruhe, Germany
traverso@fzi.de
[4] Universidad Simón Bolívar, Caracas, Venezuela
[5] ITMO University, Saint Petersburg, Russia

Abstract. Semi-structured data models like the Resource Description Framework (RDF), naturally allow for modeling the same real-world entity in various ways. For example, different RDF vocabularies enable the definition of various RDF graphs representing the same drug in Bio2RDF or Drugbank. Albeit semantically equivalent, these RDF graphs may be syntactically different, i.e., they have distinctive graph structure or entity identifiers and properties. Existing data-driven integration approaches only consider syntactic matching criteria or similarity measures to solve the problem of integrating RDF graphs. However, syntactic-based approaches are unable to *semantically* integrate heterogeneous RDF graphs. We devise SJoin, a semantic similarity join operator to solve the problem of matching *semantically equivalent* RDF graphs, i.e., syntactically different graphs corresponding to the same real-world entity. Two physical implementations are proposed for SJoin which follow blocking or non-blocking data processing strategies, i.e., RDF graphs can be merged in a batch or incrementally. We empirically evaluate the effectiveness and efficiency of the SJoin physical operators with respect to baseline similarity join algorithms. Experimental results suggest that SJoin outperforms baseline approaches, i.e., non-blocking SJoin incrementally produces results faster, while the blocking SJoin accurately matches all *semantically equivalent* RDF graphs.

1 Introduction

The support that Open Data and Semantic Web initiatives have received from the society has resulted in the publication of a large number of publicly available datasets, e.g., United Nations Data[1] or Linked Open Data cloud[2] allows for

---

[1] http://data.un.org/.
[2] http://stats.lod2.eu/.

© Springer International Publishing AG 2017
D. Benslimane et al. (Eds.): DEXA 2017, Part I, LNCS 10438, pp. 206–221, 2017.
DOI: 10.1007/978-3-319-64468-4_16

accessing billion of records. In the context of the Semantic Web, the Resource Description Framework (RDF) is utilized for semantically enriching data with vocabularies or ontologies. Albeit expressive, the RDF data model allows (e.g., due to the non-unique names assumption) multiple representations of a real-world entity using different vocabularies.

To illustrate this, consider chemicals and drugs represented in the Drugbank and DBpedia knowledge graphs. Using different vocabularies, drugs are represented from different perspectives. DBpedia contains more general information, whereas Drugbank provides more domain-specific facts, e.g., the chemical composition and properties, pharmacology, and interactions with other drugs. Figure 1 illustrates representations of two drugs in Drugbank and DBpedia. *Ibuprofen*, a drug for treating pain, inflammation and fever, and *Paracetamol*, a drug with analgesic, and antipyretic effects. Firstly, Drugbank Uniform Resource Identifiers (URIs) are textual IDs (e.g., drugbank:DB00316[3] corresponds to *Acetaminophen* and drugbank:DB01050 to *Ibuprofen*. In contrast, DBpedia utilizes human-readable URIs (e.g., dbr:Acetaminophen and dbr:Ibuprofen) to identify drugs. Secondly, the same attributes are encoded differently with various property URIs, e.g., chemicalIupacName, casRegistryNumber in Drugbank, and iupacName, casNumber in DBpedia, respectively. Thirdly, some drugs might be linked to more than one analogue, e.g., Acetaminophen in Drugbank (drugbank:DB00316) corresponds to two DBpedia resources: dbr:Paracetamol, and dbr:Acetaminophen.

Traditional join operators, e.g., *Hash Join* [2] or *XJoin* [11], are not capable of joining those resources as neither URIs nor properties match syntactically. Similarity join operators [3,5,6,8,12] tackle this heterogeneity issue, but due to the same extent of inequality string and set similarity techniques are limited in

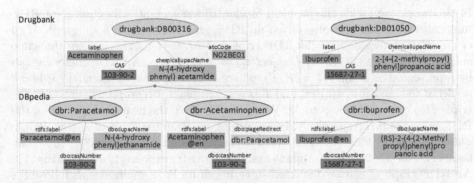

Fig. 1. Motivating Example. The Ibuprofen and Paracetamol real-world entities are modeled in different ways by Drugbank and DBpedia. Syntactically the properties and objects are different, but semantically the represent the same drugs. Drug drugbank:DB01050 matches 1-1 with dbr:Ibuprofen, while drugbank:DB00316 matches 1-2 with dbr:Paracetamol and dbr:Acetaminophen.

---

[3] Prefixes are as specified on http://prefix.cc/.

deciding whether two RDF resources should be joined or not. Therefore, we identify the need of a semantic similarity join operator able to satisfy the following requirements: **(R1)** Applicable to heterogeneous RDF knowledge graphs. **(R2)** Able to identify joinable tuples leveraging semantic relatedness between RDF graphs. **(R3)** Capable of performing perfect matching for one-to-one integration, and fuzzy conditional matching for integrating groups of N entities from one graph with M entities from another knowledge graph. **(R4)** Support of a blocking operation mode for batch processing, and a non-blocking mode for on-demand real time cases whenever results are expected incrementally.

We present *SJoin* – a semantic join operator which meets these requirements. The contributions of this article include: **(1)** Definition and description of SJoin, a semantic join operator for integrating heterogeneous RDF graphs. **(2)** Algorithms and complexity study of a blocking SJoin for $1 - 1$ integration and non-blocking SJoin for the $N - M$ similarity case. **(3)** An extensive evaluation that demonstrates benefits of SJoin in terms of efficiency, effectiveness and completeness over time in various heterogeneity conditions and confidence levels.

The article is organized as follows: The problem addressed in this work is clearly defined in Sect. 2. Section 3 presents the SJoin operator, as well as the blocking and non-blocking physical implementations, as solutions for detecting semantically equivalent entities in RDF knowledge graphs. Results from our experimental study are reported on Sect. 4. An overview of traditional binary joins and similarity joins as a related work is analyzed in Sect. 5. Finally, we sum up the lessons learned and outline future research directions in Sect. 6.

2   Problem Statement

In this work, we tackle the problem of identifying semantically equivalent RDF molecules from RDF graphs. Given an RDF graph G, we call a subgraph M of G an *RDF molecule* [4] iff the RDF triples of $M = \{t_1, \ldots, t_n\}$ share the same subject, i.e., $\forall\, i, j \in \{1, .., n\}$ $(subject(t_i) = subject(t_j))$. An RDF molecule can be represented as a pair $\mathcal{M} = (R, T)$, where R corresponds to the URI (or blank node) of the molecule subject, and T is a set of pairs $p=(prop, val)$ such that the triple *(R,prop,val)* belongs to M. We name R and T the head and the tail of the RDF molecule $\mathcal{M}$, respectively. For example, an RDF molecule of a drug *Paracetamol* is (dbr:Paracetamol, {(rdfs:label,"Paracetamol@en"), (dbo:cas Number,"103-90-2"),(dbo:iupacName,"N-(4-hydroxyphenyl)ethanamide")}). An RDF graph G can be described in terms of its RDF molecules as follows:

$$\phi(G) = \{\mathcal{M} = (R, T) | t = (R, prop, val) \in G \text{ and } (prop, val) \in T\} \tag{1}$$

Definition 1 (Problem of Semantically Equivalent RDF Graphs). *Given sets of RDF molecules $\phi(G)$, $\phi(D)$, and $\phi(F)$, and an RDF molecule $\mathcal{M}_e$ in $\phi(F)$ which corresponds to an entity e represented by different RDF molecules $\mathcal{M}_G$ and $\mathcal{M}_D$ in $\phi(G)$ and $\phi(D)$, respectively. The problem of identifying semantically equivalent entities between sets of RDF molecules $\phi(G)$ and $\phi(D)$*

consists of providing an homomorphism $\theta : \phi(G) \cup \phi(D) \to 2^{\phi(F)}$, *such that if two RDF molecules* $\mathcal{M}_G$ *and* $\mathcal{M}_D$ *represent the RDF molecule* $\mathcal{M}_e$, *then* $\mathcal{M}_e \in \theta(\mathcal{M}_G)$ *and* $\mathcal{M}_e \in \theta(\mathcal{M}_D)$; *otherwise,* $\theta(\mathcal{M}_G) \neq \theta(\mathcal{M}_D)$.

Definition 1 considers perfect 1-1 matching, e.g., determining 1-1 semantic equivalences between `drugbank:01050` and `dbr:Ibuprofen`, as well as $N - M$ matching, e.g., `drugbank:DB00316` with both `dbr:Paracetamol` and `dbr:Acetaminophen`.

3   Proposed Solution: The SJoin Operator

We propose a similarity join operator named SJoin, able to identify joinable entities between RDF graphs, i.e., SJoin implements the homomorphism $\theta(.)$. SJoin is based on the *Resource Similarity Molecule (RSM)* structure, that in combination with a *similarity function* Sim_f, and a *threshold* γ, produce a list of matching entity pairs. RSM is defined as follows:

Definition 2 (Resource Similarity Molecule (RSM)). *Given a set* $\mathbb{M}$ *of RDF molecules, a similarity function* Sim_f, *and a threshold* γ. *A Resource Similarity Molecule is a pair* $RSM=(\mathcal{M},T)$, *where:*

- $\mathcal{M} = (R,T)$ *is the head of RSM and the RDF molecule described in RSM.*
- T *is the tail of RSM and represents an ordered list of RDF molecules* $\mathcal{M}_i = (R_i, T_i)$. *T meets the following conditions:*
  - $\mathcal{M}$ *is highly similar to* $\mathcal{M}_i$, *i.e.,* $Sim_f(R, R_i) \geq \gamma$.
  - *For all* $\mathcal{M}_i = (R_i, T_i) \in T$, $Sim_f(R, R_i) \geq Sim_f(R, R_{i+1})$.

An RSM is composed of a head and tail that correspond to an RDF molecule and a list of molecules which similarity score is higher than a specified threshold γ, respectively. For example, an RSM of *Ibuprofen* (with omitted tails of *property:value* pairs) is `((dbr:Ibuprofen, `T`)[(drugbank:DB01050, `T_1`)`, `(chebi:5855, `T_2`)`, `(wikidata:Q186969, `T_3`)])` given a similarity function Sim_f, a threshold γ, and Sim_f`(dbr:Ibuprofen,drugbank:DB01050)` $\geq Sim_f$`(dbr:Ibuprofen,chebi:5855)`, and Sim_f`(dbr:Ibuprofen,chebi:5855)` $\geq Sim_f$`(dbr:Ibuprofen,wikidata:Q186969)`.

The SJoin operator is a two-fold algorithm that performs: first, *Similarity Partitioning*, and second, *Similarity Probing* to identify semantically equivalent RDF molecules. To address batch and real-time processing scenarios, we present two implementations of SJoin. **Blocking SJoin Operator** solves the 1-1 weighted perfect matching problem allowing for a batch processing of the graphs. **Non-Blocking SJoin Operator** employs fuzzy conditional matching for identifying communities of N-M entities in graphs covering the on-demand case whenever results are expected to be produced incrementally.

3.1   Blocking SJoin Operator

Figure 2 illustrates the intuition behind the blocking SJoin operator. Similarity Partitioning and Probing steps are executed sequentially. Thus, blocking SJoin operator completely evaluates both datasets of RDF molecules in the Partitioning step, and then fires the Probing step to produce the whole output.

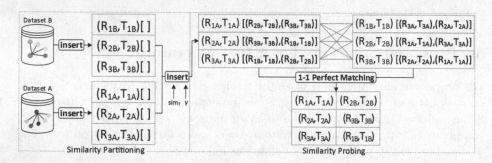

Fig. 2. SJoin Blocking Operator. Similarity Partitioning step initializes lists of RSMs and populates their tails through a similarity function Sim_f and a threshold γ. Similarity Probing step performs 1-1 weighted perfect matching and outputs the perfect pairs of semantically equivalent molecules $(\mathcal{M}_{iA}, \mathcal{M}_{jB})$.

The Similarity Partitioning step is described in Algorithm 1. The operator initializes two lists of RSMs for two RDF graphs and incoming RDF molecules are inserted into a respective list with a filled head $\mathcal{M}$ and empty tail T. To populate the tail of a RSM in the list A, SJoin resorts to a semantic similarity function for computing a similarity score between the RSM and all RSMs in the opposite list B. If the similarity score exceeds a certain threshold γ then the molecule from the list B is appended to the tail of the RSM. Finally, the tail is sorted in the descending similarity score order such that the most similar RDF molecule obtains the top position in the tail. For instance, the semantic similarity function GADES [10] is able to decide relatedness between the RDF molecules of dbr:Ibuprofen and drugbank:DB01050 in Fig. 1, and assigns a similarity score of 0.8. The algorithm supports datasets with arbitrary amounts of molecules. However, in order to guarantee 1-1 perfect matching, we place a restriction $card(\phi(D_A)) = card(\phi(D_B))$, i.e., the number of molecules in $\phi(D_A)$ and $\phi(D_B)$ must be the same. Thus, $card$(List of RSM_A) $= card$(List of RSM_B).

A 1-1 weighted perfect matching is applied at the Similarity Probing stage in the Blocking SJoin operator. It accepts the lists of RSM_A, RSM_B created and populated during the previous Similarity Partitioning step. This step aims at producing perfect pairs of semantically equivalent RDF molecules $(\mathcal{M}_{iA}, \mathcal{M}_{jB})$, i.e., $max(Sim_f(\mathcal{M}_{iA}, RSM_B)) = max(Sim_f(\mathcal{M}_{jB}, RSM_A)) = Sim_f(\mathcal{M}_{iA}, \mathcal{M}_{jB})$. That is, for a given molecule $\mathcal{M}_{iA}$, there is no molecule in the list of RSM_A which has a similarity score higher than $Sim_f(\mathcal{M}_{iA}, \mathcal{M}_{jB})$ and vice versa. Algorithm 2 describes how perfect pairs are created; Fig. 3 illustrates the algorithm.

Algorithm 1. Similarity Partitioning step for Blocking SJoin operator according to similarity function Sim_f and threshold γ

Data: Dataset $\phi(D_A)$, Sim_f, γ
Result: List of RSM_A, List of RSM_B
1 **while** *getMolecule($\phi(D_A)$)* **do**
2     $\mathcal{M}_{iA} \leftarrow$ getMolecule($\phi(D_A)$) ;
3     $R_{iA} \leftarrow head(\mathcal{M}_{iA})$;   // Get URI
4   **for** $RSM_{jB} \in$ List of RSM_B **do**
5        $RSM_{jB} = ((R_{jB}, T_{jB})[(R_{lA}, T_{lA})), \dots, (R_{kA}, T_{kA})]$;
6        $R_{jB} \leftarrow head(head(RSM_{jB}))$;   // Get URI
7   **if** $Sim_f(R_{jB}, R_{iA}) \geq \gamma$ **then**   // Probe
8           $tail(RSM_{jB}) \leftarrow tail(RSM_{jB}) + (\mathcal{M}_{iA})$;
9 **return** $sort($List of $RSM_A)$,$sort($List of $RSM_B)$

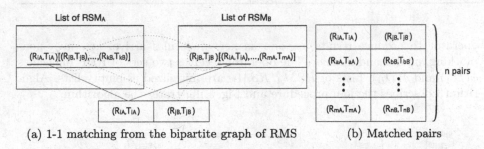

(a) 1-1 matching from the bipartite graph of RMS   (b) Matched pairs

Fig. 3. 1-1 Weighted Perfect Matching. (a) The matching is identified from the lists of RSM_A and RSM_B; RDF molecules $\mathcal{M}_{iA} = (R_{iA}, T_{iA})$ and $\mathcal{M}_{jB} = (R_{jB}, T_{jB})$ are semantically equivalent whenever R_{iA} and R_{jB} are reciprocally the most similar RDF molecules according to Sim_f.

Traversing the List of RSM_A, the algorithm iterates over each RSM_{iA}. Then, the tail of RSM_{iA}, i.e., an ordered list of *highly* similar molecules, is extracted. The first molecule of the tail RSM_{jB} corresponds to the most similar molecule from the List of RSM_B. The algorithm searches for RSM_{jB} in the List of RSM_B and examines whether the molecule (R_{iA}, T_{iA}) is the first one in the tail of RSM_{jB}. If this condition holds and (R_{iA}, T_{iA}) is not already matched with another RSM, then the pair $((R_{iA}, T_{iA}), (R_{jB}, T_{jB}))$ is identified as a perfect pair and is appended to the result list of pairs LP (cf. Fig. 3a). If false, then the algorithm finds the first occurrence of (R_{iA}, T_{iA}) in the tail of RSM_{jB} and appends the result pair to LP. When all RSMs are matched, the algorithm yields the list of perfectly matched pairs (cf. Fig. 3b).

3.2 Non-blocking SJoin Operator

The Non-Blocking SJoin operator aims at identifying $N - M$ matchings, i.e., an RSM_{iA} might be associated with multiple RSMs, e.g., RSM_{jB} or RSM_{kB}. Therefore, 1-1 weighted perfect matching is not executed which enables the

---

Algorithm 2. 1-1 Weighted Perfect Matching of RSMs bipartite graph

---

Data: List of RSM_A, List of RSM_B
Result: List of pairs $LP = ((R_{iA}, T_{iA}), (R_{jB}, T_{jB}))$

```
1  for RSM_iA ∈ List of RSM_A do
2      RSM_iA = ((R_iA, T_iA)[(R_jB, T_jB), ..., (R_kB, T_kB)]) ;          // Ordered Set
3      for (R_jB, T_jB) ∈ tail(RSM_iA) do
4          RSM_jB ← Find in the List of RSM_B ;
5          RSM_jB = ((R_jB, T_jB)[(R_lA, T_lA), ..., (R_zA, T_zA)]) ;      // Ordered Set
6          if (R_lA, T_lA) = (R_iA, T_iA) and (R_iA, T_iA) ∉ LP then
7              LP ← LP + ((R_iA, T_iA), (R_jB, T_jB)) ;                    // Add to result
8          else
9              for (R_lA, T_lA) ∈ tail(RSM_jB) do
10                 find the position of (R_iA, T_iA) ;
11 return LP
```

---

operator to produce results as soon as new molecules arrive, i.e., in a non-blocking, on-demand manner. The operator receives two sets of RDF molecules $\phi(D_A)$ and $\phi(D_B)$. Lists of RSM_A, RSM_B are initialized as empty lists. Algorithm 3 describes the join procedure and Fig. 4 illustrates the algorithm.

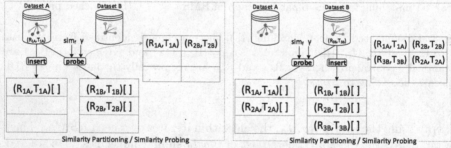

(a) Molecule (R_{iA}, T_{iA}) yields a pair $((R_{1A}, T_{1A}), (R_{2B}, T_{2B}))$

(b) Molecule (R_{3B}, T_{3B}) yields a pair $((R_{3B}, T_{3B}), (R_{2A}, T_{2A}))$

Fig. 4. SJoin Non-Blocking Operator. Identifies N-M matchings and produces results as soon as new molecule arrives. When a molecule (R_{iA}, T_{iA}) arrives, it is inserted into a relevant list and probed against another list. If the similarity score exceeds the threshold γ, a new matching is produced.

For every incoming molecule $\mathcal{M}_{iA}$ from $\phi(D_A)$, Algorithm 3 performs the same two steps: *Similarity Partitioning* and *Similarity Probing*. The URI R_{iA} of an RDF molecule extracted from the tuple (R_{iA}, T_{iA}) is probed against URIs of *all* existing $RSMs$ in the List of RSM_B (cf. Fig. 4). If the similarity score of $Sim_f(R_{iA}, \dot{R}_{jB})$ exceeds the threshold γ, then the pair $((R_{iA}, T_{iA}), (R_{jB}, T_{jB}))$ is considered as a matching and appended to the results list LP. During the *Similarity Insert* step, an RSM_{iA} is initialized, the molecule (R_{iA}, T_{iA}) becomes

---

Algorithm 3. The Non-Blocking SJoin operator executes both Similarity Partitioning and Probing steps as soon as an RDF molecule arrives from an RDF graph.

Data: Dataset $\phi(D_A)$, Sim_f, γ
Result: List of pairs $LP = ((R_{iA}, T_{iA}), (R_{jB}, T_{jB}))$

```
1  while getMolecule(φ(D_A)) do
2  │   M_iA ← getMolecule(φ(D_A)) ;
3  │   R_iA ← head(M_iA), T_iA ← tail(M_iA) ;                    // Get URI, tail
4  │   for RSM_jB ∈ List of RSM_B do
5  │   │   RSM_jB = ((R_jB, T_jB)[]) ;
6  │   │   R_jB ← head(head(RSM_jB)) ;                           // Get URI
7  │   │   T_jB ← tail(head(RSM_jB)) ;                           // Get tail
8  │   │   if Sim_f(R_iA, R_jB) ≥ γ then                        // Probe
9  │   │   │   LP ← LP + ((R_iA, T_iA), (R_jB, T_jB)) ;
10 │   head(RSM_iA) ← M_iA, tail(RSM_iA) ← [] ;
11 │   List of RSM_A ←List of RSM_A + RSM_iA ;                   // Insert
12 return LP
```

---

its head, and eventually added to the respective List of RSM_A. Algorithm 3 is applied to both $\phi(D_A)$ and $\phi(D_B)$ and able to produce results with constantly updating Lists of RSMs supporting the non-blocking operation workflow.

3.3 Time Complexity Analysis

The SJoin binary operator receives two RDF graphs of n RDF molecules each. To estimate the complexity of the blocking SJoin operator, three most expensive operations have to be analyzed. Table 1 gives an overview of the analysis. The complexity of the Data Partitioner module depends on the Algorithm 1, i.e., construction of Lists of RSM_A, RSM_B and a similarity function Sim_f. The asymptotic approximation equals to $O(n^2 \cdot O(Sim_f))$. To produce ordered tails of RSMs the similar molecules in the tail have to be sorted in the descending similarity score order. The applicable merge sort and heapsort algorithms have $O(n \log n)$ asymptotic complexity. The 1-1 Weighted Perfect Matching component has $O(n^3)$ complexity in the worst case according to the Algorithm 2. However, the Hungarian algorithm [7], a standard approach for 1-1 weighted perfect matching, converges to the same $O(n^3)$ complexity. Partitioning, sorting, and perfect matching are executed sequentially. Therefore, the overall complexity conforms to the sum of complexities, i.e., $O(n^2 \cdot O(Sim_f)) + O(n \log n) + O(n^3)$ which equals to $O(n^2 \cdot O(Sim_f)) + O(n^3)$. We thus deduce that the *SJoin* complexity depends on the complexity of a chosen similarity measure whereas the lowest achievable order of complexity is limited to $O(n^3)$.

The complexity of the non-blocking SJoin operator stems from the analysis of the Algorithm 3. The most expensive step of the algorithm is to compute a similarity score between an RSM_{iA} and RSMs in the List of RSM_B. Applied to both $\phi(D_A)$ and $\phi(D_B)$ the complexity converges to $O(n^2 \cdot O(Sim_f))$.

Table 1. The *SJoin* Time Complexity. Results for the steps of Partitioning, Sorting, and Matching, where n is the number of RDF molecules.

Stage	Blocking SJoin Complexity	Non-blocking SJoin Complexity
Partitioning	$O(n^2 \cdot O(Sim_f))$	$O(n^2 \cdot O(Sim_f))$
Sorting	$O(n \log n)$	
Matching	$O(n^3)$	
Overall	$O(n^2 \cdot O(Sim_f)) + O(n^3)$	$O(n^2 \cdot O(Sim_f))$

4   Empirical Study

An empirical evaluation is conducted to study the efficiency and effectiveness of SJoin in blocking and non-blocking conditions on RDF graphs from DBpedia and Wikidata. We assess the following research questions: **(RQ1)** Does blocking SJoin integrate RDF graphs more efficiently and effectively compared to the state of the art? **(RQ2)** What is the impact of threshold values on the completeness of a non-blocking SJoin? **(RQ3)** What is the effect of a similarity function in the SJoin results? The experimental configuration is as follows:

Benchmark: Experiment 1 is executed against a dataset of 500 molecules[4] of type Person extracted from the live version of DBpedia (February 2017). Based on the original molecules, we created two sets of molecules by randomly deleting or editing triples in the two sets. Sharing the same DBpedia vocabulary, Experiment 1 datasets have a higher resemblance degree compared to Experiment 2. Experiment 2 employs subsets of DBpedia and Wikidata of the Person class. Assessing SJoin in the higher heterogeneity settings, we sampled datasets of 500 and 1000 molecules varying triples count from 16 K up to 55 K[5]. Table 2 provides basic statistics on the experimental datasets. DBpedia D1 and D2 refer to the dumps of 500 molecules. Further, the dumps of 500 and 1000 molecules for Experiment 2 are extracted from DBpedia and Wikidata.

Table 2. Benchmark Description. RDF datasets used in the evaluation.

	Experiment 1: people		Experiment 2: people			
	DBpedia D1	DBpedia D2	DBpedia	Wikidata	DBpedia	Wikidata
Molecules	500	500	500	500	1000	1000
Triples	17,951	17,894	29,263	16,307	54,590	29,138

---

[4] https://github.com/RDF-Molecules/Test-DataSets/tree/master/DBpedia-People/20160819.

[5] https://github.com/RDF-Molecules/Test-DataSets/tree/master/DBpedia-WikiData/operators_evaluation.

Baseline: Gold standards for blocking operators comparison include the original DBpedia Person descriptions (Experiment 1) and `owl:sameAs` links between DBpedia and Wikidata (Experiment 2). We compare SJoin with a Hash Join operator. For a fair comparison, the Hash Join was extended to support similarity functions at the *Probing* stage. That is, blocking SJoin is compared against blocking similarity Hash Join and non-blocking SJoin is evaluated against non-blocking Symmetric Hash Join. The Gold standard for evaluating non-blocking operators is comprised of the precomputed amounts of pairs which similarity score exceeds a predefined threshold; gold standards are computed off line.

Metrics: We report on execution time (ET in secs) as the elapsed time required by the SJoin operator to produce all the answers. Furthermore, we measure *Precision*, *Recall* and report *F1-measure* during the experiments with blocking operators. Precision is the fraction of RDF molecules that has been identified and integrated (M) that intersects with the Gold Standard (GS), i.e., $Precision = \frac{|M \cap GS|}{|M|}$. Recall corresponds to the fraction of the identified similar molecules in the Gold Standard, i.e., $Recall = \frac{|M \cap GS|}{|GS|}$. Comparing non-blocking operators, we measure *Completeness* over time, i.e., a fraction of results produced at a certain time stamp. The timeout is set to one hour (3,600 s), the operators results are checked every second. Ten thresholds in the range [0.1 : 1.0] and step 0.1 were applied in Experiment 1. In Experiment 2, five thresholds in the range [0.1 : 0.5] were evaluated because no pair of entities in the sampled RDF datasets has a GADES similarity score higher than 0.5.

Implementation: Both blocking and non-blocking SJoin operators are implemented in Python 2.7.10[6]. Baseline improved Hash Joins are implemented in Python as well[7]. The experiments were executed on a Ubuntu 16.04 (64 bits) Dell PowerEdge R805 server, AMD Opteron 2.4 GHz CPU, 64 cores, 256 GB RAM. We evaluated two similarity functions: GADES [10] and Semantic Jaccard (SemJaccard) [1]. GADES relies on semantic descriptions encoded in ontologies to determine relatedness, while SemJaccard requires the materialization of implicit knowledge and mappings. Evaluating schema heterogeneity of DBpedia and Wikidata in Experiment 2 the similarity function is fixed to GADES.

4.1   DBpedia – DBpedia People

Experiment 1 evaluates the performance and effectiveness of blocking and non-blocking SJoin compared to respective Hash Join implementations. The testbed includes two split DBpedia dumps with semantically equivalent entities but non-matching resource URIs and randomly distributed properties; GADES and SemJaccard similarity functions. That is, both graphs are described in terms of one DBpedia ontology. Figure 5 visualizes the results obtained when applying GADES semantic similarity function in order to identify a perfect matching of graphs resources, i.e., in blocking conditions. SJoin exhibits better F1 score up to

---

[6] https://github.com/RDF-Molecules/operators/tree/master/mFuhsion.

[7] https://github.com/RDF-Molecules/operators/tree/master/baseline_ops.

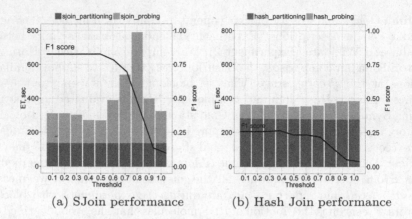

(a) SJoin performance   (b) Hash Join performance

Fig. 5. Experiment 1 (GADES) with blocking operators. The *partitioning* bar shows the time taken to partition the molecules in RSMs, *probing* indicates the time required for 1-1 weighted perfect matching. Black line chart on the right axis denotes F1 score. (a) SJoin demonstrates higher F1 score while consuming more time for perfect matching. (b) Baseline Hash Join demonstrates less than 0.25 F1 score even on lower thresholds spending less time on probing.

very high 0.9 threshold value. Moreover, the effectiveness of more than 80% is ensured up to 0.6 threshold value whereas Hash Join barely reaches 25% even on lower thresholds. The partitioning time is constant for both operators but Hash Join performs the partitioning slower due to the application of a hash function to all incoming molecules. However, high effectiveness of SJoin is achieved at the expense of time efficiency. SJoin has to complete a 1-1 perfect matching algorithm against a large 500×500 matrix whereas Hash Join performs the perfect matching *three* times but for smaller matrices equal to the size of its buckets, e.g., about 166×166 for three buckets which is faster due to the cubic complexity of the weighted perfect matching algorithm.

Figure 6 shows the results of the evaluation of non-blocking operators with GADES. SJoin outperforms the baseline Hash Join in terms of completeness over time in all four cases with the threshold in the range 0.1–0.8. Figure 6a demonstrates that the SJoin operator is capable of producing 100% of results within the timeframe whereas the Hash Join operator outputs only about 10% of the expected tuples. In Fig. 6b, SJoin achieves the full completeness even faster. In Fig. 6c both operators finish after 18 min, but SJoin retains full completeness while Hash Join reaches only 35%. Finally, with the 0.8 threshold in Fig. 6d, Hash Join performs very fast but still struggles to attain the full completeness; SJoin takes more time but sustainably achieves answer completeness. One of the reasons why Hash Join performs worse is its hash function which does not consider semantics encoded in the molecules descriptions. Therefore, the hash function partitions RDF molecules into buckets almost randomly, while it was originally envisioned to place similar entities in the same buckets.

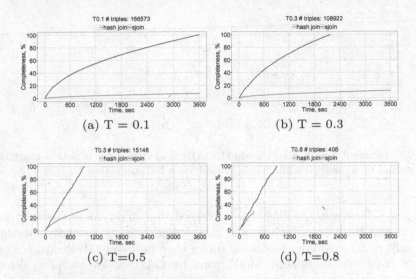

Fig. 6. Experiment 1 (GADES) with non-blocking operators. SJoin produces complete results at all threholds in contrast to Hash Join.

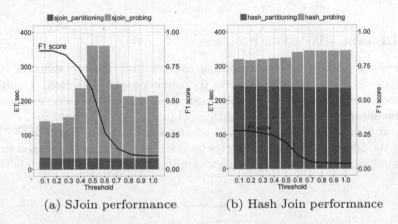

(a) SJoin performance   (b) Hash Join performance

Fig. 7. Experiment 1 (SemJaccard) with blocking operators. (a) SJoin takes less time to compute similarity scores while F1 score quickly deteriorates after threshold 0.5. (b) Baseline Hash Join in most cases consumes more time and produces less reliable matchings.

Figure 7 presents the efficiency and effectiveness of blocking SJoin and Hash Join when applying SemJaccard similarity function. As an unsophisticated measure, operators require less time for partitioning and take less time for probing stages. That is, due to the heterogeneous nature of the compared datasets, SemJaccard is not able to produce similarity scores higher than 0.4. On the other hand, SemJaccard simplicity leads to significant deterioration of the F1 score already at low thresholds, i.e., 0.3–0.4.

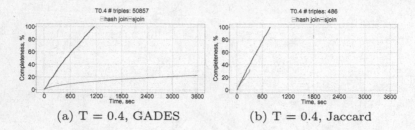

(a) T = 0.4, GADES   (b) T = 0.4, Jaccard

Fig. 8. Experiment 1 with fixed threshold. GADES identifies two orders of magnitude more results than Jaccard while SJoin still achieves full completeness.

Figure 8 illustrates the difference in elapsed time and achieved completeness of SJoin and Hash Join applying GADES or SemJaccard similarity functions. Evidently, SemJaccard outputs fewer tuples even on lower thresholds, e.g., 486 pairs at 0.4 threshold against 50,857 pairs by GADES. We therefore demonstrate that plain set similarity measures as SemJaccard that consider only an intersection of exactly same triples are ineffective in integrating heterogeneous RDF graphs.

4.2   DBpedia - Wikidata People

The distinctive feature of the experiment consists in completely different vocabularies used to semantically describe the same people. Therefore, traditional joins and set similarity joins, e.g., Jaccard, are not applicable. We evaluate the performance of SJoin employing GADES semantic similarity measure.

Figure 9 reports the efficiency and effectiveness of SJoin compared to Hash Join in the 500 molecules setup. Figure 9a justifies the range of selected thresholds as only a few number of pairs have a similarity score higher than 0.5. Block-

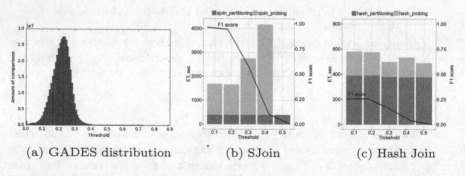

(a) GADES distribution   (b) SJoin   (c) Hash Join

Fig. 9. Experiment 2 (GADES) with blocking operators, 500 molecules. (a) The distribution of GADES similarity scores shows that there are few pairs which score exceeds 0.4 threshold. (b) SJoin requires more time but achieves more than 0.9 F1 score until T0.3. (c) Baseline Hash Join operates faster but achieves less than 0.25 F1 accuracy.

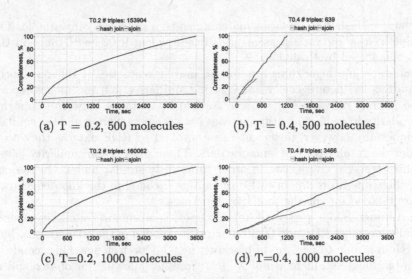

(a) T = 0.2, 500 molecules   (b) T = 0.4, 500 molecules

(c) T=0.2, 1000 molecules   (d) T=0.4, 1000 molecules

Fig. 10. Experiment 2. Non-blocking operators in different dataset sizes. In larger setups, SJoin still reaches full completeness.

ing SJoin manages to achieve higher F1 score (max 95%) up to 0.3 threshold value, but requires significantly more time to accomplish the perfect matching.

Results of non-blocking SJoin and Hash Join executed against 500 and 1000 molecules configurations are reported on Fig. 10. The observed behavior of these operators resembles the one in Experiment 1, i.e., SJoin outputs complete results within a predefined time frame, while Hash Join barely achieves 40% completeness in the case with a relatively high threshold 0.4 and small number of outputs.

Analyzing the observed empirical results, we are able to answer our research questions: (**RQ1**) Blocking SJoin consistently exhibits higher F1 scores, and the results are more reliable. However, time efficiency depends on the input graphs and applied similarity functions. (**RQ2**) A threshold value prunes the amount of expected results and does not affect the completeness of SJoin. (**RQ3**) Clearly, a semantic similarity function allows for matching RDF graphs more accurately.

5   Related Work

Traditional binary join operators require join variables instantiations to be exactly the same. For example, XJoin [11] and Hash Join [2] (chosen as a baseline in this paper) operators abide this condition. At the *Insert* step, both blocking and non-blocking Hash Join algorithms partition incoming tuples into a number of buckets based on the assumption that after applying a hash function similar tuples will reside in the same bucket. The assumption holds true in cases of simple data structures, e.g., numbers or strings. However, applying hash functions to string representations of complex data structures such as RDF molecules or RSMs tend to produce more collisions rather then efficient partitions. At the

Probe stage, Hash Join performs matching as to a specified join variable. Thus, having URI as a join variable, semantically equivalent RSMs with different URIs can not be joined by Hash Join.

Similarity join algorithms are able to match syntactically different entities and address the heterogeneity issue. String similarity join techniques reported in [3,5,12] rely on various metrics to compute a distance between two strings. Set similarity joins [6,8] identify matches between sets. String and set similarity techniques are, however, inefficient being applied to RDF data as they do not consider the graph nature of semantic data. There exist graph similarity joins [9, 13] which traverse graph data in order to identify similar nodes. On the other hand, those operators do not tackle semantics encoded in the knowledge graphs and are tailored for specific similarity functions.

In contrast, SJoin, presented in this paper, is a semantic similarity operator that fully leverages RDF and OWL semantics encoded in the RDF graphs. Moreover, SJoin is able to perform in blocking, i.e., 1-1 perfect matching, conditions or non-blocking, i.e., incremental $N - M$, manner allowing for on-demand and ad-hoc semantic data integration pipelines. Additionally, SJoin is flexible and is able to employ various similarity functions and metrics, e.g., from simple *Jaccard* similarity to complex *NED* [14] or *GADES* [10] measures, achieving best performance with semantic similarity functions.

6   Conclusions and Future Work

We presented SJoin, an operator for detecting semantically equivalent RDF molecules from RDF graphs. SJoin implements two operators: Blocking and Non-Blocking, which rely on similarity measures and ontologies to effectively detect equivalent entities from heterogeneous RDF graphs. Moreover, the time complexity of SJoin operators depends on the time complexity of the similarity measure, i.e., SJoin does not introduce additional overhead. The behavior of SJoin was empirically studied on DBpedia and Wikidata real-world RDF graphs, and on Jaccard and GADES similarity measures. Observed results suggest that SJoin is able to identify and merge semantically equivalent entities, and is empowered by the semantics encoded in ontologies and exploited by similarity measures. As future work, we plan to define new SJoin operators to compute on-demand integration of RDF graphs and address streams of RDF data.

Acknowledgments. Mikhail Galkin is supported by the project Open Budgets (GA 645833). This work is also funded in part by the European Union under the Horizon 2020 Framework Program for the project BigDataEurope (GA 644564), and the German Ministry of Education and Research with grant no. 13N13627 (LiDaKra).

References

1. Collarana, D., Galkin, M., Lange, C., Grangel-Gonzàlez, I., Vidal, M.-E., Auer, S.: FuhSen: a federated hybrid search engine for building a knowledge graph on-demand (short paper). In: Debruyne, C., et al. (eds.) OTM 2016. LNCS, vol. 10033, pp. 752–761. Springer, Cham (2016). doi:10.1007/978-3-319-48472-3_47

2. Deshpande, A., Ives, Z.G., Raman, V.: Adaptive query processing. Found. Trends Databases **1**(1), 1–140 (2007)
3. Feng, J., Wang, J., Li, G.: Trie-join: a trie-based method for efficient string similarity joins. VLDB J. **21**(4), 437–461 (2012)
4. Fernández, J.D., Llaves, A., Corcho, O.: Efficient RDF interchange (ERI) format for RDF data streams. In: Mika, P., et al. (eds.) ISWC 2014. LNCS, vol. 8797, pp. 244–259. Springer, Cham (2014). doi:10.1007/978-3-319-11915-1_16
5. Li, G., Deng, D., Wang, J., Feng, J.: Pass-join: apartition-based method for similarity joins. PVLDB **5**(3), 253–264 (2011)
6. Mann, W., Augsten, N., Bouros, P.: An empirical evaluation of set similarity join techniques. PVLDB **9**(9), 636–647 (2016)
7. Munkres, J.: Algorithms for the assignment and transportation problems. J. Soc. Ind. Appl. Math. **5**(1), 32–38 (1957)
8. Ribeiro, L.A., Cuzzocrea, A., Bezerra, K.A.A., do Nascimento, B.H.B.: Incorporating clustering into set similarity join algorithms: the *SjClust* framework. In: Hartmann, S., Ma, H. (eds.) DEXA 2016. LNCS, vol. 9827, pp. 185–204. Springer, Cham (2016). doi:10.1007/978-3-319-44403-1_12
9. Shang, Z., Liu, Y., Li, G., Feng, J.: K-join: knowledge-aware similarity join. IEEE Trans. Knowl. Data Eng. **28**(12), 3293–3308 (2016)
10. Traverso, I., Vidal, M.-E., Kämpgen, B., Sure-Vetter, Y.: Gades: a graph-based semantic similarity measure. In: SEMANTiCS, pp. 101–104. ACM (2016)
11. Urhan, T., Franklin, M.J.: Xjoin: a reactively-scheduled pipelined join operator. IEEE Data Eng. Bull. **23**(2), 27–33 (2000)
12. Wandelt, S., Deng, D., Gerdjikov, S., Mishra, S., Mitankin, P., Patil, M., Siragusa, E., Tiskin, A., Wang, W., Wang, J., Leser, U.: State-of-the-art in string similarity search and join. SIGMOD Rec. **43**(1), 64–76 (2014)
13. Wang, Y., Wang, H., Li, J., Gao, H.: Efficient graph similarity join for information integration on graphs. Front. Comput. Sci. **10**(2), 317–329 (2016)
14. Zhu, H., Meng, X., Kollios, G.: NED: an inter-graph node metric based on edit distance. PVLDB **10**(6), 697–708 (2017)

Choosing Data Integration Approaches Based on Data Source Characterization

Julio Cesar Cardoso Tesolin[✉] and Maria Cláudia Cavalcanti

Instituto Militar de Engenharia, Rio de Janeiro, RJ, Brazil
jcctesolin@gmail.com, yoko@ime.eb.br

Abstract. The Big Data era is an inevitable consequence of our capacity to generate and collect data. Therefore, data sources got a more dynamic behavior and the challenge to manage data source's integration process and the network traffic between data producers and consumers has been overwhelmed by the number of data sources and their content's volume, by the variety of their structures and formats and by the velocity of their appearance. This work presents a method to help users in choosing an appropriate approach (materialization or virtualization) for each data source in an integration environment.

Keywords: Data integration · ETL process · Hybrid architecture

1 Introduction

Our current capacity to generate and collect data changed data sources' dynamic behaviour in three important aspects: volume, variety and velocity. Therefore, new challenges came up and need our attention. For instance, a poor data integration approach selection may lead to unnecessary network data traffic or a data integration process that only uses human intervention might not be appropriated for a data intensive environment.

As stated by Doan *et al.* [1], data integration is not a trivial task and a good data source characterization is fundamental to choose the right integration approach (materialization or virtualization). Several works have already proposed solutions to select an appropriated approach for a given data source based on its characteristics [2–5]. However, some do not support a dynamic approach selection [2–4], while others do not use a broader set of characteristics [5].

In this sense, this work proposes a method to choose an appropriated data integration approach for each data source based on its own characterization and on the integration environment characteristics. It also contributes on the development of a hybrid data integration architecture, based on the proposed method, capable to select the proper approach dynamically.

2 Related Work

There were some early attempts that sought a way to select a proper integration approach for each data source in an integration environment [2,3]. Both proposed a hybrid data integration architecture, where part of the data sources is

D. Benslimane et al. (Eds.): DEXA 2017, Part I, LNCS 10438, pp. 222–229, 2017.
DOI: 10.1007/978-3-319-64468-4_17

materialized and the other part is virtualized. Some recent works [4,5], focus on data sources' characterization to select a proper integration approach in their environment. In [4], each data source is analyzed using each one of the characteristics from the proposed set, and weights are assigned to them. A final score is produced in order to select a proper integration approach for it. A similar work [5], goes a step further, using a fuzzy logic decision system to choose the appropriate integration approach. The developed system is able to dynamically change the integration approach for each data source, whenever a significant change in their characteristics set is observed.

This work differs from the others in some points: **(i)** the concept of data source's *virtualization capacity* is presented; **(ii)** some new characteristics are presented in comparison with the other works; **(iii)** the proposed integration architecture considers that the integration approach can change over the integration environment lifetime; **(iv)** it focuses on the data traffic in the integration environment and the amount of human intervention needed to manage it.

3   Data Source Characterization - Static Aspects

The hybrid integration architecture works [2–5] discuss little about data source's nature and its characteristics. Based on that, a generic data integration process cycle was used as a starting point, formed by two entities: a data provider (P) and a data consumer (C) (Fig. 1). In a traditional integration environment, the data source (DS) plays the role of a data provider and the consuming system (CS) plays the role of a data consumer. But the integration solution (IS) can play both roles: it acts as a data consumer when seen by the data source and as data provider when seen by the consuming system.

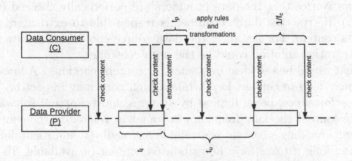

Fig. 1. Generic integration process cycle

The static aspects investigation is based on Amit Sheth's classic article [6] about system's heterogeneity. The syntactic heterogeneity is related to the format of the data provider's and how it is represented. In this case, the data provider can only be virtualized if its content is recognized (machine readable) by the data consumer. The structural heterogeneity has two aspects to be studied: the existence of a logical data model to represent the content and the schema compatibility between data providers and the data consumers. The first one deals with the data consumer's capacity to understand the logical model, a necessary condition to allow virtualization. The second one deals with the transformation complexity and if the data consumer is not able to deal with it in query execution time or it takes too long, the content has to be materialized. Finally, the semantic heterogeneity is linked to the meaning incompatibility between data provider's and data consumer's metadata. As in the structural heterogeneity, the issue is related to the capacity and the necessary processing time to adjust data provider's content. For the systemic heterogeneity, [6] gives a wide definition, while [7] treat it in a more practical way. For them, data providers have two different behaviours: they can be passive (data consumer needs to check for a new content) or active (a new content is sent by the data provider). Other aspect highlighted by them is the data provider's capacity to answer a query. Different from their behaviour, the capacity of a data provider to answer a query is tied to the data consumer's capacity to deal with it.

So, based on the previous analysis, a data provider can only be virtualized if the data consumer is capable to deal with all these heterogeneity in query execution time. This aspect is hereafter called as *virtualization capacity*.

4   Introducing Dynamic Aspects

Despite of many characteristics being brought and treated by [6, 7], some other aspects were not. Dynamic aspects, such as the data sources' volume and their content update frequency, are also important when designing a integration solution. Figure 1 represents, in a simplified way, the dynamics of the integration environment. Notice that the data providers are periodically checked for a new content (f_c). If true, the data consumer is responsible to extract, manipulate and hold its content. For instance, the data source could be the data provider and the integration solution could be the data consumer.

But, what would be the ideal frequency for content checking? A low checking frequency may lead to content loss, while a high pace may impact on the data consumer performance, being limited by systems' hardware and software capabilities. The content checking frequency is somehow related to the content's life time (t_l), i.e., how long a content is available for readings, and its update period (t_u), i.e., how long it takes for a new updated content be available. To the best of the authors' knowledge, the literature does not address it. So, to overcame this gap, this work proposes an analogy with classic experiments in statistics and probability. Considering this analogy, the binomial distribution stands out as a widely used distribution to characterize a repetition of n independent tests. It sets the probability of how many times successful k tests occur after n tests, being described by the formula $p(x = k) = \binom{n}{k}p^k q^{n-k}$.

The content's checking frequency(f_c) represents the tests that should be done to establish its existence. The success probability is the ratio between the content life time and the time between content's update, while the fail probability is the complement over the total probability. It is worth to be noted that, for this model, it is imposed that a new content checking has to be done between content's update in order not to create queues over time to process it. It is clear from Fig. 1 that it needs just one success on sampling to realize that a content exists, what can be mathematically expressed as $p(x > 0)$. As $p(x > 0)$ can be rewritten as $1 - p(x \leq 0) = 1 - p(x = 0)$, it means that by lowering $p(x = 0)$, it will increase the chances to happen at least one success on finding a content between its updates. So, k can be replaced by 0 in the original formula, which provides $p(x = 0) = \binom{n}{0} p^0 q^{n-0} = q^n$.

Replacing q as the ratio of t_l and t_u, the previous equation can be represented as $p(x = 0) = q^n = (1 - t_l * f_u)^n$, since $1/t_u$ can be expressed as the content's update frequency(f_u). Once the objective is not to discover the loss probability ($p(x = 0)$), but how many times a checking has to be made between content's update in order to realize its existence at least one time, n can be derived as

$$n = \lceil \frac{\ln(p(x = 0))}{\ln(1 - t_l * f_u)} \rceil^1 \tag{1}$$

So, once $p(x = 0)$ is given, it is possible to return the value of n that ensures that at least one sampling will be done between content's update. However, it is necessary to analyze a mathematical inconsistency introduced by the logarithmic function in order to extract n. It happens when the content life time (t_l) is equal or greater than the time between content's update (t_u). Although this condition creates negative or null logarithmic, it can be noticed that it will always be possible to find it in this scenario, i.e., the probability to find it is 100%, being enough to the checking frequency (f_c) be at least equal to the content's update frequency. Therefore, considering also that $f_c = n * f_u$, Eq. (1) can be extended and improved like this:

$$n = \begin{cases} 1 & , t_l \geqslant t_u \\ \lceil \frac{\ln(p(x = 0))}{\ln(1 - t_l * f_u)} \rceil & , t_l < t_u \end{cases} \tag{2}$$

$$f_c^{max} \geqslant f_c \geqslant n * f_u \tag{3}$$

Nevertheless, looking back at the interface between the data provider and the data consumer (Fig. 1), two other cases need to be explored: the *transport time* (t_p) and the *processing time* (t_T). The *transport time* is the amount of time needed to move the data provider's content to the data consumer and it has to be considered specially at the edge of content's lifetime (t_l). Once that is possible that the process to remove data provider's content at the same time a request reaches it, leading to an unstable state, it is reasonable to establish that the transport time has to be deducted from the content's lifetime. On the other hand, the *processing time* is the amount of time needed to make data provider's content compatible with the data consumer schema. As this operation has to be done in query time in the virtualization approach, the processing time should be taken in consideration as part of the transport time. On the other hand, in the materialization approach, the data consumer first ingests data provider's content as it is and then the transformation is done in a different step. Therefore, the processing time does not contribute to the transport time in this approach.

Measuring content's transport and processing times exposed different challenges. Although measuring the transport time looks simple, it depends on the data consumer's administration scope. When the role of the data consumer is played by a consuming system, measuring the transport time from there might not be possible. To overcome this limitation, another investigation was done and it was noticed that content's transport time can be estimated as a product of the content's volume (v_{cnt}) and the network latency (t_{lat}), leading to $t_T = v_{cnt} * t_{lat}$. On the other side, measuring processing time does not depend on the administration scope and can explain other factors like hardware and software restrictions or transformation complexities more easier using just one metric.

Taking all the previous discussions about the possible dynamic characteristics that can change the integration process into account, n can be represented as

$$n_C = \begin{cases} 1 & , (t_l^P - t_T^{PC}) \geqslant t_u^P \\ \lceil \dfrac{\ln(p(x=0)^C)}{\ln(1 - (t_l^P - t_T^{PC}) * f_u^P)} \rceil & , (t_l^P - t_T^{PC}) < t_u^P \end{cases} \tag{4}$$

$$f_c^{C,max} \geqslant f_c^C \geqslant n_C * f_u^P \tag{5}$$

where

t_l^P : content's life time, P $\in$ {DS,IS},
t_u^P : interval between content's updates, P $\in$ {DS,IS},
f_u^P : content's update frequency, P $\in$ {DS,IS},
f_c^C : checking frequency, C $\in$ {IS,CS},
t_T^{PC} : transport time between P and C, PC $\in$ {DSIS,ISCS,DSCS},
$p(x=0)^C$: loss probability, C $\in$ {IS,CS}

A data source is a candidate to be virtualized if Eq. (5) is fulfilled. Once the consuming system's checking frequency (f_c^{CS}) is considered out of the administration scope, the data source's virtualization will be dependent on its content's lifetime (t_l^{DS}), update frequency (f_u^{DS}) and volume (v_{cnt}) as well as the network latency between data source and consuming system (t_{lat}^{DSCS}) and the processing time (t_p) needed to transform the content as described in Eq. (4). If the data source has also the *virtualization capacity*, this approach can be used and the integration solution in this configuration acts only as a connection element, being transparent to the data source and the consuming system.

However, if the data source can not fulfill Eq. (5), it has to be materialized. In this case, the integration solution plays both roles - a data consumer for the data sources and a data provider for the consuming systems. As a data consumer, its checking frequency (f_c^{IS}) can be adjusted once it is in the administration scope and will be dependent on the same aspects described later, besides the processing time (t_p). On the other hand, as a data provider, its content's lifetime (t_l^{IS}) can be adjusted based on the consuming system's checking frequency (f_c^{CS}). Both adjustments can be achieved using Eq. (4).

5   FlexDI - A Hybrid Data Integration Architecture

FlexDI is a hybrid data integration architecture that supports the dynamic approach selection described in Sects. 3 and 4 for each data source in the integration environment. Inspired by other works [2,3,5], it reproduces traditional hybrid integration solutions, but also presents a control module capable to track the aspects that enable a proper integration approach selection and effectively choose it (Fig. 2).

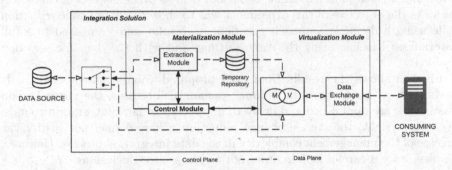

Fig. 2. Integration environment using FlexDI architecture

Tracking those aspects, for each data source and its contents, is done at several moments during the integration process. For instance, the content's virtualization capacity is defined when a data source is registered in the integration solution while the content's lifetime (t_l) aspect is set by measuring how long it is available for consumption in the data source repository. The content's update frequency (f_u), is set when a new content is discovered in the data source, measuring the time lapse since the last time a new content was found, while the checking frequency (f_c) from the consuming systems is measured by monitoring every request for a particular data source's content. The best integration approach selection for each data source is done at the end of each integration cycle. After measuring the relevant aspects, the historical collection for each aspect of each data source is represented by a unique value, that can be represented in many ways, such as historical averages, medians, moving averages etc. Once represented, all aspects are gathered and used in a decision flow that implements Eqs. (4) and (5).

Another detail that should be noticed in FlexDI is the logical union between virtualized and materialized contents of the same data source in the mediator. As it is possible that some of the contents that come from a specific data source are saved into the integration temporary repository, while others are kept in its original repository, two descriptions are built for the same data source in the mediator: one for the original data source and other for the temporary repository connection. Once it is done, these two descriptions are united to provide a unique view of the data source's content. This union not only provides a single content's presentation, but also ease the application of rules and transformations that should be done regardless the integration approach being used.

6   Evaluation

The test was planned to evaluate a thirty day integration process in a twelve hours experiment, transforming TPC-DI's data sources as described by their benchmarking process [8]. For such, all data sources and the consuming system were configured to show a dynamic behaviour using a predefined set of characteristics. As the objective of this experiment was to show the data traffic reduction while using a dynamic approach selection, its results were compared to a full materialized baseline using the same settings and with the dynamic selection disabled.

Figure 3 shows the results from each proposed volume based scenario. It shows that, as far as the total volume exchanged increases, the virtualization possibilities are diminished and delayed, happening in later integration rounds. So, at first sight, the data exchange reduction achieved may not justify the increase of the management complexity in the data integration process. However, it is necessary a careful look in order not to jump into conclusions.

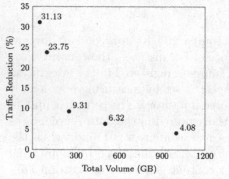
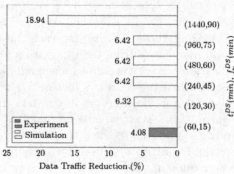

Fig. 3. Volume based scenarios   **Fig. 4.** 1000 GB simulation

Considering that the experiment should have being completed in a reasonable monitoring period (12 h), the content's lifetime (t_l) and update frequency (f_u) were set to achieve it. But these settings could have influenced the final result. In order to evaluate this assumption, a software simulation was done using the 1000 GB scenario data and a linear projection of the content's lifetime and update frequency. Figure 4 shows better results when these characteristics are greater than the first setting.

7   Conclusions

This paper proposed a method to choose the most suitable integration approach for a data source based on its characteristics and on the integration environment characteristics that they take part. Although the data traffic between peers was reduced in all volume based scenarios using a hybrid data integration architecture (FlexDI), the virtualization possibilities were diminishing as more automatic adjustments were done to keep a proper integration solution checking frequency while the data sources' volume were raising. Nevertheless, a simulation showed better results for large data volumes, when using different settings.

The collected evidences suggest that it is possible to create a method to dynamically select the best integration approach for a given data source based on its static and dynamic aspects in order to reduce data traffic and minimize human intervention in the integration process. An analysis regarding to memory and processing consumption in the integration solution and an evaluation with a different set of data sources were left for further investigations.

References

1. Doan, A., Halevy, A.Y., Ives, Z.G.: Principles of Data Integration. Morgan Kaufmann, Burlington (2012)
2. Zhou, G., Hull, R., King, R., Franchitti, J.C.: Supporting data integration and warehousing using H2O. IEEE Data Eng. **18**, 29–40 (1995)
3. Voisard, A., Jürgens, M.: Geospatial information extraction: querying or quarrying? In: Goodchild, M., Egenhofer, M., Fegeas, R., Kottman, C. (eds.) Interoperating Geographic Information Systems. ISECS, vol. 495. Springer, Boston (1999). doi:10.1007/978-1-4615-5189-8_14
4. Bichutskiy, V.Y., Colman, R., Brachmann, R.K., Lathrop, R.H.: Heterogeneous biomedical database integration using a hybrid strategy: a p53 cancer research database. Cancer Inform. **2**, 277–287 (2006)
5. Hadi, W., Zellou, A., Bounabat, B.: A fuzzy logic based method for selecting information to materialize in hybrid information integration system. Int. Rev. Comput. Softw. **8**(2), 489–499 (2013)
6. Sheth, A.: Changing focus on interoperability in information systems: from system, syntax, structure to semantics. In: Goodchild, M., Egenhofer, M., Fegeas, R., Kottman, C. (eds.) Interoperating Geographic Information Systems. ISECS, vol. 495, pp. 5–29. Springer, Boston (1999)
7. Hohpe, G., Woolf, B.: Enterprise Integration Patterns: Designing, Building, and Deploying Messaging Solutions. Addison-Wesley, Boston (2003)
8. Poess, M., Rabl, T., Jacobsen, H.A., Caufield, B.: TPC-DI: the first industry benchmark for data integration. Proc. VLDB Endow. **7**(13), 1367–1378 (2014)

Security and Privacy (I)

Towards Privacy-Preserving Record Linkage with Record-Wise Linkage Policy

Takahito Kaiho, Wen-jie Lu[⊠], Toshiyuki Amagasa, and Jun Sakuma

University of Tsukuba, Ten' nohdai 1-1-1, Tsukuba, Japan
{takahito,riku}@mdl.cs.tsukuba.ac.jp, {amagasa,jun}@cs.tsukuba.ac.jp

Abstract. We consider a situation that a large number of individuals contribute their personal data to multiple databases, and an analyst is allowed to issue a request of record linkage across any two of these databases. In such a situation, one concern is that very detailed information about individuals can be eventually obtained by analysts through sequential requests of record linkage across various database pairs. To resolve this privacy concern, we introduce a novel privacy notion for record linkage, record-wise linkage policy, with which each data contributor can individually designate database pairs that are allowed (resp. banned) to process record linkage by whitelist (resp. blacklist). We propose a secure multi-party computation to achieve record linkage with record-wise linkage policies. Also, we prove that our protocol can securely evaluate PPRL without violating given linkage policies and evaluate the efficiency of our protocol with experiments.

Keywords: Security · Privacy · Record linkage · Equijoin · Privacy policy

1 Introduction

Record linkage is recognized as an important tool for integration of information distributed over multiple data sources and has been extensively studied for decades [4]. When records contain private information of individuals, consideration of privacy is necessary to process record linkage so that it does not cause privacy violation. Suppose two databases exist and each record in the databases is associated with a single individual. Then, the aim of privacy-preserving record linkage (PPRL) is to identify pairs of records associated with the same individual without sharing identifiers or private attributes across two databases [22]. Some applications of PPRL are presented in [15,22]. In this paper, we consider privacy concerns that can appear when PPRL is performed in the following situation:

1. A large number of individuals contribute their personal data to multiple databases (e.g., transactions of bank account, medical records of hospitals), in which unique identifiers (UID, e.g., social security numbers) are agreed among all the databases, and UIDs are attached to all the records, and

© Springer International Publishing AG 2017
D. Benslimane et al. (Eds.): DEXA 2017, Part I, LNCS 10438, pp. 233–248, 2017.
DOI: 10.1007/978-3-319-64468-4_18

2. An analyst is allowed to issue a request of PPRL across any pair of two databases; databases jointly evaluate PPRL and provide the result (de-identified joined table) to the analyst.

When a large number of databases participate in this framework, the analyst might eventually learn detailed information about individuals by sequentially requesting PPRL across all pairs databases. Even if the records in the joined table are de-identified, the risk of identification increases as records from more databases are linked. The basic assumption of this study is that demand for privacy protection in the process of record linkage would be individually different from each other. For example, suppose databases of a bank, credit card company, hospital, and health insurance company agreed to provide tables that are joinable using unique identifiers to an analyst. Alice might dislike that medical records and health insurance records are linked to avoid privacy violation, but might accept record linkage across all the other pairs of databases to utilize online medical services; Bob might accept record linkage of her bank account with credit card histories to utilize online financial services, but might refuse record linkage across all the other pairs of databases (See Fig. 1). It should be noticed that existing PPRL methods do not offer such fine-grained privacy control.

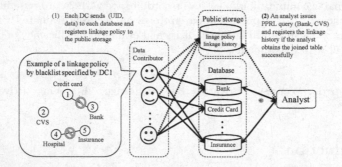

Fig. 1. Information flow among stake holders of PPRL considering linkage policies

To reconcile individually different demands for privacy protection and benefit derived from data integration, we introduce a novel privacy notion for record linkage, *record-wise linkage policy* (or linkage policy, for short), to the problem of PPRL. Each data contributor specifies pairs of databases with which record linkage is allowed (whitelist) and/or not allowed (blacklist) as her linkage policy. In the example above, to satisfy their demand for privacy, Alice can register pair (bank, credit card company) to her *blacklist* as his linkage policy; Bob can register (hospital, health insurance company) to his *whitelist* as his linkage policy. The aim of this paper is to state the problem PPRL over record-wise linkage policies and introduce a PPRL protocol with satisfying all linkage policies that are individually configured by data contributors.

1.1   Related Work

There are three major approaches for PPRL. Secure multi-party computation is a cryptographic building block that enables to perform a specified computation with private data distributed over multiple information sources without sharing them. Existing PPRL methods based on secure multiparty computation solve the problem by securely matching records without sharing identifiers or attributes, for example, by using secure set intersection [14]. Solutions with secure multiparty computation give a theoretical guarantee of security and correctness of the results at the same time; however, the computation can be relatively slow compared to non-cryptographic solutions. Inan et al. introduced a hybrid approach to attain both efficiency and theoretical guarantee of privacy by introducing k-anonymity into secure multiparty computation [11].

Similarity preserving data transformation is an alternative approach that transforms object identities or attributes into a distance space in that distance-based matching can be efficiently performed [2]. Yakout et al. presented a protocol that maps vectors into the imaginary space and performs record linkage in the space [25]. Scannapieco et al. showed a method that embeds strings into a vector space [20]. In these methods, data transformations are designed so that the original identifiers or attributes cannot be recovered from the transformed points for privacy protection. The data transformation methodology does not contain cryptographic operations, and thus achieves better performance, while the guarantee of security is not provable [23].

Given two databases with size n and m, naive PPRL, where each pair of records are examined, requires pairwise matching $O(nm)$ times. Several recent studies on PPRL consider improvements of efficiency by introducing efficient distance-based matching [25], weakening of the security model in a controlled way [16], reduction of pairwise matching using blocking [8], to name a few. In principle, the typical goal of PPRL is to obtain the result of record linkage without sharing private identities or attributes across two databases. To the best of our knowledge, no literature has discussed PPRL that considers record-wise linkage policies configured by data contributors independently.

1.2   Our Contribution

Our contribution is two-fold. First, we introduce a novel privacy notion, linkage policy, for PPRL. We consider two different types of linkage policy representation: whitelist and blacklist. Also, we define the security model of PPRL with linkage policies. Second, we propose a secure multi-party computation to compute PPRL with linkage policies and prove the security of our protocol.

The rest of this paper is organized as follows. In Sect. 2, we formalize our problem and security model. In Sect. 3, we introduce cryptographic tools that we use in our method. In Sect. 4, we propose cryptographic protocols which securely computes PPRL with linkage policies. The security proof of the proposed protocol and the asymptotic analysis are also shown. In Sect. 5, we experimentally measured the executing time and bandwidth consumption. Finally, we conclude the paper and give future work issues in Sect. 6.

2 Record Linkage Considering Linkage Policy

2.1 Preparation

Stakeholders. We assume four stakeholders exist: data contributors, analysts, databases, and a public storage. Figure 1 describes the relationship between the four stakeholders and linkage policies. All data contributors deposit their records to the databases and linkage policies to the public storage. The linkage policies describe the permission of linking the records with other databases. A database holds records collected from data contributors. An analyst issues a request of record linkage with specifying two databases to obtain linked records across the two databases. A public storage stores linkage policies and linkage histories. When receiving a request from the analyst, the two databases and the analyst jointly evaluate record linkage so that the linkage policies are satisfied. If and only if the request is valid, the databases provide the resulting linked records to the analyst and register the linkage history to the public storage.

Notation. For a positive integer Z, we write $[Z]$ to indicate the set $\{1, 2, \ldots, Z\}$. We write $z \in_r [Z]$ to mean that z is chosen uniformly at random from set $[Z]$. We use a bold upper case Roman character to denote a matrix, e.g., $\mathbf{W}$. We write $(\mathbf{W})_i$ to denote the ith row of the matrix, and write $(\mathbf{W})_{ij}$ to denote the element in the position of ith row and jth column. A vector is written in a bold lower case Roman character, e.g., $\boldsymbol{u}$. The jth element of vector $\boldsymbol{u}$ is given by u_j. A polynomial is written in an upper case calligraphic character, e.g., $\mathcal{F}$. The jth coefficient of polynomial $\mathcal{F}$ is given by $\mathcal{F}[j]$.

We write C to denote the number of data contributors and write D to denote the number of databases. We denote the number of records stored in the ith database as C_i. We assume that $C_i \leq C$ for all databases. Let x_{ik} be the record of the kth data contributor in the ith database for $k \in [C_i]$. We denote the set of records of the kth data contributor as $x_k = \{x_{ik}\}_{i \in [D]}$. Also, the set of the records of the ith database is represented by $\mathcal{DB}_i = \{x_{ik}\}_{k \in [C_i]}$.

Record Linkage. A unique identifier (UID) is assigned to each data contributor that is consistent among all databases. For example, social security numbers or pairs of name and birthday can work as the unique identifier. We leverage a mapping function $\pi : \{1, 1\}^* \to \mathcal{UID}$ where $\mathcal{UID}$ denotes the domain of the UID and $\{1, 1\}^*$ denotes an arbitrary length binary string. We suppose $\mathcal{UID}$ is a set of integers without loss of generality. Each database independently holds the mapping function π_i where $i \in [C_i]$. With this mapping, the unique identifier of x_{ij} held by the ith database is specified as $\pi_i(x_{ik})$. Intuitively, π_i works as the correspondence table for the ith database that associates records and UIDs.

Notice that x_{ik} contains the attribute value only, but does not contain UID in it. The record with the unique identifier is represented by the pair, $(\pi_i(x_{ik}), x_{ik})$. We require the mapping functions satisfy the following two conditions:

1. If the data contributor of x_{ik} and the data contributor of $x_{j\ell}$ are the same individual, $\pi_i(x_{ik}) = \pi_j(x_{j\ell})$ holds,
2. $\pi_i(x_{ik}) \neq \pi_i(x_{ik'})$ if and only if $k \neq k'$. That is, the same individual does not appear twice in every database.

We use a predicate Match which is defined as follows.

$$\texttt{Match}(x_{ik}, x_{j\ell}) = \begin{cases} \text{true} & \text{if } \pi_i(x_{ik}) = \pi_j(x_{j\ell}) \\ \text{false} & \text{otherwise.} \end{cases}$$

Match returns true only when the records x_{ik} and $x_{j\ell}$ are collected from the same data contributor. We define the join of the databases $\mathcal{DB}_i$ and $\mathcal{DB}_j$ as:

$$\mathcal{DB}_i \bowtie \mathcal{DB}_j = \{x_{ik} \| x_{j\ell} : \texttt{Match}(x_{ik}, x_{j\ell}) = \text{true}, \forall x_{ik} \in \mathcal{DB}_i, \forall x_{j\ell} \in \mathcal{DB}_j\},$$

where $\cdot \| \cdot$ denotes the concatenation of two records. Notice that tables are joinable with unique identifiers commonly used. Also, pairs of records and unique identifiers are needed to evaluate function Match to perform join.

2.2 Record Linkage with Linkage Policy

Linkage Policy. A data contributor determines a linkage policy to specify the valid linkage of the databases according to his will. The linkage policy of a data contributor with UID m is represented by an undirected graph $G_m = (\mathcal{V}, \mathcal{E}_m)$, where $\mathcal{V} = \{\mathcal{DB}_j\}_{j=1}^D$ denotes the set of databases and $\mathcal{E}_m$ is the set of undirected edges of the data contributor's linkage policy.

We consider two types of linkage policy, *whilelist* and *blacklist*. Suppose an analyst issues a request of join, $\mathcal{DB}_i \bowtie \mathcal{DB}_j$. For the case of whilelist, $e_{ij} \in \mathcal{E}_m$ means that the data contributor with UID m allows $\mathcal{DB}_i$ and $\mathcal{DB}_j$ to jointly evaluate $x_{ik} \| x_{j\ell}$ where $\pi_i(x_{ik}) = \pi_k(x_{j\ell}) = m$ and provides $x_{ik} \| x_{j\ell}$ to the analyst. On the other hand, in the case of blacklist, $e_{ij} \in \mathcal{E}_m$ indicates that two databases are not allowed to evaluate $x_{ik} \| x_{j\ell}$ where $\pi_i(x_{ik}) = \pi_j(x_{j\ell}) = m$. We designate G_m the *linkage policy graph* for UID m. Since each data contributor can individually configure her policy graph, C linkage policy graphs are given in total. We give formal definitions on the whitelist and blacklist.

Linkage Policy by Whitelist. When the whitelist representation is used, each data contributor specifies pairs of databases in which she wants to perform join. We represent the linkage policy graph G_m^W of a data contributor with UID m for whitelist with an adjacency matrix $\mathbf{W}_m$.

$$(\mathbf{W}_m)_{ij} = \begin{cases} 0 & e_{ij} \in \mathcal{E}_m^W, \text{ linkage allowed across } \mathcal{DB}_i \text{ and } \mathcal{DB}_j, \\ 1 & \text{otherwise.} \end{cases}$$

The join of $\mathcal{DB}_i$ and $\mathcal{DB}_j$ with enforcement of linkage policy $G_1^W, G_2^W, \ldots, G_C^W$ is defined as follows:

$$\mathcal{DB}_i \bowtie_W \mathcal{DB}_j = \{x_{ik} \| x_{j\ell} : \texttt{Match}(x_{ik}, x_{j\ell}) = \text{true}, e_{k\ell} \in \mathcal{E}_m^W, x_{ik} \in \mathcal{DB}_i, x_{j\ell} \in \mathcal{DB}_j\}.$$

For instance, suppose the analyst wants to join $\mathcal{DB}_1$ and $\mathcal{DB}_2$. Then, when the join follows the linkage policy, the table that the analyst obtains contains only the records that follow the specified whitelist.

Linkage Policy by Blacklist. In the blacklist method, each data contributor specifies pairs of databases that he *is unwilling to* permit linkage. Let $G_m^B = (\mathcal{V}, \mathcal{E}_m^B)$ be the linkage policy graph for blacklist of a data contributor with UID m. If $e_{ij} \in \mathcal{E}_m^B$, $\mathcal{DB}_i$ and $\mathcal{DB}_j$ are not allowed to perform record linkage of record of the data contributor with UID m. We define the adjacency matrix of G_m^B by

$$(\mathbf{B}_m)_{ij} = \begin{cases} 1 & e_{ij} \in \mathcal{E}_m, \text{ linkage } prohibited \text{ across } \mathcal{DB}_i \text{ and } \mathcal{DB}_j, \\ 0 & \text{otherwise.} \end{cases}$$

One might think two databases can always perform record linkage of the data contributor with UID m if the edge corresponds to the two databases is not contained in G_m^B. However, in the blacklist case, this is not necessarily true. Suppose there exist four databases and a data contributor with UID m prohibits record linkage across $\mathcal{DB}_1$ and $\mathcal{DB}_3$. So she adds e_{13} to her black list G_m^B. When an analyst requests join $\mathcal{DB}_1 \bowtie \mathcal{DB}_2$, the result can contain $x_{1k}\|x_{2\ell}$ where $\pi_1(x_{1k}) = \pi_2(x_{2\ell}) = m$. After that, suppose the analyst issues a request of join $\mathcal{DB}_2 \bowtie \mathcal{DB}_3$. Linkage between $\mathcal{DB}_2$ and $\mathcal{DB}_3$ is not prohibited by the blacklist; however, if it contains $x_{2\ell}\|x_{3p}$ where $\pi_2(x_{2\ell}) = \pi_3(x_{3p}) = m$ for some p, this causes violation of the blacklist for the following reason. Recall that the analyst has $x_{1k}\|x_{2\ell}$ in $\mathcal{DB}_1 \bowtie \mathcal{DB}_2$ already. If the analyst obtains $x_{2\ell}\|x_{3p}$, $x_{1k}\|x_{2\ell}$ can be readily derived by the analyst by using $x_{2\ell}$ as a key if $x_{2\ell}$ is unique in the database. This is prohibited by the blacklist of the data contributor.

Thus, to prohibit record linkage by blacklist after responding to multiple requests, we need to store the histories of linkage released previously and suppress to release linked records that can cause record linkage prohibited by the blacklist.

To this end, we introduce the *linkage history*, which represents histories of linkage requested by the analyst previously. For now, we do not explicitly explain who holds and manages the linkage histories; we here simply describe only the definition and the property of the linkage histories. We refer matrix $\mathbf{Q}_m$ as the linkage histories about the data contributor with UID m and define $\mathbf{Q}_m$ as

$$(\mathbf{Q}_m)_{ij} = \begin{cases} 1 & \text{if } (x_{ik}\|x_{j\ell}) \text{ is released previously where } \pi_i(x_{ik}) = \pi_j(x_{j\ell}) = m, \\ 0 & \text{otherwise.} \end{cases}$$

Linkage History. The linkage history $\mathbf{Q}_m$ can be interpreted as a graph in which nodes correspond to databases and edges correspond to database pairs with which record linkage is previously conducted. If two databases in the blacklist are not directly connected in $\mathbf{Q}_m$ but become reachable in $\mathbf{Q}_m$ by adding an edge specified by the request, it means that analyst obtains linkage that violates the blacklist by obtaining the response to the request.

In the example above, $\mathbf{Q}_m$ is initialized as the zero matrix, and after releasing $x_{1k}\|x_{2\ell}$ where $\pi_1(x_{1k}) = \pi_2(x_{2\ell})$ to the analyst, $(\mathbf{Q}_m)_{12}$ is set to 1. In the graph, if $(\mathbf{Q}_m)_{23}$ is set to 1, $\mathcal{DB}_1$ and $\mathcal{DB}_3$ becomes reachable, which violates the blacklist. In the blacklist method, we need to suppress possible linkage by checking reachability using the linkage history at each request from analysts.

We can check the possible violation of linkage policy $\mathbf{B}_m$ caused by join request across $\mathcal{DB}_i$ and $\mathcal{DB}_j$ as follows. First, we set $(\mathbf{Q}_m)_{ij} = 1$ and evaluate

$$\mathbf{R}_m = (\mathbf{Q}_m + \mathbf{I}) + \cdots + (\mathbf{Q}_m + \mathbf{I})^{|\mathcal{V}|} \text{ where } \mathbf{I} \text{ is the identity matrix.} \quad (1)$$

If $(\mathbf{R}_m)_{ij} \neq 0$, we can know that release of linkage between $\mathcal{DB}_i$ and $\mathcal{DB}_j$ violates the blacklist policy of data contributor with UID m

With this observation, the blacklist check function is defined by

$$\text{Check}(x_{ik}, x_{j\ell}) = \begin{cases} 0 & \text{if } \pi_i(x_{ik}) = \pi_j(x_{j\ell}) = m \text{ and } (\mathbf{R}_m)_{ij} = 0 \\ \text{false} & \text{otherwise.} \end{cases}$$

With this function, the join of $\mathcal{DB}_i$ and $\mathcal{DB}_j$ with consideration of linkage policy $G_1^B, G_2^B, \ldots, G_C^B$ is defined as follows:

$$\mathcal{DB}_i \bowtie_B \mathcal{DB}_j = \{x_{ik}\|x_{j\ell} : \text{Match}(x_{ik}, x_{j\ell}) = \text{true},$$
$$\text{Check}(x_{ik}, x_{j\ell}) = \text{true}, x_{ik} \in \mathcal{DB}_i, x_{j\ell} \in \mathcal{DB}_j\}. \quad (2)$$

Security Model. Table 1 summarizes the security model of all instances that appear in our problem. In principle, our problem is defined based on privacy-preserving record linkage. Each database privately holds its records and the mapping function. On the other hand, linkage policies, linkage histories (only when blacklist is used), and queries are treated as public information. Thus, in our protocol, we suppose these are shared by all entities. In real execution of the protocol, linkage policies and linkage histories of all databases will be stored by a public storage and provided to any entities upon requests publicly. We remark that one of the databases can play the role of the public storage. Concatenated records (linkage) $x_{ik}\|x_{j\ell}$ are released to the analyst only when the linkage policy specified by each data contributor is satisfied. Finally, we define policy-preserving record linkage by whitelist with this security model.

Definition 1. *Suppose all databases and analysts behave semi-honestly. Also, suppose linkage histories and queries are public inputs to all stakeholders. For $i = 1, \ldots, D$, let database $\mathcal{DB}_i$ and mapping π_i be the private input of the ith database. Given linkage policies for a whitelist of all data contributors, after execution of join across $\mathcal{DB}_i$ and $\mathcal{DB}_j$, the analyst learns $\mathcal{DB}_i \bowtie_W \mathcal{DB}_j$ without violating the security model of Table 1.*

The problem of policy-considering record linkage by blacklist is defined by replacing the whitelist with the blacklist and the resulting table with $\mathcal{DB}_i \bowtie_B \mathcal{DB}_j$, respectively, in Definition 1.

Table 1. Security model: "public" means the stakeholder can view this entity as a plain text. "private" means the stakeholder cannot view the value of this entity.

		$\mathcal{DB}_i$	$\mathcal{DB}_j(j \neq i)$	Analyst
Record of $\mathcal{DB}_i$	x_{ik}	public	private	conditionally public
Mapping function of $\mathcal{DB}_i$	π_i	public	private	private
Linkage policy of DC with UID m	$\mathbf{W}_m, \mathbf{B}_m$	public	public	public
Linkage history of DC with UID m	$\mathbf{Q}_m$	public	public	public
Query	(i, j)	public	public	public
Linkage	$x_{ik}\|x_{j\ell}$	private	private	conditionally public

3 Cryptographic Tools

In this section, we introduce cryptographic tools employed for our protocols.

Asymmetric Cipher. When to transfer computed results between databases and the analyst, we use an asymmetric cryptosystem to encrypt the results. In this paper, we use the RSA cryptosystem. For a message $x \in \{0,1\}^*$, we write $\mathsf{Enc}_{\mathsf{pk}^R}(x)$ to denote the RSA ciphertext given the public encryption key pk^R.

RLWE-Based Homomorphic Encryption. We propose to use the Ring Learning with Error (RLWE)-based scheme. An RLWE-based scheme such as the BGV's scheme [1] is fully homomorphic. In other words, it supports both additive and multiplicative homomorphism. Our protocols, however, only leverage the additive property. The motivation of using RLWE schemes lies on the fact that the algebraic structure of RLWE schemes enables us to perform batch homomorphic addition. This property allows us to construct scalable private record linkage protocols.

The plaintext space of the BGV's scheme is given as the polynomial ring $R_t^N = \mathbb{Z}_t[Y]/(Y^N + 1)$ where t is a prime number and $N > 0$. Intuitively, R_t^N is a set of polynomials whose coefficients are in $\mathbb{Z}_t$ and whose degree are less than n. For instance, $3 + 2Y + Y^2 \in R_5^4$ but $7 + 3Y, 1 + Y + 2Y^4 \notin R_5^4$.

Given the public encryption key of the BGV's scheme pk^B, we write $\mathsf{Enc}_{\mathsf{pk}^B}(\mathcal{M})$ to the denote the ciphertext of the polynomial $\mathcal{M} \in R_t^N$. We write $\mathsf{Dec}_{\mathsf{sk}^B}(\cdot)$ to denote the decryption of the BGV's scheme under the private decryption key sk^B. The additive homomorphism of the BGV's scheme is given as follows

$$\mathsf{Dec}_{\mathsf{sk}^B}(\mathsf{Enc}_{\mathsf{pk}^B}(\mathcal{M}_1) \oplus \mathsf{Enc}_{\mathsf{pk}^B}(\mathcal{M}_2)) = \mathcal{M}_1 + \mathcal{M}_2,$$

$$\mathsf{Dec}_{\mathsf{sk}^B}(\mathsf{Enc}_{\mathsf{pk}^B}(\mathcal{M}_1) \oplus \mathcal{M}_2) = \mathcal{M}_1 + \mathcal{M}_2, \text{ where } \mathcal{M}_1, \mathcal{M}_2 \in R_t^N.$$

Packing. When to encrypt a single integer, we can also use the BGV's scheme since an integer is a degree-0 polynomial. However, this encryption manner does not fully exploit the capacity of the BGV's scheme. For this issue, we use packing techniques. Packing techniques [17] convert a *vector of integers* to a polynomial so that more than one integer can be encrypted into a single ciphertext. In this paper, we use the following method Pack to convert a length-p $(p < N)$ integer vector v $(\forall h \ v_h \in \mathbb{Z}_t)$ to a polynomial $\mathsf{Pack}(v) = \sum_{h=0}^{p-1} v_h Y^h$. When $p < N$, we have $\mathsf{Pack}(v) \in R_t^N$. On the other hand, when $p \geq N$ we just partition v into smaller parts so that $p < N$ holds for each part and pack them separately.

By applying Pack to integer vectors u and v, we have batch homomorphic addition as follows $\mathsf{Enc}_{\mathsf{pk}^B}(\mathsf{Pack}(u)) \oplus \mathsf{Enc}_{\mathsf{pk}^B}(\mathsf{Pack}(v)) = \mathsf{Enc}_{\mathsf{pk}^B}(\mathsf{Pack}(v + u))$. Usually, we have $N \approx 10^4$ in common use of the BGV's scheme. In other words, we can encrypt thousands of integers into a single ciphertext and perform thousands of batch additions with only one homomorphic addition.

The decryption of BGV's ciphertext results in polynomials. For the sake of simplicity, we omit the definition of the unpacking function since we can access the coefficient of a polynomial directly. For instance, let polynomial $\mathcal{V}$ denote the decryption of the packed ciphertext $\mathsf{Enc}_{\mathsf{pk}^B}(\mathsf{Pack}(v))$. We simply have $\mathcal{V}[h] = v_h$.

4   Proposed Protocols

4.1   Protocol Description

In this section, we propose Linkage White protocol and Linkage Black protocol that realizes PPRL with enforcement of linkage policies represented by a whitelist and blacklist, respectively. In our protocols, the analyst sends a linkage query to two databases $\mathcal{DB}_i$ and $\mathcal{DB}_j$, and the designated databases and the analyst cooperate to each other and perform join following the policies that are specified by the data contributors. After the description of the protocols, we show the correctness of our protocols and also give a proof for the security.

Whitelist Protocol. The protocol Linkage White that addresses the whitelist policy is given in Algorithm 1. Algorithm 1 runs in a multiple rounds manner. In the protocol, we suppose the number of slots in a single ciphertext is greater than $C = \max_i C_i$, the maximum number of data contributors. If this does not hold, we use multiple ciphertexts to pack all the values. The protocol for this case can be derived by a straightforward extension of Algorithm 1.

The analyst sends a linkage query to databases $\mathcal{DB}_i$ and $\mathcal{DB}_j$ for the join of records in those databases (Line 1) and then generates key pairs used for the protocol (Line 2). Database $\mathcal{DB}_i$ firstly randomizes all the UIDs and encrypts them with BGV scheme after packing and sends the ciphertext to database $\mathcal{DB}_j$ (Line 3). $\mathcal{DB}_j$ receives the ciphertext from $\mathcal{DB}_i$ and evaluates Eq. 4 to obtain $\{\mathcal{C}_\ell\}$ (Line 4) for all $\ell \in [C_j]$. $\mathcal{DB}_j$ then sends $\{\mathcal{C}_\ell\}$ to the analyst. After receiving $\{\mathcal{C}_\ell\}$ from $\mathcal{DB}_j$, the analyst performs decryption using the secret key sk^B and obtains $\{\mathcal{F}_\ell\}$. After the decryption, the analyst sends all $\{\mathcal{F}_\ell\}$ to $\mathcal{DB}_j$ (Line 5). At Line 6, $\mathcal{DB}_j$ creates a set of ciphertext-UID pairs $S = \{(\mathsf{Enc}_{\mathsf{pk}^R}(x_{j\ell}), \pi_j(x_{j\ell})) : \mathcal{F}_{\hat{\ell}}[k] = r'_\ell\}$. At Line 7, $\mathcal{DB}_i$ receives S from $\mathcal{DB}_j$ and finds all pairs of ciphertexts

---

Algorithm 1. Linkage White

Private input of $\mathcal{DB}_i$: record $\mathcal{DB}_i = \{x_{ik}\}_{k \in [C_i]}$, map $\pi_i : \mathcal{X}_i \to \mathcal{UID}$, vector of UIDs $\boldsymbol{a}_i = (\pi_i(x_{i1}), \ldots, \pi_i(x_{iC_i}))$, adjacency matrix of linkage policy $\{\mathbf{W}_u\}_{u \in \boldsymbol{a}_i}$
Private input of analyst: None
Private output of analyst: $\mathcal{DB}_i \bowtie_W \mathcal{DB}_j$
Private output of $\mathcal{DB}_i$: $(\mathbf{Q}_u)_i$ for $u = (\boldsymbol{a}_i)_k$ and $k = 1, \ldots, C_i$
1: The analyst sends a request of join (i, j) where $C_i \leq C_j$.
2: The analyst generates a key pair of BGV scheme $(\mathsf{pk}^B, \mathsf{sk}^B)$ and a key pair of RSA cryptosystem $(\mathsf{pk}^R, \mathsf{sk}^R)$ and sends pk^B (resp. pk^R) to $\mathcal{DB}_i$ (resp. $\mathcal{DB}_j$).
3: The ith DB sends $(\mathsf{Enc}_{\mathsf{pk}^B}(\mathsf{Pack}(r_i \cdot \boldsymbol{a}_i)), r_i)$ where $\boldsymbol{a}_i = (\pi_i(x_{i1}), \ldots, \pi_i(x_{iC_i}))$ and $r_i \in_r \mathbb{Z}_n$ to $\mathcal{DB}_j$.
4: The jth DB generates $r_\ell \in_r \mathbb{Z}_n, r'_\ell \in_r \mathbb{Z}_n$ where $\ell \in [C_j]$ and vector $\mathbf{b} = (b_1, \ldots, b_{C_j})$ where

$$b_\ell = -r_i \cdot \pi_j(x_{j\ell}) + (r_\ell \cdot (\mathbf{W}_{\ell*})_{ij}) + r'_\ell. \tag{3}$$

Here, $\ell^* = \pi_j(x_{j\ell})$. Next, for all $\ell \in [C_j]$, the jth DB $\mathcal{DB}_j$ evaluates

$$\mathcal{C}_\ell = \mathsf{Enc}_{\mathsf{pk}^B}(\mathsf{Pack}(r_i \cdot \boldsymbol{a}_i)) \oplus \mathsf{Pack}(\mathbf{b}^{(\ell)}) \tag{4}$$

where $\mathbf{b}^{(\ell)} = (b_\ell, \cdots, b_{(\ell+C_i-1) \bmod C_i})$. Then, the jth DB sends $\{\mathcal{C}_\ell\}_{\ell \in [C_j]}$ to the analyst.
5: For all $\ell \in [C_j]$, the analyst decrypts $\mathcal{F}_\ell = \mathsf{Dec}_{\mathsf{sk}^B}(\mathcal{C}_\ell)$ and sends $\{\mathcal{F}_\ell\}_{\ell \in [C_j]}$ to the jth DB.
6: The jth DB sends $S = \{(\mathsf{Enc}_{\mathsf{pk}^R}(x_{j\ell}), \pi_j(x_{j\ell})) : \forall \ell \in [C_j] \text{ s.t. } \mathcal{F}_{\tilde{\ell}}[k] = r'_\ell \text{ where } k \in [C_i], \tilde{\ell} \in [C_j]\}$ to $\mathcal{DB}_i$.
7: The ith DB sends $K = \{\mathsf{Enc}_{\mathsf{pk}^R}(x_{ik}) \| \mathsf{Enc}_{\mathsf{pk}^P}(x_{j\ell}) : (\mathsf{Enc}_{\mathsf{pk}^R}(x_{j\ell}), \pi_j(x_{j\ell})) \in S, \pi_i(x_{ik}) = \pi_j(x_{j\ell})\}$ to the analyst.
8: The analyst decrypts all elements in K and obtain $\mathcal{DB}_i \bowtie_W \mathcal{DB}_j = \{x_{ik} \| x_{j\ell} : \pi_i(x_{ik}) = \pi_j(x_{j\ell})\}$.

---

such that $\pi_i(x_{ik}) = \pi_j(x_{j\ell})$ and then constructs $K = \{\mathsf{Enc}_{\mathsf{pk}^R}(x_{ik}) \| \mathsf{Enc}_{\mathsf{pk}^R}(x_{j\ell}) : \pi_i(x_{ik}) = \pi_j(x_{j\ell})\}$. The analyst receives K from $\mathcal{DB}_i$ and decrypts all elements in K to obtain the result of $\{x_{ik} \| x_{j\ell}\} = D_i \bowtie_W D_j$ (Line 8). The following theorem shows why Algorithm 1 can correctly evaluate $D_i \bowtie_W D_j$.

Theorem 1 (Correctness of Algorithm 1). *Algorithm 1 correctly solves the problem of Definition 1 with whitelist policy.*

Proof. In Line 7 of Algorithm 1, $\mathcal{DB}_j$ obtains $\{\mathcal{F}_\ell\}$. According to Eq. 4, we have $\mathcal{F}_\ell[k] = r_i \cdot (\pi_i(x_{ik}) - \pi_j(x_{j\ell})) + r_\ell \cdot (\mathbf{W}_\ell)_{ij} + r'_\ell$, where r_i is a random integer chosen by $\mathcal{DB}_i$ and r_ℓ and r'_ℓ are random integers chosen by $\mathcal{DB}_j$. Since Eq. 4 is evaluated for all ℓ, $\mathcal{F}_\ell[k]$ is evaluated over the ciphertexts for all k and ℓ. When $\pi_i(x_{ik}) = \pi_j(x_{j\ell})$ (that is, the UIDs of x_{ik} and $x_{j\ell}$ are equivalent) and $(\mathbf{W}_\ell)_{ij} = 0$ (linkage allowed by the linkage policy), we have $\mathcal{F}_{\tilde{\ell}}[k] = r'_\ell$ for some $\tilde{\ell}$. Otherwise (i.e., $\pi_i(x_{ik}) \neq \pi_j(x_{j\ell})$ or $(\mathbf{W}_\ell)_{ij} \neq 0$), $\mathcal{F}_\ell[k]$ distributes uniformly at random on $\mathbb{Z}_n$. Consequently, by checking the equality, we can extract the set S in Line 7 of Algorithm 1 that consists of $\mathcal{DB}_i \bowtie_W \mathcal{DB}_j$ where each element is encrypted with pk^R. Accordingly, the analyst obtains $\mathcal{DB}_i \bowtie_W \mathcal{DB}_j$ by decrypting each element at the end of the Algorithm 1 correctly.

Partitioning. When $C > N$, all UIDs cannot be packed into a single ciphertext in Line 3. In this case, we partition the vector into $\lceil \frac{C}{N} \rceil$ blocks so that values in each block can be packed into a single ciphertext. With this partitioning, $\mathcal{DB}_i$

Algorithm 2. CheckReachability

Input: join request (i, j), UID of data contributor u, adjacency matrix of linkage policy $\mathbf{B}_u$, linkage history $\mathbf{Q}_u$
output: Check(i, j, u)
1: jth **DB: R'** $\leftarrow$ **0**
2: **for** $d = 1$ **to** D **do**
3:     jth **DB: R** $\leftarrow$ $\mathbf{Q}_u \cdot (\mathbf{I} + \mathbf{R'})$ where **I** is the unit matrix.
4:     jth **DB: R'** $\leftarrow$ **R**
5: **end for**
6: **return** $z = r \cdot (\mathbf{R'})_{ij}$ where r is a random number distributed over $\mathbb{Z}_N$

---

sends $\lceil \frac{C}{N} \rceil$ ciphertexts in Line 3. Then, $\mathcal{DB}_j$ evaluates Eq. 4 for each ciphertext in Line 4, generates $C \cdot \lceil \frac{C}{N} \rceil$ ciphertexts as $\mathcal{C}_\ell$, and send them to the analyst.

Blacklist Protocol. The protocol for blacklist is similar to the Algorithm 1 except in the following three points:

1. Data contributor with UID m inputs a policy graph for blacklist $\mathbf{B}_m$ instead of whitelist $\mathbf{W}_m$,
2. In Eq. 3, $(\mathbf{W}_{\ell^*})_{ij}$ is replaced with Check(i, j, ℓ^*), which checks whether the join of $\mathcal{DB}_i$ and $\mathcal{DB}_j$ will violate the linkage policy by blacklist specified by the mth data contributor, given the history of joins that have been performed (see Algorithm 2),
3. After linkage happens, the linkage history matrix $\mathbf{Q}_m$ is updated. Since $\mathbf{Q}_m$ is public information, we do not specifically describe who collects and manages $\mathbf{Q}_m$ in the protocol description.

Theorem 2 (Correctness of the protocol for blacklist). *The protocol for blacklist correctly solves the problem of Definition 1 with a blacklist.*

Proof. The behavior of the protocol for blacklist is equivalent to Algorithm 1 except for Eq. 3. It thus suffices to prove that Check (Algorithm 2) returns 0 if and only if the join between $\mathcal{DB}_i$ and $\mathcal{DB}_j$ will not violate the blacklist linkage policy. The computation in the loop starting from Line 2 evaluates Eq. 1[9]. Thus, we can check the linkability between $\mathcal{DB}_i$ and $\mathcal{DB}_j$ for the data contributor with UID u by checking if $(\mathbf{R})_{ij} = 0$ holds, which is evaluated at Line 6.

4.2 Security Proof

The first requirement is that our protocol is secure against honest-but-curious databases and analyst. In other words, the analyst does not learn anything more about the data contributors beyond the joined records, i.e., $\{x_{ik} \| x_{j\ell}\}$. A honest-but-curious database does not learn anything more beyond the set of UIDs of records joined by the protocol. We give the formal security guarantee but defer the full formal proof to the full version of this paper.

Theorem 3. *If BGV scheme and RSA provide semantic security, the protocol for blacklist securely computes Eq. 2 in the presence of semi-honest adversaries. At the end of protocol for blacklist, the analyst learns $\mathcal{DB}_i \bowtie_B \mathcal{DB}_j$ but nothing else. $\mathcal{DB}_i$ and $\mathcal{DB}_j$ learn the UIDs of records in $D_i \bowtie_B D_j$ but nothing else.*

The security of protocol for whitelist is readily derived from the security proof of protocol for blacklist. We, thus, only give the security statement for blacklist protocol in Theorem 3. The formal proof is derived by following the standardized proof technology using the concept of *simulator*. Due to space limitation, we give a sketch of the proof, and defer the full argument to the full version of this paper.

Proof (Sketch). The analyst's view of the protocol execution consists only of RSA ciphertexts of the joinned records. By assumption of semantic security of RSA, we can create a simulator that can simulate the analyst's view.

The view of $\mathcal{DB}_i$ of the protocol execution consists of two components: the set of UIDs of the joint records and RSA ciphertexts. We can build up a simulator that can simulate the view of $\mathcal{DB}_i$ independently of the data providers' input, by the assumption of semantic security of RSA.

The view of $\mathcal{DB}_j$ of the protocol execution consists of three components: the set of UIDs of the joint records, a set of BGV's ciphertexts, and a set of polynomials. The first component is the output of the protocol. The second component can be simulated by the semantic security of BGV's scheme. Polynomials $\{\mathcal{F}_\ell\}$ are polynomials with uniformly distributed coefficients (except some coefficients reveal the set of UIDs of the joint records). Thereby, $\{\mathcal{F}_\ell\}$ can also be simulated (given the set of UIDs of the joint records).

4.3 Asymptotic Analysis

We analyze the asymptotic performances of Algorithm 1 and its blacklist variant regarding communication and computation. More specifically, we investigate the communication complexity of the entire protocol and the computation complexity of the following operations in the protocol. Encryption (Enc) of plaintexts (Line 3), homomorphic additions (Calc) with packing (Eq. 4 in Line 4), and decryption (Dec) of ciphertexts (Line 5).

Table 2 summarizes the computation and communication complexity of each step in Algorithm 1 (Linkage White). First, we consider the case that the number of data contributors of each database is upper bounded by $C \leq N$. In this case, since only a single ciphertext is generated in Line 3, this step is $O(1)$. In Line 4, at most C homomorphic addition is performed; this step is $O(C)$. Consequently, at

Table 2. Asymptotic communication and computational complexity of Linkage White

	Enc	Calc	Dec	Communication value
$C \leq N$	$O(1)$	$O(C)$	$O(C)$	$O(C)$
$C > N$	$O(C/N)$	$O(C^2/N)$	$O(C^2/N)$	$O(C^2/N)$

most C ciphertexts are decrypted in Line 5. Also when $C \leq N$, $O(C)$ ciphertexts of the BGV scheme is communicated. Regarding the ciphertexts of RSA, at most C ciphertexts are communicated in Line 6 and at most, $2C$ ciphertexts are communicated in Line 7. Consequently, $O(C)$ ciphertexts are communicated through the entire protocol. Here we ignored the communication of the keys because the size of the key is negligibly small.

In the case of $C > N$, we need to partition the vector into several blocks. The number of blocks of the ciphertexts, the number of homomorphic operations, and the number of decryption operations needed are $\lceil \frac{C}{N} \rceil$, $C \cdot \lceil \frac{C}{N} \rceil$, and $C \cdot \lceil \frac{C}{N} \rceil$, respectively. Thus, the computation complexity of Enc, Calc, and Dec is $O(\frac{C}{N})$, $O(\frac{C^2}{N})$, and $O(\frac{C^2}{N})$, respectively. The communication complexity when $C > N$ is $O(\frac{C^2}{N})$ because the number of the ciphertexts of the BGV scheme is $O(\frac{C^2}{N})$.

The protocol for blacklist is exactly same as that for whitelist except that it requires evaluation of Algorithm 2, which contains C dimension matrix product D times in plaintext and the computation complexity is $O(DC^3)$. Since the computation time in plaintext is ignorable compared to that in ciphertext, we can regard that the computation complexity of the blacklist protocol is almost the same as the whitelist protocol.

5 Experimental Evaluation

In this section, we present experimental results of Linkage Black. As we discussed in Sect. 4.3, the computation complexity of the protocol for blacklist is the same as that for whitelist. So, we consider the performance of the protocol for blacklist only in this section. We measured the execution time and the communications traffic for performance evaluation. All programs were implemented in C++ and run on a machine with an Intel Core-i7-2600 3.5 GHz CPU and with 8 GB RAM. For the RSA, we used 1024 bits key for providing 80 bits security level. For the BGV scheme, we leverage the HElib [6] library, using the parameters as $N = 8192, t = 7681$ for at least 80 bits security level.

We measured the computation time of operations of Enc, Calc, Dec, and KeyGen (key generation at Line 1), and the communication traffic occurred during the protocol execution among all stakeholders.

Experimental Results. The experimental results are shown in Fig. 2. According to the analysis in Sect. 4.3, the computation time of Calc is dependent on the number of the databases, D; however, the time of computation in plaintext is negligibly small compared to that in ciphertext. So, we set $D = 16, 32$.

Figure 2(a) and (b) show the computation time of each operation aspect to the number of data contributors. When $C > N$, the computation complexity of the protocol is $O(C^2/N)$. Since $N = 8192$ in our setting, the computation time grows quadratically when $C > 8192$. More specifically, when a database holds 5000 records, the analyst decrypts $5000 \cdot \lceil \frac{5000}{8192} \rceil = 5000$ records at Line 5. In the same setting, when the database holds 10^5 records, the number of ciphertexts

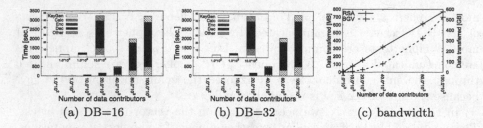

(a) DB=16   (b) DB=32   (c) bandwidth

Fig. 2. Computation time and communicated amount of Linkage Black

generated in the protocol execution is $10^5 \cdot \lceil \frac{10^5}{8192} \rceil = 1.3 \cdot 10^6$. In our experimental evaluation, decryption of 1.3×10^6 ciphertexts costs about 41 min.

Figure 2(c) represents the increase in the communication traffic of the ciphertexts of RSA (the right horizontal axis) and BGV (the left horizontal axis) when the number of the data contributors is varied. When the number of the data contributors is 10^5, the communication traffic caused by RSA is less than 1 GB whereas that by BGV is about 500 GB. This is because the size of a single ciphertext of BGV is 423 KB. Thus, when the number of the record is large, the protocol can consume seriously large bandwidth. In practical, the computation after Line 4 can be performed in a sequential manner for each ciphertext. Thus, the memory consumption can be kept constant even with large C. From these experimental evaluations, we can conclude that our protocol can work in a practical computation time and bandwidth consumption when the number of records is not too large. Reduction of bandwidth consumption with large C is an important future work to be resolved.

6   Summary and Future Works

In this paper, we introduced a novel privacy notion for PPRL, record-wise linkage policy, with which each data contributor can individually choose database pairs that are allowed (resp. banned) to process record linkage with the records of the data contributor by whitelist (resp. blacklist). We show that our protocol can correctly and securely evaluate PPRL with record-wise linkage policies represented by whitelist or blacklist. On the other hand, the network bandwidth consumption can be seriously large when the number of data contributors is very large We assumed that all databases behave semi-honestly. When databases are trusted organizations, such as banks or hospitals, this assumption is reasonable. However, the applicability of our protocol can be expanded if data contributors can enforce linkage policies even to malicious databases. Relaxation of security assumption is also one of interesting research challenges.

Acknowledgement. This research was partly supported by the program "Research and Development on Real World Big Data Integration and Analysis" of the Ministry of Education, Culture, Sports, Science and Technology, Japan and RIKEN and JST CREST Grant Number JPMJCR1302, Japan.

References

1. Brakerski, Z., Gentry, C., Vaikuntanathan, V.: (Leveled) fully homomorphic encryption without bootstrapping. ACM Trans. Comput. Theory (TOCT) **6**(3), 13 (2014)
2. Churches, T., Christen, P.: Blind data linkage using n-gram similarity comparisons. In: Dai, H., Srikant, R., Zhang, C. (eds.) PAKDD 2004. LNCS, vol. 3056, pp. 121–126. Springer, Heidelberg (2004). doi:10.1007/978-3-540-24775-3_15
3. ElGamal, T.: A public key cryptosystem and a signature scheme based on discrete logarithms. In: Blakley, G.R., Chaum, D. (eds.) CRYPTO 1984. LNCS, vol. 196, pp. 10–18. Springer, Heidelberg (1985). doi:10.1007/3-540-39568-7_2
4. Elmagarmid, A.K., Ipeirotis, P.G., Verykios, V.S.: Duplicate record detection: a survey. IEEE Trans. Knowl. Data Eng. **19**(1), 1–16 (2007)
5. Goldreich, O.: Foundations of Cryptography: Basic Applications, vol. 2. Cambridge University Press, Cambridge (2009)
6. Halevi, S., Shoup, V.: HELib. (2017). http://shaih.github.io/HELib. Accessed 10 Apr 2017
7. Hall, R., Fienberg, S.E.: Privacy-preserving record linkage. In: Domingo-Ferrer, J., Magkos, E. (eds.) PSD 2010. LNCS, vol. 6344, pp. 269–283. Springer, Heidelberg (2010). doi:10.1007/978-3-642-15838-4_24
8. Han, S., Shen, D., Nie, T., Kou, Y., Yu, G.: Scalable private blocking technique for privacy-preserving record linkage. In: Li, F., Shim, K., Zheng, K., Liu, G. (eds.) APWeb 2016. LNCS, vol. 9932, pp. 201–213. Springer, Cham (2016). doi:10.1007/978-3-319-45817-5_16
9. Higham, N.J.: Accuracy and stability of numerical algorithms (2002)
10. Hore, B., Mehrotra, S., Tsudik, G.: A privacy-preserving index for range queries. In: Proceedings of 30th International Conference on Very Large Data Bases, VLDB Endowment, vol. 30, pp. 720–731 (2004)
11. Inan, A., Kantarcioglu, M., Bertino, E., Scannapieco, M.: A hybrid approach to private record linkage. In: IEEE 24th International Conference on Data Engineering, pp. 496–505. IEEE (2008)
12. Karakasidis, A., Verykios, V.S.: Privacy preserving record linkage using phonetic codes. In: Fourth Balkan Conference on Informatics, BCI 2009, pp. 101–106. IEEE (2009)
13. Karakasidis, A., Verykios, V.S.: Secure blocking + secure matching = secure record linkage. J. Comput. Sci. Eng. **5**(3), 223–235 (2011)
14. Kissner, L., Song, D.: Private and threshold set-intersection. Technical report, DTIC Document (2004)
15. Kum, H.C., Krishnamurthy, A., Machanavajjhala, A., Reiter, M.K., Ahalt, S.: Privacy preserving interactive record linkage (PPIRL). J. Am. Med. Inform. Assoc. **21**(2), 212–220 (2014)
16. Kuzu, M., Kantarcioglu, M., Inan, A., Bertino, E., Durham, E., Malin, B.: Efficient privacy-aware record integration. In: Proceedings of 16th International Conference on Extending Database Technology, pp. 167–178. ACM (2013)
17. Naehrig, M., Lauter, K., Vaikuntanathan, V.: Can homomorphic encryption be practical? In: Proceedings of 3rd ACM Workshop on Cloud Computing Security Workshop, pp. 113–124. ACM (2011)
18. Paillier, P.: Public-key cryptosystems based on composite degree residuosity classes. In: Stern, J. (ed.) EUROCRYPT 1999. LNCS, vol. 1592, pp. 223–238. Springer, Heidelberg (1999). doi:10.1007/3-540-48910-X_16

19. Sakazaki, H., Anzai, K.: Proposal of secret computation scheme corresponding to many-to-many encryption and decryption. In: Proceedings of Computer Security Symposium 2016, vol. 2016, pp. 53–59, October 2016
20. Scannapieco, M., Figotin, I., Bertino, E., Elmagarmid, A.K.: Privacy preserving schema and data matching. In: Proceedings of 2007 ACM SIGMOD International Conference on Management of Data, pp. 653–664. ACM (2007)
21. Schnell, R., Bachteler, T., Reiher, J.: Privacy-preserving record linkage using bloom filters. BMC Med. Inform. Decis. Mak. **9**(1), 41 (2009)
22. Vatsalan, D., Christen, P., Verykios, V.S.: A taxonomy of privacy-preserving record linkage techniques. Inf. Syst. **38**(6), 946–969 (2013)
23. Verykios, V.S., Karakasidis, A., Mitrogiannis, V.K.: Privacy preserving record linkage approaches. Int. J. Data Min. Model. Manag. **1**(2), 206–221 (2009)
24. di Vimercati, S.D.C., Foresti, S., Jajodia, S., Paraboschi, S., Samarati, P.: Controlled information sharing in collaborative distributed query processing. In: 28th International Conference on Distributed Computing Systems, ICDCS 2008, pp. 303–310. IEEE (2008)
25. Yakout, M., Atallah, M.J., Elmagarmid, A.: Efficient private record linkage. In: IEEE 25th International Conference on Data Engineering, pp. 1283–1286. IEEE (2009)

Two-Phase Preference Disclosure in Attributed Social Networks

Younes Abid, Abdessamad Imine, Amedeo Napoli, Chedy Raïssi,
and Michaël Rusinowitch[✉]

Lorraine University, CNRS, Inria, 54000 Nancy, France
{younes.abid,abdessamad.imine,amedeo.napoli,
chedy.raissi,michael.rusinowitch}@loria.fr

Abstract. In order to demonstrate privacy threats in social networks we show how to infer user preferences by random walks in a multiple graph representing simultaneously attributes and relationships links. For the approach to scale in a first phase we reduce the space of attribute values by partition in balanced homogeneous clusters. Following the Deepwalk approach, the random walks are considered as sentences. Hence unsupervised learning techniques from natural languages processing can be employed in second phase to deduce semantic similarities of some attributes. We conduct initial experiments on real datasets to evaluate our approach.

Keywords: Online social network (OSN) · Attribute disclosure attacks · Privacy

1 Introduction

Social networks offer their users several means to control the visibility of their personal data and publications such as attribute values and friendship links. However even in the case when these policies are properly enforced nowadays data collection techniques and statistical correlations can provide hints on users hidden information [14]. Moreover information leaks from relatives of a user are difficult to control and, for instance, by homophily reasoning [3] an attacker can disclose and exploit sensitive data from a target. Therefore we need to anticipate such disclosure of private information from publicly available data. A way to tackle the problem is to offer users tools that rise their awareness about these privacy breaches. That is we aim to provide people algorithms that try to infer their own hidden attributes, even when social graphs are sparse or friendship links unexploitable, so that they can apply proper countermeasures when such inference are too easy.

In this work, we aim to disclose secret preferences of a social network user for instance, his/her liked movie with high probability of success. Secret preferences are either private or unspecified values of some attribute of the targeted user. The challenge is to predict the secret preference from hundreds of thousands

© Springer International Publishing AG 2017
D. Benslimane et al. (Eds.): DEXA 2017, Part I, LNCS 10438, pp. 249–263, 2017.
DOI: 10.1007/978-3-319-64468-4_19

of possible preferences in the network. Therefore to reduce the preference space the first phase consists in clustering attributes values by common likes. For instance, the values *Star Wars V* and *Star Wars IV* of the attribute movies end up having the same label (i.e. cluster identifier) since they are liked by many common users. By carefully chosing the parameters we end up with relatively homogeneous clusters of balanced sizes. The second phase consists in applying unsupervised learning techniques from natural language processing to disclose the (cluster) label of the secret preference. These techniques have proved to be quite effective for predicting missing links in sparse graphs. Finally, when the label is disclosed we can either further process the cluster content to disclose the secret preference or directly infer the preferences when the clustered values are highly similar, as for instance *Star Wars* episodes.

Let us pinpoint some noticeable features of our approach. Preferences of users for some attribute values are represented by bipartite graphs. Clustering attributes values relies only on users preferences in the considered social network and does not consider external information such as human expertise or information from other websites. We process different graphs of attributes at the same time through random walk to cross latent information about many attributes as detailed in Sect. 4. For instance, drinks preferences can play a major role to disclose the secretly liked dish of the target. To cope with over-fit problems we assign a weight to each graph in order to quantify its importance in disclosing the secret preference of the targeted user. Weights are parameters validated through off-line tests. We also exploit friendship graphs between users in order to better connect attribute graphs in the random walks.

Related Works. For space reasons we only discuss closely related works. In [11] the authors propose algorithms to detect whether a sensitive attribute value can be infered from the neighborhood of a target user in a social network. Heatherly et al. [7] seem to be the first that study how to sanitize a social network to prevent inference of social network attributes values. It relies on bayesian classification techniques. The analogous link prediction (recommendation) problem is solved in [1] by exploiting attributes to guide a random walk on graph. The random walk technique has been applied to social representations in [10]. We present here an inference technique that combines attribute clustering and random walks. The method can handle sparse social graphs and attributes with large set of values. The initial clustering allows one to obtain results in a few minutes on large graphs.

2   Social Network Model

To model social networks for privacy analysis purposes, it is important to take into account their complex structures as well as their rich contents. Infering sensitive and personal information can then be more accurate. We use graphs to model both the structure and the content of social networks. For the network structure, the link-ship networks are modeled either by directed or undirected

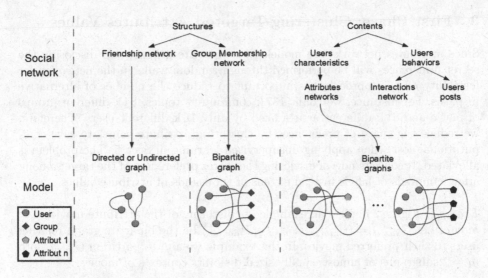

Fig. 1. Social network model for privacy analysis purposes.

graphs depending on the social network links type. For instance follow-ship on Twitter are modeled by directed graphs while friendships on Facebook are modeled by undirected graphs. Let $G_l = (U_l, L)$ be the graph of link-ship where U_l is a set of user nodes and L is a set of links between them. In the same model we use bipartite graphs to represent group membership networks. Let $G_g = (U_g, V_g, P_g)$ be the graph of memberships where U_g is a set of user nodes, V_g is a set of group nodes and P_g is a set of links between user nodes and group nodes.

For modeling the networks contents we use bipartite graphs too. In this work we focus on attributes and omit other contents. Let $G_A = (U_A, V_A, P_A)$ be the graph depicting the preferences of users concerning the attribute A where U_A is a set of user nodes, V_A is a set of nodes representing the different possible values of A and P_A is a set of edges between user nodes U_A and attribute values nodes V_A. P_A represents the preferences (or "likes") of users in U_A for the different attribute values. Figure 1 depicts the detailed model above.

Some attributes such as gender have a small set of values. Some others such as music, book and politics have a huge set of possible values. Predicting the favorite book titles or music tunes of a user among scaling thousands of possibilities is hard in a single step. To cope with this problem we decompose the set of possibilities into a few clusters as detailed in Sect. 3. Our objective will be to predict an attribute value by first predicting the cluster that contains this value. For instance, we aim to predict first the favorite music genres and favorite book genres instead of the favorite music tunes and the favorite book titles directly. Then once such a cluster is determined, infering a prefered item will be easier thanks to the smaller size of clusters compared to the whole set of values. Even a random selection strategy in this last step generates interesting results as shown by our experiments.

3   First Phase: Clustering Targeted Attributes Values

Since attributes networks are modeled by bipartite graphs, and disclosing the secret preferences will be performed through random walks in the networks, for feasibility of the approach it is important to reduce the number of alternative in paths. For instance, we count $137\,k$ community topics, $84\,k$ different groups of music and $31\,k$ different artists liked by only $15\,k$ different users. Therefore, we reduce the space of preferences by clustering attribute values to save computational cost when applying unsupervised learning in Sect. 5. The problem is alleviated. It consists now of disclosing the secret preferences of the target among a few hundreds of labels instead of tens of thousands of attribute values.

Example. Figure 2 depicts an example of clustering of the attribute movie, $A = movie$. Let $G_{movie} = (U_{movies}, V_{movies}, P_{movies})$ be the bipartite graph relating users to their preferred movies. In this example we aim to partition G_{movie} into $n_l = 2$ subgraphs of almost equally sized disjoints contexts of movies.

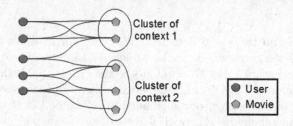

Fig. 2. Example of clustering the values of the attribute movies into disjoint clusters of context.

Clustering all attributes values into contexts requires huge up-to-date knowledge about many fields and many cultures. For instance, Eddie Murphy movies are linked to comedy in 2017 but in 2007 his name was correlated to drama for his role in Dreamgirls for which he picked up his only Oscar nomination. To cope with this problem we cluster the attributes values based on users preferences. However we do not cluster users simultaneously since it is obvious that they can have very different preferences at the same time. For instance, the same user can like both horror and documentary movies. Furthermore, we aim to disclose secret preferences by exploiting information from different graphs, including the friendship graph, as detailed in Sect. 4.

The problem can be related to a k-way graph partitioning problem since the goal is to divide the set of attribute values into k subset of about equal size. Since we also aim to maximize similarities between attribute values belonging to the same subgraphs, different approaches of dense subgraph discovery could be applied to iteratively seek and cut the densest subgraph from the original graph [8]. However, due to the sparsity of the social graph we consider, the dense subgraphs are usually small and the algorithms mentioned in [8] end up partitioning the graph into a large number of not equally-sized subgraphs with decreasing densities.

To cope with this issue, we propose a greedy algorithm that adds constraints on the size of subgraphs and the similarity between attribute values of each subgraph. In the following we denote by $|S|$ the cardinal of a set S.

Objective Function. Our aimed objective is to find a partition π_l of attribute values in n_l clusters that maximize the similarity between values inside each cluster. We define the similarity between two attribute values v and v' to be the Jaccard coefficient that measures the ratio of their common likes to the union of their likes, where the likes of an attribute value f (say, a movie) is by definition $|\{u \in U_{movies} \text{ s.t.} (u, f) \in P_{movies}\}|$ and denoted by $likes(f)$. That is,

$$similarity(v, v') = \frac{likes(v) \cap likes(v')}{likes(v) \cup likes(v')} \qquad (1)$$

For computational efficiency the number of clusters n_l must be small. But if n_l is too small the neural network detailed in Sect. 5 will be doomed to learn from insufficient data. On the other hand, if n_l is too large the neural network predictions will not be reliable due to over-fitting. Moreover, clusters must be almost equally-sized to avoid fostering a particular label. Therefore we only consider partitions $(c_1, \ldots, c_{n_l})$ of the attribute values satisfying $\sqrt{m} \leq |c_k| \leq 2\sqrt{m}$ for $1 \leq k \leq n_l$, where m is the number of all attribute values, that is the number of movies in our running example. Consequently, the number n_l of clusters satisfies $\frac{\sqrt{m}}{2} \leq n_l \leq \sqrt{m}$. The set of partitions satisfying the constraints above is denoted by Π_l. A good criteria for a candidate cluster c is to maximize the average similarity $similarity(c)$ between all couples of attribute values inside this cluster. Hence the objective function is given by Expression 2.

$$\max_{(c_1, \ldots, c_{n_l}) \in \Pi_l} \frac{1}{n_l} \left(\sum_{k=1}^{n_l} similarity(c_k) \right) \qquad (2)$$

Algorithm. Computing the average similarity of a cluster c is expensive due to the quadratic number of couples of values in c. Moreover, the algorithm needs to find the cluster of maximal average similarity among the numerous ones of size between $\sqrt{m}$ and $2\sqrt{m}$. To get around with this problem, we propose a greedy algorithm that computes only the similarity between a cluster of movies and an unlabeled attribute value (that is a value not assigned yet to a cluster). Therefore we define:

$$similarity(c, v) = \frac{\sum_{v' \in c} similarity(v', v)}{|c|} \qquad (3)$$

The idea now is to seek, from a set of unlabeled attribute values, an attribute value with maximal similarity with the cluster c (function seek_max_similar). Then add the chosen attribute value (max_similar) to c. The algorithm keeps adding attribute value to c until it reaches the stop conditions. It then defines next clusters sequentially the same way as detailed in Algorithm 1 until all attribute values are labeled.

Stop Conditions. The algorithm stops adding attribute values to the current cluster c when the size of the cluster c is equal to $\mathtt{int}(2\sqrt{m})$ or is in $[\sqrt{m}, 2\sqrt{m}-1]$ and one of the two following additional conditions is fulfilled: (i) the similarity between c and any of unlabeled attribute values is less than $\frac{1}{2}$; (ii) the number of unlabeled attribute values is higher than $\sqrt{m}$. In other word, there exists no sufficiently similar attribute value to add to the current cluster and there is enough unlabeled attribute values to create new clusters. There is also a stopping condition (line 11) when the number of unlabeled attribute values is $\mathtt{int}(\sqrt{m})$ to guarantee that the size of the last cluster will be at least $\sqrt{m}$. Finally, the main loop stops when all attribute values are labeled.

Data: $G_A = (U_A, V_A, P_A)$,
Result: π_l ▷ decomposition of V_A into l clusters
1   $B \leftarrow \sqrt{|V_A|}$
2   $V \leftarrow V_A$ ▷ V contains values not assigned to a cluster
3   **while** $|V| > 0$ **do**
4     $c \leftarrow \mathtt{one\_most\_liked}(V)$ ▷ initialisation of a new cluster with one element
5   **while** $|c| < 2B$ *and* $|V| > 0$ **do**
6   **if** $B \le |c|$ **then**
7   **if** $\mathtt{max\_similarity}(c, V) < \frac{1}{2}$ **and** $|V| > B$ **then**
8 | break
9   **end**
10   **if** $|V| = \mathtt{int}(B)$ **then**
11 | break
12   **end**
13   **end**
14       $\mathtt{max\_similar} \leftarrow \mathtt{seek\_max\_similar}(c, V)$
15       $c \leftarrow c \cup \mathtt{max\_similar}$
16       $V \leftarrow V \setminus \mathtt{max\_similar}$
17   **end**
18     $\pi_l \leftarrow \pi_l \cup c$
19 **end**

Algorithm 1. Partition of a set of attribute values into clusters.

Size of Partitions. We have analyzed the performance of the proposed algorithm with respect to the minimal size of computed clusters, where no cluster can have twice the size of other cluster from the same partition. Tests depicted by Fig. 3 show that the choice of the minimal size to be the root square of the size of the set of attribute values yields good results for both very sparse graphs like Users-FastFoods graph (density $= 0.0018$) and less sparse graphs like the Users-Actors graph (density $= 0.012$). We note that this choice yields some clusters of high similarity (≥ 0.7), few subgraphs (less than the square root of the number of attribute values) and relatively high mean similarity compared to all partitions similarities (larger than the mean of the means of all similarities).

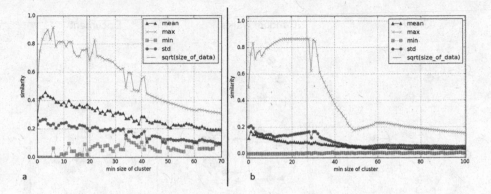

Fig. 3. Variation of partitioned bipartite subgraph similarities with respect to the minimal size of subgraphs, (a) Users-Actors graph: 15 k users, 364 actors, (b) Users-FastFoods graph: 15 k users, 777 fast foods.

4   Random Walks in a Social Attributed Network

In this section we aim to express the latent information in the graphs modeling both the structure and the content of the network into a document that will be processed in Sect. 5 to disclose secret preferences as detailed in Sect. 6.

As illustrated in Fig. 4, the document is constructed by connecting all graphs through random jumps between them and random walk between their nodes (see also [10]). Since the values of the analyzed attributes are labeled, they are represented by their clusters in the final document. For instance, the first walk depicted by Fig. 4 is $[u_1, u_4, v_{2,3}, u_4]$. But for efficiency the walk $[u_1, u_4, c_{2,2}, u_4]$ is stored instead in the document since the value $v_{2,3}$ belongs to the cluster $c_{2,2}$.

Let n be the total number of graphs that model the social network, comprising a link-ship graph $G_1 = G_l = (U_l, L)$ and $n-1$ attribute graphs $G_x = (U_x, V_x, P_x)$. Let U be the set of users in all graphs and n_1 its cardinality. Jumps between two graphs, G_x and G_y, are possible if the current walker state is a user node, say u_z, that belongs to both graphs ($u_z \in U_x \cap U_y$). The walker is allowed to jump from the user node u_z to the graph G_y with a probability $p_{z,y}$. The probability $p_{z,y}$ is defined in Eq. 4 where weights are parameters used to quantify the importance of each graph in disclosing secret preferences (e.g., value of some sensitive attribute) of the target.

$$p_{z,y} = \begin{cases} \dfrac{weight(G_y)}{\sum_{\{1 \leq x \leq n | u_z \in U_x\}} weight(G_x)} & \text{if } u_z \in U_y \\ 0 & \text{otherwise} \end{cases} \tag{4}$$

For each graph $G_y = (U_y, V_y, P_y)$ we define two line stochastic adjacency matrices, $T_{U \times V_y}$ and $T_{V_y \times U}$, and a jump matrix, J_y, that leads to G_y as detailed in 5.

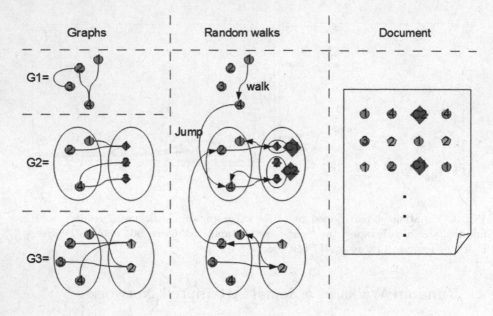

Fig. 4. Example of multi graph random walk.

$$J_y = diag(p_{z,y}|u_z \in U)$$

$$T_{U \times V_y}(i,j) = \begin{cases} \frac{1}{deg_y(u_i)} & \text{if}(u_i, v_j) \in P_y \\ 0 & \text{otherwise} \end{cases}$$

$$T_{V_y \times U}(i,j) = \begin{cases} \frac{1}{deg(v_i)} & \text{if}(u_j, v_i) \in P_y \\ 0 & \text{otherwise} \end{cases} \tag{5}$$

where U is the set of all users in all graphs and $deg_y(u_i)$ is the degree of user u_i in graph G_y.

For the link-ship graph $G_1 = G_l = (U_l, L)$ we define a jump matrix J_1 in the same way as in Eq. 5 but only one line stochastic adjacency matrix $T_{U \times U}$ as detailed in Eq. 6.

$$T_{U \times U}(i,j) = \begin{cases} \frac{1}{deg_l(u_i)} & \text{if}(u_j, u_i) \in L \\ 0 & \text{otherwise} \end{cases} \tag{6}$$

We define now a first order random walk where the next steps probabilities depend only on the current location. Given a source node S we perform a multi-graph random walk of fixed length l. Steps are generated by the distribution detailed in Expressions 7:

$$\forall k \in [2,l], P(s_k|s_{k-1}) = \begin{cases} p_{z,y} \times \frac{1}{deg_y(s_{k-1})} & \text{if } (s_{k-1}, s_k) \in P_y \\ & \text{and } s_{k-1} = u_z \text{ and } s_k \in V_y \\ p_{z,l} \times \frac{1}{deg_l(s_{k-1})} & \text{if } (s_{k-1}, s_k) \in L \\ & \text{and } s_{k-1} = u_z \text{ and } s_k \in U \quad (7) \\ \frac{1}{deg_y(s_{k-1})} & \text{if } (s_{k-1}, s_k) \in P_y \\ & \text{and } s_{k-1} \in V_y \text{ and } s_k \in U \\ 0 & \text{otherwise} \end{cases}$$

The transition matrix is defined by blocks as follows:

$$T = \begin{bmatrix} J_1 \times T_{U \times U} & J_2 \times T_{U \times V_2} & \cdots & J_i \times T_{U \times V_i} & \cdots & J_n \times T_{U \times V_n} \\ \hline T_{V_2 \times U} & & & & & \\ \cdots & & & & & \\ T_{V_i \times U} & & & 0 & & \\ \cdots & & & & & \\ T_{V_n \times U} & & & & & \end{bmatrix}$$

For the example in Fig. 4 the jump matrices and the right stochastic adjacency matrices are as following (assuming $weight(G_1) = weight(G_2) = weight(G_3)$): $J_1 = diag(\frac{1}{3}, \frac{1}{3}, \frac{1}{2}, \frac{1}{3})$, $J_2 = J_3 = diag(\frac{1}{3}, \frac{1}{3}, 0, \frac{1}{3})$ and

$$T_{U \times U} = \begin{bmatrix} 0 & 0 & 0 & 1 \\ 0 & 0 & \frac{1}{2} & \frac{1}{2} \\ 0 & 1 & 0 & 0 \\ \frac{1}{2} & \frac{1}{2} & 0 & 0 \end{bmatrix} \quad T_{U \times V_2} = \begin{bmatrix} 1 & 0 & 0 \\ 1 & 0 & 0 \\ 0 & 0 & 0 \\ 0 & \frac{1}{2} & \frac{1}{2} \end{bmatrix} \quad T_{U \times V_3} = \begin{bmatrix} 0 & 1 \\ 1 & 0 \\ 0 & 1 \\ 1 & 0 \end{bmatrix}$$

$$T_{V_2 \times U} = \begin{bmatrix} \frac{1}{2} & \frac{1}{2} & 0 & 0 \\ 0 & 0 & 0 & 1 \\ 0 & 0 & 0 & 1 \end{bmatrix} \quad T_{V_3 \times U} = \begin{bmatrix} 0 & \frac{1}{2} & 0 & \frac{1}{2} \\ \frac{1}{2} & 0 & \frac{1}{2} & 0 \end{bmatrix}$$

Hence, the transition matrix is deduced as following:

	u_1	u_2	u_3	u_4	$v_{2,1}$	$v_{2,2}$	$v_{2,3}$	$v_{3,1}$	$v_{3,2}$
u_1	0	0	0	$\frac{1}{3}$	$\frac{1}{3}$	0	0	0	$\frac{1}{3}$
u_2	0	0	$\frac{1}{6}$	$\frac{1}{6}$	$\frac{1}{3}$	0	0	$\frac{1}{3}$	0
u_3	0	$\frac{1}{2}$	0	0	0	0	0	0	$\frac{1}{2}$
u_4	$\frac{1}{6}$	$\frac{1}{6}$	0	0	0	$\frac{1}{6}$	$\frac{1}{6}$	$\frac{1}{3}$	0
$v_{2,1}$	$\frac{1}{2}$	$\frac{1}{2}$	0	0					
$v_{2,2}$	0	0	0	1					
$v_{2,3}$	0	0	0	1			0		
$v_{3,1}$	0	$\frac{1}{2}$	0	$\frac{1}{2}$					
$v_{3,2}$	$\frac{1}{2}$	0	$\frac{1}{2}$	0					

5 Second Phase: Applying Natural Language Learning

In Sect. 4 we performed multi-graph random walk to translate both the structure and the content of the social network into walks. Walks collected in the final document can be interpreted as sentences, where the words are network nodes. Hence, inferring a link between a user node and an attribute value node is similar to the natural languages processing (NLP) problem of estimating the likelihood of words co-occurrence in a corpus.

Here we use a word2vec NLP model [5,9] with skip-gram model and hierarchical Softmax to encode the steps in embeddings. Embeddings where first introduced in 2003 by Bengio et al. [2]. The basic idea is to map one-hot encoded vectors that represent words in a high-dimensional vocabulary space to a continuous vector space with lower dimension. This approach has the virtue of storing the same information in a low-dimensional vector. The skip-gram model for NLP aims to compute words embeddings in order to predict the context of a given word. The input of the neural network is a high-dimensional one-hot vector which represents the target word and its output is a real low-dimensional vector, the *embedding* of the target word, that holds contextual information. The neural network is shallow with one hidden layer and the objective function given by Equation (4) in [10] maximizes the probability of appearance of the target word within a context of w words. This model has the advantage of generating good words representations [9] and it shows good results when it comes to learning structural representations of vertices in a social network [6,10].

Here we adapt this model to the disclosure of secret preferences of users in social networks. To that end, we perform weighted random walks on social graphs representing both friendship structures and attribute preferences of users. In contrast to [10] and [6] where users vertices which have similar friends will be mapped to similar embeddings, in our case both friends and preferences play a role in calibrating embeddings. The relative importance of friends and preferences in computing embeddings are quantified by the graphs weights. With this in mind, profiles that share the most important preferences (of highly weighted graphs) can have similar embeddings even if they do not have similar neighborhood. Moreover, vertices of different types, for instance movies, musics and users, are represented by vectors belonging to the same euclidean space. Hence, secret preferences will be easily predicted through linear algebra as detailed in Sect. 6. Additionally, by analyzing the variation of accuracy with respect to graph weights we deduce correlations between attributes as detailed in Sect. 7

6   Ranking Attribute Values for Predicting Preferences

Users, clusters of targeted attributes and values of other attributes, are encoded by vectors. The vectors are ranked according to a similarity measure with the target user vector. The inference algorithm will disclose as prefered attribute value one with the smallest rank or highest similarity.

In [12] Schakel and Wilson show that word2vec unsupervised learning algorithm encodes word semantics by affecting vectors in the same direction for co-occurrent words during training. Besides, the magnitude of a vector reflects both the frequency of appearance of related words in the corpus and the homogeneity of contexts.

In fact, words that appear in different contexts are represented by vectors that average vectors pointing in different contexts directions. Hence, the final vector magnitude generally decreases with respect to contexts. With that in mind, words used only in few contexts have generally higher magnitude than other words that have the same frequency but are used in more contexts. And the higher the word frequency is, the higher the chance it has to be used in different contexts.

To measure semantic similarity between vertices we apply cosine similarity which is widely used in NLP. This metric measures the cosine of the angle formed by two vectors which represent two different vertices. It yields values in the interval $[-1, 1]$ that quantify the topical similarity between vertices regardless their *centrality*. We discuss why cosine similarity is better adapted than euclidean distance for our purpose. For instance, *Star Wars* and *Titanic* are two famous movies that attract a large audience. The vertices which represent them are connected to many user vertices in the social network. Consequently, their embeddings weight average the embeddings of many dissimilar embeddings of many users. Hence, the embeddings magnitude of these two famous movies are lower than the embeddings magnitude of the other less famous movies. Therefore, the euclidean distance between the vectors which encode *Titanic* and *Star Wars* is lower than the euclidean distance between any of them and the rest of non famous movies. However, the angle between the vectors which represent *Star Wars* and *Titanic* is large due to the fact that they point in different context. With that in mind, if a given user likes Celine Dion song's *My Heart Will Go On*, his encoding vector will points in closer direction to the direction in which points the vector of *Titanic* because the vectors encoding *Titanic* and *My Heart Will Go On* points in similar context. Hence, we can predict that this user might like the *Titanic* movie even if the euclidean distance between the vector which encodes his vertex and the vector which encodes the *Titanic* is large. We note that if a user has few friends and few preferences his vector magnitude will be high. On the other hand, vectors encoding hub users have low magnitude. So their euclidean distances to the vectors encoding *Titanic* and *Star Wars* are low but they do not necessary like them.

7   Experimental Results

Dataset. Our dataset contains 15012 Facebook profiles of students and their direct friends. The sample is connected to more than 5 millions Facebook profiles from all over the world. Thus, we take up the challenging task of disclosing secret preferences of users from a highly diverse community with rich background from all over the globe. Our sampled graph of 15012 Facebook users is connected to 1022847 different liked objects. Objects are pages created on Facebook or any other object on the Internet connected to Facebook through Open Graph protocol (OGp). The OGp is a Facebook invention that enables any web page to become an object in a social graph[1]. Facebook labels objects by types. We

---

[1] https://developers.facebook.com/docs/sharing/opengraph.

counted 1926 different types of object in our sample. Those types of objects are considered as attributes and modeled by bipartite graphs in our model. For instance the most liked type of object in the sample is *community topics* with 137338 different liked objects.

Experimental Setup. We detail the example of disclosing the secret travel agency from which the target user books his vacation. We first select target users in the bipartite graph of Users-TravelAgencies and the hide their preferences. The selection algorithm seeks users who like at least λ travel agencies and removes $r\%$ of their preferred ones. We also add a constraint on the travel agencies graph connectivity in order to guarantee that the random walk detailed in Sect. 4 can reach any travel agency and the neural network detailed in Sect. 5 can learn about all the travel agencies.

Then we have performed random walks on 7 graphs (6 bipartite graphs for attributes and 1 friendship graph) as in Sect. 4 and where the travel agencies are labeled (w.r.t. clusters). In this example, we have selected graphs with similar sizes and densities, and various subjects to focus our tests on the subjects rather than the mathematical properties of the graph. Details about the analyzed graphs are given in Table 1. The results of the first clustering phase are also detailed in Table 2.

Table 1. Details about the graphs used for learning

Graphs	Sizes	Densities
Users-Users	15012 users	8.94×10^{-6}
Users-TravelAgencies	3370 users, 4827 travel agencies	6×10^{-4}
Users-ConsultingAgencies	2288 users, 4176 counsulting agencies	7×10^{-4}
Users-LocalBusiness	2386 users, 4350 local business	5×10^{-4}
Users-Politicians	2554 users, 4589 politicians	9×10^{-4}
Users-AppPages	4396 users, 4244 app pages	8×10^{-4}
Users-Causes	2547 users, 4410 causes	6×10^{-4}

Hyper-Parameters. We have tuned the hyper-parameters of the neural network as recommended in [10]. That is, the size of the skip-gram window is 10. The length of the walks is 80. The number of repetitions of walks is 10. And the dimension of the embeddings is 128. We rather focus on validating the weights of the different graphs. We used Bayesian optimization as depicted in [13] to automatically tune weights.

Results. We use the area under the ROC curve (AUC) as defined in [4] to measure the accuracy of the infered links. The amount that AUC exceeds 0.5 tells how much the inference algorithm is better than random guessing. The AUC for link prediction problem is computed as following:

$$\frac{nr_{(nel>esl)} + 0.5 \times nr_{(nel=esl)}}{n_{nel} \times n_{esl}}$$

where n_{nel} is the number of not existing links, n_{esl} is the number of existing but secret links, $nr_{(nel>sl)}$ is the number of couples of a not existing link and a secret link of smaller rank, $nr_{(nel=esl)}$ is the number couples of a not existing link and a secret link of the same rank. Note that AUC value will be 0.5 if the ranks are independent and identically distributed.

In our model links between the targeted user and the travel agencies which belong to the same cluster will have the same rank. Assuming that all clusters have different ranks (69 different cosines coded on 2 bytes in an euclidean space of dimension 128 where vectors are coded on 256 bytes) the AUC can be computed as following:

$$AUC = AUC_1 + AUC_2 \times \frac{nr_{(nel=esl)}}{n_{nel} \times n_{esl}}$$
$$AUC_1 = \frac{nr_{(nel>esl)}}{n_{nel} \times n_{esl}}$$

where AUC_1 is the accuracy of ranking clusters and AUC_2 is the accuracy of ranking values inside the selected cluster c_s (that should contain the secretly prefered value). Due to graph sparsity (only 2 travel agencies are liked by a user in average) we can make the following approximations when the goal is to predict one given secret link at a time ($n_{esl} = 1$).

$$n_{nel} \times n_{sl} \simeq m - 1$$
$$nr_{(nel=esl)} \simeq |c_s| - 1$$
$$AUC \simeq AUC_1 + AUC_2 \times \frac{|c_s|-1}{m-1}$$

Since $m = |TravelAgencies| = 4827$ and $\sqrt{m} - 1 \leq |c_s| - 1 \leq 2\sqrt{m} - 1$ we have

$$0.0146 = \frac{1}{\sqrt{m}+1} = \frac{\sqrt{m}-1}{m-1} \leq \frac{|c_s|-1}{m-1} \leq 2\frac{\sqrt{m}-1}{m-1} = 2\frac{1}{\sqrt{m}+1} = 0.0292$$

For the results depicted in Table 3 the rank inside clusters is generated by independent and identical distribution ($AUC_2 = 0.5$).Therefore $AUC_2 \times \frac{|c_s|-1}{m-1}$ is negligible w.r.t. AUC_1 in that case and does not affect the global accuracy of the prediction. We can observe that the obtained AUC in Table 3 are clearly above 0.5 showing a satisfactory performance from the proposed method. Computation times are in the order of a few minutes. Increasing the number of steps in random walks improves accuracy but affects efficiency.

Table 2. Processing of Users-TravelAgencies graph

Full Users-TravelAgencies graph			
Number of travel agencies	4827		
Number of user to travel agency links	9804		
Removed links			
Minimal degree of targets: λ	10		
Percentage of removed links per target: r	10	20	30
Number of targets	69	45	31
Total number of removed links	80	101	106
Graph partitions			
Number of clusters	68	68	68
Maximal number of Travel Agencies in a cluster	138	138	138
Minimal number of Travel Agencies in a cluster	69	69	69
Best cluster similarity between Travel Agencies	0.857	0.614	0.71
Worst cluster similarity between Travel Agencies	0.03	0.003	0.002
Mean of clusters similarities between Travel Agencies	0.12	0.112	0.114
Std of clusters similarity between Travel Agencies	0.14	0.114	0.12

Table 3. AUC and the best weights configurations.

Graphs	Best weights configurations		
Users-Users	0.2498	0.2154	0.168
Users-TravelAgencies	0.2498	0.138	0.209
Users-CounsultingAgencies	**0.0002**	**0.0002**	0.164
Users-LocalBusiness	**0.0002**	0.2154	**0.111**
Users-Politicians	0.2498	0.2154	0.117
Users-AppPages	**0.0002**	**0.0002**	**0.114**
Users-Causes	0.2498	0.2154	0.117
Percentage of removed links per target: r	10	20	30
Best Mean Accuracy Result (AUC)	0.6836	0.6715	0.6724
	69 targets	45 targets	31 targets

8   Conclusion

We have proposed a new method for infering hidden attribute values or preferences in social networks. The method relies on first clustering attribute values and then applying efficient machine learning technique from natural language processing. The method has been fully implemented and the first experiments are encouraging. However we need to perform larger scale experiments which is not easy due to the restrictions in crawling social networks and the needs to anonymize properly the collected data. The next step is to develop online tools for users so that they can control privacy leaks from their footprints in social networks.

References

1. Backstrom, L., Leskovec, J.: Supervised random walks: predicting and recommending links in social networks. CoRR, abs/1011.4071 (2010)
2. Bengio, Y., Ducharme, R., Vincent, P., Janvin, C.: A neural probabilistic language model. J. Mach. Learn. Res. **3**, 1137–1155 (2003)
3. Elkabani, I., Khachfeh, R.A.A.: Homophily-based link prediction in the Facebook online social network: a rough sets approach. J. Intell. Syst. **24**(4), 491–503 (2015)
4. Gao, F., Musial, K., Cooper, C., Tsoka, S.: Link prediction methods and their accuracy for different social networks and network metrics. Sci. Program. **2015**, 172879:1–172879:13 (2015)
5. Goldberg, Y., Levy, O.: word2vec Explained: deriving Mikolov et al.'s negative-sampling word-embedding method. CoRR, abs/1402.3722 (2014)
6. Grover, A., Leskovec, J.: node2vec: scalable feature learning for networks. In: Proceedings of the 22nd ACM SIGKDD International Conference on Knowledge Discovery and Data Mining, San Francisco, CA, USA, 13–17 August 2016, pp. 855–864 (2016)
7. Heatherly, R., Kantarcioglu, M., Thuraisingham, B.: Preventing private information inference attacks on social networks. IEEE Trans. Knowl. Data Eng. **25**(8), 1849–1862 (2013)
8. Lee, V.E., Ruan, N., Jin, R., Aggarwal, C.: A survey of algorithms for dense subgraph discovery. In: Aggarwal, C., Wang, H. (eds.) Managing and Mining Graph Data, pp. 303–336. Springer, Boston (2010)
9. Mikolov, T., Sutskever, I., Chen, K., Corrado, G., Dean, J.: Distributed representations of words and phrases and their compositionality. CoRR, abs/1310.4546 (2013)
10. Perozzi, B., Al-Rfou, R., Skiena, S.: Deepwalk: online learning of social representations. In: The 20th ACM SIGKDD International Conference on Knowledge Discovery and Data Mining, KDD 2014, New York, NY, USA, 24–27 August 2014, pp. 701–710 (2014)
11. Ryu, E., Rong, Y., Li, J., Machanavajjhala, A.: Curso: protect yourself from curse of attribute inference: a social network privacy-analyzer. In: Proceedings of the 3rd ACM SIGMOD Workshop on Databases and Social Networks, DBSocial 2013, New York, NY, USA, 23 June 2013, pp. 13–18 (2013)
12. Schakel, A.M.J., Wilson, B.J.: Measuring word significance using distributed representations of words. CoRR, abs/1508.02297 (2015)
13. Snoek, J., Larochelle, H., Adams, R.P.: Practical Bayesian optimization of machine learning algorithms. In: Advances in Neural Information Processing Systems 25: 26th Annual Conference on Neural Information Processing Systems 2012, Proceedings of a Meeting Held, 3–6 December 2012, Lake Tahoe, Nevada, USA, pp. 2960–2968 (2012)
14. Zheleva, E., Terzi, E., Getoor, L.: Privacy in social networks. In: Synthesis Lectures on Data Mining and Knowledge Discovery. Morgan & Claypool Publishers (2012)

Access Control Policies for Relational Databases in Data Exchange Process

Adel Jbali[✉] and Salma Sassi

Economics and Management of Jendouba/VPNC Laboratory, Faculty of Law,
University of Jendouba, Jendouba, Tunisia
adeljbali@mail.com, salma.sassi@fsjegj.rnu.tn

Abstract. Nowadays, many organizations rely on database systems as
the key data management technology for a large variety of tasks. This
wide use of such systems involved that security breaches and unau-
thorized disclosures threat those systems especially when the data is
exchanged between several parts in a distributed system. Consequently,
access control must adapt to this exchange process to maintain data privacy. In this paper, the challenge is to design an approach to deal with
access control policies in a context of data exchange between relational
databases. In fact, the main problem that we are dealing with is that
given a set of policies attached to a source schema and a set of mapping
rules to a target schema, the question is how the policies will pass from
the source schema to the target schema and what are the policies that
will be attached to the target schema to comply with the set of source
policies. For that purpose, we propose in this paper our methodology
called Policies-generation.

Keywords: Access control · Data exchange · Authorization view ·
Database security

1 Introduction

Databases may contain sensitive and critical government or enterprise information like medical document, personal health information or customer credit information cards. We focus in this paper on the security challenge that arises when
this data is exchanged between distributed parts. Yet, although the importance
of taking into consideration access control in a context of a data exchange, we
have remarked during our bibliographic study the absence of works that tackles directly this issue. Data exchange is the process of taking data structured
under a source schema and materializing an instance of a target schema that
reflects as accurately as possible the source data [12]. Hence, the challenge in a
such situation is to determine the way in which the access control policies will
be translated from the source database to the target database, and the set of
rules that will preserve the target policies to remain comply with source policies.
Complying with the source policies means that a prohibited access at the source

© Springer International Publishing AG 2017
D. Benslimane et al. (Eds.): DEXA 2017, Part I, LNCS 10438, pp. 264–271, 2017.
DOI: 10.1007/978-3-319-64468-4_20

level should also be prohibited at the target level to avoid policy violation. Thus, we propose a methodology that defines in the first time the set of access control policies attached to the source database and we give our reasons for the access control model choice, and in the second time we give our algorithm *Policies-generation* in which we specify how the access control policies are translated from the source to target database. In fact, our algorithm treats three steps: the first step is the policies filtering decision in which the algorithm determines the policies that will pass to the target database and the ignored policies. The second step is the policies modification decision in which our algorithm determines the set of policies that pass to the target database without modification and the policies that will be regenerated according to the mapping rules. In the last step, the algorithm generates the new policies respecting the mapping rules if it is possible. The remainder of the paper is organized as follows: Sect. 2 gives an overview of research effort related to our work. In Sect. 3 we describe our approach. We conclude in Sect. 4.

2   Background

In this section, we will discuss the different works which are related in any way to our problem. We highlight that, in actual database research literature, there are not significant contributions focusing on the relevant issue of accees control in relation with data exchange. This further confirms to us the novelty of the research we propose.

2.1   Logical Foundations of Relational Data Exchange

According to [12] data exchange is the problem of materializing an instance that adheres to a target schema, given an instance of a source schema and a specification of the relationship between the source schema and the target schema. This problem arises in many tasks requiring data to be transferred between independent applications that do not necessarily adhere to the same data format (or schema).

Definition 1. *A schema mapping is a triple $M = (S, T, \sum)$, where S and T are the disjoint source and target schema respectively, and $\sum$ is a finite set of sentences of some logical language over the schema $S \cup T$ [1]. We can define a schema mapping from a semantic view as a set of all pairs (I, J) where I is the source instance, J is the target instance and (I, J) must satisfies $\sum$. We said that the target instance J is a solution for I with respect to M if (I, J) satisfies $\sum$. The set of solution to I with respect to M is denoted by $Sol_M(I)$.*

- $\sum_{st}$ *consists of a set of source-to-target dependencies (stds), of the form:*

$$\forall \left(\phi(x) \longrightarrow \exists y \psi(x, y) \right)$$

Where $\phi(x)$ and $\psi(x, y)$ are conjunctions of atomic formulas in S and T respectively.

– $\sum_t$ *the set of target dependencies. It represents the union between a set of equality generating dependencies (egds) of the form:* $\forall\,(\phi(x) \longrightarrow xi = xj)$ *for* xi, xj *variables in* x, *and a set of tuple generating dependencies of the form* $\forall\,(\phi(x) \longrightarrow \exists y\psi(x,y))$ *where* $\phi(x)$ *and* $\psi(x,y)$ *are conjunctions of atomic formulas in* T.

2.2 Access Control Model for Relational Databases with Authorization Views

Definition (Authorization View) 2. *A set of authorization views specifies what information a user is allowed to access. The user writes the query in terms of the database relations, and the system tests (by considering the authorization views) the query for validity by determining whether it can be evaluated over the database* [4].

The view based access control model was emerged to enforce access control at tables, columns and even cell level (fine granularity). This model can be enforced with the definition of a secure context for each request that encapsulate the information related to the query [4]. Another feature of this model is its ability to define permission based on the data content. Declarative languages like SQL query language, facilitated the development of this specification using the concept of view and query rewriting technique. According to [6] the author classify view based access control model in two categories. The first one is Truman Model which applies the query rewriting technique to the user query to provide only the answers that are authorized to the user. The second model is Non-Truman Model, in a such model a query rewriting technique is also executed, but the difference according to the previous model is in the interpretation of the query. Thereby, In our methodology, no query rewrite is performed also we mention the use of authorization views based on conjunctive queries (the clause where) to represent both the sets of source policies and target policies.

3 Access Control and Data Exchange: A View Based Approach

In this section we introduce our approach and we show through the Fig. 1 an overview about our proposal. In fact, this approach is split in four major step: The first step incarnates the process of data exchange. The second step talks about the adaptation of the View Based Access Control model to define the authorization views. The third step talks about the views transformation and we introduce in the last step the algorithm Policies-generation.

3.1 Specification of the Mapping Rules (Phase A)

As we had mentioned previously, our methodology aims to translate the authorization views from the source to target database in a context of data exchange

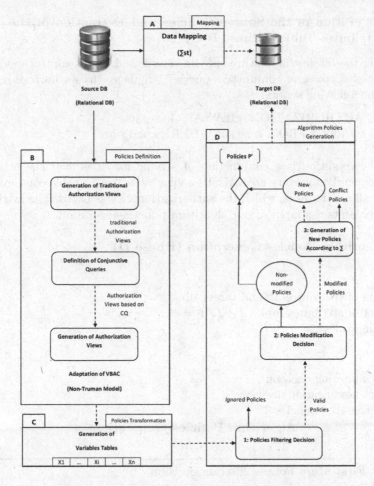

Fig. 1. Policies-generation: system architecture

respecting the mapping rules. The first type of tgds is a full tgds which there are no existentially quantified variables. It corresponds to a simple data migration from source-db to target-db:

$$\forall x \left(\phi(x) \longrightarrow \psi(x) \right)$$

The second tgds treats the case of exchanging the attributes of the tables through an integration mapping (used in data integration system) [9]. In fact, for the set of distinguish attributes A_i that was translated from the source-db to the target-db, we associate a set of distinguish authorization views for these attributes and we show through this mapping how to generate a global authorization view that synthesizes all the source authorization views related to those attributes through the intersection of conjunctive queries.

3.2   Generation of the Source Policies and Extraction of the Attributes-Tables (Phase B-C)

Admitting the interesting features of the view based access control model and the expressive power of conjunctive queries language, the authorization views admits the following syntax:

CREATE AUTHORIZATION VIEW AS view-name
SELECT attributes FROM relation WHERE condition

Where the condition is a conjunction of atomic formulas over the attributes A_i. Subsequently, for every authorization view we extract a table containing the names of all attributes on which the authorization view is based. The attributes-tables represents the input of our algorithm policies-generation.

3.3   Algorithm Policies-Generation (Phase D)

Input

– Schema of the source-db and target-db
– The set of attributes-tables T_j: $T_j[i] = A_i$
– Mapping rules

Output

– Policies filtering decision
– Policies modification decision
– New generated policies

Algorithm Policies-Generation

First step: Policies filtering decision

    Begin
1 $i \longleftarrow 0$
2 *solexistance* $\longleftarrow$ *true*
3 $n = T_j.length$
4 *while* $((solexistance = true) \,\&\, (i < n))\, do$
5 *if* $(\phi(T_j[i]) \longleftarrow \emptyset)$
6 *solexistance* $\longleftarrow$ *false*
7 *write(policy will be ignored)*
8 *else*
9 $i++$
10 *endif*
11 *endwhile*
12 *if* $(i = n)\, then$
13 *write("policy will pass to the target schema)*
14 *endif*
15 *Return*

This first step of our algorithm treats the problem of policy passing decision. Indeed, the algorithm run through the table T which represents the attributes A_i of the authorization view. Then for each attribute A_i in the table, it tests if this attribute admits a solution in the target schema, if it is not then the policy will be ignored since there is an attribute in the authorization view that had not been pass to the target schema. If we finish with a counter equal to the table length, it means that all the attributes of the authorization view had been passed to the target schema and the authorization view will also pass.

Second step: Policies modification decision
Begin
1   $i \longleftarrow 1$
2   $k \longleftarrow 1$
3   *for i from* 1 *to n do*
4   *if* $(\phi\,(T_j\,[i]) \longleftarrow \psi\,(Y_i))\,\&\,(A_i = Y_i)\,then$
5     $k + +$
6   *endif*
7   *endfor*
8   *if* $(k = n)\,then$
9   *write*("*policy will be pass to target schema without modification*)
10   *else*
11   *write*("*policy will be regenerated according to the mapping rules*)
12   *endif*
13   *Return*

In this step, our algorithm performed the following treatment. It compares the attribute A_i in the source schema with their solution Y_i in the target schema cell by cell. If it founds that all the attribute A_i in the source schema are identical to those in the target schema it avers that the authorization view will pass to the target schema without modification. Otherwise, the policy will be regenerated according to the mapping rules.

Third step: Generation of global view
Begin
1   $i \longleftarrow 1$
2   $k \longleftarrow 1$
3   $m \longleftarrow T_{j+1}.length$
4   $equal \longleftarrow true$
5   *while* $(k \leq m\,\&\,equal = true)\,do$
6   *for i from* 1 *to n do*
7   *if* $(\phi\,(T_j\,[i]) = \phi\,(T_{j+1}\,[k]))\,then$
8   *if* $(T_j\,[i]\,intersect\,T_{j+1}\,[k] <> \emptyset)\,then$

```
9    write ("the global view is the intersection of the source attributes")
10   else
11     equal ⟵ false
12     write ("conflict policies")
13   endif
14 ⟍ else
15   k ⟵ k + 1
16   endif
17   endwhile
18   if (k = m + 1) then
19   write ("the global view is the conjuction among all the source attributes")
20   endif
21   Return
```

In this last step, our algorithm generates a global authorization view according to the mapping rules mention in Sect. 3.1. Hence, in this step we consider two source authorization views represented by two attributes-tables T_j and T_{j+1}, our algorithm run through the tables and if it detects that the authorization views admit common attributes images by the mapping rules then for every common attributes images it performs an intersection between those ones. If the two common attributes have common elements means that $\phi(T_j) \cap \phi(T_{j+1}) \neq \emptyset$, then the global authorization view is defined based on the new intersection and by adding attributes that are not in common. If it is not the case $(\phi(T_j) \cap \phi(T_{j+1}) = \emptyset)$ and the two attributes doesn't have any common elements then we have a conflict policies in this case and the generation of a global authorization view may conduct to a policy violation. In the case where the source authorization views don't admit common attributes then the global authorization view is defined by the conjunction of all attributes of the conjunctive queries.

4   Conclusion

In this work we have investigated an interesting problem in databases access control, we have focused on the access control policies in a context of data exchange between relational databases. We proposed an approach which exploits the view based access control model and the conjunctive query language to define the set of source policies called authorization views. Then by reasoning about the attributes, we have extracted an algorithm able to produce the set of target policies that should be attached to the target database respecting to the mapping rules in order to comply with source policies. As perspectives of our work, we aim to deal with the case where our algorithm ignores the policy, we will try to find an insight that assures the passage of the policy from the source schema to target schema with the minimum missing data selection. Another perspective is to make our system more flexible to deal with different data representation model like XML and RDF.

References

1. Fuxman, A., Kolaitis, P.G., Miller, R.J., Tan, W.C.: Peer data exchange. ACM Trans. Database Syst. (2006)
2. Ten Cate, B., Chiticariu, L., Kolaitis, P., Tan, W.: Laconic schema mappings: computing the core with SQL queries. Proc. VLDB Endow
3. Marnette, B., Mecca, G., Papotti, P.: Scalable data exchange with functional dependencies. Proc. VLDB Endow. (2010)
4. Bertino, E., Ghinita, G., Kamra, A.: Access control for databases: concepts and systems. Found. Trends Databases
5. Gertz, M., Jajodia, S.: Handbook of Database Security Applications and Trends (2007)
6. Haddad, M., Hacid, M.-S., Laurini, R.: Data integration in presence of authorization policies. In: 11th IEEE International Conference on Trust, Security and Privacy in Computing and Communications, TrustCom (2012)
7. Arenas, M., Barceló, P., Reutter, J.: Query Languages for Data Exchange: Beyond Unions of Conjunctive. Springer Science+Business Media, LLC (2010)
8. Sellami, M., Hacid, M.S., Gammoudi, M.M.: Inference control in data integration systems. In: Debruyne, C. (ed.) OTM 2015. LNCS, vol. 9415, pp. 285–302. Springer, Cham (2015). doi:10.1007/978-3-319-26148-5_17
9. Sellami, M., Gammoudi, M.M., Hacid, M.S.: Secure data integration: a formal concept analysis based approach. In: Decker, H., Lhotská, L., Link, S., Spies, M., Wagner, R.R. (eds.) DEXA 2014. LNCS, vol. 8645, pp. 326–333. Springer, Cham (2014). doi:10.1007/978-3-319-10085-2_30
10. Nait Bahloul, S.: Inference of security policies on materialized views. Report master 2 (2009). http://liris.cnrs.fr/snaitbah/wiki
11. Kolaitis, P.G.: Schema mappings and data exchange. In: Proceedings of the Twenty-fourth ACM SIGMOD-SIGACT-SIGART Symposium on Principles of Database Systems (2012)
12. Fagin, R., Kolaitis, P.G., Miller, R.J., Popa, L.: Data exchange: semantics and query answering. Theor. Comput. Sci. (2002)
13. Miller, R.J.: Retrospective on Clio: schema mapping and data exchange in practice. In: Proceedings of the 2007 International Workshop on Description Logics (2007)
14. Rizvi, S., Mendelzon, A, Sudarshan, S., Roy, P.: Extending query rewriting techniques for fine-grained access control. In: International Conference on Management of Data (2004)

Web Search

Orion: A Cypher-Based Web Data Extractor

Edimar Manica[1,2(✉)], Carina F. Dorneles[3], and Renata Galante[1]

[1] II - UFRGS, Porto Alegre, RS, Brazil
{edimar.manica,galante}@inf.ufrgs.br
[2] Campus Ibirubá - IFRS, Ibirubá, RS, Brazil
[3] INE/CTC - UFSC, Florianópolis, SC, Brazil
dorneles@inf.ufsc.br

Abstract. The challenges in Big Data start during the data acquisition, where it is necessary to transform non-structured data into a structured format. One example of relevant data in a non-structured format is observed in entity-pages. An entity-page publishes data that describe an entity of a particular type (e.g. a soccer player). Extracting attribute values from these pages is a strategic task for data-driven companies. This paper proposes a novel class of data extraction methods inspired by the graph databases and graph query languages. Our method, called **Orion**, uses the same declarative language to learn the extraction rules and to express the extraction rules. The use of a declarative language allows the specification to be decoupled from the implementation. **Orion** models the problem of extracting attribute values from entity-pages as Cypher queries in a graph database. To the best of our knowledge, this is the first work that models the problem of extracting attribute values from template-based entity-pages in this way. Graph databases integrate the alternative database management systems, which are taking over Big Data (with the generic name of NoSQL) because they implement novel representation paradigms and data structures. Cypher is more robust than XPath (a query language that is common used to handle web pages) because it allows traversing, querying and updating the graph, while XPath is a language specific for traversing DOM trees. We carried out experiments on a dataset with more than 145k web pages from different real-world websites of a wide range of entity types. The **Orion** method reached 98% of F1. Our method was compared with one state-of-the-art method and outperformed it with a gain regarding F1 of 5%.

Keywords: Data extraction · Cypher · Graph database · Graph query language · NoSQL

1 Introduction

Heterogeneity, scale, timeliness, complexity, and privacy problems with Big Data impede progress at all the stages of the pipeline that creates knowledge from data. The problems start during data acquisition because much data is not

© Springer International Publishing AG 2017
D. Benslimane et al. (Eds.): DEXA 2017, Part I, LNCS 10438, pp. 275–289, 2017.
DOI: 10.1007/978-3-319-64468-4_21

Table 1. An output example of the data extraction.

Attribute	Value
Name	Lionel Messi
Position	FORWARD
Number	10
Place of birth	Rosario
Date of birth	24/06/1987
Weight	72 kg
Height	170 cm

Fig. 1. A real-world entity-page.

natively in a structured format and transforming such content into a structured format for later analysis is a major challenge [2]. One example of relevant data in a non-structured format is observed in template-based entity-pages. A template-based entity-page is a page, generated by a server-side template, which publishes data that describe an entity of a particular type. Figure 1 shows a fragment of a real-world entity-page of the Official FC Barcelona Web Site[1]. This entity-page publishes some attributes (*name, position, number*, among others) of an entity of the *soccer player* type. Extracting data from entity-pages is a strategic task for data-driven companies. For example, Google and Bing use data about entities to provide direct answers. Siri Voice Assistant uses data about entities to answer user queries. Other examples include the use of data about entities to analyze social media, search the deep web, and recommend products [15].

This paper focuses on the problem of extracting attribute values published in template-based entity-pages. This problem involves several challenges because: (i) the presentation formats and the set of attributes that form an entity are subject to a set of variations [12]; (ii) there are several distinct ways of publishing attribute values in template-based entity-pages (horizontal tables, vertical tables, horizontal lists, vertical lists, DIVs, SPANs, etc.); and (iii) the task of discovering the template-based entity-pages in the sites is not entirely precise [21], then the set of entity-pages of a site provided as input to a data extraction method can contain noise pages. Table 1 presents an output example of a data extraction method that receives, as input, the entity-page shown in Fig. 1.

The literature provides several methods to extract attribute values from template-based entity-pages. A significant portion [5,9,18,19,28] of these methods exploit traversal graphs of DOM (Document Object Model) trees that are mapped to XPath expressions for data extraction. Despite the recent popularization of the NoSQL (Not only SQL) databases and the advancement in research and development of graph databases and graph query languages, we did not find

---

[1] https://www.fcbarcelona.com/football/first-team/staff/players/2016-2017/messi/.

in the literature a method that uses a graph query language over a graph database for data extraction. Graph query languages are more robust than XPath because they allow traversing, querying and updating the graph, while XPath is a language specific for traversing DOM trees.

In this paper, we propose a novel class of data extraction methods inspired by the graph databases and the graph query languages. Our method, called **Orion**, uses a declarative graph query language to learn how to extract the attribute values in the template-based entity-pages of a site and carry out the extraction, so its specification is decoupled from its implementation. In particular, **Orion** automatically generates a set of Cypher queries over a graph database so that each query extracts the values of an attribute in all the entity-pages of a site.

We decided to model the problem as queries in a graph database because current graph databases are parallelizable, have an efficient distribution of index and memory usage, and have efficient graph algorithms [4,11]. We express the queries in the graph database using the Cypher language [1], a declarative graph query language that allows for expressive and efficient querying and updating of the graph store. Holzschuher and Peinl [20] argue that Cypher is a promising candidate for a standard graph query language because: (i) from a readability and maintainability perspective, Cypher seems well suited since its syntax is quite easy to understand for developers familiar with SQL; and (ii) for data with inherent graph-like structures, the code is additionally more compact and easier to read than SQL code.

We carried out experiments on a dataset with more than 145k web pages from different real-world websites of a wide range of entity types. **Orion** reached 98% of F1. We compared our method with one state-of-the-art method: *Trinity* [26]. Our method outperformed *Trinity* with a gain in terms of F1 of 5%.

This study serves at least the following purposes: (i) to provide a concise specification of the Cypher queries that extract the attribute values from template-based entity-pages of a site; (ii) to define an effective and automatic technique that uses Cypher instructions to generate the Cypher queries that extract the attribute values; and (iii) to show the effectiveness of our method through experiments. In our point of view, one important advantage of our method is to use the same declarative language (Cypher) for all the stages that are involved in the data extraction process.

This paper is structured as follows. Section 2 discusses related work. Section 3 defines the key concepts that are on the basis of our method. Section 4 presents the **Orion** method. Section 5 carries out experiments to determine the effectiveness of the **Orion** method and compares it with the baseline. Section 6 concludes the paper and makes recommendations for further studies in this area.

2   Related Work

Three of the most popular data extraction methods are *RoadRunner* [13], *ExAlg* [6], and *DEPTA* [27]. *RoadRunner* and *ExAlg* learn a regular expression that models the template used to generate the input documents, while *DEPTA* exploits a partial tree alignment algorithm. Inspired by these methods, other methods were proposed aiming to fulfill specific gaps. A recent example is the *Trinity* method, which receives a set of template-based entity-pages of a site and returns the attribute values published in these pages. Whenever *Trinity* finds a shared pattern among the entity-pages, it partitions them into the prefixes, separators, and suffixes and analyses the results recursively, until no more shared patterns are found. Prefixes, separators, and suffixes are arranged in a tree that is traversed to build a regular expression with capturing groups that represents the template that was used to generate the entity-pages. *RoadRunner*, *ExAlg*, *DEPTA* and *Trinity* are unsupervised, i.e., they learn rules that extract as much prospective data as they can, and the user then gathers the relevant data from the results. **Orion** is also unsupervised, but it uses declarative languages while *RoadRunner*, *ExAlg*, *DEPTA* and *Trinity* use imperative languages. Declarative languages are more intuitive than imperative languages because declarative languages focus on what the computer is to do, and the imperative languages focus on how the computer should do it, i.e., the declarative languages get away from "irrelevant" implementation details [25].

The most methods that use declarative languages for data extraction exploit traversal graphs of DOM trees that are mapped to XPath expressions [5,9,18,19,28]. **Orion** also exploits traversal graphs of DOM trees, but it uses Cypher expressions. Cypher is more robust than XPath because it allows traversing, querying and updating the graph, while XPath is a language specific for traversing DOM trees. These features of the Cypher language allow it to be used in all the stages of the process of data extraction. Badica *et al.* [7] propose a data extraction method (*L-wrapper*) that has declarative semantics. *L-wrapper* employs inductive logic programming systems for learning extraction rules, and XSLT technology for performing the extraction. **Orion** employs Cypher for both leaning extraction rules (that are Cypher queries) and performing the extraction. The *W4F* toolkit [24] consists of a retrieval language to identify Web sources, a declarative extraction language (the HTML Extraction Language) to express robust extraction rules and a mapping interface to export the extracted information into some user-defined data-structures. Users guide the generation of the extraction rules using a visual extraction wizard. On the other hand, **Orion** does not require the user intervention to learn the extraction rules.

Other works have goals or strategies that are similar to those of **Orion**, but they are not directly comparable. *WebTables* [10] and *ListExtract* [16] extract data published in HTML tables and lists, respectively. *DeepDesign* [22] extracts data from web pages while browsing. Users guide the extraction process of *Deep-Design* by performing a labeling of fields within an example record and can fine-tune the process as it runs based on an incremental, real-time visualization of results. *ClustVX* [17] extracts attribute values from a page that describes several entities. *SSUP* [21], which is our previous work, discovers entity-pages on the web by combining HTML and URL features. *WADaR* [23] is a post-processing strategy to eliminate noise in the extracted values. **Orion** is more generic applicable than *WebTables* and *ListExtract* since it does not require the data to be arranged in a limited number of specific patterns that frequently occur on the Web. Moreover, CSS has enabled designers to move away from table-based layouts and many of the structural cues used in table-based methods have been eradicated from HTML [22]. **Orion** is unsupervised, as opposed to *DeepDesign* that requires the user to label the web page and monitor the entire process of data extraction. **Orion** is effective in extracting attribute values from pages that describe a single entity (entity-pages), as opposed to *ClustVX*. A study [14] showed that entity-pages are among the most common forms of structured data available on the web. *SSUP* and *WADaR* are complementary to **Orion** since *SSUP* finds the entity-pages, **Orion** extracts their attribute values, and *WADaR* eliminates noise in the extracted values.

3   Preliminary Definitions

An `entity` is an object, which is described by its attributes (e.g. *name*, *weight*, and *height*). There are many categories of entities, called `entity types`. A `template-based entity-page` is a web page that describes an entity of a particular type and is dynamically generated by populating fixed HTML templates with content from structured databases. For example, Fig. 1 presents a template-based entity-page that describes an entity of the *soccer player* type.

S-Graph. Since we are using a graph database, we represent a set of template-based entity-pages of a site as a disconnected graph, called `S-graph`. The S-graph is composed of trees, one for each entity-page. The tree of an entity-page is a modified version of the DOM tree. Our modified version differs from the DOM tree because it only includes tag nodes and textual nodes, as well as, it stores the following properties of each node: (i) *value* - the name of a tag node or the text of a textual node; (ii) *url* - the URL of the entity-page; (iii) *type* - the literal "*TAG*" for tag nodes and the literal "*TEXT*" for textual nodes; (iv) *position* - the information of the position of the node related to its siblings; and (v) *function* - the literal "*TEMPLATE*" for textual nodes that are part of the template of the entity-pages, the literal "*DATA*" for other textual nodes and the absence of this property for tag nodes.

Definition 1. *An S-graph is a pair (N, E), where N is a set of nodes and E is a set of relationships (edges) between nodes. There is a node in the S-graph for each tag node or textual node of each template-based entity-page. There is a relationship between two nodes (n_x and n_y) in the S-graph if n_y is a child of n_x in the DOM tree of the entity-page. Each node has the properties: value, url, type, position, and function (except for tag nodes).*

For example, Fig. 2 shows the HTML of the entity-page presented in Fig. 1. Figure 3 presents the visual representation of the S-graph of the site that contains the entity-page given in Fig. 1. We highlighted data nodes using bold face and template nodes using italic.

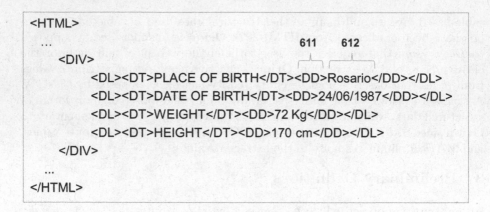

Fig. 2. HTML of the entity-page presented in Fig. 1.

Cypher Wrapper. In web data extraction, a `wrapper` is a program that extracts structured data from a collection of web pages [23]. From a collection of web pages P, a wrapper produces a relation R with schema Σ. A `relation` R is a set of n-tuples $\{U_1, .., U_n\}$ where n is the number of tuples. A `schema` $\Sigma = \{A_1, .., A_m\}$ is a tuple representing the attributes of the relation, where m is the number of attributes. In our context, the pages in P are entity-pages of a site that follow the same template. We model the wrapper through Cypher queries over the S-graph.

Definition 2. *A `Cypher wrapper` W is a set of queries $\{Q_1, .., Q_m\}$, where each Q_i is a Cypher query over the S-graph that extracts the values of the A_i attribute from all the entity-pages in the S-graph. The application of a Cypher wrapper means to execute its queries over the S-graph and associate the values returned by the queries to the corresponding attribute. The application of a Cypher wrapper W to a collection of entity-pages P produces a relation with schema $\{A_1, .., A_m\}$.*

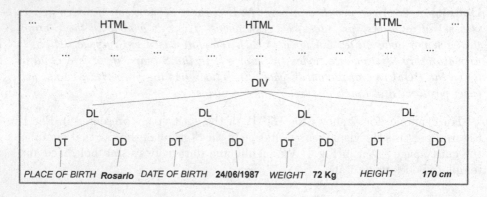

Fig. 3. S-graph of the site that contains the entity-page presented in Fig. 1.

For example, Fig. 7 shows the Cypher query that extracts the *height* attribute in the S-graph given in Fig. 3. The application of a Cypher wrapper to the entity-page presented in Fig. 1 returns the tuple shown in Table 1.

This paper focuses on the challenge of inducing a Cypher wrapper from a collection of template-based entity-pages of a site. The **Orion** method carries out this induction automatically, as described in the next section.

4   The Orion Method

This section describes **Orion**, a Cypher-based Web Data Extractor that extracts attribute values from template-based entity-pages. Figure 4 presents an overview of the **Orion** method. The input is a set of template-based entity-pages of a site. The output is the set of attribute values published in these entity-pages. The **Orion** method has four main stages: (1) *Graph Building* builds an S-graph from the entity-pages provided as input to the method; (2) *Template Recognition* identifies the nodes in the S-graph that are part of the template behind the entity-pages; (3) *Query Induction* induces Cypher queries from the template nodes; (4) *Value Extraction* extracts the attribute values published in the template-based entity-pages by executing the induced Cypher queries over the S-graph. Next subsections describe these stages in details.

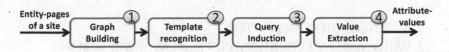

Fig. 4. Overview of the **Orion** method.

4.1   Graph Building

This stage parses the HTML of the entity-pages provided as input to the **Orion** method and builds the S-graph, without the *function* property, which is added in the next stage (Template Recognition). The S-graph is built using the Cypher language. For example, Fig. 5 shows the Cypher instructions that insert in the S-graph: (a) a node representing the tag identified in Fig. 2 by ID 611; (b) a node representing the textual content identified in Fig. 2 by ID 612; and (c) a relationship representing that the 612 node is a child of the 611 node.

```
(a)   CREATE (n {value: "DD", url: "http://... ", type: "TAG", position:2}) RETURN ID(n)     //611
(b)   CREATE (n {value: "Rosario", url: "http://...", type: "TEXT", position:1}) RETURN ID(n) //612
(c)   MATCH (a), (b) WHERE ID(a) = 611 AND ID(b) = 612 CREATE (a) - [r:contains] -> (b)
```

Fig. 5. Examples of Cypher instructions to create a S-graph. These instructions insert into the S-graph: (a) a tag node; (b) a textual node; and (c) a relationship.

For example, if the input set is composed of the entity-pages of the site that contains the entity-page given in Fig. 1, this stage builds the S-graph shown in Fig. 3. However, the textual nodes are not classified as template (italic) or data (bold) at this stage, since the *function* property is not defined yet.

4.2   Template Recognition

This stage evaluates the nodes of the S-graph and finds the ones that are part of the template behind the entity-pages. These nodes are called template nodes. A `template node` is a textual node with a value that occurs in, at least, δ entity-pages. Other textual nodes are called `data nodes`. Figure 6(a) presents the Cypher instruction that finds the template nodes and sets their *function* property. Line 1 finds all the textual nodes. Line 2 obtains the values (of the textual nodes) that occur in, at least, δ distinct entity-pages. Line 3 adds the values obtained through Line 2 to a list L. Line 4 finds all the textual nodes with a value that belongs to the L list. Line 5 assigns the literal "*TEMPLATE*" to the *function* property of the nodes obtained through Line 4. Figure 6(b) shows the Cypher instruction that finds the data nodes and sets their *function* property. This instruction must be executed after the Cypher instruction that finds the template nodes. In Fig. 6(b), Line 1 finds the textual nodes without the *function* property (at this moment, only the template nodes have the *function* property). Line 2 assigns the literal "*DATA*" to the *function* property of the nodes obtained through Line 1.

(a)
```
1. MATCH (a) WHERE a.type= "TEXT"
2. WITH a.value AS v, COUNT(DISTINCT a.url) AS qt WHERE qt >= δ
3. WITH COLLECT(v) AS L
4. MATCH (b) WHERE b.type="TEXT" AND ANY(x IN L WHERE x=b.value)
5. SET b.function= "TEMPLATE"
```

(b)
```
1. MATCH (a) WHERE a.type= "TEXT" AND a.function IS NULL
2. SET a.function= "DATA"
```

Fig. 6. Cypher instructions to set *function* property.

For example, if the input set is composed of the entity-pages of the site that contains the entity-page presented in Fig. 1, this stage classifies the textual nodes of the S-graph (illustrated in Fig. 3) as template (highlighted in italic) or data (highlighted in bold). After this stage, the S-graph is complete.

4.3   Query Induction

This stage induces the Cypher queries (over the S-graph) that allow us to extract the attribute values published in the entity-pages. For each data node, a breadth-first search finds the closest template node. The distance is the number of nodes between the template node and the data node. The path from the closest template node to the data node is used to build a Cypher query. The query filters: (i) the *value* property of all the nodes in the path, except the data node because the *value* property of the data node is what we want to extract; (ii) the *position* property of all the nodes in the path; (iii) the *type* property of all the nodes in the path. The query returns the *value* and *url* properties of the data node. The premise behind this stage is that data nodes from different entity-pages that contain the value of the same attribute generate the same query. The queries that are not generated by, at least, γ data nodes (from different entity-pages) are filtered out.

For example, Fig. 7 presents the Cypher query generated from the data node with *value= "170 cm"* (in the S-graph illustrated in Fig. 3). The closest template node to this data node is the template node with *value= "HEIGHT"* (there are three nodes between them). This Cypher query specifies a path in the S-graph, which: (i) starts at a node with *value= "HEIGHT"*, *type= "TEXT"*, and *position=1*; (ii) follows to a node with *value= "DT"*, *type= "TAG"*, and *position=1*; (iii) follows to a node with *value= "DL"*, *type= "TAG"*, and *position=4*; (iv) follows to a node with *value= "DD"*, *type= "TAG"*, and *position=2*; and (v) ends at a node with *type= "TEXT"* and *position=1*. This Cypher query returns the *value* and *url* properties of the data node.

```
MATCH (a)--(b)--(c)--(d)--(e)
WHERE a.value="HEIGHT"  AND a.type="TEXT"  AND a.position=1
   AND b.value="DT"     AND b.type="TAG"   AND b.position=1
   AND c.value="DL"     AND c.type="TAG"   AND c.position=4
   AND d.value="DD"     AND d.type="TAG"   AND d.position=2
                        AND e.type="TEXT"  AND e.position=1
RETURN e.url, e.value
```

Fig. 7. An example of a Cypher query.

4.4 Value Extraction

This stage executes the Cypher queries (induced in the previous stage) over the *S-graph* to extract the attribute values published in the entity-pages. The Cypher queries can be applied over the S-graph used to generate them or over another S-graph with other entity-pages of the same template (same site).

For example, the execution of the query given in Fig. 7 over the entity-page shown in Fig. 1 returns *"170 cm"*. The application of all the Cypher queries (induced from the S-graph that is presented in Fig. 3) over the entity-page shown in Fig. 1 produces the output presented in Table 1.

5   Experimental Evaluation

The experiments were carried out to evaluate the efficacy of **Orion** and compare it with one state-of-the-art method for data extraction from entity-pages.

Table 2. Summary of the datasets.

Dataset	Entity type	Sites	Entity-pages	Attributes
DATASET_S	Autos	10	17,923	Model, price, engine, fuel_economy
	Books	10	20,000	Title, author, isbn_13, publisher, publication_date
	Cameras	10	5,258	Model, price, manufacturer
	Jobs	10	20,000	Title, company, location, date_posted
	Movies	10	20,000	Title, director, genre, mpaa_rating
	NBA players	10	4,405	Name, team, height, weight
	Restaurants	10	20,000	Name, address, phone, cuisine
	Universities	10	16,705	Name, phone, website, type
DATASET_W	Books	10	1,315	Author, title, publisher, ISBN13, binding, publication date, edition
	Stock quotes	10	4,646	Last value, day high, day low, 52 wk high, 52 wk low, change %, open, volume, change $
	Soccer players	10	5,745	Position, birthplace, height, national team, club, weight, birthdate, nationality, number
	Videogames	10	12,329	Publisher, developer, ESRB, genre
Total		120	148,326	

5.1 Experimental Setup

Datasets. The experiments were carried out in two datasets ($DATASET\_S^2$ and $DATASET\_W^3$). Each dataset is a collection of real-world web entity-pages categorized per entity type. These datasets include a ground truth that describes the attribute values in each entity-page. We chose these datasets because they have been used in the literature [9,19,23] to evaluate different web data extraction methods. Table 2 summarizes the datasets.

Metrics. We used standard measures such as precision, recall, and F1 [8]. The extracted values were compared with the values in the ground truth to compute the number of true positives (TP), false negatives (FN), and false positives (FP), since this allowed us to assess precision as $P = \frac{VP}{VP+FP}$, recall as $R = \frac{VP}{VP+FN}$,

---

[2] Available on http://swde.codeplex.com/.
[3] Available on http://www.dia.uniroma3.it/db/weir/.

and F1 as $F1 = 2\frac{P \times R}{P+R}$. The Student's t-test [3] with the standard significance level ($\alpha = 0.05$) was employed to determine whether the difference between the results of two methods was statistically significant.

Baselines. We chose *Trinity* as our baseline because it is: (i) a state-of-the-art method for data extraction from template-based entity-pages; (ii) unsupervised; (iii) not restricted to specific HTML patterns (e.g. HTML tables and lists); (iv) carried out experiments comparing it with four baselines (including RoadRunner [13]) and outperformed all them in terms of effectiveness.

Methodology. The results of *Trinity* were obtained through the prototype provided by the authors[4]. *Trinity* has two parameters (*min* and *max*), which represent the minimum and maximum size respectively of the shared patterns for which the algorithm searches. We used the parameter configuration defined in [26], i.e., $min = 1$ and $max = 0.05 \times m$, where m denotes the size in tokens of the smallest entity-page of the site. The **Orion** method has two parameters: (i) δ - the number of entity-pages that a value must occur to be considered part of the template; and (ii) γ - the pruning threshold of the generated queries. We tested the values 10%, 20%, .., 100% to the δ and γ parameters on the *book* type of *DATASET_W*. The configuration that delivered the highest F1 was using $\delta = 60\%$ and $\gamma = 30\%$. This configuration was employed in all the experiments because it was able to achieve the highest degree of efficacy.

We used the Neo4j 3.0.6 [1] to store the S-graph because it is a highly scalable native graph database. However, another graph database that supports Cypher can be used. Another graph query language can also be used. In this case, the queries presented in this paper must be mapped to the new language. We replace the literals "TEXT", "TAG", "TEMPLATE", and "DATA" with integers.

Orion and *Trinity* are unsupervised techniques, which means that it is the user who has to assign a semantic label to each extraction rule. An extraction rule is a Cypher query in **Orion** and a capturing group in *Trinity*. We adopted the same methodology as *Trinity* [26] to discover the extraction rule that extracts each attribute in the ground truth. Specifically, we found the extraction rule that is the closest to each attribute. To do so, we compared the values extracted by each extraction rule to every attribute in the ground truth. Given an attribute in the ground truth, we considered that the precision and recall to extract them corresponds to the extraction rule (Cypher query or capturing group) with the highest F1 measure.

5.2 Orion *versus* Trinity

In this experiment, we aim to answer the following question: *which method is more effective:* **Orion** *or Trinity*? We evaluate the efficacy through the F1 metric. The higher the F1 value, the greater the degree of efficacy. Table 3 presents the recall, precision, and F1 produced by the methods in each entity type of the

---

[4] Available at http://www.tdg-seville.info/Download.ashx?id=341.

two datasets. On average, **Orion** and *Trinity* reached an F1 of 0.98 and 0.93, respectively. The mean gain of **Orion** in terms of F1 was 5%. The scale of the web translates this percentage into a large number of attribute values that can be correctly extracted. The t-test shows that this gain is statistically significant because the p-value (3.12×10^{-7}) is less than the level of significance $(\alpha = 0.05)$.

Table 3. Comparison between **Orion** and *Trinity*.

Dataset	Entity type	Recall		Precision		F1		
		Trinity	Orion	Trinity	Orion	Trinity	Orion	%
DATASET_S	Autos	0.96	0.99	0.96	1.00	0.96	0.99	3
	Books	0.90	0.97	0.92	1.00	0.90	0.98	**9**
	Cameras	0.96	0.91	0.98	1.00	0.97	0.95	−2
	Jobs	0.96	0.92	0.97	0.98	0.96	0.94	−3
	Movies	0.97	0.97	0.99	1.00	0.97	0.98	1
	NBA players	0.95	1.00	0.95	1.00	0.95	1.00	5
	Restaurants	0.91	0.96	0.91	1.00	0.91	0.98	7
	Universities	0.93	0.97	0.93	1.00	0.93	0.98	5
DATASET_W	Books	0.86	0.98	0.86	0.99	0.86	0.99	**15**
	Stock quotes	0.92	1.00	0.95	1.00	0.92	1.00	8
	Soccer players	0.92	0.96	0.91	0.97	0.91	0.97	6
	Videogames	0.93	0.97	0.94	1.00	0.93	0.98	5
Mean		0.93	0.97	0.94	0.99	0.93	0.98	5

Orion outperformed *Trinity* because, as opposed to *Trinity*, **Orion** is not affected by: (i) missing attributes (i.e. attributes where the values are only published in some of the entity-pages of the site) that are published in simple elements (i.e. when the element has only one textual node); (ii) attributes published in horizontal tables (i.e. the header is in the first line of the table); and (iii) noise pages (i.e. pages in the input set that follow a template different from the most entity-pages in the set). *Trinity* mistakenly assumes that each value of a missing attribute is a part of the value of the immediately preceding attribute in the HTML of the page. *Trinity* is not able to extract correctly the attribute values that are published in the middle columns of the horizontal tables because these attribute values have the same prefix and suffix. In this case, *Trinity* erroneously regards values from different attributes as being the values of a same multivalued attribute. A noise page in the input set makes *Trinity* creates a wrong ternary tree and, consecutively, generates an erroneous regular expression, which affects its efficacy. This behavior occurs because *Trinity* is based on patterns shared by all the entity-pages provided as input. However, the task of collecting the entity-pages of a site is not trivial and, eventually, some non-entity-pages are collected together. Another situation verified in the datasets is the occurrence of entity-pages (that describe entities of the same type) within the same site, but that follow a different template. For example, in the *job* type, there is a site where the attributes values are published: (i) on the left in some entity-pages; (ii) on the right in some entity-pages; and (iii) on the center in other entity-pages.

Orion outperformed *Trinity* in 83% of the entity types. **Orion** obtained the highest gain regarding F1 in the *book* type (15% in the $DATASET\_W$ dataset) and (9% in the $DATASET\_S$ dataset). This type contains several missing attributes. *Trinity* outperformed **Orion** in the *camera* and *job* types. However, the difference did not exceed 3%. Section 5.3 explains the cases of failure of the **Orion** method that caused this negative difference.

It is important to note that we performed exhaustive experiments with **Orion** and *Trinity*, which included 148,326 entity-pages from 120 real-world websites from different entity types. On average, each site has 1,236.00 entity-pages. The experiments performed by the authors of *Trinity* included, on average, 37.89 entity-pages per site and the maximum number of entity-pages of a site was 252.

The results show that **Orion** has an efficacy that is significantly higher than *Trinity* because on average it achieved the highest F1 value. The reason for this is that **Orion**, as opposed to *Trinity*, is not affected by the noise pages, the attribute values published in the horizontal tables or the missing attributes. Hence **Orion** is able to extract correct values in more entity-pages and to reduce the number of values that are erroneously extracted. Moreover, the implementation of the **Orion** method is based on a declarative language while the implementation of the *Trinity* method is based on an imperative language. Declarative languages are more intuitive than imperative languages [25].

5.3 Analysis of the Cases of Failure

To better understand our results, we have analyzed the cases of failure of the **Orion** method. This analysis might be very useful for developers of new methods to extract attribute values from template-based entity-pages because it shows the difficulties of handling the particularities of this kind of pages.

The first case of failure is when the values of an attribute are published in the entity-pages of a site in textual nodes that also contain other information. In some sites, the label and the value of the attribute are published in the same textual node. In other sites, the values of two attributes are published in the same textual node. **Orion** has textual node granularity, i.e., it does not segment the content of a textual node. Therefore, **Orion** extracts noise with the values of the attribute. A possible solution is to apply a post-processing to segment the extracted values and eliminate noise, as carried out by Ortona *et al.* [23].

The second case of failure occurred in sites where the attribute values were published in mixed elements (elements with two or more textual nodes). **Orion** extracted incorrect values when the attribute values were published in mixed elements and one of the following variations occurred: (i) missing attributes - attributes where the values are only published in some of the entity-pages of the site; (ii) multivalued attributes - more than one value is published for the same attribute of the same entity; and (iii) multiordering attributes - the attributes are published in a different order in the entity-pages of the site. These variations did not affect the efficacy of the **Orion** method when the attribute values were published in simple elements (i.e. elements with only one textual node). A possible solution is to apply a pre-processing to mixed elements.

The third case of failure occurred when the values of an attribute are published in the entity-pages of a site using different tags. Let v be the node that contains the value of an attribute and w be the closest template node to v. If the variation in the template occurs in the subtree rooted at the lowest common ancestor of v and w, then the recall of **Orion** is affected. A possible solution to these cases is to select complementary Cypher queries for each attribute.

6   Conclusion

In this paper, we proposed **Orion**, a novel method to extract attribute values from template-based entity-pages, which is inspired by the graph databases and graph query languages. Our method represents the template-based entity-pages as a graph, named S-graph. The **Orion** method automatically induces a set of Cypher queries over the S-graph where each query extracts the values of an attribute from all the entity-pages of a site. The **Orion** method has declarative semantics, i.e., its specification is decoupled from its implementation. Moreover, the Cypher language is more robust than XPath because it allows traversing, querying and updating the graph, while XPath is a language specific for traversing DOM trees. Our experiments, performed on 120 real-world websites, proved that the **Orion** method achieves very high precision, recall, and F1. Our method outperformed a state-of-the-art method (*Trinity*) regarding F1. As future work, we plan to: (i) define a pre-processing to handle with mixed elements; (ii) a post-processing to eliminate noise from the extracted values; (iii) evaluate the processing time of our method; and (iv) compare the **Orion** method with other methods based on declarative languages (e.g. [5,7,9,18,19,24,28]).

References

1. Neo4j Tecnology. The Neo4j Manual v2.3.3, March 2016. http://neo4j.com/docs/stable/index.html
2. Agrawal, D., et al.: Challenges and Opportunities with Big Data – A Community White Paper Developed by Leading Researchers Across the United States, March 2012. http://cra.org/ccc/docs/init/bigdatawhitepaper.pdf
3. Anderson, T., Finn, J.: The New Statistical Analysis of Data. Springer Texts in Statistics. Springer, Heidelberg (1996)
4. Angles, R., Gutierrez, C.: Survey of graph database models. ACM Comput. Surv. **40**(1), 1:1–1:39 (2008)
5. Anton, T.:. Xpath-wrapper induction by generating tree traversal patterns. In: Bauer, M., Brandherm, B., Fürnkranz, J., Grieser, G., Hotho, A., Jedlitschka, A., Kröner, A. (eds.) Lernen, Wissensentdeckung und Adaptivität (LWA) 2005, GI Workshops, Saarbrücken, 10th–12th October 2005, pp. 126–133. DFKI (2005)
6. Arasu, A., Garcia-Molina, H.: Extracting structured data from web pages. In: Proceedings of the 2003 ACM SIGMOD International Conference on Management of Data, SIGMOD 2003, pp. 337–348. ACM, New York (2003)
7. Badica, C., Badica, A., Popescu, E., Abraham, A.: L-wrappers: concepts, properties and construction. Soft. Comput. **11**(8), 753–772 (2007)

8. Baeza-Yates, R., Ribeiro-Neto, B.: Modern Information Retrieval: The Concepts and Technology Behind Search. Addison Wesley Professional, Boston (2011)
9. Bronzi, M., et al.: Extraction and integration of partially overlapping web sources. VLDB Endow. **6**(10), 805–816 (2013)
10. Cafarella, M.J., Halevy, A., Wang, D.Z., Wu, E., Zhang, Y.: Webtables: exploring the power of tables on the web. VLDB Endow. **1**(1), 538–549 (2008)
11. Cattell, R.: Scalable SQL and NoSQL data stores. SIGMOD Rec. **39**(4), 12–27 (2011)
12. Chang, C.-H., Kayed, M., Girgis, M.R., Shaalan, K.F.: A survey of web information extraction systems. IEEE Trans. Knowl. Data Eng. **18**(10), 1411–1428 (2006)
13. Crescenzi, V., Mecca, G.: Automatic information extraction from large websites. J. ACM **51**(5), 731–779 (2004)
14. Crestan, E., Pantel, P.: Web-scale table census and classification. In: Proceedings of the Fourth ACM International Conference on Web Search and Data Mining, WSDM 2011, pp. 545–554. ACM, New York (2011)
15. Deshpande, O., et al.: Building, maintaining, and using knowledge bases: a report from the trenches. In: SIGMOD (2013)
16. Elmeleegy, H., Madhavan, J., Halevy, A.: Harvesting relational tables from lists on the web. VLDB **20**(2), 209–226 (2011)
17. Grigalis, T.: Towards web-scale structured web data extraction. In: WSDM (2013)
18. Gulhane, P., Rastogi, R., Sengamedu, S.H., Tengli, A.: Exploiting content redundancy for web information extraction. In: WWW (2010)
19. Hao, Q., Cai, R., Pang, Y., Zhang, L.: From one tree to a forest: a unified solution for structured web data extraction. In: SIGIR (2011)
20. Holzschuher, F., Peinl, R.: Performance of graph query languages: comparison of cypher, gremlin and native access in Neo4j. In: Proceedings of the Joint EDBT/ICDT 2013 Workshops, EDBT 2013, pp. 195–204. ACM, New York (2013)
21. Manica, E., Galante, R., Dorneles, C.F.: SSUP – a URL-based method to entity-page discovery. In: Casteleyn, S., Rossi, G., Winckler, M. (eds.) ICWE 2014. LNCS, vol. 8541, pp. 254–271. Springer, Cham (2014). doi:10.1007/978-3-319-08245-5_15
22. Murolo, A., Norrie, M.C.: Revisiting web data extraction using in-browser structural analysis and visual cues in modern web designs. In: Bozzon, A., Cudre-Maroux, P., Pautasso, C. (eds.) ICWE 2016. LNCS, vol. 9671, pp. 114–131. Springer, Cham (2016). doi:10.1007/978-3-319-38791-8_7
23. Ortona, S., Orsi, G., Furche, T., Buoncristiano, M.: Joint repairs for web wrappers. In: 32nd IEEE International Conference on Data Engineering, ICDE 2016, Helsinki, Finland, 16–20 May 2016, pp. 1146–1157. IEEE Computer Society (2016)
24. Sahuguet, A., Azavant, F.: Wysiwyg web wrapper factory (W4F). In: Proceedings of WWW Conference (1999)
25. Scott, M.: Programming Language Pragmatics. Elsevier Science, Amsterdam (2015)
26. Sleiman, H.A., Corchuelo, R.: Trinity: on using trinary trees for unsupervised web data extraction. TKDE **26**(6), 1544–1556 (2014)
27. Zhai, Y., Liu, B.: Web data extraction based on partial tree alignment. In: Proceedings of the 14th International Conference on World Wide Web, WWW 2005, pp. 76–85. ACM, New York (2005)
28. Zhang, J., Zhang, C., Qian, W., Zhou, A.: Automatic extraction rules generation based on xpath pattern learning. In: Chiu, D.K.W., Bellatreche, L., Sasaki, H., Leung, H., Cheung, S.-C., Hu, H., Shao, J. (eds.) WISE 2010. LNCS, vol. 6724, pp. 58–69. Springer, Heidelberg (2011). doi:10.1007/978-3-642-24396-7_6

Search and Aggregation in XML Documents

Abdelmalek Habi[✉], Brice Effantin, and Hamamache Kheddouci

Université de Lyon, Université Lyon 1, CNRS LIRIS, UMR 5205, 69622 Lyon, France
{abdelmalek.habi,brice.effantin-dit-toussaint,
hamamache.kheddouci}@univ-lyon1.fr

Abstract. Information retrieval encounters a migration from the traditional paradigm (returning an ordered list of responses) to the aggregate
search paradigm (grouping the most comprehensive and relevant answers
into one final aggregated document). Nowadays extensible markup language (XML) is an important standard of information exchange and representation. Usually the tree representation of documents and queries is
used to process them. It allows to consider the XML documents retrieval
as a tree matching problem between the document trees and the query
tree. Several paradigms for retrieving XML documents have been proposed in the literature but only a few of them try to aggregate a set
of XML documents in order to provide more significant answers for a
given query. In this paper, we propose and evaluate an aggregated search
method to obtain the most accurate and richest answers in XML fragment search. Our search method is based on the Top-k Approximate Subtree Matching (TASM) algorithm and a new similarity function is proposed to improve the returned fragments. Then an aggregation process
is presented to generate a single aggregate response containing the most
relevant, exhaustive and non-redundant information given by the fragments. The method is evaluated on two real world datasets. Experimen-
tations show that it generates good results in terms of relevance and
quality.

1 Introduction

Structured documents focus on relevant information, they contain heterogeneous
contents which are organized with structural information. The structure of a
document can be used to process textual information with more granularity
than the entire document. Most of the proposed XML Information Retrieval
(XML IR) systems aim to return, as answer to the user's queries, the most relevant and exhaustive elements (subtrees) possible. For instance in [7] authors
use XML fragments for expressing the information needs and present an extension of vector space model for XML information retrieval.The results are ranked

This work is partially funded by the French National Agency of Research project:
Contextual and Aggregated Information Retrieval (ANR-14-CE23-0006).

D. Benslimane et al. (Eds.): DEXA 2017, Part I, LNCS 10438, pp. 290–304, 2017.
DOI: 10.1007/978-3-319-64468-4_22

by their relevance. In [9] authors present a probabilistic model for representation of documents which allows to use the structural information to improve the returned results. In [16] an approach is described for searching, scoring and ranking the relevant components in XML information retrieval. The XFIRM system presented in [24] is based on a relevance propagation method for both keywords and keywords-and-structural constraints queries. In XFIRM the scores are propagated from the leaf nodes to the inner nodes. And in [25,27] authors propose an approach that combines the result ranking with the structured queries search to improve the precision of the returned results. Thus XML IR systems return relevant information units instead of returning the entire document. The answer contains a list of organized elements according to a given criteria, often their relevance towards the query. The list of the returned results (elements) is organized from the most relevant element to the least one.

The vision of XML IR systems is limited. A relevant information in a document may not be contiguous in the list of results, as it may be scattered across several documents. In such case the returned answer is just a group of elements that cannot match the desired answer. In fact the desired answer can be a composition of many elements, a sort of a summary or a grouping, which made these paradigms less efficient [18]. Figure 1 presents an example of a list R of results returned by relevant subtrees search system for the query Q, $R = \{T_1, T_2, T_3, T_4\}$. This list contains elements (subtrees) of the same interrogated document which can be successive or dispersed according to the used similarity function.

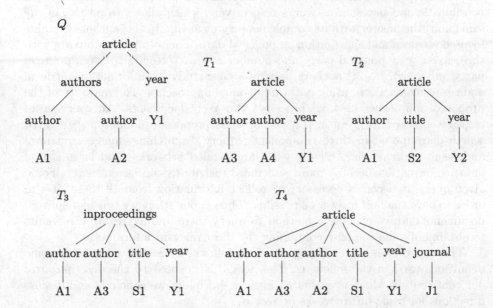

Fig. 1. Example of query and its relevant subtrees returned by top-k approximate subtree matching algorithm, with $k = 4$.

An aggregation of the results solves this problem of dispersed and scattered responses, by changing the traditional search paradigms. The aggregated search proposes to group in the same document (aggregate), all the relevant and non redundant information that can satisfy a user's query. Currently, most search engines perform some levels of aggregation [14].

In [28, 29] authors propose a model that allows merging XML data streams. It is based on a merge template that defines the result document's structure, and a set of functions to be applied to the XML data streams. Authors in [31] propose a merging operation over XML documents. This merging is directed by known conditions which makes it possible to identify the elements to be merged. Different systems are proposed to provide a multi documents summary that gives an overview of the documents containing the response. In [8] authors present the Query, Cluster, Summarize (QCS) system for document retrieval. QCS regroups the retrieved documents into clusters and creates a cluster's multidocuments summary and a document's summary. In [17] a method is proposed to extract a single-document summarization based on genetic operators. In [22] authors introduce a summarization method for the web-search approach by generating a concise/comprehensive summary for each cluster of documents retrieved. And in [23] a system is presented to summarize multi-document news.

In [10] authors propose a system that generates a resulting snippets from the results of an existing XML keyword search system. If we consider all these works from an aggregation view, we find them far from the problem of aggregation that we consider. Authors in [6, 20] present an XML keyword search model based on possibilistic and bayesian networks respectively, which allows to return, in the same unit, the answer with its complementary elements. In [11] authors examine focused retrieval and aggregation on political data. The result is a summary with three axes: year, political party and number of search results (person, political party and year). In [21] authors provide a case study for information retrieval and result aggregation, using natural language approaches. The majority of the proposed approaches [2–4, 30] processes the Web documents. An overview of aggregated search is given in [12] and authors propose a model for aggregated search decomposed in three components: query dispatching, nuggets retrieval and result aggregation. Actually with large scaled networks (and large size of data), information become more and more distributed on several sites. For a given query, it becomes necessary to collect information from all these sites in order to have the best answer as possible. Thus in our study we consider splitted documents and we propose a method to query them and aggregate the results to obtain only one response containing the most relevant information.

The rest of this paper is organized as follows. In Sect. 2 we present some definitions used in the following. The Sect. 3 is devoted to the description of our contribution which is evaluated in Sect. 4. Finally we conclude and discuss directions for some future works in Sect. 5.

2   Preliminaries

In graph theory, a *tree* $T = (V, E)$ is a connected graph without cycles where any two nodes are connected by exactly one path. If you suspend a tree by a node $r \in V(T)$, you obtain a *rooted tree* on r. In a rooted tree, a hierarchical relationship exists between any node and its neighbors. Then each node x (except the root r) has exactly one *parent* in the tree (*i.e.* its predecessor in the path from r to x), denoted $parent(x)$, and the remaining of its neighbors are its *children*. If a node $x \neq r$ has no child, it is called a *leaf*. Thus *sibling* nodes x, y satisfy $parent(x) = parent(y)$ and an *ancestor* of a node x is a node in the path from r to x. Note that an *independent ancestor* of v is an ancestor that conveys only the v's information. In the remaining we consider only rooted trees. Moreover a tree is said *labeled* if each node is given a label. A tree is said *ordered* if the order between siblings is important, otherwise the tree is *unordered*.

The strength of XML lies in its ability to describe any data domain through its extensibility. XML tags describe the hierarchical structure of the content. An XML element is delimited by an opening and a closing tag and it describes a semantic unit or an hierarchy of the document. An XML document can be represented as a labeled tree [1]. This last contains a root node that represents the entire document, internal nodes that organize the structure and leaf nodes describing the semantic information. Note that we keep the notion of semantic to refer the contents of the leaf nodes.

The user's need in XML documents can be formulated through queries that describe both content and structure of the desired information. Authors in [7] argue that the use of queries in form of XML fragment offer more flexibility in terms of representation and approximation of information needs. Thus we use this type of queries (Content-and-Structure queries). This allows to process the search phase with a tree matching. Using a tree representation of an XML document we can find two types of search. The first one is based on the exact tree matching and the second one is based on approximate tree matching. In our case since the user does not have a complete view on the entire document structure, the approximate tree matching seems the most appropriate. One of the most used metrics for approximate tree matching is the *tree edit distance*. In [26] the authors formalized the edit distance between strings of character used in [15] for trees. The edit distance between two trees is defined as the minimal cost of edit operations that transform a tree into another tree.

Let T, Q be two labeled trees. The tree edit distance between T and Q is defined as:

$$TED(T,Q) = \min_{e_1,\ldots,e_n \in \gamma(T,Q)} \sum_{i=1}^{n} cst(e_i),$$

where $\gamma(T,Q)$ is the set of edit operations and $cst()$ represents the edit operation cost. The edit operations are:

- Insertion: insert a node v as a child of w in T, w's children become v's children.
- Deletion: delete a node v in T, v's children become children of v's parent.
- Relabeling: replace the label of a node by another label.

Figure 2 represents an example of the three edit operations.

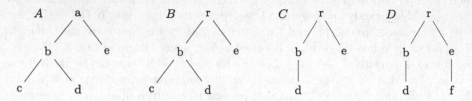

Fig. 2. Tree edit distance operations: $(A) \longrightarrow (B)$: relabeling of node a by r; $(B) \longrightarrow$ (C): deletion of node c; $(C) \longrightarrow (D)$: insertion of node f.

3   Our Contribution

Recall that our aim is to construct an aggregated document from the results of a query Q sent to a collection $C = \{D_1, D_2, \ldots\}$ of documents. This process is done in two phases. The first one is the retrieval process from the collection C. For each document D_i, this phase gives a list of partial subtrees of D_i (called *fragments*). The second phase is the aggregation process where one fragment of each list is selected to be aggregated in the final document.

3.1   Tree Matching for Information Retrieval

The answer of a user's query may be dispersed in the same document. In this case the search algorithm must return several fragments ordered by a similarity function. In the literature several search algorithms aim to find the best answers for a given query. In this work, we consider the *TASM* algorithm *(Top-k Approximate Subtree Matching problem)* [5] since it is one of the most effective solutions to search the k-best answers for queries. From a given query tree, TASM allows to identify the top-k subtree in a data tree with smallest edit distance. It achieves a space complexity that is independent of the data tree size. Moreover it does not require any predefined rules for relaxation on subtrees that are more likely to respond to the query, *i.e.* all document subtrees are considered as a candidate answer, evaluated by an edit distance function.

Definition 1 ([5], **Top-k Approximate Subtree Matching problem**). *Let Q (query) and T (document) be ordered labeled trees, n be the number of nodes of T, T_i be the subtree of T that is rooted at node t_i and includes all its descendants, $d(.,.)$ be a distance function between ordered labeled trees, and $k \leq n$ be an integer. A sequence of subtrees, $R = (T_{i_1}, T_{i_2}, \ldots, T_{i_k})$, is a top-k ranking of the subtrees of the document T with respect to the query Q iff*

1. *the ranking contains the k subtrees that are closest to the query:* $\forall T_j \notin R$: $d(Q, T_{i_k}) \leq d(Q, T_j)$, *and*
2. *the subtrees in the ranking are sorted by their distance to the query:* $\forall 1 \leq j < k : d(Q, T_j) \leq d(Q, T_{i_{j+1}})$.

The *top-k approximate subtree matching (TASM) problem is the problem of computing a top-k ranking of the subtrees of a document T with respect to a query Q.*

In *TASM*, authors used *TED* as a similarity measure based on a cost model for edit operations. An insertion (deletion) operation has a cost equal to the cost of an inserted (deleted) node. The cost of a relabeling operation is the average between a deletion and an insertion operation. This cost model does not distinguish between the two types of similarity (content and structure). Since in our case we need to distinguish them, we propose a new cost model. In this model the relabeling operation is considered as two operations, a deletion and an insertion operations. Moreover the deletion of a leaf and its independent ancestors have cost 0 since they are considered as enrichment for the answer.

Definition 2. *Let T and Q be labeled trees, $t_i \in V(T)$, $q_j \in V(Q)$. The cost assigned to a node is cst, an insertion operation of node t_i is denoted $(\epsilon \to t_i)$, a deletion operation of node t_i is denoted $(t_i \to \epsilon)$, and a relabeling operation of node q_j by the label of node t_i is denoted $(q_j \to t_i)$. We denoted by leaf a leaf node and IndAnc an independent ancestor of a deleted leaf node. The edit operation costs are defined by*

$$\gamma(q_j, t_i) = \begin{cases} cst & \text{if } (\epsilon \longrightarrow t_i), \\ 2 \times cst & \text{if } (q_j \longrightarrow t_i), \\ 0 & \text{if } (t_i \longrightarrow \epsilon) \wedge (t_i = leaf \vee t_i = IndAnc), \\ cst & \text{if } (t_i \longrightarrow \epsilon) \wedge (t_i \neq leaf \wedge t_i \neq IndAnc). \end{cases}$$

From this TED computation, we need to define a similarity function in order to consider the fragments as semantic entities delimited by hierarchical structures.

Definition 3. *Let T be a document, Q be a query, $T_v \in T$ be a subtree in T rooted on v and $TED(T_v, Q)$ be the tree edit distance between T_v and Q, according to the cost model described in Definition 2. We define the similarity function $Sim(T_v, Q)$ by*

$$Sim(T_v, Q) = \frac{1}{TED(T_v, Q) \times \left(1 + \frac{|Rel_l| + |Ins_l|}{|leaf_{query}|}\right)},$$

where $|Rel_l|$ is the number of relabeled leaves, $|Ins_l|$ is the number of inserted leaves and $|leaf_{query}|$ represents the number of leaves in Q.

Thus the fragment with the highest similarity is the most relevant fragment in comparison to the query. In a fragment, if a leaf is relabeled (*i.e.* an information changed) or a leaf is inserted (*i.e.* a missing information), the fragment is penalized and its accuracy becomes less important (so it loses its relevance obtained by the *TED*). Example 1 shows the difference between our similarity function and that used in *TASM*.

Example 1. *We consider the example of Fig. 1, the subtrees T_1, T_2, T_3, T_4 are the results of the query Q for $k = 4$. With the TASM's similarity, the subtree T_1 is on top of the list because it has the best TED, while T_4 is the last in the list ($TED(T_1, Q) = 3$, $TED(T_2, Q) = 4$, $TED(T_3, Q) = 5$ and $TED(T_4, Q) = 7$). This ranking is correct according only to the structure and it is not if structure and semantics are considered. With our similarity function, the order T_4, T_3, T_1, T_2 is given (with $Sim(T_4, Q) = 1$, $Sim(T_3, Q) = 0.15$, $Sim(T_1, Q) = 0.12$ and $Sim(T_2, Q) = 0,086$).*

This retrieval process is applied on every document of the initial collection and it gives, for each document, an ordered list of *candidate fragments* according to the similarity function given in Definition 3. The candidate fragments are heterogeneous, each having a different granularity. Filtering these lists, to eliminate irrelevant fragments according to the query, is necessary. The filtering aims at evaluating each candidate fragment separately with structural and semantical criteria. Let Q be the query and F_i be a candidate fragment. The fragment F_i is considered irrelevant and it will not be considered as a response if it verifies the following condition:

$$(TED(F_i, Q) \geq (2 \times |Q|)) \quad \lor \quad \left(\frac{|Rel_l| + |Ins_l|}{|leaf_{query}|} \geq \tau \right).$$

The first part of the condition is used to structurally distinguish Q and F_i (indeed $2 \times |Q|$ is reached if all the elements of Q are relabeled). In the second part of the condition, parameter τ with ($0 \leq \tau \leq 1$) presents a threshold of evaluation for the fragment's content. If the number of inserted leaves and relabeled leaves is close to the number of leaves in the query, the fragment is considered irrelevant. The threshold τ is determined experimentally.

In the next phase, some of these fragments will be aggregated to produce a final document.

3.2 Aggregation of the Fragments

The previous phase gave, for each document of the collection, a list of fragments answering to the query Q, ordered by the similarity function given above. In this section we describe a method to construct one document from these fragments. The resulting document should be as exhaustive and relevant as possible. It should not contain redundancy and it must be semantically correct with respect to the returned responses. This aggregation process is decomposed in different steps:

1. Selection of the fragments to be aggregated. At the first iteration, a base fragment is selected and denoted F_{base}. Then for each iteration, we need to identify from the response lists, a fragment F to be aggregated to F_{base}. This identification is done among the lists not yet used.
2. Construction of the primary aggregate. We introduce the composition operators to aggregate F with F_{base}.
3. If the final aggregate is not yet obtained, the primary aggregate is considered as the new base fragment and the process is repeated from step 1.

Step 1: Selection. The selection process aims at selecting the fragments to be aggregated. In the first iteration, it also identifies the base fragment. For every iteration, we select a fragment in a response list L to have the most additional relevant information to F_{base}. Note that since we take, in L, the most interesting fragment, we consider that no other fragment of L has relevant information and we remove L from the remaining of the process.

Base Fragment: Among all the fragments of the optimized lists, a fragment is chosen to be the aggregation base. It must be the most relevant possible that gives a partial view of the aggregated result structure. The base fragment F_{base} is the fragment with the highest similarity among the best fragments of all response lists.

Fragment Selection: To select a fragment F, a comparison is done between F_{base} and all the fragments of response lists. This comparison is computed by a path decomposition of each tree (fragment). A path decomposition of a tree T is the set of paths from the root to each of its leaves (note that we keep the notations $root$ and $leaf$ for the first and last nodes of each path). Then, for each fragment F we construct a similarity matrix between F_{base} and F. We denote by $M(F_{base}, F)$ the similarity matrix of size $n = \max\{|F_{base}|, |F|\}$, where the rows of $M(F_{base}, F)$ represent the paths p_i of F_{base} and the columns represent the paths f_j of F. The matrix elements are computed as follows:

$$M(p_i, f_j) = \begin{cases} lcs(p_i, f_j) & \text{if } leaf(p_i) = leaf(f_j), \\ 0 & \text{otherwise.} \end{cases}$$

The *longest common subsequence* between two paths, denoted $lcs(p, p')$, is computed by the algorithm given in [19] which has $O(ND)$ time and space complexity, where N is the sum of the lengths of p and p' and D is the size of the minimum edit operations for p and p'. After constructing the similarity matrix we use the Hungarian algorithm [13] to compute the matching on it. The Hungarian algorithm is considered as one of the most efficient matrix matching algorithms, with an $O(n^3)$ time complexity. The fragment for which the matching is maximized is selected for the next step.

Step 2: Construction of the Aggregate. The Hungarian algorithm gives the set of matchings between path decompositions of F_{base} and F. Note that the information given by the paths (on their leaves) intersect (*i.e.* are similar). Let $\mathcal{R} = \{(p_i, f_j)\}$ be the set of these matchings where $p_i \in F_{base}$, $f_j \in F$ and $lcs(p_i, f_j) \geq 1$. We consider each pair (p_i, f_j) independently, and for each pair (p_i, f_j) we define a subtree of F to be aggregated to F_{base}. Thus for each pair (p_i, f_j):

- The *inclusion subtree* $F(f_j)$ of F is the subtree of F rooted on the first ancestor that does not carry two intersection information. To do that we identify each node of F in a post-order. Thus $lcs(p_i, f_j)$ can be represented by the identifiers of its nodes. Let L be the list of identifiers of the $lcs(p_i, f_j)$ nodes. We also define R as the list of identifiers of all the leaves in $\mathcal{R} \setminus \{(p_i, f_j)\}$. The root's identifier of the inclusion subtree is then given by $id = \max\{l \in L | \forall r \in R, r > l\}$.
- The composition of this inclusion subtree with F_{base} is done by merging the common nodes and by adding the new nodes of $F(f_j)$ in F_{base} (connected as in $F(f_j)$).
- We do $F = F \setminus \{F(f_j)\}$. We remove the subtree $F(f_j)$ from the fragment F before the new pair (p_i, f_j) (to avoid duplication of information in the next iterations). Note that F stays a connected tree after the deletion.

Finally if the fragment F still contains information (after considering all pairs of $\mathcal{R}$), the remaining tree F is added to F_{base} by adding a fictive node as root between F_{base} and F.

3.3 Complexity

In this section we analyze the running time of our method. Let Q be a query of size $|Q| = m$. A document D, of size $|D| = n$, is splitted into p documents $D_1, D_2, \ldots, D_p$. The computation of the time complexity can be divided into two parts. The first one is the complexity of the retrieval process and the second one is the complexity of the aggregation phase. The retrieval process is based on the TASM algorithm on p documents. Authors prove in [5] that the runtime complexity of TASM is $O(m^2n)$ for a document. Thus for the entire part, the complexity of the retrieval process is $O(pm^2n)$. In the other part, it is important to note that the number of the returned fragments for each D_i is k (with $k \leq n$) and the size of a returned fragment is bounded by $(2m + k)$ (by [5]). Since the computation of the similarity function is done in the process of retrieving, the filtering step is done in $O(pk)$ time. Moreover, the base fragment F_{base} is chosen in a constant time. Then for all the remaining D_i, all their fragments are considered. In the fragment selection, the cost to construct the similarity matrix is $O(m^2)$ and the cost of the Hungarian matching is $O(m^3)$. For each selected fragment, the inclusion subtree is computed in $O(m^2)$ and the composition (insertion, union, deletion) in $O(n)$. Thus for all the fragments of all documents the aggregation phase is done in $O(p^2km^3)$ time. We can remark that the complexity is independent of the size of documents. It depends on the number of partitions p and the size of the query m. Note that typically m is small and the size of document is much larger than the size of query ($n \gg m$).

4   Experiments

The evaluation of the aggregated search is a difficult challenge, it can be considered as an issue that is so far from being solved [12]. It may be related to several criteria like the complementarity of the returned elements which constitute the final aggregate, the non-redundancy of information and completeness.

In this section we describe and discuss experimental results to evaluate our method. However, to our knowledge there is no other method for aggregation based on structural and semantical constraints. Thus we compare our method with the TASM algorithm.

4.1   Test Collection

Our evaluation is based on a test collection decomposed into three parts: a set of documents to be interrogated, a set of queries to be searched in the document collection and associated relevance judgments for each query. In our study we construct our collection of documents. We use two real world datasets (with XML format). The DBLP[1] dataset is an XML representation of a bibliographic information on major computer science journals and proceedings. The IMDb[2] dataset gives a representation of Internet Movie Database in an XML document. For each dataset, the whole document (*global document*) D_b is partitioned into p partitions which verify: $\bigcup_{i=1}^{p} P_i = D_b$ and $\bigcap_{i=1}^{p} P_i \neq \emptyset$. Each partition is considered as an independent document. In our work, queries are also given with an XML structure and for each dataset we construct a set of queries from the global document. Each set contains 20 queries with different sizes, structures and contents. Each of them is evaluated by a set of experts (researchers) to construct the relevant judgment sets that gathers all relevant answers from the global document to a given query.

4.2   Evaluation Method

On each of the test collections, three sets of experiments are conducted. The first one called *global TASM* is an implementation of TASM with its cost model on the global documents D_b. Algorithm *TASM* returns for each document a list of fragments with size k that are deemed relevant by the *TED* function. The second one called *global TASM aggregation* is an implementation of TASM with our cost model on the global documents D_b, augmented with the proposed aggregation approach applied on the list of fragments sorted by the similarity function *Sim*. The last one called *partition TASM aggregation* is an implementation of our algorithm on the p partitions. In our experiments we evaluated our approach with several values of k varying from 1 to 15. The best results for the three experimental scenarios are for $k = 6$ and $\tau = 0.7$. Thus in the remaining, we fix $k = 6$ and $\tau = 0.7$.

---

[1] http://dblp.uni-trier.de/xml/.
[2] http://research.cs.wisc.edu/niagara/data/.

4.3   Result Discussions

To evaluate the results of each experiment, we use two criteria. The first one is the *recall* which represents the ratio between the number of relevant returned elements and the number of relevant elements of the response in the judgment set. The second criteria is the *precision* which represent the ratio between the relevant returned elements and the number of returned elements.

Figures 3 and 4 show the recall of responses of the three algorithms for IDMb and DBLP datasets respectively. It reports the overall performance of our method in terms of relevant responses. In the IDMb dataset, *partition TASM*

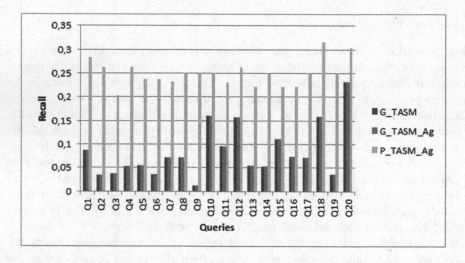

Fig. 3. Recall for the three algorithms on IMDb dataset.

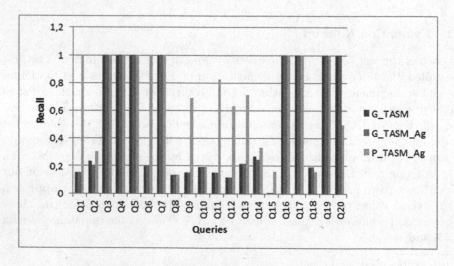

Fig. 4. Recall for the three algorithms on DBLP dataset.

aggregation carries 30% (for $Q20$) to 90% (for $Q9$) of relevant information more than both *global TASM* and *global TASM aggregation*. In contrast, there is no gain in terms of enrichment of the response for *global TASM aggregation*. This is due to the fact that all the returned fragments (using *global TASM* and *global TASM aggregation*) are of small sizes because they are related to the size of the query and the value k, and the subtrees in the global document are large. We can consider the queries used to interrogate this dataset as aggregative queries when the result is obtained by interrogating a set of documents to assemble relevant fragments. Each fragment contributes partially to the response but all together constitute a complete response. The same results appear for DBLP database for aggregative queries ($Q1, Q2, Q6, Q9, Q11$–$Q15$). For the other queries, the experiments show less efficiency compared to IDMb. This is due to the nature of the used queries. They do not have a response that carries a gain in terms of information enrichment. They are very generic and their judgment response sets are very large. This leads to consider all responses of *global TASM* and *global TASM aggregation* as relevant. On the other hand the results of this type of queries in *partition TASM aggregation* are a construction and an enrichment of one fragment. This makes its recall small compared to the size of the judgment responses.

These results are evaluated without considering whether a response contains irrelevant and/or redundant elements. The redundancy of elements in the same response is considered as an irrelevant information. For each query's answer, we consider the precision of returned elements to evaluate their qualities. Figures 5 and 6 show the precision of returned elements of the three algorithms for IDMb and DBLP datasets respectively. In both IDMb and DBLP, *partition TASM aggregation* shows a high precision of results for the majority of queries. It returns only the relevant elements and eliminates the redundancies. On the other hand,

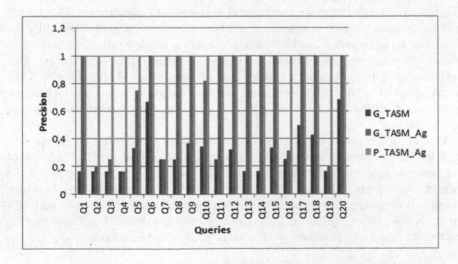

Fig. 5. Precision for the three algorithms on IMDb dataset.

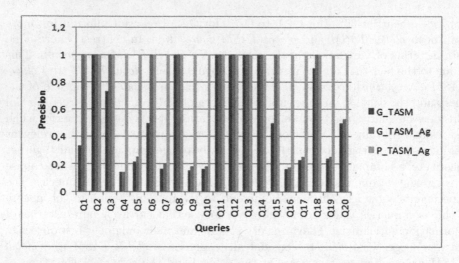

Fig. 6. Precision for the three algorithms on DBLP dataset.

global TASM aggregation shows an improvement in terms of precision compared to *global TASM*. This is due to the elimination of redundancies of fragments returned.

5   Conclusion

In this paper we discuss search and aggregation in the context of the retrieval in XML documents. Firstly we introduce a search model based on an approximate tree matching algorithm, the Top-k Approximate Subtree Matching. It is adapted to distinguish between structural and semantic similarities. This search model provides a set of fragments candidate to answer a query. Secondly we introduce an aggregation model which generates a single result document containing the most relevant elements and their complementary information. Our method shows its performance in terms of relevance and quality for the XML fragment search where the information are distributed over several documents. Our method can be readily extended to support any data format having a tree representation.

Our approach could be extended or adapted to many other problems. We consider the problem of search and aggregation in general graphs. This problem becomes more difficult when considering the complex representation of the data in graph form. However, it is more challenging for the distributed graphs. Another promising problem is that of querying multiple data sources with different representations (different collections, pictures, video, ...) and aggregating the results into a single answer with a standardized representation.

References

1. W3C XML web page. http://www.w3.org/XML/
2. Arguello, J.: Improving aggregated search coherence. In: Hanbury, A., Kazai, G., Rauber, A., Fuhr, N. (eds.) ECIR 2015. LNCS, vol. 9022, pp. 25–36. Springer, Cham (2015). doi:10.1007/978-3-319-16354-3_3
3. Arguello, J., Capra, R.: The effect of aggregated search coherence on search behavior. In: Proceedings of the 21st ACM International Conference on Information and Knowledge Management, CIKM 2012, New York, NY, USA, pp. 1293–1302. ACM (2012)
4. Arguello, J., Diaz, F., Callan, J., Carterette, B.: A methodology for evaluating aggregated search results. In: Clough, P., Foley, C., Gurrin, C., Jones, G.J.F., Kraaij, W., Lee, H., Mudoch, V. (eds.) ECIR 2011. LNCS, vol. 6611, pp. 141–152. Springer, Heidelberg (2011). doi:10.1007/978-3-642-20161-5_15
5. Augsten, N., Barbosa, D., BÃűhlen, M., Palpanas, T.: TASM: top-k approximate subtree matching. In: 2010 IEEE 26th International Conference on Data Engineering (ICDE 2010), pp. 353–364, March 2010
6. Bessai-Mechmache, F.Z., Alimazighi, Z.: Aggregated search in XML documents. J. Emerg. Technol. Web Intell. 4(2), 181–188 (2012)
7. Carmel, D., Maarek, Y.S., Mandelbrod, M., Mass, Y., Soffer, A.: Searching XML documents via XML fragments. In: Proceedings of the 26th Annual International ACM SIGIR Conference on Research and Development in Information Retrieval, SIGIR 2003, pp. 151–158. ACM, New York (2003)
8. Dunlavy, D.M., OâĂŹLeary, D.P., Conroy, J.M., Schlesinger, J.D.: QCS: a system for querying, clustering and summarizing documents. Inf. Process. Manag. 43(6), 1588–1605 (2007)
9. Géry, M., Largeron, C., Thollard, F.: Probabilistic document model integrating xml structure. In: Proceedings in INEX, pp. 139–149 (2007)
10. Huang, Y., Liu, Z., Chen, Y.: Query biased snippet generation in xml search. In: Proceedings of the 2008 ACM SIGMOD International Conference on Management of Data, SIGMOD 2008, pp. 315–326. ACM, New York (2008)
11. Kaptein, R., Marx, M.: Focused retrieval and result aggregation with political data. Inf. Retrieval 13(5), 412–433 (2010)
12. Kopliku, A., Pinel-Sauvagnat, K., Boughanem, M.: Aggregated search: a new information retrieval paradigm. ACM Comput. Surv. 46(3), 41:1–41:31 (2014)
13. Kuhn, H.W.: The hungarian method for the assignment problem. Naval Res. Logistics Q. 2(1–2), 83–97 (1955)
14. Lalmas, M.: Aggregated search. In: Melucci, M., Baeza-Yates, R. (eds.) Advanced Topics in Information Retrieval. The Information Retrieval Series, vol. 33, pp. 109–123. Springer, Heidelberg (2011). doi:10.1007/978-3-642-20946-8_5
15. Levenshtein, V.I.: Binary codes capable of correcting deletions, insertions, and reversals. In: Soviet Physics Doklady, vol. 10, pp. 707–710 (1966)
16. Mass, Y., Mandelbrod, M.: Retrieving the most relevant xml components. In: INEX 2003 Workshop Proceedings, p. 58. Citeseer (2003)
17. Mendoza, M., Bonilla, S., Noguera, C., Cobos, C., León, E.: Extractive single-document summarization based on genetic operators and guided local search. Expert Syst. Appl. 41(9), 4158–4169 (2014)
18. Murdock, V., Lalmas, M.: Workshop on aggregated search. SIGIR Forum 42(2), 80–83 (2008)

19. Myers, E.W.: An O(ND) difference algorithm and its variations. Algorithmica **1**(1), 251–266 (1986)
20. Naffakhi, N., Faiz, R.: Aggregated search in XML documents: what to retrieve? In: 2012 International Conference on Information Technology and e-Services, pp. 1–6, March 2012
21. Paris, C., Wan, S., Thomas, P.: Focused and aggregated search: a perspective from natural language generation. Inf. Retrieval **13**(5), 434–459 (2010)
22. Qumsiyeh, R., Qumsiyeh, R., Ng, Y.-K., Ng, Y.-K.: Searching web documents using a summarization approach. Int. J. Web Inf. Syst. **12**(1), 83–101 (2016)
23. Radev, D., Otterbacher, J., Winkel, A., Blair-Goldensohn, S.: Newsinessence: summarizing online news topics. Commun. ACM **48**(10), 95–98 (2005)
24. Sauvagnat, K., Hlaoua, L., Boughanem, M.: XFIRM at INEX 2005: ad-hoc and relevance feedback tracks. In: Fuhr, N., Lalmas, M., Malik, S., Kazai, G. (eds.) INEX 2005. LNCS, vol. 3977, pp. 88–103. Springer, Heidelberg (2006). doi:10.1007/978-3-540-34963-1_7
25. Schlieder, T., Meuss, H.: Result ranking for structured queries against xml documents. In: DELOS Workshop Information Seeking, Searching and Querying in Digital Libraries, Zurich, Switzerland (2000)
26. Tai, K.-C.: The tree-to-tree correction problem. J. ACM **26**(3), 422–433 (1979)
27. Theobald, M., Schenkel, R., Weikum, G.: TopX and XXL at INEX 2005. In: Fuhr, N., Lalmas, M., Malik, S., Kazai, G. (eds.) INEX 2005. LNCS, vol. 3977, pp. 282–295. Springer, Heidelberg (2006). doi:10.1007/978-3-540-34963-1_21
28. Tufte, K., Maier, D.: Aggregation and accumulation of XML data. IEEE Data Eng. Bull. **24**(2), 34–39 (2001)
29. Tufte, K., Maier, D.: Merge as a lattice-join of xml documents. In: 28th International Conference on VLDB (2002)
30. Turpin, L., Kelly, D., Arguello, J.: To blend or not to blend? Perceptual speed, visual memory and aggregated search. In: Proceedings of the 39th International ACM SIGIR Conference on Research and Development in Information Retrieval, SIGIR 2016, pp. 1021–1024. ACM, New York (2016)
31. Wei, W., Liu, M., Li, S.: Merging of XML documents. In: Atzeni, P., Chu, W., Lu, H., Zhou, S., Ling, T.-W. (eds.) ER 2004. LNCS, vol. 3288, pp. 273–285. Springer, Heidelberg (2004). doi:10.1007/978-3-540-30464-7_22

Representing and Learning Human Behavior Patterns with Contextual Variability

Paula Lago[1]([⊠]), Claudia Roncancio[2], Claudia Jiménez-Guarín[1],
and Cyril Labbé[2]

[1] Universidad de Los Andes, Bogotá, Colombia
{pa.lago52,cjimenez}@uniandes.edu.co
[2] Univ. of Grenoble Alpes, CNRS, Grenoble INP, LIG, 38000 Grenoble, France
{claudia.roncancio,cyril.labbe}@univ-grenoble-alpes.fr

Abstract. For Smart Environments used for elder care, learning the inhabitant's behavior patterns is fundamental to detect changes since these can signal health deterioration. A precise model needs to consider variations implied by the fact that human behavior has an stochastic nature and is affected by context conditions. In this paper, we model behavior patterns as usual activity start times. We introduce a Frequent Pattern Mining algorithm to estimate probable start times and their variations due to context conditions using only one single scan of the activity data stream. Experimentation using the Aruba CASAS and the ContextAct@A4H datasets and comparison with a Gaussian Mixture Model show our proposition provides adequate results for smart home environments domains with a lower computational time complexity. This allows the evaluation of behavior variations at different context dimensions and varied granularity levels for each of them.

1 Introduction

Smart Homes used for elderly home care, can increase independent living time and reduce long-term care costs. By monitoring Activities of Daily Living (ADL), behavior patterns, their changes and their evolution, it is possible to infer a person's health status and her ability to live independently [14].

One element to consider about behavior patterns is the probable start time of each monitored activity. It is possible to model start times since we tend to do the same activities at *approximately* the same time [14]. Changes in the start time of an activity can signal health disturbances. Nonetheless, the time when an activity is done may vary depending on context conditions. For example, one may wake up later on weekends or take a bath earlier when it is sunny outside to go out. Learning and modeling these variations avoids false alarms and improves routine understanding for both the elder and her caregivers.

If context conditions are not considered *during* learning, some patterns may go undetected [12]. But the large number of context dimensions and attribute values increases the complexity of the analysis specially since not all activities

© Springer International Publishing AG 2017
D. Benslimane et al. (Eds.): DEXA 2017, Part I, LNCS 10438, pp. 305–313, 2017.
DOI: 10.1007/978-3-319-64468-4_23

are affected by the same dimensions nor by the same scale of them (hour, day, month...). While the probable start time of waking up may vary on weekends, that of cooking may vary only on Fridays and that of going out may vary when it rains. Although we can model start times using normal distributions, analyzing their possible variations at different scales is difficult.

Current approaches either don't consider possible variations in the start time of each activity, or consider the same attributes and the same scale for every activity (Sect. 2). In this paper, we propose the TIMe algorithm (Sect. 3) to automatically learn probable start times of daily living activities and variations due to context conditions based on frequent pattern mining. We compare TIMe's start time intervals to the intervals found by fitting a Gaussian Mixture Model using two real-life datasets (ContextAct@A4H [7] and Aruba CASAS [2]). Our method, though less accurate than fitting a normal distribution, has a lower cost and is processed in a stream setting. TIMe evaluates raw data simultaneously with multiple context dimensions and scales to detect variations, without storing raw observation data, to analyze patterns. We can analyze changes and evolution in real-time, protecting user privacy (Sect. 4).

2   Behavior Pattern Learning in Smart Homes

Analyzing the temporal dimension of behavior can improve the accuracy of anomaly detection [6,9]. This includes both relative ordering of activities and absolute temporal characteristics such as activity durations, start times and the temporal gaps between activities. The main approaches for modeling behavior, knowledge-driven and data-driven [1,4], emphasize the relative sequential structure while absolute temporal characteristics are overlooked.

We model behavior start times since we tend to do the same activities at approximately the same time. Proposed approaches regard time as either a continuous or a discrete dimension. When considered as a continuous dimension, start time is mostly modeled using a Gaussian Distribution (GD) [5]. Most accurately, with a linear combination of GD (a Gaussian Mixture Model, GMM) multiple occurrences of the same activity during a day are modeled [14]. When seen as a discrete dimension, a day is segmented into chunks and the behavior occurring each interval is modeled [10,11]. While the Gaussian Distribution method models the tightness or flexibility of a routine with its variance, discretizing the temporal space is faster and allows easier computations. Dawadi et al. [3] use m equal-sized windows to find activity frequency on each but model each activity distribution over the entire day-long period, thus finding a trade-off between both representations.

The usual start time of a behavior is modeled as the interval(s) with the highest observation frequencies or as $[\mu - \sigma_i, \mu + \sigma]$. But these times vary according to context conditions without them being anomalies [8]. To model these variations some authors create a different model for each possible situation [9,10]. This approach is not scalable with the number of dimensions defining a situation. Other authors include context as an activity feature, thus every variation is a different concept [6]. This hinders routine understanding since the semantic relationship between variations of the same activity is lost. Finally, other approaches model each context dimension separately [5]. While this method is scalable, it does not find frequent behaviors on rare contexts [12].

In sum, finding the probable start time(s) of an activity while considering context dimensions and different granularities (scales) for each of them with a scalable algorithm is still an open issue. We propose a frequent pattern mining method to find a suitable *trade-off between continuous and discrete representations that is computationally efficient* and precise enough to model behavior.

3   Discovering Most Probable Start Time Intervals and Probable Context Variations

In this section we formalize the problem of mining most probable start time(s) for an activity and its possible variations in light of context conditions as a frequent pattern mining problem (Sect. 3.1). By introducing the notion of a maximum start time interval and the expected number of observations in such interval (Sect. 3.2), our approach to detect start times does not previously need neither the number of possible occurrences in a day of each activity nor the context dimension granularity. TIMe (Sect. 3.3), allows not only the analysis of different context dimensions and scales but also the analysis of pattern evolution in real-time and can be run on small computers.

3.1   Mining Activity Start Time Intervals Problem

Let $\mathcal{T} = (T, \leq)$ be a time domain. A time instant $t_x \in T$ is called a *timestamp*. Since we are interested in finding *periodic* patterns, we define a periodic granularity in which each granule groups W time instants of T. There exists a mapping $f_W(t) = t \bmod W$ for each timestamp $t \in \mathcal{T}$ allowing to represent the timestamp in a *circular time domain*. We denote this circular domain as $\mathcal{T}'$.

Example 1. Let $\mathcal{T}$ be a time domain whose time instants are represented as a Unix epoch. Let $W = 86400\,\mathrm{s}$ (24 h). Let, $t_1 = 1488656696$ (2017-03-04 19:44:56) and $t_2 = 1489520696$ (2017-03-14 19:44:56), then $f_W(t_1) = 71096$ (19:44:56) and $f_W(t_2) = 71096$ (19:44:56). Both t_1 and t_2 represent the same point within the defined period.

Let $\mathcal{A} = \{a_1, a_2, \ldots a_n\}$ be a set of activity labels representing the activities to be monitored. For example, $A = \{\text{SLEEP, COOK, GO\_OUT}\}$

An activity observation is a tuple $o_i = (e_i, t_j)$ with $e_i \in \mathcal{A}, t_j \in \mathcal{T}$. e_i represents the label of the observed activity and t_j represents its start time for the current observation. A contextualized activity observation is a tuple $o_i = (e_i, t_j, [c_1, c_2, \ldots, c_k])$, where $c_1, c_2, \ldots, c_k$ define context attribute values of the observation such as day of the week, weather description, temperature, noise level, etc. Context attributes are a vector in which each item takes value in a specific context domain. A stream of activity observations is an unbounded sequence $\mathcal{S} = <o_1, o_2, o_3, \ldots >$, where o_i is a contextualized activity observation.

Finding the probable start time of each monitored activity a_i is equivalent to finding all the time intervals $[s, e]$, where $s, e \in \mathcal{T}'$, at which the probability that an observation of the activity belongs to the interval is greater than a user chosen parameter φ. Given a sequence $\mathcal{S}$, the problem can be expressed as follows:

$$\forall a_i \in \mathcal{A} \text{ find the time intervals } [s, e] \text{ where} s, e \in \mathcal{T}' \text{such that:}$$
$$P(\exists o_i = (a_i, t_j) \in \mathcal{S} \mid f_W(t_j) \in [s, e]) > \varphi \quad (1)$$

The time interval $[s, e]$ can be interpreted as: *"the time interval when there is a high probability of starting activity a_i"* or as *"the time interval around which activity a_i usually starts"*. Notice that for each activity a_i, there may be 0 or more probable start time intervals, each of which may have a different size $e - s$.

Similarly, we find variations due to a given context value c_k as the time interval(s) with a probability greater than the given parameter (see Eq. 2). We assume all c_k are independent from each other, so they are considered separately.

$$\frac{P(\exists o_i = (a_i, t_j, c_k) \in \mathcal{S} | f_W(t_j) \in [s, e])}{P(\exists o_i = (a_i, t_l, c_k) \in \mathcal{S})} > \varphi \quad (2)$$

3.2 Maximum Start Time Interval Size

As said before, we define the total size of the timespace $\mathcal{T}'$ by W, expressed in a corresponding temporal unit. For example, if the space is a day, W has 24 h or 1440 min. This is the period of analysis. To mine probable start time intervals, we divide this space into smaller, w-sized chunks called *slots*.

Let *ntotal* be the number of observations of an activity seen thus far and φ be the minimum probability for a time interval to be considered frequent (see Eq. 1). Following (see Eq. 2), and assuming events are evenly distributed, the number y of expected observations in a frequent time interval is given by: $y = ntotal * \varphi$. The maximum size T of a frequent interval is given by the minimum number of slots that would be needed to have a frequent interval under a uniform distribution. Thus T is defined by: $T = \frac{W}{w} * \varphi$. From y and T, the number n of expected observations in a slot belonging to a frequent interval is given by

$$n = y * \frac{1}{T} \quad (3)$$

A probable start time interval is thus, a set of contiguous slots that have each more than n observations.

3.3 TIMe: An Algorithm for Mining Start Time Intervals

The algorithm relies on counting the total number of activity observations and the number of observations per activity per slot. Since there are a total of $\frac{W}{w}$ slots in $\mathcal{T}'$, a matrix of counters, $M_{|\mathcal{A}| \times \frac{W}{w} + 1}$ is used for this. In the matrix, $m_{i,0}$

counts the frequency of activity a_i in the stream and $m_{i,j}$ for each $j > 0$ counts the frequency of activity a_i in slot s_j.

TIMe runs in two phases. The first (Algorithm 1), runs when a new activity observation is made. It updates both the total count of activity observations and that of the corresponding slot and activity (Algorithm 1 lines 3 to 5). The second (Algorithm 2), runs at the end of each period. It finds the frequent intervals for each activity by first finding the slots with counts greater than n (Eq. 3). These are the candidate slots. All contiguous candidate slots are merged into a single interval. If its total count is greater than the count of the activity times φ, then the interval is considered a probable start time for the activity. To find contextual variability of the most probable start times, we use a tree structure representing different scales of each context dimension. Each node has a matrix keeping counts for observations at the specific context value represented by the node. The root of the tree keeps the total number of observations.

Algorithm 1 TIME($w, \mathcal{S}$)

```
    total := 0, counts[0...|𝒜| − 1][0...W/w]
2:  for oᵢ ∈ 𝒮 do
        total := total + 1
4:      count[oᵢ.a][0] := count[oᵢ.a][0] + 1
        count[oᵢ.a][oᵢ.t] := count[oᵢ.a][oᵢ.t] + 1
6:      if  Reached end of period  then
            callmerge_intervals(counts, total)
8:      end if
    end for
```

Algorithm 2
merge_intervals($c[][], total$)

```
    𝕀 ← ∅ , min := total * w/W
2:  for i ∈ [0...|𝒜 − 1|]  do
        s, e := −1
4:      for  j ∈ [1...W/w] do
            n := c[i][j]/c[i][0]
6:      if n > min then
            if s == −1 then
8:          s, e := j
            else
10:         e := j
            end if
12:     else if s! = −1 then
            𝕀 ← 𝕀 ∪ (i, s, e), s, e := −1
```

4   Empirical Evaluation

In this section, we prove three properties of TIMe by empirical evaluation: (1) It can find multiple start times in a day for an activity with a precision similar to that of a GMM (Sect. 4.1), (2) It detects contextual variability at different scales for such intervals (Sect. 4.2) and, (3) Its patterns may be used to increase routine awareness from sensor data (Sect. 4.3). We used the Aruba dataset (222 days) from CASAS project [2] and the ContextAct@A4H dataset (28 days) from the Amiqual4Home Lab [7]. From the former dataset, we focus on the SLEEP (401 observations), EATING (259 obs.), MEAL PREPARATION (1605 obs.) and WASH DISHES (67 obs.) activities. From the latter, we focus on the SLEEP (26 obs.), SHOWER (24 obs.), COOK (77 obs.) and WASH DISHES (52 obs.) activities. Both datasets contain real-life sensor data of activities of daily living annotated by the inhabitant.

4.1   Comparing Intervals to GMM

This experiment assessed the precision of the intervals found by TIMe by comparing them to those found by fitting a GMM on the Aruba dataset. We found the optimal number of components for the GMM using the BIC criterion [13] and fit the model using the scikit-learn library[1]. Taking $[\mu - \sigma, \mu + \sigma]$ of each component as start interval, we compared them to those found by TIMe graphically (Fig. 1). Notice how both methods find different number of probable intervals each with a different size for each activity. Eating shows the highest difference because it is too irregular around midday so TIMe cannot find any usual time but GMM finds a low probability interval.

As a quantitative measure, we calculated the Jaccard similarity of the intervals whose size is measured in minutes (Table 1). Eating has the lowest similarity due to the interval between 9:30 and 17:00, not recognized by TIMe but found by GMM. Even though, the average similarity for every other activity is suitable.

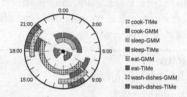

Fig. 1. GMM v.s. TIMe start time intervals

Table 1. Jaccard similarity of GMM and TIMe intervals

Activity	Average similarity
Sleep	76.7%
Eating	38.8%
Cook	87.1%
Wash dishes	59.9%

4.2 Finding Pattern Variability

This experiment evaluated how TIMe can detect variability with respect to different context conditions. For the Aruba dataset we consider the day of the week as a context dimension and evaluated differences in week and weekend patterns (Fig. 2) and in day to day variability (Figs. 3 and 4). These results show how an activity can start at different times each day.

To explore variability with respect to other context dimensions, we used ContextAct@Home considering day of the week and weather conditions (Figs. 2, 3 and 4). Wake up corresponds to the end time of sleep. The variations

Fig. 2. Week v.s. weekend time variations

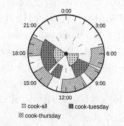

Fig. 3. Cooking time variations

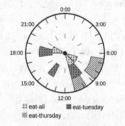

Fig. 4. Eating time variations

---

[1] http://scikit-learn.org/.

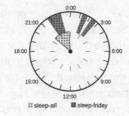

:: bath-all   ■ bath-weekend   :: sleep-all   ■ sleep-friday   :: wakeup-all   ■ wakeup-rain

Fig. 5. Bath variations   **Fig. 6.** Sleep variations   **Fig. 7.** Wake variations

found help to better reason about routines. For example, if only patterns on the whole dataset were mined, taking a bath after midday (Fig. 5) or going to sleep later on Fridays (Fig. 6) or waking up later when it rains (Fig. 7) would be considered an anomaly but those changes are common given their context.

4.3   Understanding Sensor Data Using Patterns

To show how sensor events patterns increase understanding, we compared start time intervals of the bed pressure, stove state, shower faucet state and dishwasher faucet state sensor events to the intervals of sleeping, cooking, showering and washing dishes activity annotations respectively (Fig. 8). We chose these activities because the sensors highly correlate to doing the activity. Intervals are very similar for all activities (Table 2).

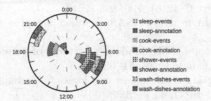

:: sleep-events
■ sleep-annotation
// cook-events
■ cook-annotation
::: shower-events
■ shower-annotation
// wash-dishes-events
■ wash-dishes-annotation

Fig. 8. Activity v.s. sensor events intervals (A4H dataset)

Table 2. Average similarity for activity and sensor events intervals

Activity	Jaccard similarity
Sleep	84.6%
Shower	81.2%
Cook	89.3%
Wash dishes	77.1%

Showering and washing dishes have lower similarities because these activities started a while after stating it and because the faucet is used for other activities (like washing vegetables). Still, patterns resemble the inhabitant's routine greatly.

5   Discussion, Conclusion and Perspectives

In this paper, we presented TIMe, a frequent pattern mining algorithm on stream data to learn start times and their variations without manual data slicing. With

GMM as a baseline, our experiments show that probable start time intervals mined with TIMe are suitable (Sect. 4.1) but at a significant lower cost since TIMe uses a single data scan to analyze incoming data in a streaming mode.

TIMe can analyze variability due to multiple context dimensions each at multiple scales without manual data slicing. Our results show that there are in fact variations in patterns when analyzed in the light of different context conditions, but that not all patterns change with respect to the same context condition nor scale (Sect. 4.2). Given this, it is specially important to analyze each pattern separately and that each dimension is evaluated independently from others.

Being able to find pattern variations can help to better understand behavior and better analyze events so that false alarms are reduced. We have shown that patterns from sensor events are highly similar to those of annotated activities (Sect. 4.3).

TIMe maintains the relationship between an activity and its variations. This not only allows a separation of concerns but also creates a richer semantic model of behavior. As such, activity recognition methods can focus on global characteristics and personal characteristics such as start time are dealt with by a behavior analysis method.

As future work we will apply TIMe to analyze pattern evolution and also study how to learn other properties of behavior (expected frequency, frequent sequences). These aspects complete knowledge about personal behavior patterns.

References

1. Chen, L., Hoey, J., Nugent, C.D., Cook, D.J., Yu, Z.: Sensor-based activity recognition. IEEE Trans. Syst. Man Cybern. Part C **42**(6), 790–808 (2012)
2. Cook, D., Crandall, A., Thomas, B., Krishnan, N.: CASAS: a smart home in a box. IEEE Comput. **46**(7), 62–69 (2013)
3. Dawadi, P.N., Cook, D.J., Schmitter-Edgecombe, M.: Modeling patterns of activities using activity curves. Pervasive Mob. Comput. **28**, 51–68 (2015)
4. Rodríguez, N.D., Cuéllar, M., Lilius, J., et al.: A survey on ontologies for human behavior recognition. ACM Comput. Surv. **46**(4), 43:1–43:33 (2014)
5. Forkan, A., et al.: A context-aware approach for long-term behavioural change detection and abnormality prediction in AAL. Pattern Recognit. **48**(3), 628–641 (2015)
6. Jakkula, V.R., Crandall, A.S., Cook, D.J.: Enhancing anomaly detection using temporal pattern discovery. In: Kameas, A., Callagan, V., Hagras, H., Weber, M., Minker, W. (eds.) Advanced Intelligent Environments, pp. 175–194. Springer, Heidelberg (2009). doi:10.1007/978-0-387-76485-6_8
7. Lago, P., Lang, F., Roncancio, C., Jiménez-Guarín, C., Mateescu, R., Bonnefond, N.: The ContextAct@A4H real-life dataset of daily-living activities. In: Brézillon, P., Turner, R., Penco, C. (eds.) CONTEXT 2017. LNCS, vol. 10257, pp. 175–188. Springer, Cham (2017). doi:10.1007/978-3-319-57837-8_14
8. Lago, P., Jiménez-Guarín, C., Roncancio, C.: A case study on the analysis of behavior patterns and pattern changes in smart environments. In: Pecchia, L., Chen, L.L., Nugent, C., Bravo, J. (eds.) IWAAL 2014. LNCS, vol. 8868, pp. 296–303. Springer, Cham (2014). doi:10.1007/978-3-319-13105-4_43

9. Monekosso, D.N., Remagnino, P.: Behavior analysis for assisted living. IEEE Trans. Autom. Sci. Eng. **7**(4), 879–886 (2010)
10. Moshtaghi, M., Zukerman, I., Russell, R.A.: Statistical models for unobtrusively detecting abnormal periods of inactivity in older adults. User Model. User-Adap. **25**(3), 231–265 (2015)
11. Nait Aicha, A., Englebienne, G., Kröse, B.: Unsupervised visit detection in smart homes. Pervasive Mob. Comput. **34**, 157–167 (2016)
12. Rabatel, J., Bringay, S., Poncelet, P.: Contextual sequential pattern mining. In: 2010 IEEE ICDM Workshops, pp. 981–988 (2010)
13. Schwarz, G.: Estimating the dimension of a model. Ann. Stat. **6**(2), 461–464 (1978)
14. Soulas, J., Lenca, P., Thépaut, A.: Unsupervised discovery of activities of daily living characterized by their periodicity and variability. Eng. Appl. Artif. Intell. **45**, 90–102 (2015)

Truthfulness of Candidates in Set of t-uples Expansion

Ngurah Agus Sanjaya Er[1,4]($\boxtimes$), Mouhamadou Lamine Ba[2],
Talel Abdessalem[1,3,4], and Stéphane Bressan[3]

[1] Télécom Paristech, Paris, France
sanjaya.agus@telecom-paristech.fr
[2] Université Alioune Diop de Bambey, Bambey, Senegal
[3] National University of Singapore, Singapore, Singapore
[4] UMI IPAL, CNRS, Paris, France

Abstract. Set of t-uples expansion refers to the task of building a set of t-uples from a corpus based on some examples, or seed t-uples. Set of t-uples expansion requires a ranking mechanism to select the relevant candidates. We propose to harness and compare the performance of different state-of-the-art truth finding algorithms for the task of set of t-uples expansion. We empirically and comparatively evaluate the accuracy of these different ranking algorithms. We show that truth finding algorithms provide a practical and effective solution.

Keywords: Set expansion · Seeds · t-uples · Ranking · Truth finding · Evaluation

1 Introduction

A set expansion system provides a different approach to extracting information from the World Wide Web. It starts with some examples of the information of interest and automatically extracts matching candidates from the Web. The term *seeds* is often used to denote examples in set expansion system. For example, if one wants to retrieve the list of all the presidents of the United States, she/he can start by giving *<Barrack Obama>*, *<Bill Clinton>* and *<Donald Trump>* as three seeds. The system then will try to figure out and return all the possible other US President names. DIPRE [8], SEAL [27] and STEP [13] are three examples of set expansion systems which extract either atomic entities or n-ary (n being the number of elements) relations from the Web.

Motivations. A given set expansion system may, however, extract *false candidates* due to the inherent imprecision of any extraction process; for instance, a system could extract *<Hillary Clinton>* as a US President. In order to mitigate this problem, these systems implement a ranking mechanism as a manner to verify the level of the truthfulness of the extracted candidates. As an example,

© Springer International Publishing AG 2017
D. Benslimane et al. (Eds.): DEXA 2017, Part I, LNCS 10438, pp. 314–323, 2017.
DOI: 10.1007/978-3-319-64468-4_24

SEAL [27] uses a weighted graph to rank candidates. It gives more weight to candidates which are extracted from many Web pages. Similarly, STEP [13] uses a weighted graph where the weight of each node is the result of running PageRank [21] function on the graph. The performance of the set expansion process is, therefore, impacted by the precision of the implemented ranking mechanism. Unfortunately, currently implemented mechanisms do not always guarantee high performance because they can return candidates that are found to be incorrect in some scenarios. The ranking of SEAL system is biased in the presence of Web pages with different reliability level, e.g. the candidate <*Hillary Clinton*> could come from the majority of Web pages, but these are untrustworthy sources. We argue that incorporating truth finding in the set of tuples expansion routine may help to solve such a kind of issue. Identifying true and false values in an automated way, usually known as truth finding, is a challenging problem that has been extensively studied in several contexts such as data integration and fact checking. Numerous truth finding algorithms, each of them exploring specific aspects behind the truth or targeting particular applications, have been proposed in this perspective; for respectively a short presentation and a detailed picture of existing truth finding approaches we defer to [1,5].

Contribution. In this work, we propose to incorporate truth finding algorithm as a mean to justify the level of the truthfulness of the extracted candidates. Integrating a truth finding step in the set of tuples expansion is natural, as we show it in Sect. 3. We harness and compare the performance of different state-of-the-art truth finding algorithms for the new task of set of t-uples expansion, by empirically and comparatively evaluating the precision, recall and F-measure of these different ranking algorithms with several sets of t-uples expansion cases. As the main result, we show that the proposed approach to set of t-uples expansion leveraging truth finding algorithms is practical and effective. We identify some of the challenges to be overcome in difficult cases.

The remaining of the paper is organized as follows. In Sect. 2, we summarize the related works on set expansion and truth finding. We formally introduce the translation of the truth finding problem in our set of t-uples expansion approach in Sect. 3. The performance evaluation is presented in Sect. 4, whereas Sect. 5 ends the paper with some research perspectives.

2   Related Work

Set expansion systems try to extract elements of a particular semantic class from a given data source and some examples of its members. More precisely, given a set of seeds of a particular semantic class and a collection of documents or corpus, the set expansion systems extract more elements of this class from the input collection. Set expansion systems are useful for many applications such as vocabulary or dictionary construction [9,24], question answering [29], and knowledge extraction [19]. These systems fetch relevant documents containing the given seeds. Sources of documents are the World Wide Web [8,13,27,28], encyclopedias [6], search logs [20,31]. The systems locate occurrences of the seeds in the documents and infer patterns (wrappers). The generated wrappers are then used to extract candidates from the documents. Finally, the systems rank the extracted candidates using such ranking mechanisms as random walk [27], PageRank [13,28], Bayesian sets [28], iterative thresholding [16], nearest neighbour [20]. Brin [8] proposed DIPRE. DIPRE retrieves documents containing the seeds and infers wrappers to extract new candidates. The same process is repeated using the newly found seeds and ended when there are no more candidates. DIPRE [8] does not employ any ranking strategies. SEAL is proposed by Wang and Cohen [27]. It expands sets of atomic entities from a collection of semi-structured documents and generates a pattern for a specific page. The ranking mechanism in SEAL is based on a weighted graph model and a random walk on the graph. In [28], Wang and Cohen extend SEAL to support the extraction of binary relations. Er et al. [13] extend the set expansion problem to set of t-uples expansion and introduce STEP.

Areas such as data integration [10,32] and information retrieval [2] are very active on studying the issue of finding object true statements from multiple conflicting sources. Such a problem is called truth finding [2], truth discovery [11,30,32,33], fact checking, data fusion [7,12,17,23], etc., as highlighted by surveys and comparative analysis in [1,5,12,18,25]. A usual truth finding setting is an iterative process that deterministically merges conflicting statements and automatically computes a confidence value for each statement based on the trustworthiness level of its sources: source trustworthiness are themselves unknown apriori. The truth consists of the statements with the highest confidence scores. According to the core of the implemented iterative function, one can classify current truth finding approaches as follows: (i) **agreement-based** methods such as Majority voting, TruthFinder [30], Cosine, 2-Estimates, and 3-Estimates [15]; (ii) **MAP estimation based** methods such as MLE [26], LTM [32], LCA [22]; and (iii) **Bayesian Inference-based** methods such as Depen and its variants [10]. The truth presents several facets in real applications, as shown in [10,15,22,30]. Ba et al. present, in [3], possible structural correlations among data attributes to present a complementary study of the proposed models. None amongst the existing truth finding algorithms outperforms the others in all cases. Combiner [4,25] is an ensembling model, enabling to exploit and integrate the best of several concurrent models in order to outperform each of them. VERA [2] implements Combiner and proposes a Web platform to check the veracity of Web facts of numerical types.

3 Set Expansion with Truth Finding

We formally present hereafter our use of truth finding (instead of PageRank) on candidate t-uples extracted by STEP which is introduced and detailed in [13]. We fix the set $\mathcal{T}$ of extracted candidate t-uples representing instances of real world objects characterized by some attributes (or properties). We restrict ourselves in this work to the common *one-truth* framework where any attribute of every object has only *one correct value* and *many possible wrong values*. We consider fixed finite sets of *attribute labels* $\mathscr{A}$ and *values* $\mathscr{V}$, as well as a finite set of *object identifiers* $\mathscr{I}$. Then, we have formally:

Definition 1. *A* t-uple *t is a pair* (i, v) *where* $i \in \mathscr{I}$ *and* v *a mapping* $v : \mathscr{A} \to \mathscr{V}$.

STEP inspects documents using wrappers to extract a collection of t-uples as illustrated in Example 1. In this setting, we describe a *data source* as modeled by a given document (where the t-uple has been extracted) and a wrapper (the extractor of the t-uple). At attribute level, we define indeed a data source as follows:

Definition 2. *A* source *is a partial function* $S : \mathscr{I} \times \mathscr{A} \to \mathscr{V}$ *with non-empty domain. A (candidate)* ground truth *is a total function* $G : \mathscr{I} \times \mathscr{A} \to \mathscr{V}$.

Given a source S and two distinct attributes a, $a' \in \mathscr{A}$ we state that $S(i_1, a)$ and $S(i_2, a')$ refer to statements about attributes a and a' of the same t-uple if $i_1 = i_2$ where i_1 and i_2 are object identifiers in $\mathscr{I}$.

Example 1. Consider the following t-uples *<Alger, Dinar Algerian>*, *<Alger, Leka Algerian>*, and *<Alger, Dinar Algerian>* extracted from three different data sources *source_wrap1*, *source_wrap2* and *source_wrap3* by a set expansion system. One can observe these are about the real world object "Country" with their properties "Capital City" and "Currency".

The main target of the truth finding issue is to determine the actual ground truth based on the statements of a number of sources. Two sources S and S' are *conflicting* whenever there exists $i \in \mathscr{I}$, $a \in \mathscr{A}$ such that $S(i, a)$ and $S'(i, a)$ are both defined and $S(i, a) \neq S'(i, a)$. A source is *correct* with respect to the ground truth G on $i \in \mathscr{I}$, $a \in \mathscr{A}$ when $S(i, a) = G(i, a)$, and *wrong* when $S(i, a)$ is defined but $S(i, a) \neq G(i, a)$. A truth finding process is formally defined as follows:

Definition 3. *A truth finding algorithm* F *is an algorithm that takes as input a set of sources* $\mathscr{S}$ *and returns a candidate ground truth* $F(\mathscr{S})$, *as well as an estimated* accuracy. *Accuracy*$_F(S)$ *for each source* $S \in \mathscr{S}$ *and confidence score Confidence*$_F(S(i, a))$ *for every statement* $S(i, a)$ *made by a source* S *on any attribute* a *of the object* i.

Truth finding processes are mostly fixpoint iteration algorithms which compute a candidate ground truth based on source accuracy (or trustworthiness) values and confidence scores. In the special case of majority voting, one can respectively define source accuracy and confidence for F simply as the proportion of true statements in the ground truth and the average accuracy scores of the providers: see Example 2.

Example 2. Applying majority voting on candidate t-uples in Example 1, we obtain accuracy values of 1, 1/2 and 1 for *source_wrap1*, *source_wrap2*, and *source_wrap3* respectively. Indeed the true values are chosen by *voting* for the attributes "Capital City" and "Currency" will be "*Alger*" and "*Dinar Algerian*". According to the confidence scores of the statements "*Alger*", "*Dinar Algerian*" and "*Leka Algerian*" will be set to 5/6, 1 and 1/2. Obviously "*Dinar Algerian*" from top ranked sources, i.e. *source_wrap1* and *source_wrap3*, is chosen instead of "*Dinar Algerian*".

In this paper, we aim at improving the ranking of the STEP system using truth finding algorithms. To do so, we will consider and experiment on most popular state-of-the-art truth finding models which support the *one-truth* setting. We will then compare their performance against that of PageRank, as we shall detail in Sect. 4.

4   Performance Evaluation

We proceed here to a comparative evaluation of eleven truth finding algorithms and PageRank on different t-uple datasets extracted by STEP. We show that our proposed set of t-uples expansion approach leveraging truth finding algorithms is effective.

4.1   Experimentation Setting

Input Datasets. We intensively performed our tests on seven datasets consisting of a set of t-uples extracted by STEP for distinct domains. We refer our reader to [14] for a complete description of the datasets. Table 1a provides a summary of these latter.

Table 1. Description of datasets

(a) Description of Datasets and Ground Truth

Dataset	#Web_Page	#Wrapper	#Candidate_T-uples	#Distinct_T-uples	Ground_Truth
DT1 (Calling Code)	16	98	5572	1001	239
DT2 (Nobel)	5	7	446	237	107
DT3 (Chat Abbr.)	12	11	5758	5167	2697
DT4 (Currency)	9	39	2473	927	244
DT5 (FIFA Player)	10	13	260	141	61
DT6 (Miss Universe)	8	3	130	82	65
DT7 (Oscar)	6	5	368	337	87

(b) Excerpt of input dataset in AllegatorTrack

Object	Attribute	Value	Source
Angola	Capital_City	Luanda	http://1min.in/content/international/currency-codes_1
Angola	Currency	Kwanza	http://1min.in/content/international/currency-codes_1
Angola	Currency	Kuanza	http://1min.in/content/international/currency-codes_2

DAFNA Input. To evaluate with state-of-the-art truth finding algorithms we rely on the DAFNA platform. DAFNA provides AllegatorTrack [25], a Web application setting up an implementation of most of the state-of-the-art truth finding algorithms within the same setting. AllegatorTrack accepts as input a set of entries whose main elements consists of an object key, an object attribute (or property), an attribute value, and a data source. As a consequence, we then technically decompose the objects and its attributes in the t-uples extracted by our system. An object is represented by the first element in the t-uple, which is denoted as the *object key*, while the other remaining elements serve as the attributes. We further use this key element to decompose a t-uple into *key-attribute* t-uples. We do this for all the t-uples in our different datasets. In general, if a t-uple is composed of n elements, then after decomposition we obtain n-1 key-attribute t-uples.

STEP may extract the identical or distinct t-uples from this page using different wrappers. As a result, we refer within our one-truth setting to the source of a given candidate t-uple by indexing the name of the inspected Web page with the number of the used wrapper as shown with input example to AllegatorTrack in Table 1b.

Evaluated Metrics. We evaluate and compare the precision, recall, and F-measure (a.k.a. the harmonic mean) of PageRank and eleven truth finding models according to a certain ground truth. We use the following equations to compute these metrics: $p = \frac{\sum_{i=1}^{|R|} Entity(i)}{|R|}; r = \frac{\sum_{i=1}^{|R|} Entity(i)}{|G|};$ F-measure $= 2 * \frac{p*r}{p+r}$ where $Entity(i)$ is a binary function that returns true if the i-th candidate is found in the ground truth and false otherwise, $|R|$ is the number of distinct extracted candidates, and $|G|$ is the size of the ground truth. $|R|$ and $|G|$ is the fifth and last column in Table 1a.

4.2 Experiments

We started by experimenting with truth finding algorithms via AllegatorTrack. We considered each of our seven datasets and performed on it the following models: Cosine (CO), 2-Estimates (2E), 3-Estimates (3E), Depen (DP), Accu (AC), AccuSim (AS), AccuNoDep (AN), TruthFinder (TF), SimpleLCA (SL), GuessLCA (GL), and Combiner (CM). We ran those tests without changing the default values of the mandatory parameters for each algorithm, i.e. the *prior source trustworthiness* and *convergence threshold* which are set by default to 0.8 and 0.001 respectively: these are default values proven to be optimal. The output of every truth finding algorithm is, for every object attribute, a confidence score and a truth label (true or false). To evaluate the performance of the truth finding algorithms w.r.t. the ground truth, we will only consider attribute values whose truth labels are set to true for each t-uple. We then continued our experimentation by performing PageRank (PR) on the same datasets; we used the same implementation of PageRank, as well as the identical experimentation setting, as in our previous paper [13]. For performance evaluation, we will only consider t-uples with the highest score for each object returned by PageRank algorithm. We conducted those experiments in two stages as detail now in the following.

Per-attribute Experiments. In the first phase, we were interested in doing tests on datasets per attribute, i.e., if a given dataset contains objects with several attributes we have considered sub-datasets per attributes and ran experiments on each sub-dataset separately. We did such a splitting of each tested dataset because we made the focus on the performance of the algorithms per attribute, as we will later explain with the results in Table 2a and b. In this perspective, we will use "Attr", the abbreviation for an attribute, and add an index to differentiate the different attributes belonging to the same given dataset when the latter has been split into several sub-datasets.

Overall Dataset Experiments. In the second phase, we conducted tests using every overall dataset, i.e. without doing the splitting according to the attributes. The goal was to study the performance of the ranking algorithms on any given entire input dataset.

4.3   Result Analysis

Per-attribute Precision and Recall Analysis. Measures in Table 2a and b show that at least one truth finding algorithm is better or equal in performance to PageRank, except for DT4 (Attr2), DT5 (Attr2), and DT6 (Attr1). In dataset DT4, five truth finding algorithms (2E, AC, AS, AN, and SL) receive 0 in precision, while the highest score is achieved by Cosine (CO). They achieve the minimum precision score because these algorithms affect a false true label to the second attribute (Attr2) of the object.

Table 2. Precision and recall per-attribute

(a) Precision measures per-attribute

		CO	2E	3E	DP	AC	AS	AN	TF	SL	GL	CM	PR
DT1	Attr1	0.172	0.361	0.368	0.360	0.368	0.368	0.368	0.360	0.368	0.360	0.361	0.368
DT2	Attr1	0.137	0.147	0.168	0.147	0.153	0.153	0.153	0.147	0.153	0.153	0.158	0.153
DT3	Attr1	0.246	0.194	0.201	0.241	0.231	0.236	0.231	0.283	0.214	0.220	0.255	0.064
DT4	Attr1	0.620	0.667	0.702	0.647	0.676	0.676	0.676	0.632	0.702	0.673	0.647	0.620
	Attr2	0.084	0	0.017	0.003	0	0	0	0.005	0	0.003	0.005	0.020
DT5	Attr1	0.730	0.888	0.873	0.841	0.888	0.888	0.888	0.793	0.888	0.793	0.841	0.888
	Attr2	0.873	0.841	0.809	0.888	0.873	0.873	0.873	0.873	0.873	0.873	0.873	0.920
DT6	Attr1	0.924	0.878	0.924	0.924	0.924	0.924	0.924	0.924	0.924	0.924	0.924	0.939
	Attr2	0.818	0.787	0.818	0.787	0.838	0.838	0.848	0.803	0.818	0.848	0.787	0.803
DT7	Attr1	0.442	0.863	0.915	0.536	0.915	0.915	0.915	0.547	0.915	0.884	0.915	0.410
	Attr2	0.410	0.810	0.831	0.452	0.831	0.831	0.831	0.494	0.915	0.778	0.831	0.336

(b) Recall measures per-attribute

		CO	2E	3E	DP	AC	AS	AN	TF	SL	GL	CM	PR
DT1	Attr1	0.423	0.887	0.904	0.883	0.904	0.904	0.904	0.883	0.904	0.883	0.887	0.895
DT2	Attr1	0.252	0.271	0.308	0.271	0.280	0.280	0.280	0.271	0.280	0.280	0.290	0.280
DT3	Attr1	0.247	0.195	0.201	0.241	0.232	0.236	0.232	0.284	0.215	0.221	0.255	0.064
DT4	Attr1	0.873	0.939	0.988	0.910	0.951	0.951	0.951	0.889	0.988	0.947	0.910	0.873
	Attr2	0.119	0	0.025	0.004	0	0	0	0.008	0	0.004	0.008	0.029
DT5	Attr1	0.754	0.918	0.902	0.869	0.918	0.918	0.918	0.820	0.918	0.820	0.869	0.918
	Attr2	0.902	0.869	0.836	0.918	0.902	0.902	0.902	0.902	0.902	0.902	0.902	0.951
DT6	Attr1	0.938	0.892	0.938	0.938	0.938	0.938	0.938	0.938	0.938	0.938	0.938	0.954
	Attr2	0.831	0.800	0.831	0.800	0.862	0.862	0.862	0.815	0.831	0.862	0.800	0.815
DT7	Attr1	0.442	0.863	0.916	0.537	0.916	0.916	0.916	0.547	0.916	0.884	0.916	0.411
	Attr2	0.411	0.811	0.832	0.453	0.832	0.832	0.832	0.495	0.916	0.779	0.832	0.337

Overall Performance Analysis. According to Table 3a and b, PageRank achieves the lowest average for the overall precision and recall. The best truth finding algorithms, in this case, are Accu and TruthFinder respectively (indicated by bold in Table 3a and b). However, in terms of the trade-off between precision and recall, Fig. 1 proves that Accu is an obvious choice. We conclude our analysis by stating that the set expansion problem can significantly gain effectiveness with truth finding, as demonstrated by the different results of our intensive experiments on various datasets.

Table 3. Overall precision and recall

(a) Overall precision measures

	CO	2E	3E	DP	AC	AS	AN	TF	SL	GL	CM	PR
DT1	0.172	0.361	0.368	0.360	0.368	0.368	0.368	0.360	0.368	0.360	0.361	0.366
DT2	0.137	0.147	0.168	0.147	0.153	0.153	0.153	0.147	0.153	0.153	0.153	0.153
DT3	0.261	0.208	0.213	0.256	0.245	0.244	0.245	0.459	0.226	0.233	0.242	0.069
DT4	0	0	0.003	0	0	0	0	0.003	0.000	0.003	0.055	0.128
DT5	0.841	0.809	0.809	0.777	0.809	0.809	0.809	0.730	0.809	0.809	0.460	0.079
DT6	0.787	0.742	0.818	0.757	0.818	0.818	0.818	0.772	0.787	0.818	0.742	0.696
DT7	0.378	0.842	0.915	0.442	0.925	0.915	0.915	0.484	0.915	0.884	0.915	0.010
Average	0.368	0.444	0.471	0.391	**0.474**	0.472	0.473	0.422	0.465	0.470	0.325	0.214

(b) Overal recall measures

	CO	2E	3E	DP	AC	AS	AN	TF	SL	GL	CM	PR
DT1	0.423	0.891	0.908	0.887	0.908	0.908	0.908	0.887	0.908	0.891	0.787	0.891
DT2	0.252	0.271	0.308	0.271	0.280	0.280	0.280	0.271	0.280	0.280	0.280	0.280
DT3	0.262	0.209	0.213	0.257	0.245	0.245	0.245	0.460	0.226	0.233	0.242	0.069
DT4	0	0	0.004	0	0	0	0	0.004	0	0.004	0.078	0.180
DT5	0.869	0.836	0.836	0.803	0.836	0.836	0.836	0.754	0.836	0.836	0.475	0.082
DT6	0.800	0.754	0.831	0.769	0.831	0.831	0.831	0.785	0.800	0.831	0.754	0.692
DT7	0.379	0.842	0.916	0.442	0.916	0.916	0.916	0.484	0.916	0.916	0.263	0.011
Average	0.434	0.493	0.517	0.498	0.517	0.517	0.517	**0.527**	0.508	0.513	0.436	0.366

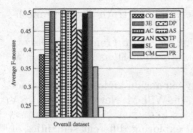

Figure 1 compares average F-measure of truth finding algorithms and PageRank given the F-measure scores of every approach over our seven datasets. Obviously, truth finding algorithms outperforms PageRank with an improvement ranging from 10% to 26% in terms of both precision and recall. AC is the best truth finding model for ranking w.r.t. computed F-measure scores.

Fig. 1. Average F-measure scores

Difficult Cases. Our evaluation has revealed some cases where truth finding becomes hard, that is: (i) for particular attributes, e.g. Attr2 of DT4 and; (ii) for particular datasets, e.g. DT2 and DT4. We further investigated this problem at dataset level for DT2 and DT4. We first computed the precision and recall of Majority Voting on these datasets and observed that they are very low which prove that DT4 and DT2 suffer from the fact the majority of their sources are constantly wrong, yielding a biased truth finding process. We also manually inspected the datasets and discovered a lot of noises, e.g. nonnormalized values such a "Singapore Dollar" and "Dollar". A better data preparation before the ranking or the use of ontology into the set of t-uples expansion may be helpful.

5 Conclusion

In this paper, we have proposed to leverage truth finding algorithm to candidates extracted by the set of t-uples expansion. We empirically showed that truth finding algorithms can significantly improve the truthfulness of the extracted candidates. We intend to use truth finding algorithms into our STEP system, by also tackling difficult cases. An extended presentation of this work with more details and extended discussions is available in [14].

Acknowledgment. This work has been partially funded by the Big Data and Market Insights Chair of Télécom ParisTech and supported by the National University of Singapore under a grant from Singapore Ministry of Education for research project number T1 251RES1607.

References

1. Waguih, D.A., Berti-Équille, L.: Truth discovery algorithms: an experimental evaluation. CoRR, abs/1409.6428, September 2014
2. Ba, M.L., Berti-Equille, L., Shah, K., Hammady, H.M.: VERA: a platform for veracity estimation over web data. In: WWW (2016)
3. Ba, M.L., Horincar, R., Senellart, P., Wu, H.: Truth finding with attribute partitioning. In: WebDB SIGMOD Workshop, Melbourne, Australia, May 2015

4. Berti-Equille, L.: Data veracity estimation with ensembling truth discovery methods. In: IEEE Big Data Workshop (2015)
5. Berti-Equille, L., Borge-Holthoefer, J.: Veracity of Big Data: From Truth Discovery Computation Algorithms to Models of Misinformation Dynamics. Morgan & Claypool, San Rafael (2015)
6. Bing, L., Lam, W., Wong, T.L.: Wikipedia entity expansion and attribute extraction from the web using semi-supervised learning. In: WSDM, New York, NY, USA (2013)
7. Bleiholder, J., Draba, K., Naumann, F.: FuSem: exploring different semantics of data fusion. In: VLDB, Vienna, Austria (2007)
8. Brin, S.: Extracting patterns and relations from the world wide web. In: WWW and Databases Workshop (1998)
9. Chen, Z., Cafarella, M., Jagadish, H.V.: Long-tail vocabulary dictionary extraction from the web. In: WSDM, New York, NY, USA (2016)
10. Dong, X.L., Berti-Equille, L., Srivastava, D.: Integrating conflicting data: the role of source dependence. PVLDB **2**(1), 550–561 (2009)
11. Dong, X.L., Berti-Equille, L., Srivastava, D.: Truth discovery and copying detection in a dynamic world. PVLDB **2**(1), 562–573 (2009)
12. Dong, X.L., Naumann, F.: Data fusion: resolving data conflicts for integration. PVLDB **2**(2), 1654–1655 (2009)
13. Er, N.A.S., Abdessalem, T., Bressan, S.: Set of t-uples expansion by example. In: iiWAS, New York, NY, USA (2016)
14. Er, N.A.S., Ba, M.L., Abdessalem, T., Bressan, S.: Truthfulness of candidates in set of t-uples expansion. Technical report, National University of Singapore, School of Computing, TRA5/17, May 2017
15. Galland, A., Abiteboul, S., Marian, A., Senellart, P.: Corroborating information from disagreeing views. In: WSDM, New York, USA, February 2010
16. He, Y., Xin, D.: SEISA: set expansion by iterative similarity aggregation. In: WWW, New York, NY, USA (2011)
17. Li, Q., Li, Y., Gao, J., Zhao, B., Fan, W., Han, J.: Resolving conflicts in heterogeneous data by truth discovery and source reliability estimation. In: SIGMOD, Snowbird, Utah, USA, May 2014
18. Li, X., Dong, X.L., Lyons, K., Meng, W., Srivastava, D.: Truth finding on the deep web: is the problem solved? PVLDB **6**(2), 97–108 (2012)
19. Moens, M., Li, J., Chua, T. (eds.): Mining User Generated Content. Chapman and Hall/CRC, Boca Raton (2014)
20. Paşca, M.: Weakly-supervised discovery of named entities using web search queries. In: CIKM, New York, NY, USA (2007)
21. Page, L., Brin, S., Motwani, R., Winograd, T.: The pagerank citation ranking: bringing order to the web. Technical report (1999)
22. Pasternack, J., Roth, D.: Latent credibility analysis. In: WWW, Rio de Janeiro, Brazil, May 2013
23. Pochampally, R., Das Sarma, A., Dong, X.L., Meliou, A., Srivastava, D.: Fusing data with correlations. In: SIGMOD, Snowbird, Utah, USA, May 2014
24. Sarker, A., Gonzalez, G.: Portable automatic text classification for adverse drug reaction detection via multi-corpus training. J. Biomed. Inform. **53**, 196–207 (2015)
25. Waguih, D.A., Goel, N., Hammady, H.M., Berti-Equille, L.: AllegatorTrack: combining and reporting results of truth discovery from multi-source data. In: ICDE, Seoul, Korea (2015)
26. Wang, D., Kaplan, L., Le, H., Abdelzaher, T.: On truth discovery in social sensing: a maximum likelihood estimation approach. In: IPSN, Beijing, China, April 2012

27. Wang, R.C., Cohen, W.W.: Language-independent set expansion of named entities using the web. In: ICDM (2007)
28. Wang, R.C., Cohen, W.W.: Character-level analysis of semi-structured documents for set expansion. In: EMNPL, Stroudsburg, PA, USA (2009)
29. Wang, R.C., Schlaefer, N., Cohen, W.W., Nyberg, E.: Automatic set expansion for list question answering. In: EMNLP, Stroudsburg, PA, USA (2008)
30. Yin, X., Han, J., Yu, P.S.: Truth discovery with multiple conflicting information providers on the web. In: IEEE TKDE, June 2008
31. Zhang, Z., Sun, L., Han, X.: A joint model for entity set expansion and attribute extraction from web search queries. In: AAAI (2016)
32. Zhao, B., Rubinstein, B.I.P., Gemmell, J., Han, J.: A Bayesian approach to discovering truth from conflicting sources for data integration. PVLDB 5(6), 550–561 (2012)
33. Zhao, Z., Cheng, J., Ng, W.: Truth discovery in data streams: a single-pass probabilistic approach. In: CIKM, Shangai, China, November 2014

Data Clustering

Interactive Exploration of Subspace Clusters for High Dimensional Data

Jesper Kristensen[1], Son T. Mai[1,3(✉)], Ira Assent[1], Jon Jacobsen[1], Bay Vo[2], and Anh Le[3]

[1] Department of Computer Science, Aarhus University, Aarhus, Denmark
{jesper,mtson,ira,jon}@cs.au.dk
[2] Faculty of Information Technology, Ho Chi Minh City University of Technology, Ho Chi Minh City, Vietnam
bayvodinh@gmail.com
[3] Department of Computer Science, University of Transport, Ho Chi Minh City, Vietnam
anh@hcmutrans.edu.vn

Abstract. PreDeCon is a fundamental clustering algorithm for finding arbitrarily shaped clusters hidden in high-dimensional feature spaces of data, which is an important research topic and has many potential applications. However, it suffers from very high runtime as well as lack of interactions with users. Our algorithm, called AnyPDC, introduces a novel approach to cope with these problems by casting PreDeCon into an anytime algorithm. It quickly produces an approximate result and iteratively refines it toward the result of PreDeCon at the end. This scheme not only significantly speeds up the algorithm but also provides interactions with users during its execution. Experiments conducted on real large datasets show that AnyPDC acquires good approximate results very early, leading to an order of magnitude speedup factor compared to PreDeCon. More interestingly, while anytime techniques usually end up slower than batch ones, AnyPDC is faster than PreDeCon even if it run to the end.

Keywords: Subspace clustering · Interactive clustering

1 Introduction

Clustering is a major data mining task that separates points into groups called clusters so that points inside each group are more similar to each other than between different groups. Traditional clustering algorithms only look for clusters in full dimension (attribute) of data. However, when the dimensionality increases, the data space becomes sparse [7] and thus meaningful clusters are no longer exist in full dimensional space, thus making these techniques inappropriate for processing high dimensional data.

Due to this *curse of dimensionality* problem [7,10], finding subspace clusters hidden in high dimensional feature spaces of data has recently become one of the

D. Benslimane et al. (Eds.): DEXA 2017, Part I, LNCS 10438, pp. 327–342, 2017.
DOI: 10.1007/978-3-319-64468-4_25

most important research topics with many potential applications in many fields [10]. While most techniques focus on finding clusters related to fixed subsets of attributes, e.g., SUBCLU [11], PreDeCon [5] is a fundamental technique which is specifically designed to find arbitrarily shaped clusters where their points are specific on a certain number of dimensions. In PreDeCon, each point is assigned a so-called *subspace preference vector* indicating the distribution of its local neighborhood along each dimension. These vectors are then used for calculating the distance among points in a weighted scheme. And the results are employed for finding areas with high point density (or clusters).

One major drawback of PreDeCon is that it needs to calculate the neighborhoods of all points under the weighted distance functions, thus resulting in $O(mn^2)$ time complexity where m and n are the dimension of data and number of points, respectively. In our era of big data, this quadratic time complexity obviously is a computational bottleneck that limits its applicability.

In this paper, we aim at an efficient extension of the algorithm PreDeCon, called *anytime PreDeCon* (AnyPDC), with some unique properties as follows.

– **Anytime.** AnyPDC is an interactive algorithm. It quickly produces an approximate result and then continuously refines it until it reaches the same result as PreDeCon. During its runtime, it can be suspended for examining intermediate results and resumed for finding better results at any time. This scheme is very useful when coping with large datasets under arbitrary time constraints. To the best of our knowledge, AnyPDC is the first *anytime* extension of PreDeCon proposed in the literature so far.
– **Efficiency.** Most indexing techniques for speeding up the range query processing suffers from performance degeneration on high dimensional data due to the *curse of dimensionality* problem described above. Reducing the total number of range queries as well as range query times therefore play a crucial role for enhancing the efficiency of PreDeCon. By using the Euclidean distance as a filter for the weighted distance, AnyPDC significantly reduces the time for calculating the neighborhoods of points. Moreover, it repeatedly learns the cluster structure of data at each iteration, actively chooses only a small subset of points for producing clusters, and ignores points that do not contribute to the determination of the final result. Consequently, it produces the same result as PreDeCon with much less number of range queries. These makes AnyPDC much efficient than PreDeCon.

2   Backgrounds

2.1   The Algorithm PreDeCon

Given a dataset D with n points in m dimensions. Let $A = \{A_1, \cdots, A_m\}$ be the set of attributes of points in D. Given four parameters $\mu \in \mathbb{N}$, $\epsilon \in \mathbb{R}$, $\delta \in \mathbb{R}$, and $\lambda \in \mathbb{N}$. The general idea of PreDeCon [5] is finding areas of high density points, where their local ϵ-neighborhoods are specific in a predefined number of dimensions, as clusters.

Definition 1 *(ϵ-neighborhood). The ϵ-neighborhood of a point p, denoted as $N_\epsilon(p)$, under the Euclidean distance d is defined as $N_\epsilon(p) = \{q \mid d(p,q) \leq \epsilon\}$.*

In PreDeCon, each point p is assigned a so-called *subspace preference vector* which captures the main direction of the local neighborhood of p.

Definition 2 *(Subspace preference vector). The subspace preference vector of a point p is denoted as $\overline{w}_p = \{w_1, \ldots, w_d\}$ where*

$$w_i = \begin{cases} 1 & \text{if } Var_{A_i}(N_\epsilon(p)) > \delta \\ \kappa & \text{otherwise} \end{cases}$$

where $\kappa \gg 1$ is a predefined constant, $VAR_{A_i}(N_\epsilon(p))$ is the variance of $N_\epsilon(p)$ along an attribute A_i and is defined as:

$$Var_{A_i}(N_\epsilon(p)) = \frac{\sum_{q \in N_\epsilon(p)} (d(\pi_{A_i}(p), \pi_{A_i}(q))^2}{|N_\epsilon(p)|}$$

where $\pi_{A_i}(p)$ is the projection of p onto an attribute A_i.

PreDeCon uses preference vectors as weights for different dimensions when calculating the distance among points.

Definition 3 *(Preference weighted distance). The preference weighted distance between two points p and q is defined as:*

$$d_{pref}(p,q) = max\{d_p(p,q), d_q(p,q)\}$$

where $d_p(p,q)$ is the preference weighted distance between p and q wrt. the preference weighted vector $\overline{w}_p$.

$$d_p(p,q) = \sqrt{\sum_1^d w_i \cdot d(\pi_{A_i}(p), \pi_{A_i}(q))^2}$$

Definition 4 *(Preference dimensionality). Preference dimensionality of a point p, denoted as $Pdim(p)$, is the number of attributes A_i with $Var_{A_i}(N_\epsilon(p)) \leq \delta$.*

Definition 5 *(Preference weighted neighborhood). Preference weighted neighborhood of a point p, denoted as $P_\epsilon(p)$, is the set of point q, where $d_{pref}(p,q) \leq \epsilon$.*

Definition 6 *(Core property). A point p is a core point, denoted as $core(p)$, if $|P_\epsilon(p)| \geq \mu \wedge Pdim(p) \leq \lambda$. If p is not a core but one of its preference weighted neighbors is a core and $Pdim(p) \leq \lambda$ then it is a border point. Otherwise, it is a noise point.*

Definition 7 *(Density-reachability). Object p is directly density-reachable from q, denoted as $p \triangleleft q$, if $core(q) \wedge p \in P_\epsilon(q) \wedge Pdim(p) \leq \lambda$.*

Definition 8 *(Density-connectedness).* *Two points p and q are density-connected, denoted as $p \bowtie q$, iff there is a sequence of points $(x_1, \ldots, x_m)$, where $\forall x_i : core(x_i)$ and $p \triangleleft x_1 \triangleleft \cdots \triangleright x_m \triangleright q$.*

Definition 9 *(Cluster).* *A cluster is a maximal set C of density-connected points in D.*

In PreDeCon, a cluster is a set of points that are specific in a predefined number of dimensions and are density-connected together. For constructing clusters, it performs all range queries on all objects for finding their neighbors, thus resulting in $O(dn^2)$ time complexity overall.

2.2 Anytime Algorithms

Most existing algorithms are *batch* ones, i.e., they only provide a single answer at the end. However, in many modern intelligent systems, users might need to terminate the algorithms earlier prior to completion due to resource constraints such as computational times. Or they want to examine some results earlier while the algorithms continue to find other better ones. Therefore, *anytime* techniques [20] are introduced. They can return approximate results whose qualities are improved overtime. Obviously, they are a very useful approach for coping with time consuming problems in many fields, e.g., [8, 12–15, 18].

3   The Algorithm AnyPDC

3.1   General Ideas

Range Query Processing. Since PreDeCon depends on the neighborhood queries for constructing clusters, enhancing their performance is thus crucial for speeding up PreDeCon.

Lemma 1. *For each pair of points p and q, $d(p,q) \leq d_{pref}(p,q)$.*

Lemma 1 can be straightforwardly proven from Definitions 3 and 2 since $\forall w_i : w_i \geq 1$. Consequently, any indexing technique for Euclidean distance (ED) can be employed for calculating $N_\epsilon(p)$ as a filter. $P_\epsilon(p)$ can be directly obtained from $N_\epsilon(p)$ by calculating the preference weighted distance $d_{pref}(p,q)$ for each $q \in N_\epsilon(p)$. If $d_{pref}(p,q) \leq \epsilon$, q will belong to $P_\epsilon(p)$. This *filter-and-refinement* scheme helps to reduce the neighborhood query time, thus speeding up the algorithm.

Summarization. PreDeCon expands a cluster by examining each point and its neighbors sequentially. This scheme incurs many redundant distance calculations, which decrease the performance of algorithm. AnyPDC, in contrast, follows a completely different approach extended from the anytime algorithm AnyDBC [12] for coping with the local subspace definitions of PreDeCon. It first summarizes points into homogeneous groups called *local clusters*. Each consists of a core point p (as a representative) and its directly density-reachable points.

Definition 10 *(Local cluster). A set S of points is called a local cluster of a point p, denoted as $lc(p)$ if p is a core and for each point q in S, p and q are density-connected. If all points inside $lc(p)$ are directly density-reachable from p, we also call $lc(p)$ a primitive cluster.*

Lemma 2. *All points p in a local clusters of a core point q belong to the same cluster.*

Lemma 2 is directly inferred from Definitions 7 to 9. Thus, we only need to connect these local clusters for producing the final clustering result of PreDeCon.

Lemma 3. *If $a \in lc(p) \cap lc(q)$ and $core(a)$, all points in $lc(p) \cup lc(q)$ belongs to the same cluster.*

Proof. For every point $x \in lc(p)$ and $y \in lc(q)$, we have $p \bowtie x$ and $y \bowtie q$. Since a is a core point and $a \in lc(p) \cap lc(q)$, we have $a \bowtie x$ and $a \bowtie y$. Thus, x and y belong to the same cluster (Definitions 8 and 9).

Lemma 4. *Given two primitive clusters and their representative p and q, if $d(p,q) > 3\epsilon$, p and q will never be directly connected via theirs members under the d_{pref} distance.*

Proof. Let u and v be arbitrary points inside $lc(p)$ and $lc(q)$, respectively. We have $d(p,q) \leq d(p,u) + d(q,u)$ and $d(q,u) \leq d(q,v) + d(u,v)$ (triangle inequality). Therefore, $d(u,v) \geq d(p,q) - d(p,u) - d(q,v)$. Since $d(q,u) \leq \epsilon$ and $d(q,v) \leq \epsilon$, we have $d(u,v) > \epsilon$. Due to Lemma 1, $d_{pref}(u,v) > \epsilon$, i.e., u and v are not directly density-reachable following Definition 7.

Active Query Selection. Instead of performing all queries like PreDeCon, AnyPDC actively and iteratively chooses some most promising points to connect local clusters together to build clusters and ignores ones that do not contribute to the final results. Consequently, it produces the same result as PreDeCon with much less number of queries. Thus, it is significantly faster than PreDeCon.

3.2   The Algorithm AnyPDC

The algorithm AnyPDC consists of some major steps described below. As described above, the general idea of AnyPDC is reducing the number of range queries for enhancing the efficiency. Thus, at the end, many queries may not be performed. Therefore, we assign for each point a state that describes its current status as follows.

Definition 11 *(Object state). The state of a point p represents its current status during the runtime of AnyPDC and is denoted as $state(p)$. If a query has been performed on p, $state(p)$ is processed. Otherwise, it is unprocessed. Beside that, depending on its core property. The state of p is additionally assigned as core, border, or noise. The untouched state means that p is unprocessed and it is not inside the neighborhoods of any processed points.*

Step 1: Summarization. In the beginning, all points are marked as *untouched*. For each point q, we additional store for it the number of neighbors it currently has, denoted as $nei(q)$. If a range query is executed on a point p and $q \in P_\epsilon(p)$ then we increase $nei(q)$ by 1.

AnyPDC iteratively and randomly chooses α points with *untouched* states, performs neighborhood queries in parallel, and checks if they are core points. If a point p is a core, it is marked as *processed-core* and all of its directly density-reachable points will be put to a local cluster (primitive cluster) with p as a representative. For each points $q \in P_\epsilon(p)$, if $state(q)$ is *untouched*, it will be marked as *unprocessed-border* or *unprocessed-core* depending on the number of neighbors $nei(q)$ it currently has. If $state(q)$ is *processed-noise*, it is marked as *processed-border* since it is surely a border of a cluster. If p is a core point either in *processed* or *unprocessed* state, its state remains. If p is not a core, it is marked as *processed-noise* and is stored in the so-called noise list L for a post processing step. Note that, if $Pdim(p) > \lambda$, p is surely a noise point following Corollary 1. In this case, we do not need to calculate $P_\epsilon(p)$ for saving runtime.

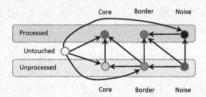

Fig. 1. Object transition state schema

The summarization process stops when all points are put into local clusters or the noise list L. Note that, in contrast to [5], $Pdim$ calculation is performed only when it is required for calculating the preference weighted distances and all calculated preference weighted vectors will be stored for assessing the other preference weighted distances to reduce runtime.

Corollary 1. *If $Pdim(p) > \lambda$, p is surely a noise point.*

Corrolary 1 is directly inferred from Definition 6. Figure 1 summarizes the transitions of the states of points during all Steps of AnyPDC. Overall, the state of a point p can only be changed from *noise* to *border* to *core* and from *unprocessed* to *processed* state.

Lemma 5. *For every point p, $state(p)$ only changes following the transition state schema described in Fig. 1.*

All state transitions of Lemma 5 can be verified through Definitions 6 and 11. As described in Sect. 3.1, we need to connect these local clusters to form the final clustering result. To do so, at the end of step 1, AnyPDC builds a graph $G = (V, E)$ that represents the current connectivity statuses of all local clusters.

G will then be used by AnyPDC for analyzing the potential cluster structures and selecting points for performing queries in the next Steps.

Definition 12 *(Local cluster connection graph). The connection graph $G = (V, E)$ represents the current connectivity statuses of all local clusters, where each vertex $v \in V$ represents a local cluster and each edge $e = (u, v)$ of G represent the relationship between two vertices u and v.*

For every pair of local clusters (u, v) if $d(u, v) > eps$, u and v will never be directly connected via their members following Lemma 4. Therefore, u and v will not be connected in the graph G. Otherwise, we connect u and v in G and assign for this edge (u, v) a state as follows.

Definition 13 *(Edge state). Each edge $e = (u, v)$ of G is assigned one of the three states, denoted as $state(e)$ or $state(u, v)$, including yes, weak, and unknown depending on the relationship between its local clusters u and v as follows:*

- *$state(u, v) = yes$ if $lc(u)$ and $lc(v)$ share a core point*
- *$state(u, v) = weak$ if $|lc(u) \cap lc(v)| \neq \emptyset$ and $state(u, v) \neq yes$*
- *$state(u, v) = unknown$ if $|lc(u) \cap lc(v)| = \emptyset$*

Note that, if $(u, v) \notin E$, we also denote that $state(u, v) = no$. The $state(u, v)$ therefore reflexes the possibility that all points $lc(u)$ and $lc(v)$ belong to the same cluster. If it is *weak*, u and v have higher change to be in the same cluster than if it is *unknown*. These states will be exploited by AnyPDC to choose points for performing queries in the next steps. Moreover, the states of edges will be changed following the changes of the core properties of points.

Lemma 6. *Given two arbitrary points $p \in lc(x)$ and $q \in lc(y)$, if p and q belong to the same cluster in DBSCAN, there exist a path of edges in G that connects $lc(x)$ and $lc(y)$.*

Proof. Assume that there is no path in G that connect $lc(x)$ and $lc(y)$. Let $S = s_i$ be an arbitrary path of points that connect p and q. Let LS be a set of local clusters that contain all s_i. Due to Lemma 4, there must exist two adjacent points s_i and s_{i+1} where $d(s_i, s_{i+1}) > \epsilon$ or $d_{pref}(s_i, s_{i+1}) > \epsilon$ (Lemma 1). This lead to a contradiction.

Due to Lemma 6, we only need to examine the graph G to build clusters. At any time, clusters can be extracted by finding connected components of *yes* edges in the graph G following Lemma 7. All points inside local clusters of a connected component will be assigned same cluster label.

Lemma 7. *Each connected component of yes edges of G indicates a subspace preference weighted cluster.*

Proof. Let $p \in lc(x)$ and $q \in lc(y)$ be two arbitrary points in C. Let $lc(x) = lc(c_1), \ldots, lc(c_t) = lc(y)$ be a path of local clusters that connects $lc(x)$ and $lc(y)$. Due to Definition 13, there is a set of core points that connects c_1 and c_t together. Thus, p and q are density-connected.

Lemma 8. *When all edges of G are in the yes states, the overall cluster structure will not change anymore.*

Proof. Due to Lemma 5, if a point p is a core either in *unprocessed* or *processed* states, it will never be changed back to *border* or *noise*. Thus, for two arbitrary vertices u and v in G, if $state(u, v) = yes$, it will never change despite of any additional queries. And so is the clustering result following Lemma 7.

Step 2: Checking Stop Condition. At the beginning of step 2, AnyPDC finds all connected components of the graph G. Then, it changes all *weak* and *unknown* edges inside each connected component to the *yes* state since all vertices inside them will belong to the same cluster as stated in Lemma 7.

If there is no *weak* and *unknown* edges in G, there will be no more change in the clustering result following Lemma 8. Thus, we can safely stop AnyPDC for saving runtime. This makes AnyPDC much efficient than PreDeCon.

If there still exist *weak* or *unknown* edge, AnyPDC must continue to perform range queries to determine the core properties of *unprocessed* points and to connect local clusters together. However, before going to the next Steps, AnyPDC will merge all local clusters in each connected component into a single local cluster. One of the representatives of these local clusters will be selected as the representative for the whole group. The general purpose is to reduce the total number of graph nodes. Fewer vertices means better performance and smaller overhead for maintaining the graph G, e.g., smaller time for finding connected components or merging clusters in the next Steps. After merging local clusters, AnyPDC rebuilds the graph G following Lemma 9.

Lemma 9. *Given two connected component $C = \{lc(c_1), \ldots, lc(c_a)\}$ and $D = \{lc(d_1), \ldots, lc(d_b)\}$ of G. We have:*

- *Case A: $\forall lc(c_i) \in C, lc(d_j) \in D : state(lc(c_i), lc(d_j)) = no \Rightarrow state(C, D) = no$*
- *Case B: $\exists lc(c_i) \in C, lc(d_j) \in D : state(lc(c_i), lc(d_j)) = weak \Rightarrow state(C, D) = weak$*
- *Case C: otherwise $state(C, D) = unknown$*

Proof. Since there is no connection between $lc(c_i)$ and $lc(d_j)$, C and D will obviously have no connection. Thus, Case A is hold. Moreover, according to Lemma 6, C and D will surely belong to different clusters. Following Definition 13, $C \cap D \neq \emptyset$ since $\exists lc(c_i) \in C, lc(d_j) \in D : lc(c_i) \cap lc(d_j) \neq \emptyset$. Moreover, for all point p in $C \cap D$, p is not a core otherwise C and D will belong to the same connected component. Thus, case B is hold. Case C is proven similarly.

Step 3: Select Points. Selecting right points for querying is very important if we want to reduce the number of queries. Thus, AnyPDC introduces an *active clustering* scheme that iteratively analyzes the current result and scores all unprocessed points based on some statistical information and connectivity statuses of G and their positions in G. Then, AnyPDC iteratively chooses α top scored points to perform the range queries for updating the cluster structures. Here we use the same heuristic proposed in [12] to select points for updating.

Definition 14 *(Node statistic [12]). The statistical information of a node $u \in V$, denoted as $stat(u)$, is defined as follows:*

$$stat(u) = \frac{usize(u)}{|pclu(u)|} + \frac{|pclu(u)|}{n}$$

where $usize(u)$ is the number of unprocessed points inside $pclu(u)$ and n is the total number of points.

Definition 15 *(Node degree [12]). Given a node u and its adjacent nodes $N(u)$ in the graph G. The degree of u, denoted as $deg(u)$, is defined as follows:*

$$deg(u) =$$
$$w(\textstyle\sum_{v \in N(u) \wedge state(u,v)=weak} stat(v)) + \sum_{v \in N(u) \wedge state(u,v)=unknown} stat(v) - \psi(u)$$

where $\psi(u) = 0$ if u does not contain border points otherwise it is the total number of weak and unknown edges of u.

The node degree of a node u indicates the relationship between u and its adjacent ones. If u is inside an uncertain area, it should have many undetermined connections. Thus, it has higher $deg(u)$ and vice versa. Due to the merging scheme in step 4 and the Lemma 8, u should have lower priority to reach to an earlier termination condition. The score of each point is calculated as in [12].

Definition 16 *(Point score [12]). The score of an unprocessed point p, denoted as $score(p)$, is defined as follows:*

$$score(p) = \textstyle\sum_{u \in V \wedge p \in pclu(u)} deg(u) + \frac{1}{nei(p)}$$

where $nei(p)$ is the current number of neighbors of p.

Step 4: Update Cluster Structures. For each selected point p, if p is not a core point, it will be marked as *processed-border* since it is already inside the neighborhood of a core point. Otherwise, p is marked as a *processed-core* point. All of its neighbors are reassigned new states following the transition state scheme in Fig. 1. Then, $P_\epsilon(p)$ will be merged into all local clusters that contain p.

Lemma 10. *Assume that all point p's neighbors are merged into local clusters $lc(q)$, they and $lc(q)$ belong to the same cluster.*

Proof. For each arbitrary point $x \in P_\epsilon(p)$ and $y \in lc(q)$, we have $p \bowtie x$. Moreover, $q \bowtie p$ following Definition 10. Since p is a core point, we have $q \bowtie x$ due to Definition 8. Therefore, $P_\epsilon(p) \cup lc(q)$ is a local cluster and belongs to a cluster.

Since the states of points have been changed, the connections among vertices of G will be changed. Thus, at the end of step 3, G is updated locally, i.e., only edges related to changed clusters are examined. Their status will be changed following Definition 13. Moreover, according to Lemma 11, if one of the two local clusters $lc(p)$ and $lc(q)$ has been fully processed, i.e., all their points are in *processed* states, and $state(lc(p), lc(q)) \neq yes$, then $lc(p)$ and $lc(q)$ will belong to different clusters. And, their edge can be safely removed from G.

Lemma 11. *Given two local clusters $lc(p)$ and $lc(q)$, if the edge (p, q) is still in weak or unknown state but $lc(p)$ or $lc(q)$ are fully processed, i.e., queries are executed for all points inside them, $lc(p)$ and $lc(q)$ will never be directly connected via their members.*

Proof. Assume that a node $lc(p)$ is fully processed and there exists a chain of density-connected core points $X = \{p, \ldots, d, e, \ldots, q\}$ that connects the two core points p and q (Definition 8). However, if e is a core point, all of its neighbors including d are merged into $lc(p)$. Thus, $state(lc(p), lc(q))$ must be *yes*. This leads to a contradiction.

Step 2, 3, and 4 are repeated until the algorithm is stopped (by Lemma 8). We now go to the final step 5.

Step 5: Determining Noise and Border Points. The goal of this step is to examine the noise list L to see if points are true noise or border ones. To do so, all points p inside L are examined. If $Pdim(p) > \lambda$, p is surely a noise according to Corollary 1. Otherwise, AnyPDC needs to check whether p is a border of a cluster. If p is currently in *processed-border* state, it is already inside a cluster. Otherwise, AnyPDC check all the preference weighted neighbors $q \in P_\epsilon(p)$. If q is a core, then p will surely be a border. If there is no core point inside $P_\epsilon(p)$, additional range queries must be performed if one of p's neighbors is still in *unprocessed* state. If a core is found, p is a border. Otherwise, p is surely a noise.

3.3 Algorithm Analysis

Lemma 12. *The final result of AnyPDC is identical to that of PreDeCon.*

Proof. AnyPDC produces clusters following the notion of PreDeCon as proven in Lemmas 2 to 11. Thus, when all queries are processed, it will produce the same results as PreDeCon. Moreover, when all the edges of G are in the *yes* state. There will be no change in the graph G and thus the final cluster structure. Thus, if we stop the algorithm, the acquired result is still similar to the final one, though many queries have not been processed.

Note that, shared border points will be assigned the label of one of its clusters based on the examining order of points in PreDeCon. And so is AnyPDC. Thus, they are the only difference between AnyPDC and PreDeCon.

Lemma 12 shows an interesting property of AnyPDC. It has the powers of both approximation techniques (in terms of its anytime scheme) as well as exact techniques at the end.

Let n, v, and l are the number of points, number of initial graph nodes, and the size of the noise list L. Let b is the total number of iterations. The time complexity of AnyPDC is $O(b(v^2 + vn) + l\mu n)$ in the worst case. It consumes $O(v^2 + vn + l\mu)$ memory for storing the graph G and local clusters. Note that, v and b are much smaller than n and n/α for all real datasets in our experiments, respectively.

4 Experiments

All experiments are conducted on a Linux workstation with two 3.1 GHz Xeon CPUs and 128 GB RAM. We compare AnyPDC with the original PreDeCon algorithm on real datasets including Electric Device (ED)[1], Brain-Computer Interfaces (BCI)[2], and Weather Temperature data (WT)[3]. They consist of 16637, 41680, and 566268 points in 96, 96 and 12 dimensions, respectively. Unless otherwise stated, we use default parameters $\kappa = 4$, $\delta = 0.0001$, $\lambda = 5$ (following suggestions in [5]). Default values for μ is 5. The default block size is $\alpha = 128$.

4.1 Anytime PreDeCon

Anytime Performance. For assessing the anytime properties of AnyPDC, we use the results of PreDeCon as a ground truth. At each iterations of AnyPDC, we measure how similar its intermediate result is compared to that of PreDeCon using the Normalized Mutual Information (NMI) score [19]. NMI results are in $[0, 1]$ where 1 indicates perfect clustering results, i.e., AnyPDC produces the same results as PreDeCon.

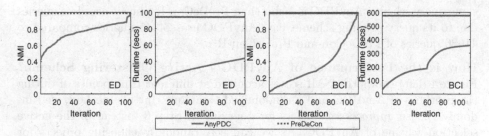

Fig. 2. NMI scores and runtimes of AnyPDC for the datasets ED (left) and BCI (right) during their execution in comparison with PreDeCon (represented by horizontal lines)

Figure 2 shows the result of AnyPDC and PreDeCon for the ED ($\epsilon = 0.6$) and BCI ($\epsilon = 0.8$) datasets. The NMI score increases at each iteration and comes

---

[1] http://www.cs.ucr.edu/~eamonn/time_series_data/.

[2] http://www.bbci.de/competition/iii/.

[3] http://www.cru.uea.ac.uk/data/.

to 1 at the end. That means the results of AnyPDC gradually come closer to those of PreDeCon and are identical at the end. For ED as an example, if we stop the algorithm at the first iteration, AnyPDC requires only 9.4 s compared to 93.3 s of PreDeCon, which is 10 times faster. Even if it runs till the end, AnyPDC requires only 43.5 s (around 2.1 times faster than PreDeCon). It is very interesting since the final runtimes of anytime algorithms are usually larger than those of the batch ones.

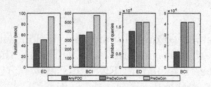

Fig. 3. Runtimes (left) and numbers of queries (right) of different algorithms for the dataset ED and BCI

Why Does AnyPDC Faster than PreDeCon? There are two reasons: (1) it has a more efficient neighborhood query scheme as described in Sect. 3.1 and (2) it uses much less queries than PreDeCon due to its efficient query pruning scheme as also described in Sect. 3.1. For the former reason, we slightly modify the algorithm PreDeCon replacing its old query scheme with the new one and name the algorithm PreDeCon-R. Figure 3 shows the runtimes and numbers of queries of AnyPDC, PreDeCon, and PreDeCon-R for the two datasets ED and BCI with the same parameters as in Fig. 2. Due to its efficient query scheme, PreDeCon-R is more efficient than PreDeCon though they use the same numbers of queries, e.g., 50.7 s vs. 93.3 s for the dataset ED. For the latter reason, with the same query scheme, AnyPDC is faster than PreDeCon-R since it uses less queries due to its query pruning scheme, e.g., AnyPDC uses 14345 queries compared to 41680 queries of PreDeCon and PreDeCon-R.

How is the Performance of AnyPDC's Active Clustering Scheme? Figure 4 (left) shows the NMI scores acquired at different time points of during the runtimes of AnyPDC and a randomized technique. In this method, we randomly choose *unprocessed* objects for querying at Step 3 instead of the active selection scheme of AnyPDC. As we can see, randomly choosing objects for

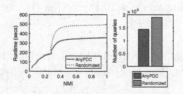

Fig. 4. Performance of AnyPDC's active clustering scheme for the dataset BCI ($\epsilon = 0.8$)

querying requires much more time to reach to a certain NMI score than the *active clustering* scheme of AnyPDC. Moreover, it requires more queries than AnyPDC to finish as shown in Fig. 4 (right). While AnyPDC requires only 14345 queries at the end, the randomized one consumes 18947 queries.

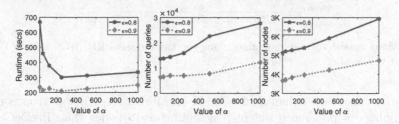

Fig. 5. The effect of the block size α on the performance of AnyPDC on the dataset BCI with different parameter ϵ

The Effect of Blocksize α. Figure 5 shows the effect of the block size α (from 32 to 1024) on various aspects of AnyPDC on the dataset BCI with different values of ϵ. As we can see, when α increases, the number of initial graph nodes v will increase since there are more overlap among nodes at Step 1 of AnyPDC. Similarly, large α means more objects will be selected at each iteration of AnyPDC. The more objects we select in a block, the more redundant queries may happen since the cluster structure will not be updated frequently. This increases the number of used queries to produce clusters. However, the changes on both the number of queries and number of nodes are small compared to the values of α. For example, with $\epsilon = 0.8$, when α is changed from 32 to 1024 (32 times), the number of nodes increases from 5139 to 6925 (1.34 times) and the number of queries comes from 13511 to 27648 (2.04 times).

Due to its anytime scheme, AnyPDC suffers from overheads for scoring objects and updating the cluster structure at each iteration. Thus, when α increases, the final cumulative runtime of AnyPDC decreases since it does not have to re-evaluating objects frequently. Thus, the overhead is reduced. However, when α is large enough, redundant queries will happen, thus making it a little bit slower as we can see from Fig. 5. For $\epsilon = 0.8$ as an example, it takes AnyPDC 667.7 and 300.8 s when α comes from 32 to 256, respectively. However, when α is increased to 512 and 1024, the final runtime of AnyPDC slightly increases to 311.1 and 334.7 s, respectively.

Performance Comparison. Figure 6 further compares the final cumulative runtimes of AnyPDC and other techniques for different real datasets and values of ϵ. Note that we choose ϵ so that all objects are noise in the beginning and increase it until most objects are inside a single cluster. AnyPDC is much faster than the original PreDeCon algorithm, especially on large datasets. Taking the dataset WT as an example, with $\epsilon = 0.4$, PreDeCon requires 15019.1 s, while AnyPDC needs only 466.1 s, which is 32.2 times faster. Moreover, AnyPDC is also

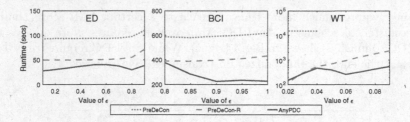

Fig. 6. Performance of different algorithms for the datasets ED, BCI, and WT with different values of ϵ

up to 4.31 times faster than PreDeCon-R on all datasets, though it uses much less numbers of preference weighted neighborhood queries than PreDeCon-R. The reason is that both AnyPDC and PreDeCon must perform the neighborhood queries on Euclidean distance to calculate the preference weighted vectors for all objects. Since this is also very expensive, it reduces the performance gap between AnyPDC and PreDeCon-R.

5   Related Works and Discussion

Subspace Clustering. While traditional clustering algorithms focus on finding clusters in full feature attributes of data, subspace clustering algorithms aim at finding homogeneous groups of objects hidden inside subspaces of high-dimensional datasets [10]. There exist many different approaches. For example, CLIQUE [4] divides the data space into grids and connects dense cells containing significant number of objects to build clusters. SUBCLU [11] try to find density-based clusters hidden inside subsets of dimensions. While the former techniques potentially output overlapped clusters, PROCLUS [3] follows a k-medoid-like clustering scheme for produce non-overlapped clusters. Providing a complete survey of these techniques is out of scope of this paper. Interested readers please refer to other surveys for more information, e.g., [10,17].

PreDeCon and its Variants. Similar to PROCLUS, PreDeCon also produces non-overlapped density-based clusters using a specialized distance measure that captures the subspace of each cluster. However, it follows the density-based cluster notions of DBSCAN [6]. It can detect clusters with arbitrary shapes and is able to detect outliers in contrast to PROCLUS.

There exist some extensions of PreDeCon proposed in the literature. IncPreDeCon [9] is an incremental version of PreDeCon for dynamic data. It relies on the locality of the cluster structure changes for efficiently updating the results wrt. each inserted or deleted points in a database. HiSC [1] is a hierarchical extension of PreDeCon which produces a reachability plot containing an order of points and subspace distances, which is calculated based on the intersection of two subspace preference vectors of two points p and q. Clusters can be built by cutting through the reachability plot with an arbitrary value of ϵ. Each separated part forms a cluster. An extension of HiSC, DiSH [2], allows one cluster to be embedded in many other super-clusters thus leading to complex hierarchical relationship among clusters. HDDSTREAM [16] is another extension of PreDe-Con designed to cope with streaming data. To the best of our knowledge, none of these algorithms is an anytime algorithm like AnyPDC. Moreover, AnyPDC focuses on reducing the number of neighborhood queries by extending the any-time clustering algorithm in [12] to cope with the subspace notions of PreDeCon. Thus, it is more efficient than others.

6   Conclusion and Discussion

In this paper, we introduce for the first time an anytime extension of the fundamental clustering algorithm PreDeCon, called AnyPDC. The major ideas of AnyPDC are (1) processing queries in block, (2) maintaining and updating an underlying graph structure representing local clusters, and (3) actively studying the current cluster structure and choosing only meaningful objects for examining their neighbors. Experiments on various real datasets show very interesting and promising results. AnyPDC approximates the final clustering results of PreDe-Con very well at intermediate iterations. Even if it runs to the end, it is also faster than PreDeCon.

Acknowledgments.. This research is funded by Vietnam National Foundation for Science and Technology Development (NAFOSTED) under grant number 102.05-2015.10.

References

1. Achtert, E., Böhm, C., Kriegel, H.-P., Kröger, P., Müller-Gorman, I., Zimek, A.: Finding hierarchies of subspace clusters. In: Fürnkranz, J., Scheffer, T., Spiliopoulou, M. (eds.) PKDD 2006. LNCS (LNAI), vol. 4213, pp. 446–453. Springer, Heidelberg (2006). doi:10.1007/11871637_42
2. Achtert, E., Böhm, C., Kriegel, H.-P., Kröger, P., Müller-Gorman, I., Zimek, A.: Detection and visualization of subspace cluster hierarchies. In: Kotagiri, R., Krishna, P.R., Mohania, M., Nantajeewarawat, E. (eds.) DASFAA 2007. LNCS, vol. 4443, pp. 152–163. Springer, Heidelberg (2007). doi:10.1007/978-3-540-71703-4_15
3. Aggarwal, C.C., Procopiuc, C.M., Wolf, J.L., Yu, P.S., Park, J.S.: Fast algorithms for projected clustering. In: SIGMOD, pp. 61–72 (1999)
4. Agrawal, R., Gehrke, J., Gunopulos, D., Raghavan, P.: Automatic subspace clustering of high dimensional data for data mining applications. In: SIGMOD, pp. 94–105 (1998)
5. Böhm, C., Kailing, K., Kriegel, H.P., Kröger, P.: Density connected clustering with local subspace preferences. In: ICDM, pp. 27–34 (2004)
6. Ester, M., Kriegel, H.P., Sander, J., Xu, X.: A density-based algorithm for discovering clusters in large spatial databases with noise. In: KDD (1996)
7. Hinneburg, A., Aggarwal, C.C., Keim, D.A.: What is the nearest neighbor in high dimensional spaces? In: VLDB, pp. 506–515 (2000)

8. Kobayashi, T., Iwamura, M., Matsuda, T., Kise, K.: An anytime algorithm for camera-based character recognition. In: ICDAR, pp. 1140–1144 (2013)
9. Kriegel, H.-P., Kröger, P., Ntoutsi, I., Zimek, A.: Density based subspace clustering over dynamic data. In: Bayard Cushing, J., French, J., Bowers, S. (eds.) SSDBM 2011. LNCS, vol. 6809, pp. 387–404. Springer, Heidelberg (2011). doi:10.1007/978-3-642-22351-8_24
10. Kriegel, H.P., Kröger, P., Zimek, A.: Clustering high-dimensional data: a survey on subspace clustering, pattern-based clustering, and correlation clustering. TKDD 3(1), 1–58 (2009)
11. Kröger, P., Kriegel, H.P., Kailing, K.: Density-connected subspace clustering for high-dimensional data. In: SDM, pp. 246–256 (2004)
12. Mai, S.T., Assent, I., Storgaard, M.: AnyDBC: an efficient anytime density-based clustering algorithm for very large complex datasets. In: KDD (2016)
13. Mai, S.T., He, X., Feng, J., Böhm, C.: Efficient anytime density-based clustering. In: SDM, pp. 112–120 (2013)
14. Mai, S.T., He, X., Feng, J., Plant, C., Böhm, C.: Anytime density-based clustering of complex data. Knowl. Inf. Syst. 45(2), 319–355 (2015)
15. Mai, S.T., He, X., Hubig, N., Plant, C., Böhm, C.: Active density-based clustering. In: ICDM, pp. 508–517 (2013)
16. Ntoutsi, I., Zimek, A., Palpanas, T., Kröger, P., Kriegel, H.: Density-based projected clustering over high dimensional data streams. In: SDM (2012)
17. Sim, K., Gopalkrishnan, V., Zimek, A., Cong, G.: A survey on enhanced subspace clustering. Data Min. Knowl. Discov. 26(2), 332–397 (2013)
18. Ueno, K., Xi, X., Keogh, E.J., Lee, D.J.: Anytime classification using the nearest neighbor algorithm with applications to stream mining. In: ICDM (2006)
19. Zaki, M.J., Meira Jr., W.: Data Mining and Analysis: Fundamental Concepts and Algorithms. Cambridge University Press, New York (2014)
20. Zilberstein, S.: Using anytime algorithms in intelligent systems. AI Mag. 17(3), 73–83 (1996)

Co-clustering for Microdata Anonymization

Tarek Benkhelif[1,2]([✉]), Françoise Fessant[1], Fabrice Clérot[1],
and Guillaume Raschia[2]

[1] Orange Labs, 2, Avenue Pierre Marzin, 22307 Lannion Cédex, France
{tarek.benkhelif,francoise.fessant,fabrice.clerot}@orange.com
[2] LS2N - Polytech Nantes,
Rue Christian Pauc, BP50609, 44306 Nantes Cédex 3, France
guillaume.raschia@univ-nantes.fr

Abstract. We propose a methodology to anonymize microdata (i.e. a table of n individuals described by d attributes). The goal is to be able to release an anonymized data table built from the original data that protects against the re-identification risk. The proposed solution combines co-clustering with synthetic data generation to produce anonymized data. Co-clustering is first used to build an aggregated representation of the data. The obtained model, which is also an estimator of the joint distribution of the data, is then used to generate synthetic individuals, in the same format as the original ones. The synthetic individuals can now be released. We show through several experiments that these synthetic data preserve sufficient information to be used in place of the real data for various data mining tasks (supervised classification, exploratory data analysis). Finally, the protection against the re-identification risk is evaluated.

Keywords: Co-clustering · Synthetic individual data · Re-identification risk · Anonymity

1 Introduction

There is an increasingly social and economic demand for open data in order to improve planning, scientific research or market analysis. In particular, the public sector via its national statistical institutes, healthcare or transport authorities, is pushed to release as much information as possible for the sake of transparency. Private companies are also implicated in the valorization of their data through exchange or publication like Orange that has recently made available to the scientific community several mobile communication datasets collected from its networks in Senegal and Ivory Coast [1]. When published data refer to individual respondents and involve personal information, they must be anonymized so that an attacker is unable to re-identify a precise individual in the released data or infer some knew sensitive knowledge about it [8]. In this article we focus on microdata. The recent literature about privacy preservation for microdata is mainly organized around two privacy concepts (i) group anonymization

© Springer International Publishing AG 2017
D. Benslimane et al. (Eds.): DEXA 2017, Part I, LNCS 10438, pp. 343–351, 2017.
DOI: 10.1007/978-3-319-64468-4_26

techniques such as k-anonymity and (ii) random perturbation methods with in particular the concept of Differential Privacy (DP). K-anonymity seeks to prevent re-identification of records by making each record indistinguishable within a group of k or more records with respect to the non-confidential attributes (the probability of re-identification is at most $1/k$) and allows the release of data in its original form [14]. Most of the algorithms proposed to sanitize data with respect to k-anonymity are based on the use of *generalization* and *suppression* [10]. Solutions based on clustering have also been proposed [15]. The k-anonymity fails to prevent the disclosure of confidential attributes (or inference attacks). Different techniques have been developed to overcome these limitation as for instance l-diversity [12] or t-closeness [11]. There are yet few practical examples of the application of k-anonymity beyond $k = 10$ and for sets of more than 10 attributes [13]. When the number of dimensions increases, aggregation damages the information contained in the data with consequences on its utility. The notion of protection defended by DP is the strong guarantee that the presence or absence of an individual in a dataset will not significantly affect the result of aggregated statistics computed from this dataset. DP works by adding some controlled noise to the computed function [7]. Most of the existing works on differentially private data publishing focus on releasing some low dimensional statistical information about the dataset (counts, marginal tables or histograms). When applied to microdata release, DP techniques suffer from the curse of dimensionality, they cannot achieve either reasonable scalability or desirable utility. Special treatment is needed to overcome the challenges incurred by high dimensionality [6]. In addressing these challenges, one of the most promising ideas is to decompose in a differentially private way high-dimensional data into a set of low-dimensional marginal tables, along with an inference mechanism that infers the joint data distribution from these marginal tables [5]. The joint distribution is then used as a generative model to sample synthetic individual records. This idea that consists in generating synthetic datasets is already explored by some national statistics institutions for the publication of detailed individual datasets while ensuring the protection of individual's privacy [16].

In the anonymization scenario we propose, we assume that the released data is intended to be used in many data mining contexts so the anonymization process can't be driven by a specific data mining task. The challenge in designing and implementing the anonymization solution is to achieve a good balance between privacy and utility [9]. We present an approach that combines co-clustering, an unsupervised data mining analysis technique, and synthetic data generation. Co-clustering is used to partition the space of individuals into clusters of at least k individuals (as in regular k-anonymity, but with high values of k and many attributes). The co-clustering gives a summary of the raw data [2]. It can be exploited at various granularities depending on the required level of privacy while minimizing the loss of information. The co-clustering model can also be considered as a generative model to sample synthetic individual records in the same format as original data and which are expected to preserve similar statistical properties. In this work we address the risk of record re-identification.

The paper is organized as follows. Section 2 first reviews background on co-clustering, then the proposed anonymization approach is described. Section 3 reports experiments on a real-life dataset. The utility of the produced synthetic dataset is evaluated through a supervised classification task and an exploratory analysis one. We also evaluate the level of protection brought by the released synthetic data. The final section gathers some conclusions and future lines of research.

2 Description of the Anonymization Methodology

In this section, we introduce the key technique of co-clustering on which is based our anonymization methodology. Then we detail how it is used for the generation of synthetic data from raw data.

2.1 Co-clustering

Co-clustering works by rearranging the rows and columns of a data matrix in order to highlight blocks that have homogeneous density. In the anonymization task, rows are individuals and columns are variables that can be of mixed-types, categorical as well as numerical. To infer the best partition of the matrix, we choose the approach implemented in the data mining software named Khiops [4], from Orange Labs (www.khiops.com). This approach (i) is non-parametric and does not require user intervention or fine-tuning, (ii) is easily scalable and can, consequently, be used to analyze large datasets, (iii) is able to deal with several mixed variables. Khiops algorithm is based on data grid models which partition the sets of values of each variable into intervals for numerical variables and into groups of values for categorical variables. The cross-product of the univariate partitions forms a multivariate partition of the representation space into a set of cells. This multivariate partition, called data grid, constitutes the summary of the data and is a piecewise constant non-parametric estimator of the joint probability of all the variables. The detailed formulation of the optimization criterion as well as the optimization algorithms and the asymptotic properties are detailed in [4] for a co-clustering with mixed variables. [4] has demonstrated that the approach behaves as a universal estimator of joint density asymptotically tending towards the true underlying distribution. A post-processing technique that can be considered as an agglomerative hierarchical clustering has been associated to the co-clustering. This allows exploring the retrieved patterns at any granularity, up to the finest model, without any user parameter. The number of sub-intervals, that is the level of details kept in the functions, is automatically adjusted in an optimal way when the number of clusters is reduced [3].

2.2 Technical Solution Description

In this section we detail how the co-clustering technique, stated in the above section, allows us to build a k-anonymized model of a multidimensional table (of n individuals and d attributes that can be both numerical and categorical) and how this model is exploited to generate a table of synthetic individuals. The whole process can be organized into four steps: data preparation, co-clustering, co-clustering simplification and synthetic data generation. Let us notice that our solution is not suitable in situations where the amount of available instances is too small to produce an informative co-clustering model.

Data Preparation Step. This first step turns the d-dimensional co-clustering problem into a 2-clustering one. This is achieved via a preparation step in which all the attributes of the table are recoded as categorical attributes. If the attribute is already categorical, it remains unchanged. If the attribute is numerical, it is discretized into intervals of equal frequencies and binarized. An individual from the original table is now described by the list of the d values it takes on the attributes.

Co-clustering Step. A co-clustering model is then built from this new representation (i.e. between the individuals and the binary variables) resulting in the formation of clusters of individuals and clusters of parts of variables (hereafter referred as id-clusters and vp-clusters respectively). Individuals in the same cluster have a similar distribution on vp-clusters and vice-versa.

Co-clustering Simplification Step. The clustering aggregation ability mentioned above offers the user a way to control the precision kept in data and thus the level of information she wants to disclose by decreasing the number of clusters in a greedy optimal way. We choose the level of granularity that ensures that each id-cluster is populated enough with k or more individuals. From now: (i) each individual is represented by its cluster, an (ii) id-cluster is represented by a probability distribution onto the vp-clusters. At this stage of the process we have a k-anonymous representation of the id-clusters in the sense that an individual in a cluster is indistinguishable within a group of k individuals. An individual is only described by a probability distribution onto the vp-clusters.

Synthetic Data Generation Step. The purpose in this last step is to generate individual records based on the co-clustering model. The co-clustering can be viewed as an estimator of the joint density between individuals and parts of variables. It is used as a generative model to generate synthetic individuals in the same format as the original ones. To generate a synthetic individual:

– randomly select an id-cluster among all available clusters, focus on the part of the data grid which corresponds to this cluster.
– assign a value to the first attribute:
      select a vp-co-cluster among the co-clusters that hold some parts matching the attribute, according to the distributions of the values of the attribute in these co-clusters. We are now in a co-cluster in which we can randomly draw a value for the attribute,

– iterate over all the attributes until the synthetic individual is complete. We have to comply with the constraint that a simulated individual must have as many values as attributes and can receive only one value by attribute.

Iterate for all the synthetic individuals to create. We create as many synthetic individuals as there are in the original set. Thus we get a set of synthetic individuals which are no longer empirical individuals.

3   Experimentation

In this section we conduct several experiments on a real-life microdata set in order to illustrate the efficiency of our proposition on a practical case. The objective is to explore the utility of synthetic data for solving various data mining tasks as well as to evaluate the level of protection offered against re-identification of individuals.

3.1   Experimental Settings

We experiment with the Adult database[1]. We used 13 variables {age, workclass, education, education num, marital status, occupation, relationship, race, sex, capital gain, capital loss, hours per week, native country}. We randomly select 80% of the 48882 observations in order to build the co-clustering and generate the synthetic data, the remaining 20% are used for the evaluations.

Data Preparation. In the data preparation phase, the numerical variables are discretized in deciles and binarized, the categorical variables are left unchanged.

Co-clustering and Co-clustering Simplification. The finest co-clustering grid is obtained for 34 id-clusters and 58 vp-clusters. At this level, the least populated id-cluster contains 500 individuals whereas the most populated cluster has size 950. One can go back to the co-clustering hierarchy until all the clusters are populated by the desired number of individuals k. Thus, to obtain $k = 1300$ it is necessary to go up to 15 id-clusters. A more coarse representation of id-clusters and vp-clusters is obtained.

Generation of Synthetic Datasets. For the purposes of the experimentation, a table of synthetic individuals is generated at different levels of co-clustering aggregation. The chosen levels are: {34, 30, 25, 20, 15, 10, 5, 1} id-clusters. At each level, the generated table contains as many individuals as there were in the initial table (i.e. 39100 records).

3.2   Synthetic Data Usefulness Evaluation

We are now interested in using the generated synthetic data. The question that arises here is: are the synthetic data representative of the original data and can they be used for mining in the same way as the latter?

---

[1] https://archive.ics.uci.edu/ml/.

Supervised Classification. For this task, a variable is selected as the target, the other variables being the explanatory variables. Two classifiers are learned: the first one with the set of synthetic data generated at a given level of co-clustering, the second one with the original data. Then the two models are successively deployed on the actual 20% test data that had been set aside previously. Thus, we can evaluate the performances of the classifiers on the «real individuals». The criteria that are evaluated are the classification accuracy (ACC) and the area under the ROC curve (AUC). We used the supervised learning tool of the Khiops software suite which implements a naive Bayesian classifier with variable selection and averaging of models (downloadable at khiops.predicsis.com). The experimental results obtained with the target variable *education* are presented on Table 1. The target's values have been divided into 2 groups «higher education» {yes, no}. The other variables are used as explanatory variables for the learning of the models (except *education num* correlated to the target). The performances measured when the real data are used to learn the model (no privacy) are respectively 0.839 for AUC and 0.798 for ACC.

Table 1. AUC and ACC measured on the test data. For a given level of co-clustering (given by the number of id-clusters) the model was learned from the synthetic data generated at this level

Number of id-clusters	1	2	5	6	10	15	20	25	30	34
AUC	0.5	0.621	0.766	0.771	0.828	0.830	0.833	0.833	0.833	0.834
ACC	0.714	0.714	0.715	0.725	0.788	0.788	0.793	0.795	0.794	0.795

One can observe that the classification performances obtained on the actual test data from the models learned on the synthetic datasets are close to those measured when the real data are used. The classification performances break down at higher levels of co-clustering aggregation. Different target variables were evaluated in the same way (gender and marital status). Classification behaviors similar to those presented above were observed. We concluded that the synthetic data preserve the properties of the actual data in a way that allows to use them for supervised classification.

Exploratory Analysis. The question that arises here is: what level of knowledge can be extracted from a set of synthetic data and how close is this knowledge to the one which would be extracted from the actual data? The experimental protocol is the following: (i) build a co-clustering from the synthetic data, (ii) deploy this co-clustering on the test data, the deployment consists in assigning to each record of the test set a cluster, (iii) deploy the co-clustering used to build the synthetic data set on the same test data (co-clustering built from the original data and aggregated until it reaches the appropriate level). The two co-clusterings, the one built from the real data and the one built from the synthetic data, can be compared with the confusion matrix crossing the two co-clusterings. We have analyzed these confusion matrices at several aggregation levels. For example, the synthetic set coming from an aggregation level of 34 *id*-clusters allows to build also a «synthetic»co-clustering of 34 synthetic id-clusters. For an aggregation level of 5 *id*-clusters this number is 6. Regardless of the considered level of aggregation, the two «synthetic»and «real»co-clusterings are consistent, with mainly diagonal confusion matrices. The level of information in the original data is well-preserved in the synthetic data. However, as expected, synthetic data does not preserve information at a finer level than the data from which they were generated. They do not allow building a finer level co-clustering. Thus, new information is not learned through analysis since it is not possible to fall below the level predefined by the synthetic data set.

3.3   Protection of Individuals

In this section we experimentally evaluate the risk of record re-identification provided by the synthetic data. The experimental attack we propose is the following: we assume that the synthetic records and also all the original records, except one, are available to the attacker. The question that arises is: how this last record can remain hidden? For the experimentation we first isolate a record corresponding to an individual from the raw data, this record is hidden from the attacker. The attacker can now make the mapping between the two sets synthetic and raw, and count to how many remaining synthetic records the original unknown record can be attributed. Iterated over all the original records, this gives a measure of the attacker's uncertainty associated to records re-identification. We report here some numerical results for the experimented data. For a synthetic data set generated with the data grid at the finest level (with 34 *id*-clusters), we measure that, on average, an unknown original record can be imputed to the 31550 synthetic records that don't match with the original ones (among 39100, the size of the experimented set, i.e. to 80.6% of the synthetic records). For a synthetic dataset generated with a co-clustering aggregated with 10 *id*-clusters, this value reaches 86.4%. The risk of record re-identification is thus very low.

4   Conclusion

This work presents a generic methodology for the anonymization of microdata sets. We focus on the re-identification risk. The goal was to be able to produce anonymized individual data that preserve sufficient information to be used in place of the real data. Our approach involves combining k-anonymity with synthetic data generation. K-anonymity is achieved based on co-clustering, a data joint distribution estimation technique, providing the analyst with a summary of the data in a parameter-free way. From the finest co-clustering, the user can choose, thanks to an aggregation functionality, the appropriate granularity level to constitute clusters of at least k individuals. This step doesn't require a costly preparation of data as in classical k-anonymization methods. The co-clustering is also seen as a generative model that is used to build synthetic individuals in the same format as the original individuals. We have shown that synthetic data generated in this way retain the statistical properties of raw data. In terms of privacy, the re-identification risk has been experimentally evaluated. Our method proposes an interesting balance between privacy and utility, thus using the synthetic data for various data mining tasks can be envisaged. We plan now to extend the methodology to address the risk of sensitive attribute disclosure.

References

1. Blondel, V.D., Esch, M., Chan, C., Clérot, F., Deville, P., Huens, E., Morlot, F., Smoreda, Z., Ziemlicki, C.: Data for development: the D4D challenge on mobile phone data. arXiv preprint arXiv:1210.0137 (2012)
2. Boullé, M.: Universal approximation of edge density in large graphs. Technical report, arXiv, arXiv:1508.01340 (2015)
3. Boullé, M., Guigourès, R., Rossi, F.: Nonparametric hierarchical clustering of functional data. Adv. Knowl. Discov. Manag. (AKDM-4) **527**, 15–35 (2014)
4. Boullé, M.: Functional data clustering via piecewise constant nonparametric density estimation. Pattern Recogn. **45**(12), 4389–4401 (2012)
5. Chen, R., Xiao, Q., Zhang, Y., Xu, J.: Differentially private high-dimensional data publication via sampling-based inference. In: Proceedings of the 21th ACM SIGKDD International Conference on Knowledge Discovery and Data Mining, KDD 2015. ACM, New York, pp. 129–138 (2015)
6. Cormode, G.: The confounding problem of private data release (Invited Talk). In: 18th International Conference on Database Theory (ICDT 2015), Leibniz, LIPIcs, vol. 31, pp. 1–12 (2015)
7. Dwork, C.: Differential privacy: a survey of results. In: Agrawal, M., Du, D., Duan, Z., Li, A. (eds.) TAMC 2008. LNCS, vol. 4978, pp. 1–19. Springer, Heidelberg (2008). doi:10.1007/978-3-540-79228-4_1
8. G29, WG: Opinion 05/2014 on anonymization techniques. Technical report, EC (2014). http://ec.europa.eu/
9. Hundepool, A., Domingo-Ferrer, J., Franconi, L., Giessing, S., Nordholt, E.S., Spicer, K., De Wolf, P.P.: Statistical Disclosure Control. Wiley, Hoboken (2012)
10. LeFevre, K., DeWitt, D.J., Ramakrishnan, R.: Incognito: efficient full-domain k-anonymity. In: Proceedings of the 2005 ACM SIGMOD International Conference on Management of data, pp. 49–60. ACM (2005)
11. Li, N., Li, T., Venkatasubramanian, S.: t-closeness: privacy beyond k-anonymity and l-diversity. In: 2007 IEEE 23rd International Conference on Data Engineering, pp. 106–115. IEEE (2007)
12. Machanavajjhala, A., Kifer, D., Gehrke, J., Venkitasubramaniam, M.: L-diversity: privacy beyond k-anonymity. ACM Trans. Knowl. Discov. Data (TKDD) **1**(1), 3 (2007)
13. Prasser, F., Bild, R., Eicher, J., Spengler, H., Kohlmayer, F., Kuhn, K.A.: Lightning: utility-driven anonymization of high-dimensional data. Transa. Data Priv. **9**(2), 161–185 (2016)
14. Samarati, P.: Protecting respondents identities in microdata release. IEEE Trans. Knowl. Data Eng. **13**(6), 1010–1027 (2001)

15. Torra, V., Navarro-Arribas, G., Stokes, K.: An overview of the use of clustering for data privacy. In: Celebi, M.E., Aydin, K. (eds.) Unsupervised Learning Algorithms, pp. 237–251. Springer, Cham (2016). doi:10.1007/978-3-319-24211-8_10
16. Vilhuber, L., Abowd, J.M., Reiter, J.P.: Synthetic establishment microdata around the world. Stat. J. IAOS **32**(1), 65–68 (2016)

STATS - A Point Access Method for Multidimensional Clusters

Giannis Evagorou[✉] and Thomas Heinis

Imperial College London, London SW7 2AZ, UK
{g.evagorou15,t.heinis}@imperial.ac.uk

Abstract. The ubiquity of high-dimensional data in machine learning and data mining applications makes its efficient indexing and retrieval from main memory crucial. Frequently, these machine learning algorithms need to query specific characteristics of single multidimensional points. For example, given a clustered dataset, the *cluster membership* (*CM*) query retrieves the cluster to which an object belongs.

To efficiently answer this type of query we have developed *STATS*, a novel main-memory index which scales to answer *CM* queries on increasingly big datasets. Current indexing methods are oblivious to the structure of clusters in the data, and we thus, develop STATS around the key insight that exploiting the cluster information when indexing and preserving it in the index will accelerate look up. We show experimentally that STATS outperforms known methods in regards to retrieval time and scales well with dataset size for any number of dimensions.

Keywords: High-dimensional indexing · Clustering

1 Introduction

Machine learning algorithms have received considerable attention from the research community, demonstrating their usefulness of extracting knowledge from data. Applications of machine learning are vast and include image segmentation, object or character recognition and many more. In image segmentation, for example, machine learning is used to identify tumors and other pathologies to determine the best surgical plan or diagnosis [5]. Executing machine learning efficiently is hence key to analyzing big amounts of data.

Not only do machine learning algorithms require a vast number of iterations to converge, but in each iteration, they execute a vast number of queries on high-dimensional data. Machine learning algorithms [13] thus, crucially depend on the efficient execution of CM (cluster membership) queries– testing to which cluster an object belongs—during their execution. This is particularly true for algorithms that assume background knowledge, i.e., semi-supervised learning algorithms like constrained K-means [14] which uses partial or complete cluster membership (CM) information to improve the accuracy of the clustering.

With these applications in mind, we develop STATS, a new index to answer *CM queries* efficiently. Formally, STATS takes a point query $Q = (q_1, \ldots, q_n)$

© Springer International Publishing AG 2017
D. Benslimane et al. (Eds.): DEXA 2017, Part I, LNCS 10438, pp. 352–361, 2017.
DOI: 10.1007/978-3-319-64468-4_27

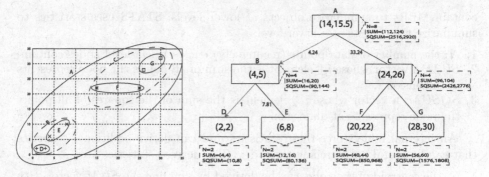

Fig. 1. STATS indexing points

and, if Q belongs to any cluster C, it returns the id of C. Known indexes for CM queries store the cluster ID separately in the index which increases the size of the data structure, making retrieval slower.

STATS is a main-memory index that uses statistics to index multidimensional points along with their cluster ID. The cluster ID is an integral part of the index, which makes STATS more efficient when answering CM queries. STATS uses bottom-up clustering to produce a hierarchical data structure, which is the equivalent of a dendrogram in agglomerative clustering approaches [9]. The dendrogram guides the search to a particular cluster (which contains the points).

Our experimental evaluation shows that STATS outperforms other approaches in terms of retrieval time and it scales well for datasets of increasing size. Furthermore, its performance does not deteriorate as dimensionality increases.

2   STATS

In this paper, we develop STATS, a novel dimension agnostic main-memory index that uses *statistics* to represent clustered points in space. STATS uses *agglomerative clustering* (i.e. bottom-up clustering) to cluster data points in a recursive manner. In case of clustered data, a large number of points can be effectively indexed using a representative point - the cluster's centroid.

During its building phase, STATS uses agglomerative clustering to produce a hierarchical structure of statistics that can be used to calculate distances and guide the search. The result of the building phase is an *unbalanced binary tree*. An internal node contains statistics that summarize the content of its descendants. An inner node is a cluster without the data but with statistics that represent the constituent points at lower levels. Leaf-level nodes contain the data points.

In this project, we propose the use of statistics, instead of other approximations (e.g. Minimum Bounding Rectangles used in R-Tree variants [2,3,12]), to effectively summarize points in k-dimensional space. These statistics can efficiently support query operations. Like the internal nodes of R-tree variants contain MBRs to summarize objects at lower levels, STATS uses statistics to summarize the objects at a lower level:

1. N: the number of data points (recursively) enclosed leaf-level nodes contain,
2. SUM: a vector whose i_{th} element is the sum of all points the node represents at dimension i,
3. $SQSUM$: a vector whose i_{th} element is the sum of the squares of all points the node represents at dimension i.

An internal node stores these statistics and can derive additional statistics that can be useful for searching the index. The derived statistics are:

1. C, the centroid of the node, can be derived by dividing the SUM vector with the scalar value N. C represents the node and is needed for queries,
2. VAR, a vector whose i_{th} element is the variance of the coordinates over all points that the node represents at dimension i. VAR_i is computed by calculating the expression: $\frac{SQSUM_i}{N} - (\frac{SUM_i}{N})^2$, for every dimension i,
3. STD, a vector whose i_{th} element is the standard deviation of the coordinates at dimension i, over all points that the node represents. Standard deviation is estimated by performing an element-wise square root of the vector VAR.

Let us assume the four clusters in 2D space described in Fig. 1 (left); the resulting data structure is depicted in Fig. 1 (right). STATS starts building the index structure based on the clusters in the dataset. Each cluster in the dataset is summarized into a leaf node of STATS. To build the upper levels of STATS nodes are recursively merged starting at the leaf level, i.e., the statistics of low-level nodes are combined into a higher-level node. Merging the nodes and their statistics continues bottom up until the root of the tree is created.

The example in Fig. 1 illustrates this: the ancestor of the leaf nodes $(2, 2)$ and $(6, 8)$ contains the centroid $(4, 5) = (\frac{SUM_1}{N}, \frac{SUM_2}{N}) = (\frac{16}{4}, \frac{20}{4})$. The required information is shown in the dashed boxes, while (some of) the derived information is shown in the solid boxes. To avoid calculation of the derived information during the querying phase, we store both the derived and the required information. When main memory is scarce, however, derived information is not stored.

Leaf nodes also need to store information about single points. A Red-Black tree maintains the distance of each point from the cluster's centroid. To rapidly locate the point within a leaf node, we index the distance of each point from the centroid of its enclosing cluster/leaf-node. In particular, at the leaf-level we maintain a Red-Black tree that holds the following key-value pairs: $[d(p_i, c), p_i]$, where p_i is a point that belongs to the cluster/leaf-node c and $d(p_i, c)$ is the distance of p_i from c's centroid. Put differently, $d(p_i, c)$ is used as the key to index a Red-Black tree to retrieve p_i and thus, impose a total order on all points in the leaf node. The distance metric used (i.e., d) to index the points at the leaf level must be the same as the one used during the querying phase.

Building Phase

The building phase uses agglomerative clustering to create an unbalanced binary tree. At each step of the algorithm, the two nearest clusters are merged generating a new cluster, which is added to the collection of available clusters to be merged. Importantly, each cluster is represented by its centroid. Algorithm 1 explains the building phase (see Table 1 for the notation used).

The building phase starts by estimating the pairwise distances between each pair of clusters in CL (lines 1 to 5). The most important steps of the algorithm are explained next: (1) Line 5 inserts the pairwise distances between all clusters into the heap, H. (2) Line 7 retrieves from the heap the two clusters which have the closest centroids to each other. (3) If the two clusters have already been merged, they are ignored and removed from the set of all clusters (lines 8 to 10). (4) Line 11 merges the nearest clusters creating a new cluster, which only contains statistics summarizing points or clusters. (5) The heap is updated with the distances between the new cluster and the remaining clusters (line 13). (6) Nodes are created bottom-up to form an unbalanced binary tree (line 15).

Algorithm 1. Building phase	**Algorithm 2.** Querying algorithm
1: **for** c_i in CL **do**	**Input:**qp ▷ The query point
2: **for** c_j in CL **do**	**Output:**$clusterId$
3: **if** $c_i \neq c_j$ **then**	1: node = root
4: Calculate $d(c_i, c_j)$	2: **while** *node* **do**
5: Insert(d,H)	3: left = node.getLeft()
6: **while** H is **not** empty **do**	4: right = node.getRight()
7: H.**top**()	5: leftDist = distance(left,qp)
8: **if merged**(c_i,c_j) **then**	6: rightDist = distance(right,qp)
9: H.**pop**()	7: **if** $leftDist < rightDist$ **then**
10: continue	8: node = left
11: c_{ij}=**merge**(c_i, c_j)	9: **else**
12: $CL = CL - c_i - c_j$	10: node = right
13: **update**(c_{ij},CL,H)	11: **if** node.isLeaf() **then**
14: $CL = CL \cup c_{ij}$	12: $leafDis = dist(node, qp)$
15: **createNodes**($c_i, c_j, c_{i,j}$)	13: point = RBtree(leafDis)
16: H.**pop**()	14: **report** point.clusterId()

As an example consider the index in Fig. 1. The algorithm begins by estimating the distance between each pair of clusters (i.e., the leaf nodes) with a given distance metric. The initial distance calculation is shown in Table 2 (**i**). It finds nodes D and E closest to each other, and they are thus merged, creating node B - the parent of D and E. The distance between B and all other nodes has to be calculated and inserted into the heap (see line 13 of Algorithm 1). Table 2 (**ii**) (where we use eager removal of elements from the heap to make the example more concise) shows the updated heap after the merge of D and E. At the next step of the algorithm, nodes F and G are merged, creating node C. Finally, C and B are closest to each other and thus merged to create the root of tree A.

The merging process (line 11 in Algorithm 1) does not consider all the points of the two clusters to be merged. If an internal node, n, at a high level of the tree represents l_n leaf nodes and the size of the clusters at the leaf level is s, then the complexity for merging nodes n and m is $\mathcal{O}(l_n * l_m * s)$. For bigger nodes (and bigger datasets) merging nodes can become prohibitively expensive and merging close to the root of the tree is more expensive because nodes involve more points. To mitigate this expensive merging procedure we merge the required statistics instead. Therefore, if we want to create cluster c by merging clusters a and b, then we estimate the expressions in Eq. 1, for every $i \in [1, \ldots, d]$ where d is the number of dimensions. As an example consider nodes D and E in Fig. 1. STATS can efficiently merge them and create node B by calculating the aforementioned expressions, as in Eq. 2.

$$N_c = N_a + N_b$$
$$SUM_i^c = SUM_i^a + SUM_i^b \quad (1)$$
$$SQSUM_i^c = SQSUM_i^a + SQSUM_i^b$$

$$N_b = 2 + 2$$
$$SUM_b = \begin{bmatrix} 4 \\ 4 \end{bmatrix} + \begin{bmatrix} 12 \\ 16 \end{bmatrix} = \begin{bmatrix} 16 \\ 20 \end{bmatrix} \quad (2)$$
$$SQSUM_b = \begin{bmatrix} 10 \\ 8 \end{bmatrix} + \begin{bmatrix} 80 \\ 136 \end{bmatrix} = \begin{bmatrix} 90 \\ 144 \end{bmatrix}$$

Distance Calculation and the Building Phase

Throughout this work, we represent a cluster with a *centroid*. A centroid is a point in a k-dimensional space, where its i_{th} dimension is the average over all points in the same dimension. Assuming a cluster with N points in a k-dimensional space, we can formally define the centroid to be: $[c_1, \ldots, c_k]$, where $c_i = \frac{1}{N} \sum_{j=1}^{N} x_{ji}$, with x_{ji} as the value of the i_{th} dimension of a random point j.

During the building phase, we estimate the distance between two clusters, a and b, by using the *euclidean* distance metric: $d(a, b) = \sqrt{\sum_{i=1}^{d} (c_i^a - c_i^b)^2}$. Although the *DED* (i.e. Default Euclidean Distance) metric is not appropriate for high dimensional data [1], especially when executing nearest neighbour queries, it is suitable when the centroids of the clusters are well separated.

STATS Querying

STATS uses a point as the key to retrieve the ID of the cluster the point belongs to. Algorithm 2 describes the procedure followed to find a query point. The search

Table 1. Building phase notation

Symbol	Description
H	Heap
CL	All clusters
c_{ij}	Merged cluster
c_i	Random cluster i
c_j	Random cluster j
$d(c_i, c_j)$	Distance between clusters i & j

Table 2. Building phase - heap contents

i			ii		
p_i	p_j	$d(p_i, p_j)$	p_i	p_j	$d(p_i, p_j)$
D	E	$\sqrt{52}$	F	G	$\sqrt{128}$
F	G	$\sqrt{128}$	F	B	$\sqrt{545}$
E	F	$\sqrt{392}$	G	B	$\sqrt{1201}$
D	F	$\sqrt{724}$	iii		
E	G	$\sqrt{968}$	p_i	p_j	$d(p_i, p_j)$
D	G	$\sqrt{1460}$	C	B	$\sqrt{841}$

for a point, qp, starts at the root, where the distance between qp and the root's descendants is calculated (lines 5–6 of Algorithm 2). The algorithm then follows the path with the least distance (lines 7–10), until STATS reaches the leaf node where the point resides.

We illustrate the process with an example where we look for the point $(1, 2)$ that resides in leaf node D (see Fig. 1 - left). The search algorithm follows the path $A \rightarrow B \rightarrow D$ shown in Fig. 1 (the branch with the smallest distance).

When the search algorithm reaches the leaf node, the distance between qp and the leaf node's centroid is calculated. The distance is used to locate the point in logarithmic time (guaranteed by the Red-Black tree) without performing a linear search over all points in the cluster. Algorithm 2 uses Eq. 3 to estimate the distance between qp and a cluster b. In Eq. 3, c_i^b is the value of the centroid and σ_i^b is the standard deviation of a cluster b at dimension i. We can use the DED as illustrated in Eq. 4 to estimate the distance between a query point and each node in the hierarchy. Nevertheless, the query accuracy of Algorithm 2 deteriorates when using the DED metric.

$$d(qp, b) = \sqrt{\sum_{i=1}^{d} \frac{(qp_i - c_i^b)^2}{(\sigma_i^b)^2}} \quad (3) \qquad d(qp, b) = \sqrt{\sum_{i=1}^{d} (qp_i - c_i^b)^2} \quad (4)$$

The default Euclidean distance is not suitable for high-dimensional data. The work in [1] has explored the behavior of the L_k norm for various values of k (including the L_2 norm, i.e., Euclidean distance) and has proven that the L_2 norm is *not meaningful* for high dimensional data. The problem with DED and high dimensional data is that the ratio of the distances of the farthest and nearest objects to a query object is almost 1 [1], i.e., the distance between a query object and its farthest neighbor is as big as the distance between the same query object and its nearest neighbor. This blurry distinction between farthest and nearest objects renders the default Euclidean distance *unsuitable* for the querying Algorithm 2. Normalized Euclidean Distance (NED) is suitable for clustered data as it considers the spread of a cluster in each dimension.

Table 3. Datasets

Series 1		Series 2		Series 3	
# of clusters	Size (GB)	Dimensions	Size	# of clusters	Cluster size
128	0.071	10	0.354 MB	100	5K
256	0.143	20	0.7 MB	200	5K
400	0.226	30	1.1 GB	300	5K
512	0.290	40	1.4 GB	400	5K
1000	0.564	50	1.7 GB	500	5K
1500	0.847	60	2.1 GB	600	5K
2000	1.1	70	2.4 GB	700	5K
3000	1.7	80	2.8 GB	800	5K
3500	2.0	90	3.1 GB	900	5K
4000	4.5	-	-	-	-

Fig. 2. Clustered data - structure

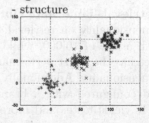

3 Experimental Evaluation

We compare the retrieval time of STATS with an R*-Tree [4] (bulkloaded R*-Tree using STR [7] and a plain R*-Tree [2]) and a KD-Tree [10]. We additionally evaluate the accuracy of the Querying Algorithm 2 for several distance measures. We assume a static clustered dataset and a workload of CM queries.

3.1 Experimental Setup

We run the experiments on a virtual machine (one core at 3.6 GHz, 14 GB of RAM and 100 GB of disk) with a 64-bit Debian-based OS - Linux Mint with a kernel version of 4.4.0.21. We use numactl [6] on Linux to ensure that experiments run on one processing core. The R-Tree [4] uses a branching factor of 100 and a fill factor of 70%. For both trees, we use the parameters providing the best performance. All the implementations are in-memory.

3.2 Experimental Methodology

To measure the retrieval time, we use synthesized datasets up to 90 dimensions containing well-separated clusters produced with ELKI [11]. Figure 2 shows an example of our synthesized datasets. Cluster A has a normal distribution with a mean of 0, and a standard deviation of 10, the two dimensions of cluster B have a normal distribution with a mean of 50 and a standard deviation of 10, etc.

Retrieval Time and Dataset Size. In this series, we compare the scalability of STATS with the KD-Tree and the R*-Tree (see Fig. 3a). This series assumes constant cluster size at 10'000 points, three dimensions and a varying number of clusters (see Table 3). Not only does the R-tree follow an upward trend, but its retrieval time for the smallest datasets is also high. STATS has the lowest retrieval time and scales well with dataset size. The KD-tree has similar performance to STATS due to its very good behavior in low dimensions. Nevertheless, both experience a slight increase when querying the two biggest datasets.

Retrieval Time and Dimensionality. This series measures the impact of dimensionality on retrieval time (Fig. 3b). The number of clusters is 200 and with a size of 10'000, while we increase dimensionality (see Table 3). The R*-Tree has the worst performance with a mean retrieval time of 285 μs across any dimensionality. Additionally, we compare with an R*-Tree loaded with STR [7] and configured with a branching factor of 100 and a fill factor of 0.8. Using STR yields faster and more stable execution times. STATS and the KD-Tree experience a steady increase in retrieval time as dimensionality increases.

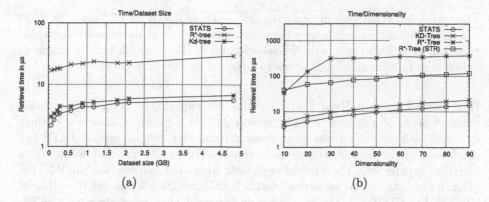

Fig. 3. Cluster membership query time (μs) (logscale)

Distance Metric and Querying Accuracy. STATS can calculate both *NED* and *DED* metrics. To illustrate the impact of choice of distance metric, we execute 50'000 queries to measure the percentage of false answers for three metrics, the DED metric, the NED metric and the Manhattan Distance (MD) metric. Generally, when using the L_k norm, the contrast between the farthest and nearest neighbours is bigger for small values of k rather than for big values of k. For this reason, we use the L_1 norm for Algorithm 2.

For this experiment (see Fig. 4a), we use a dataset of dimensionality 30 with a varying number of clusters. Increasing the number of clusters results in more leaf nodes, which increases the number of paths in the tree. We show that choosing the wrong distance metric can create ambiguity in the choice of path.

The results show that the MD and the DED metric have almost the same accuracy. STATS' building phase creates the inner nodes of the hierarchy bottom-up. More centroids are thus created in the space apart from the centroids of the *actual clusters/leaf nodes*, i.e., the clusters that contain the points.

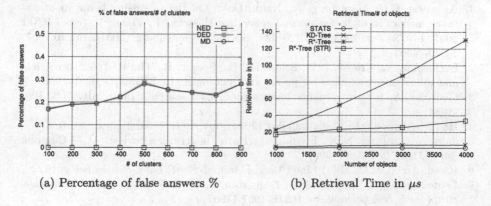

(a) Percentage of false answers %   (b) Retrieval Time in μs

Fig. 4. Queering accuracy and performance on a real dataset

These high-level centroids increase the ambiguity in the choice of path. The NED metric alleviates this problem by considering the extent of the clusters and thus, Algorithm 2 follows the cluster whose centroid is close enough but its extent is big enough to accommodate the query point in its circumference.

Retrieval Time on Real Data. In the last experiment (Fig. 4b), we index measurements of abalones [8]. The cluster ID for a group of abalones is their age, and each dimension is a measurement (diameter, height, etc.). STATS has the lowest retrieval time. When using the R*-Tree without bulk-loading, time increases rapidly with the number of objects indexed. Bulk-loading the R*-Tree yields a more stable and lower time, but it is still considerably higher than that of STATS. The KD-Tree is the only indexing approach that competes with STATS.

4   Conclusions

With STATS, we depart from the norm of approximating points using bounding based approximations, and we suggest the use of statistics to approximate multidimensional points. STATS stores cluster membership as an integral part of the index making it an efficient method to answer CM queries.

As our experiments show, STATS performs very well regardless of dimensionality and dataset size. It can index well separated clustered data and can efficiently answer CM queries. STATS is sensitive to the choice of distance metric but as we showed, NED provides very good performance on a range of datasets.

Acknowledgements. This work is supported by the EU's Horizon 2020 grant 650003 (Human Brain project), EPSRC's PETRAS IoT Hub and HiPEDS grant reference EP/L016796/1).

References

1. Aggarwal, C.C., Hinneburg, A., Keim, D.A.: On the surprising behavior of distance metrics in high dimensional space. In: Bussche, J., Vianu, V. (eds.) ICDT 2001. LNCS, vol. 1973, pp. 420–434. Springer, Heidelberg (2001). doi:10.1007/3-540-44503-X_27
2. Beckmann, N., Kriegel, H.P., Schneider, R., Seeger, B.: The R*-tree: an efficient and robust access method for points and rectangles. In: SIGMOD 1990 (1990)
3. Guttman, A.: R-trees: dynamic index structure for spatial data. In: SIGMOD 1984 (1984)
4. Hadjieleftheriou, M.: libspatialindex (2014). https://libspatialindex.github.io/
5. Jain, A.K., Murty, M.N., Flynn, P.J.: Data clustering: a review. ACM Comput. Surv. **31**(3), 264–323 (1999)
6. Kleen, A.: NUMACTL(1) Linux User's Manual. SuSE Labs, September 2016
7. Leutenegger, S.T., Lopez, M.A., Edgington, J.: STR: a simple and efficient algorithm for R-tree packing. In: ICDE 1997 (1997)
8. Lichman, M.: UCI Machine Learning Repository (2013)

9. Maimon, O., Rokach, L.: Data Mining and Knowledge Discovery Handbook. Springer-Verlag New York, Inc., New York (2005)
10. Muja, M., Low, D.G.: nanoflann (2016). https://github.com/jlblancoc/nanoflann
11. Ludwig Maximilian University of Munich: ELKI data mining library (2016)
12. Sellis, T.K., Roussopoulos, N., Faloutsos, C.: The R+-tree: a dynamic index for multi-dimensional objects. In: VLDB 1987 (1987)
13. Suthaharan, S.: Machine Learning Models and Algorithms for Big Data Classification: Thinking with Examples for Effective Learning, 1st edn. Springer Publishing Company Incorporated, Heidelberg (2015)
14. Wagstaff, K., Cardie, C., Rogers, S., Schrödl, S.: Constrained K-means clustering with background knowledge. In: ICML 2001 (2001)

LinkedMDR: A Collective Knowledge Representation of a Heterogeneous Document Corpus

Nathalie Charbel[1(✉)], Christian Sallaberry[2], Sebastien Laborie[1],
Gilbert Tekli[3], and Richard Chbeir[1]

[1] University Pau & Pays Adour, LIUPPA, EA3000, 64600 Anglet, France
`{nathalie.charbel,sebastien.laborie}@univ-pau.fr`, `rchbeir@acm.org`
[2] University Pau & Pays Adour, LIUPPA, EA3000, 64000 Pau, France
`christian.sallaberry@univ-pau.fr`
[3] University of Balamand (UOB), 100, Tripoli, Lebanon
`gilbert.tekli@fty.balamand.edu.lb`

Abstract. The ever increasing need for extracting knowledge from heterogeneous data has become a major concern. This is particularly observed in many application domains where several actors, with different expertise, exchange a great amount of information at any stage of a large-scale project. In this paper, we propose LinkedMDR: a novel ontology for Linked Multimedia Document Representation that describes the knowledge of a heterogeneous document corpus in a semantic data network. LinkedMDR combines existing standards and introduces new components that handle the connections between these standards and augment their capabilities. It is generic and offers a pluggable layer that makes it adaptable to different domain-specific knowledge. Experiments conducted on construction projects show that LinkedMDR is applicable in real-world scenarios.

Keywords: Heterogeneous documents · Document Representation · Ontologies · Information Retrieval

1 Introduction

Modern large-scale projects adopt a systematic approach to project planning, where several actors of different expertise are involved. Throughout the project, they contribute and exchange a wide variety of technical and administrative knowledge depending on their background and role in the project. As an example, in the construction industry, actors (i.e., owners, consultants and contractors) exchange contracts, technical specifications, administrative forms, technical drawings and on-site photos throughout the different stages of a construction process [13]. The interchanged documents, originated from different sources, do not usually have a common standard structure. Also, they show heterogeneity in their formats (e.g., pdf, docx, xlsx, jpeg, etc.), contents (e.g., architecture,

© Springer International Publishing AG 2017
D. Benslimane et al. (Eds.): DEXA 2017, Part I, LNCS 10438, pp. 362–377, 2017.
DOI: 10.1007/978-3-319-64468-4_28

electrical, mechanical, structure, etc.), media types (e.g., image, text, etc.) and versions. In addition, the documents may have implicit and explicit inter and intra-document links[1] and references.

In the literature, several works have been undertaken to define metadata on documents and their contents. These annotation models can be classified according to whether they deal with text-based content (e.g., TEI [22]), image-based content (e.g., EXIF [7]) or multimedia-based content [1–3,8,16]. Nevertheless, none of the existing works considers (i) the various inter and intra-document links, (ii) semantic annotations of both texts and images on a content and structural levels, (iii) documents multimodality, and (iv) particularities of domain-specific documents (such as technical drawings).

In this context, our aim is to provide a clear and a complete picture of the collection of heterogeneous documents exchanged in multidisciplinary and large projects. It is of a key importance to have the needed data easily available and adapted to the specific actor at any point in time based on his domain and requirements. To do so, there is a need to define a common and generic data model representing all the documents and their possible connections by means of semantic concepts and relations. Such linked data network would provide a collective knowledge that allows better indexing and semantic information processing and retrieval. This minimizes actors workload, cognitive efforts and human errors.

In this paper, we present LinkedMDR: a novel ontology for Linked Multimedia Document Representation. LinkedMDR combines existing standards that address metadata, structure and content representation (i.e., DC [6], TEI [22] and MPEG-7 [21]). In addition, it introduces new components augmenting the capabilities of these standards and allowing adaptation to different domain-specific knowledge. Our proposal is evaluated in the context of various construction related documents. Also, on-going evaluations in other application domains (such as the energy management domain) are still in progress.

The remainder of this paper is organized as follows. Section 2 illustrates the motivations through a real-world scenario encountered in the construction industry. Section 3 reviews related works in metadata, structure and content representation. Section 4 describes our LinkedMDR data model. Finally, Sect. 5 presents experimental results, while Sect. 6 concludes the paper and shapes some future works.

2 Motivations

To motivate our work, we investigate a representative scenario in the context of construction projects, provided by Nobatek[2], a French technological resource center involved in the sustainable construction domain.

---

[1] Inter-document links are relations between documents. Intra-document links are relations between elements of the same document.

[2] http://www.nobatek.com.

2.1   Context

Nobatek is a consultancy office that assists investors in managing services related to sustainable construction projects. Its main role is to ensure the compliance of a project with the environmental standards and quality performance. Throughout the different construction stages, Nobatek communicates with other consultancy offices, contractors and specialized subcontractors. Across all of the involved actors, there exist members with different interests and profiles. Figure 1 illustrates some of the various heterogeneous documents[3] that are involved in such projects. For instance, there are several documents that describe technical specifications (d_1 and d_5), thermal properties (d_2) and acoustic characteristics (d_3) of the building. Also, there is an excerpt of a technical drawing related to the ground floor (d_4) and an image (d_6). Currently, the engineers at Nobatek have to manually search for information in all these documents. In order to ensure the compliance of the building's exterior facades with the environmental standards, one needs to search for the related information contained in these documents produced by several specialized consultants (e.g., offices for acoustic and thermal analysis services). Given the large volume of information, the search becomes tedious and misleading as one can miss useful information.

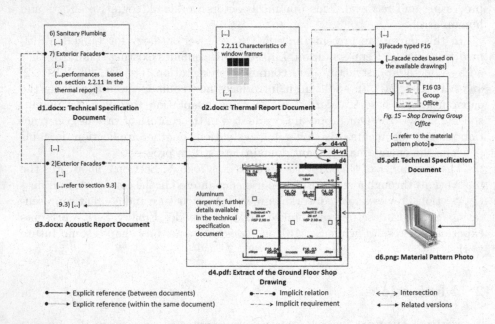

Fig. 1. Example of heterogeneous documents exchanged within a construction project.

---

[3] For the sake of simplicity, we only present 6 documents. However, other documents could be also involved such as videos, audios and 3D drawings.

2.2 Challenges

- **Challenge 1: Representing various inter and intra-document links -** One has to search for the relevant information in a collective knowledge built on a multitude of documents. Figure 1 shows that a document d_i has various relations (references, mutual topics, versions, etc.) with other documents. These relations can be implicit or explicit. For instance, some textual sections of d_1 and d_3 have an implicit relation since they both describe building's exterior facades. The document d_5 has explicit reference to the image d_6 and the technical drawing d_4, which itself has several versions. This raises the need to index these documents and analyze their metadata information in order to build the relations between them as in Fig. 1.
- **Challenge 2: Handling information semantics on content and structural levels -** One needs to locate, on different depth levels, the relevant information contained in several documents recalling the same topic. For instance, d_5 describes general information regarding exterior facades, while d_1 contains detailed information of these facades together with other building services, such as plumbing and electricity. d_2 focuses particularly on the thermal properties of exterior and interior facades. This raises the need to reason over the document's content, its general metadata, and structural metadata that describes different depth levels (e.g., page, section, paragraph or sentence), thus to associate them with semantic concepts.
- **Challenge 3: Handling documents multimodality -** One has to work with heterogeneous documents: d_1, d_2 and d_3 are Word documents, d_4 is a CAD drawing in a PDF format, d_5 a PDF document and d_6 a PNG image. Therefore, it is convenient to handle different document types and formats.
- **Challenge 4: Ensuring extensibility -** One may encounter other types of construction related documents involving different structures, types of media, formats and document links. For example, an extension of the scenario illustrated in Fig. 1 may involve an audio file that analyzes the noise impact before and after the cladding of the facades. This raises the need to handle evolution of the information.

In the following section, we demonstrate that existing models are only capable of partially addressing the aforementioned challenges within a given information system.

3 Related Work

In this section, we describe existing standards and models addressing metadata and content representation (Sect. 3.1). We conclude with a discussion and a comparative study (Sect. 3.2).

3.1   Metadata and Content Representation Standards and Models

Dublin-Core - The Dublin Core Metadata Initiative (DC) is a Metadata standard for describing a wide range of online multimedia resources [6]. The Dublin Core Metadata Element Set consists of 15 Elements describing the content of a document (e.g., title, description, etc.), the intellectual property (e.g., creator, rights, etc.), and its instantiation (e.g., date, format, etc.) [25]. This standard also offers a set of qualifiers which aim to modify the properties of the Dublin Core statements.

MPEG-7 - The Multimedia Content Description Interface is an ISO/IEC standard developed by MPEG (Moving Picture Experts Group) [21]. Its aim is to provide a rich set of complex standardized tools describing low and high level features while also covering different granularities of data (e.g., collection, image, video, audio, segment) and different areas (content description, management, organization and navigation). The three main standardized components of MPEG-7 are: Descriptors (Ds), Description Schemes (DSs) and a Description Definition Language (DDL). While the MPEG-7 Ds are representations of features, the MPEG-7 DSs support complex descriptions and specify the structure and the semantics of the relationships among its constituents: Ds and DSs [17]. The MPEG-7 DDL is a standardized language based on XML schema. It allows the extension of existing Ds and DSs as well as the introduction of new components for specific domains [10].

Ontology-Based Models - Many initiatives have been taken for the purpose of building multimedia ontologies, such as [1,16,24], or transforming existing formats into ontologies, such as [8]. The aim of these studies is to bridge the gap between low level features with automatically extractable information by machines and high level human interpretable features of the same information [19]. There is often the need to combine several standards in order to meet the requirements of complex multimedia applications [18]. This boosts the efforts for building such ontologies. For instance, the Core Ontology for MultiMedia (COMM) [1] is built to satisfy multimedia annotation requirements. It is designed based on the MPEG-7 standard, DOLCE as a foundational ontology, and two ontology design patterns. In contrast to COMM, other MPEG-7 based ontologies are designed using a one-to-one automatic mapping of the entire standard to OWL (e.g., MPEG-7 Rhizomik [8]). Other ontologies, such as the Multimedia Metadata Ontology (M3O) [16], aim to provide abstract pattern-based models for multimedia metadata representation. M3O is centered on the formal upper-level ontology DOLCE+DnS Ultralight. One of the recent ontologies for describing multimedia content is the Media Resource Ontology developed by the W3C Media Annotation Working Group [24]. It is subject to multiple alignments with several existing multimedia metadata, such as MPEG-7, Dublin Core and EXIF.

LINDO Metadata Model - In the context of distributed multimedia information systems, such as the video surveillance applications that use different indexing engines, interoperability problems arise when metadata of different formats are combined together. The LINDO project (Large scale distributed INDexation of multimedia Objects) [12] took up the challenge of handling different metadata standards within its distributed information system, such as Dublin Core, EXIF and MPEG-7. Thus, the authors in [3] define a unified XML-based metadata model that encapsulates these standards based on two levels: general metadata information describing the entire document and metadata related to multimedia contents (image, text, video and audio).

XCDF Format - Problems such as over-segmentation, additional noisy information, and lack of structure preservation produced by PDF generators are often associated with the PDF format. Great effort has been put into overcoming these issues [2]: XCDF is a canonical format which purpose is to represent the results of the physical structure extraction of the PDF documents in a single and structured format. It is based on the XML format and its DTD provides basic elements for general document representation (page, fonts, image, graphic, textblock, textline and token).

EXIF - EXIF (Exchangeable Image File Format) standard is a widely used standard for describing digital images [7]. It mainly supports a set of tags related to image-specific features (e.g., width, height, pixel composition, color), and many other general tags (e.g., image title, date and time, creator, version).

TEI - In this paper, we focus on the TEI (Text Encoding Initiative) [22] as it is a commonly adopted text-driven descriptive standard. It is based on XML format and provides a way to describe information and meta-information within a textual document. TEI offers a representational form of a text (e.g., chapters, sections, paragraphs, lists, items, page break, table, figures and graphics) as well as a set of semantically rich elements (e.g., cross reference links, title, name and address) that could improve information retrieval.

3.2 Discussion

We evaluated the existing standards and models that describe metadata and content related to a heterogeneous document corpus based on the challenges previously mentioned in Sect. 2.2. The results are depicted in Table 1.

It is obvious that the standards that focus only on text or image description are limited because they cannot handle the different types of multimedia documents (Challenge 3). Nevertheless, these standards, in particular the TEI for the text, provide relevant components that can be reused in our proposal. For example, the *<ref>* element allows us to represent document links but is limited to cross-references. As for the multimedia-based standards, the MPEG-7

Table 1. Evaluation of the existing standards and models with regard to the identified challenges.

Challenges	Properties (Approaches)		Metadata and Content Representation Standards and Models									
			Multimedia-based								Image-based	Text-based
			Dublin Core	MPEG-7	COMM	M3O	MediaOnt	Mpeg-7 Rhizomik	LINDO Metadata Model	XCDF Format	EXIF	TEI
Challenge 1	Relations Description	Intra-document Link	Partial	Partial	Partial	Partial	Partial	Partial	Partial	x	x	Partial
		Inter-document Link	Partial	Partial	Partial	Partial	Partial	Partial	Partial	x	x	Partial
Challenge 2	Descriptive Metadata Representation		√	√	√	√	√	√	√	x	√	√
	Content Description		x	√	√	√	x	√	√	x	x	√
	Structural Metadata Representation	Image	x	√	√	√	x	√	√	x	x	x
		Text	x	Partial	Partial	Partial	x	Partial	Partial	Partial	x	√
Challenge 3	Multimodality		√	x	x	x	x	x	√	x	x	x
Challenge 4	Extensibility		Partial	Partial	Partial	Partial	Partial	Partial	Partial	Partial	x	Partial

defines $<StillRegion>$ and $<TextAnnotation>$ elements, which could be used to describe different parts of a technical drawing and its associated text legends. However, these two elements do not allow the description of relevant details that are sometimes encountered within text legends (Challenges 2 and 4), nor possible relations with descriptors from other standards, such as TEI or DC (Challenge 1).

The multimedia-based ontologies and models are also limited. Most of them aggregate descriptors of existing standards but do not offer a solution to the challenges mentioned above (specifically, Challenges 1 and 2). For instance, they are not capable of relating descriptors from MPEG-7, DC and TEI. Hence, these models inherit the deficiencies of their constituent standards.

To our knowledge, there is currently no available representation of a semantic network of linked data that can describe the collective knowledge of a heterogeneous document corpus. Our proposal, illustrated in the following section, answers this need through the design of a novel ontology relying on the most adopted metadata standards. We consider the ontologies as a reliable and an efficient solution to support Semantic Information Retrieval from heterogeneous multimedia data [9].

4   LinkedMDR: A Novel Ontology for Linked Multimedia Document Representation

We propose a novel ontology for Linked Multimedia Document Representation, entitled LinkedMDR[4]. Our ontology models the knowledge embedded in a corpus of heterogeneous documents. This approach is based on (i) the combination

---

[4] LinkedMDR is an OWL ontology created on Protégé. Details on the LinkedMDR ontology, the overall concepts and relations are available at http://spider.sigappfr. org/linkedmdr/.

of the relevant standards for metadata representation and content description (e.g., DC [6], TEI [22] and MPEG-7 [21]), and (ii) the introduction of new semantic concepts and relations that augment the capabilities of these standards and link them through a semantic network. Our proposed ontology is made of three main layers: (i) the core layer serving as a mediator, (ii) the standardized metadata layer comprising elements from existing standards, and (iii) the pluggable domain-specific layer that is adaptable to domain-specific applications. The multiple layers of LinkedMDR ensure its genericity and extensibility (See Fig. 2).

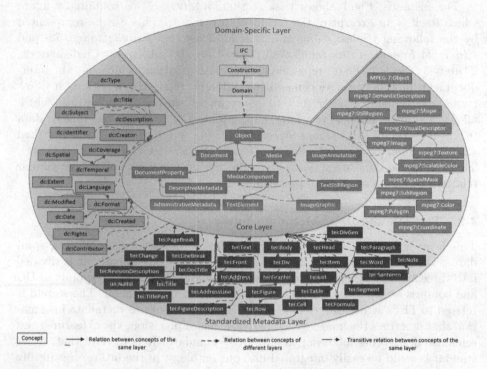

Fig. 2. LinkedMDR: the semantic model representing a corpus of multimedia documents.

4.1   The Core Layer

This layer introduces new concepts and relations that enrich the existing standards, mainly: (i) concepts that model the global composition of a given document and the metadata properties associated to it (e.g., *Document*, *Media*, *MediaComponent* and *DocumentProperty*), (ii) *Object* entity which abstracts *Document*, *Media* and *MediaComponent* while introducing a rich set of relations between them (i.e., *hasPart*, semantic, temporal and spatial relations), (iii) concepts that subsume other concepts involved in an adjacent layer (e.g., *DescriptiveMetadata*, *AdministrativeMetadata* and *TextElement* are super concepts generalizing metadata information of DC and TEI respectively), and (iv) concepts that extend some elements of MPEG-7 and TEI (e.g., *TextStillRegion* extends *tei:Text* and *mpeg7:StillRegion* to describe text metadata inside an image, *Image-Graphic* extends *mpeg7:Image* and *tei:Graphic* to represent the image metadata inside a text, and *ImageAnnotation* extends *tei:Text* and *mpeg7:Image* to annotate an image with another image or a structured text).

In the remainder of the paper, we use RDF [23] statements to illustrate instances of our ontology (with *lmdr* as prefix) as RDF is the most widely used data model for representing a semantic and extensible graph.

For instance, Fig. 1 shows that section 3 (*div3*) of d_5 contains a figure which itself is an excerpt of the technical drawing d_4. This can be represented by the following triples: *<lmdr:d5, lmdr:hasPart, tei:d5.div3.figure15>* and *<lmdr:d4, lmdr:contains, tei:d5.div3.figure15>*. It addresses Challenges 1, 2 and 3 (Sect. 2.2). It represents a spatial relation which is an inter-document link between documents of different types. In addition, it illustrates the semantics associated to concepts of various granularities, which allows to infer further relations and enrich the network of linked data (e.g., if *<tei:d5.div3.figure15, lmdr:contains, tei:dn.div1.figure1>*, then *<lmdr:d4, lmdr:contains, tei:dn.div1.figure1>* is deduced by transitivity). This cannot be done with existing standards since they do not handle similar relations on different levels of precision.

4.2   The Standardized Metadata Layer

This layer is made of selected metadata information defined by existing standards: Dublin Core [6], TEI [22] and MPEG-7 [21]. Consequently, it is divided into three sub-layers, each dedicated to a standard. The first corresponds to DC and comprises metadata information of the document in general. The second is related to TEI's structural text metadata. The third contains metadata information that describes the image with different levels of precision, visual features and semantic descriptors following the MPEG-7 standard. It is to note that, other standards could be easily integrated into our ontology, in the future, specifically into this layer.

This layer also involves relations between its sub-layers. For instance, we added the *isRevisedBy* relation in order to link the *tei:Change* concept (a set of changes made during the revision of a document) to the corresponding *dc:Contributor* concept (a person or organization responsible for making these changes). Further, each sub-layer is connected to the core layer (Sect. 4.1) through relations between their respective concepts. For example, the subsumption relation in *<tei:Text, lmdr:isA, lmdr:Media>*, *<mpeg7:StillRegion, lmdr:isA, lmdr:MediaComponent>*, *<dc:Title, lmdr:isA, lmdr:DescriptiveMetadata>* and *<lmdr:TextElement, lmdr:isOn, tei:PageBreak>* allow concepts of TEI, MPEG-7 and DC to inherit common properties from the core layer.

For instance, we can represent the documents of Fig. 1, such as d4, using triples related to the DC metadata: e.g., *<lmdr:d4, lmdr:hasProperty, dc:d4.title>* and *<dc:d4.title, rdf:value, "ShopDrawing">*. Also, d_4 contains several technical drawings, each related to a specific building floor and described on different pages of the document. The MPEG-7 metadata helps in describing the different regions of the drawings but without any information on the corresponding pages. Similarly, the TEI metadata represents the pages of each drawing but without any description of their content. Hence, considering Challenge 1 (Sect. 2.2), the links between the metadata of these different standards is only possible via the LinkedMDR concepts and relations of the core layer: $<lmdr:d4, lmdr:hasPart, lmdr:d4.imagegraphic>$, $<lmdr:d4.imagegraphic,$ $lmdr:isOn, tei:d4.page1>$, $<lmdr:d4.imagegraphic, lmdr:hasPart, mpeg7:d4.$ $stillregion>$.

4.3   The Pluggable Domain-Specific Layer

The previously mentioned layers are generic and independent of the type of the linked multimedia documents. However, we aim to provide a generic ontology for any domain-specific application. We introduce this pluggable layer as a means to make LinkedMDR adaptable to any domain-specific knowledge. To do so, we present a new concept entitled *Domain* and we link it to the *Object* concept of the core layer. This way domain-specific concepts can be added under the *Domain* while relating to sub-concepts of *Object* (i.e., *Document, Media* and *MediaComponent*).

In this paper, we introduce an example showing how we can make this layer adaptable to the construction domain. We add *Construction* concept as a sub-concept of *Domain*. We also relate the latter to *IFC* concept which comprises concepts from ifcOWL[5], which is the conversion of the IFC (Industry Foundation Classes) [4] schema into ontology. The IFC standard is the complete and fully stable open and international standard for exchanging BIM (Building Information Modeling) data, independently of the different software applications. It involves building and construction objects (including physical components, spaces, systems, processes and actors) and relationships between them [11].

As an example, section 7 (*div7*) in d_1 (Fig. 1) describes the exterior facades. It is now possible to link section 7 with the related IFC object (e.g., window4): $<lmdr:d1, lmdr:isA, lmdr:Document>$, $<lmdr:d1, lmdr:hasPart, tei:d1.div7>$, $<tei:d1.div7, tei:isA, tei:Div>$, $<tei:d1.div7, lmdr:isRelated, ifc:window4>$, and $<ifc:window4, ifc:isA, ifc : BuildingElement>$. This particularly answers Challenges 1 and 4.

5   Experimental Evaluation

We have conducted several experiments to evaluate the quality of the annotation of heterogeneous corpora based on LinkedMDR, in different scenarios, w.r.t. the previously discussed objectives (See Sect. 2.2). We show the results of 2 experiments applied on real-world construction projects provided by Nobatek.

---

[5] Available at http://ifcowl.openbimstandards.org/IFC4_ADD2.owl.

Note that, an on-going evaluation of LinkedMDR in the context of HIT2GAP[6], a European H2020 funded project, is still in progress.

5.1   Context

Test Data - We hand-picked 5 heterogeneous corpora of real construction projects from Nobatek. Table 2 shows the document composition of each test corpus.

Table 2. Document composition of the test corpora.

Corpus	No. of documents	Document formats	Corpus size (MB)
1	10	3 docx, 4 pdf, 3 png	20
2	10	7 pdf, 2 png, 1 jpeg	27.4
3	17	5 docx, 10 pdf, 2 png	54.8
4	15	5 xslx, 1 docx, 7 pdf, 2 png	112.3
5	12	1 xslx, 10 pdf, 1 png	38.2

Prototype - We have developed LMDR Annotator, a java-based prototype, which purpose is to automatically annotate a document corpus based on (i) DC (using Apache Tika API), (ii) TEI (using Oxgarage Web service), (iii) MPEG-7 (using MPEG-7 Visual Descriptors API), and (iv) LinkedMDR. As for the latter, it uses the GATE (General Architecture for Text Engineering) API for the automatic generation of explicit inter and intra-document link annotations based on regular expressions encountered in text. The output XML files generated by the GATE module and the other annotation modules (related to DC, TEI and MPEG-7) are automatically transformed into LinkedMDR instances using tailored XSLT processors. The final output is an RDF file describing the entire corpus. Further details on the prototype's architecture and its different modules are available online[7].

Experiment 1 - The aim of this experiment is to compare LinkedMDR with its alternatives (i.e., DC [6], TEI [22], MPEG-7 [21], and the three combined) regardless of the annotation tools. We manually annotated one corpus (Corpus 1) using each standard separately, the three standards combined, and LinkedMDR. The manual annotations ensure the best possible document representation that can be generated from each of the data models.

We then evaluate the conciseness of each data model. More particularly, for each data model we look at the total number of annotated documents, the number

---

[6] HIT2GAP (Highly Innovative Building Control Tools) is a large-scale project that involves 21 partners and provides an energy management platform for managing building energy behavior. Further details are available at: http://www.hit2gap.eu/.

[7] http://spider.sigappfr.org/linkedmdr/lmdr-annotator.

of annotation elements[8], redundancies (overlapping metadata), and the number of annotation files required for covering the maximum number of relevant criteria. For the latter, we calculate F_2-scores[9] which weight the Recall measure higher than the Precision, to emphasize missed relevant annotations. Note that, we define the relevant criteria (e.g. semantic intra-document links, topological intra-document links, general metadata, text/image-specific metadata, etc.) as to address the requirements that we aim to satisfy (See Sect. 2.2). The considered factors are particularly important for future Information Retrieval application.

Experiment 2 - In a real-world application, manually annotating a document corpus is a tedious job that requires much effort and technical knowledge. In this experiment, we evaluate the effectiveness of our LMDR Annotator in automatically annotating different corpora (Corpus 2 to 5). We calculate Precision, Recall and F_2-measure. One RDF annotation file ($Corpus_m.rdf$) is generated per corpus, such that

$$Corpus_m.rdf = Corpus_m[DC].rdf \cup Corpus_m[TEI].rdf \cup$$
$$Corpus_m[MPEG7].rdf \cup Corpus_m[xLinks].rdf$$

where $Corpus_m[DC].rdf$, $Corpus_m[TEI].rdf$ and $Corpus_m[MPEG7].rdf$ are the set of generated LinkedMDR instances related to the standardized metadata layer (See Sect. 4.2) for the $Corpus_m$; $Corpus_m[xLinks].rdf$ is the set of generated LinkedMDR instances, related to the Core layer (See. Sect. 4.1), representing explicit inter and intra-document links involved within $Corpus_m$; $Corpus_m$ is the set of heterogeneous documents of the mth project.

5.2 Experimental Results

Experiment 1: Evaluating the Conciseness of LinkedMDR - Table 3 shows that using DC, the three standards combined, and LinkedMDR, we were able to annotate the 10 source documents involved in Corpus 1. In contrast, using TEI and MPEG-7, the annotations were not exhaustive. This is due to the incapacity of annotating images and technical drawings in TEI and textual documents in MPEG-7. The Annotation files that we generate using the DC standard contain a very small number of annotation elements since DC covers generic metadata and neglects structure and content representation of the documents (weakest f_2-score: 0.25). The combination of the three standards produces a significant number of annotation elements since the TEI and MPEG-7 standards are very verbose while not covering all of the expected annotation

---

[8] The number of XML tags in the XML annotation files that we generated based on the existing standards and the number of RDF triples that we generated in the LinkedMDR ontology.

[9] F_2-measure: $(5 \times P \times R) / (4 \times P + R)$

Recall: No. of covered relevant criteria/Total No. of expected criteria.

Precision: No. of covered relevant criteria/Total No. of annotated criteria.

elements. Also, it involves many redundancies caused by common metadata elements between the DC, TEI and MPEG-7 standards (f_2-score: 0.70). Hence, the current experiment shows that the annotation based on LinkedMDR is the most concise since it provides the highest f_2-score (0.94) with a relatively small number of annotation elements, all in a single annotation file, and without any redundancy.

Table 3. Results of experiment 1: the evaluation of the conciseness of the existing standards and LinkedMDR in annotating corpus 1.

Annotation Model	No. of Annotated Documents	Cumulative No. of Annotation Elements	No. of Annotation Files	No. of Overlapping Metadata	F_2-Scores
DC [6]	10	131	10	0	0.25
TEI [22]	5	807	5	0	0.53
MPEG-7 [21]	5	495	5	0	0.29
Combined Standards	10	1433	20	128	0.70
LinkedMDR	10	604	1	0	0.94

Experiment 2: Evaluating the Effectiveness of LMDR Annotator - Figure 3 shows the f_2-scores evaluating the outputs of each annotation module separately (i.e., $Corpus_m[DC].rdf$, $Corpus_m[TEI].rdf$, $Corpus_m[MPEG7].rdf$, and $Corpus_m[xLinks].rdf$) then their union ($Corpus_m.rdf$). The results of the automatic annotation of the 4 corpora have f_2-scores that range from 0.48 ($Corpus_4.rdf$) to 0.63 ($Corpus_2.rdf$).

Looking over the individual annotation modules, the f_2-scores slightly change from one corpus to another. $Corpus_m[DC].rdf$ is in general the most effective since it involves LinkedMDR instances generated from documents' meta-tags (e.g., title, creator, date, format, etc.) which are easily extracted automatically using the Apache Tika API. On the other hand, $Corpus_m[MPEG7].rdf$ produces the lowest scores since relevant concepts, such as $mpeg7:StillRegion$, are relatively difficult to generate automatically. In fact, the MPEG-7 Visual Descriptors library is limited to the automatic extraction of low level features, such as color and texture characteristics. Advanced feature extraction requires sophisticated computer vision and machine learning algorithms (such as [14]) which, so far, are not adopted in our prototype. As for $Corpus_m[xLinks].rdf$, one can see that the GATE java API is reliable in automatically extracting explicit inter and intra-document references from regular expressions encountered in the textual documents. However, in some cases, such as in Corpus 4, we obtain a relatively lower score. This is due to some expressions presenting ambiguous references that could not be handled automatically without the use of advanced semantic disambiguation techniques [5,20].

Figure 4 illustrates the recall values w.r.t. to the total expected LinkedMDR instances per corpus. Intuitively, the recall scores decrease when the number of documents increases since more complex inter and intra-document links are involved, which cannot yet be resolved. This emphasizes that our LMDR Annotator, at its current state, offers relatively low recall scores when compared to

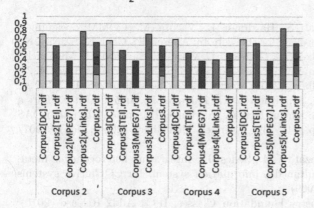

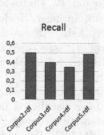

Fig. 3. Results of experiment 2: F_2-scores measuring the effectiveness of our LMDR Annotator for corpus 2 to 5.

Fig. 4. Recall scores based on the total expected Linked-MDR instances per corpus.

the annotation potential proposed by the LinkedMDR ontology. Our current effort focuses on furthering the annotation capabilities of LMDR Annotator as experiments have shown promising results for its adoption in real world projects.

6 Conclusion and Future Work

This paper introduces LinkedMDR: Linked Multimedia Document Representation, a novel ontology for representing the collective knowledge embedded in a heterogeneous document corpus. LinkedMDR is based on the combination of the DC [6], TEI [22], and MPEG-7 [21] standards with the introduction of new concepts and relations augmenting the capabilities of these standards and allowing adaptation to different domain-specific knowledge. Experiments were conducted on real-world construction projects from Nobatek using our LMDR Annotator, a java-based prototype for automatic annotation of a heterogeneous document corpus based on LinkedMDR.

Building on our findings, we are working on extending our research towards the energy management domain, in the context of the European H2020 funded project HIT2GAP. Furthermore, we are extending the LMDR Annotator to handle more inter and intra-document links (such as semantic, topological, spatial) and improve the effectiveness of the automatic annotation based on LinkedMDR. We also aim at providing an interactive interface for non expert users in order to reason over the LinkedMDR ontology using non technical queries such in [15].

References

1. Arndt, R., Troncy, R., Staab, S., Hardman, L., Vacura, M.: COMM: designing a well-founded multimedia ontology for the web. In: Aberer, K., Choi, K.-S., Noy, N., Allemang, D., Lee, K.-I., Nixon, L., Golbeck, J., Mika, P., Maynard, D., Mizoguchi, R., Schreiber, G., Cudré-Mauroux, P. (eds.) ASWC/ISWC -2007. LNCS, vol. 4825, pp. 30–43. Springer, Heidelberg (2007). doi:10.1007/978-3-540-76298-0_3
2. Bloechle, J.-L., Rigamonti, M., Hadjar, K., Lalanne, D., Ingold, R.: XCDF: a canonical and structured document format. In: Bunke, H., Spitz, A.L. (eds.) DAS 2006. LNCS, vol. 3872, pp. 141–152. Springer, Heidelberg (2006). doi:10.1007/11669487_13
3. Brut, M., Laborie, S., Manzat, A.M., Sedes, F.: Integrating heterogeneous metadata into a distributed multimedia information system. In: COGnitive systems with Interactive Sensors (2009)
4. buildingSMART: IFC-Industry Foundation Classes, IFC4 Add2 Release (2016). http://www.buildingsmart-tech.org/specifications/ifc-releases/ifc4-add2-release
5. Charbel, N., Tekli, J., Chbeir, R., Tekli, G.: Resolving XML semantic ambiguity. In: EDBT, pp. 277–288 (2015)
6. Dublin Core Metadata Initiative: DCMI Metadata Terms (2012). http://dublincore.org/documents/dcmi-terms/
7. EXIF: Exchangeable Image File Format for digital still cameras (2002). http://www.exif.org/Exif2-2.PDF
8. Garcia, R., Celma, O.: Semantic integration and retrieval of multimedia metadata. In: 5th International Workshop on Knowledge Markup and Semantic Annotation, pp. 69–80 (2005)
9. Guo, K., Liang, Z., Tang, Y., Chi, T.: SOR: an optimized semantic ontology retrieval algorithm for heterogeneous multimedia big data. J. Comput. Sci. (2017)
10. Hunter, J.: An overview of the MPEG-7 description definition language (DDL). IEEE Trans. Circuits Syst. Video Technol. 11(6), 765–772 (2001)
11. Huovila, P.: Linking IFCs and BIM to sustainability assessment of buildings. In: Proceedings of the CIB W78 2012: 29th International Conference (2012)
12. ITEA: LINDO-Large scale distributed INDexation of multimedia Objects (2010). https://itea3.org/project/lindo.html
13. Klinger, M., Susong, M.: Chapter, phases of the contruction project. In: The Construction Project: Phases, People, Terms, Paperwork, Processes. American Bar Association (2006)
14. OpenCV: Open Source Computer Vision Library (2011). http://opencv.org
15. Pankowski, T., Brzykcy, G.: Data access based on faceted queries over ontologies. In: Hartmann, S., Ma, H. (eds.) DEXA 2016 Part II. LNCS, vol. 9828, pp. 275–286. Springer, Cham (2016). doi:10.1007/978-3-319-44406-2_21
16. Saathoff, C., Scherp, A.: Unlocking the semantics of multimedia presentations in the web with the multimedia metadata ontology. In: Proceedings of the 19th International Conference on World Wide Web, pp. 831–840. ACM (2010)
17. Salembier, P., Smith, J.R.: MPEG-7 multimedia description schemes. IEEE Trans. Circuits Syst. Video Technol. 11(6), 748–759 (2001)
18. Scherp, A., Eissing, D., Saathoff, C.: A method for integrating multimedia metadata standards and metadata formats with the multimedia metadata ontology. Int. J. Semant. Comput. 6(01), 25–49 (2012)
19. Suarez-Figueroa, M.C., Atemezing, G.A., Corcho, O.: The landscape of multimedia ontologies in the last decade. Multimed. Tools Appl. 62(2), 377–399 (2013)

20. Tekli, J., Charbel, N., Chbeir, R.: Building semantic trees from XML documents. Web Semant.: Sci. Serv. Agents World Wide Web **37**, 1–24 (2016)
21. The Moving Picture Experts Group: MPEG7-Multimedia Content Description Interface (2001). http://mpeg.chiariglione.org/standards/mpeg-7
22. The Text Encoding Initiative Consortium: TEI-Text Encoding Initiative (1994). http://www.tei-c.org/release/doc/tei-p5-doc/en/Guidelines.pdf
23. W3C: Resource Description Framework (2004). https://www.w3.org/RDF/
24. W3C: Ontology for Media Resources 1.0 (2012). http://www.w3.org/TR/mediaont-10/
25. Weibel, S., Kunze, J., Lagoze, C., Wolf, M.: Dublin Core metadata for resource discovery. Technical report 2070-1721 (1998)

Top-K and Skyline Queries

Skyline-Based Feature Selection for Polarity Classification in Social Networks

Fayçal Rédha Saidani[1](✉), Allel Hadjali[2], Idir Rassoul[1], and Djamal Belkasmi[3]

[1] LARI Laboratory, University of Mouloud Mammeri, Tizi-Ouzou, Algeria
faycal.saidani@ummto.dz, idir_rassoul@yahoo.fr
[2] LIAS/ENSMA, University of Poitiers 1,
Avenue Clement Ader, Futuroscope Cedex, France
allel.hadjali@ensma.fr
[3] DIF-FS/UMBB, Boumerdès, Algeria
belkasmi.djamal@gmail.com

Abstract. This paper deals with the feature selection in sentiment analysis for the purpose of polarity classification. We propose a method for selecting a subset of non-redundant and discriminating features, providing better performance in classification. This method relies on the skyline paradigm often used in multi criteria decision and Database fields. To demonstrate the effectiveness of our method with regard to dimensionality reduction and classification rate, some experiments are conducted on real data sets.

Keywords: Feature selection · Sentiment analysis · Skyline queries · Feature weighting

1 Introduction

With the emergence of social networks and more specifically Twitter sphere, the subjective textual data grow exponentially. In front of this important and constant flow of subjective data, it would be interesting to transform this conveyed information into useful knowledge in order to better analyze the needs, experiences and expectations of internet users are very significant at the same time for both customers and professionals. Also, given that social networks are increasingly used as a means of political debate and the ease which opinion are exchanges is such that they arouse the interest of the policies and inspire more research to find the way to integrate them into traditional political polls. Thus, the sentiment analysis, also called polarity classification, aims precisely at disambiguating automatically that opinionated information by translating it by a valence of positive or negative polarity.

Several works have been proposed in the literature for the sentiment analysis. There are two main approaches: (i) those based on lexicons [1, 2] and (ii) those based on machine learning [3]. The former rely on a list (dictionaries, thesaurus) of subjective words either created manually or automatically. If a document includes them, it is considered as subjective document. One of the pioneering works is presented in [4]. The authors proposed a method to estimate the orientation of a word or phrase, by

© Springer International Publishing AG 2017
D. Benslimane et al. (Eds.): DEXA 2017, Part I, LNCS 10438, pp. 381–394, 2017.
DOI: 10.1007/978-3-319-64468-4_29

comparing its similarity to seed words using Pointwise Mutual Information (PMI) and Information Retrieval. Other works focused on using adjectives or verbs as pointers of the semantic orientation [1, 5].

The machine learning based approaches use different types of classifiers such as Support Vector Machines (SVM) [6] and Naïve Bayes (NB) [7] which are trained on a given data set of documents. In order to process the latter, text features are extracted and machine learning algorithm is trained with a labelled data set in order to construct in a supervised manner the classification model. The main strength of these approaches lies in their ability to analyse the text of any domain and produces classification models that are tailored to the problem. In addition, they can be applied to multiple languages [3].

Lexicon-based approaches suffer however in most cases from the lack of specialized lexicons for specific tasks that generally require a vocabulary adapted to the field of predilection, unlike machine learning methods that rely mainly on data without considering human interaction. Therefore, they constitute a good alternative for our research task.

However, the performance of machine learning algorithm is heavily dependent on the features (terms) used during the learning phase; therefore a preliminary task of feature selection is generally essential before any learning. The complexity of selection depends on the type of objective to be achieved. Indeed, simple thematic classification generally requires few evaluation criteria or preferences to be taken into consideration. But, for more complex tasks such as sentiment analysis where several factors must be taken into account, it is generally difficult to obtain a relevant subset based solely on a single measure whereas each of them could bring some benefit to others. Unfortunately with feature selection methods commonly used in sentiment analysis, very few of them take into account the interactions between these different factors and thus tend to select features with redundant rather than complementary information.

In this paper, we investigate a novel idea that consists of leveraging skyline paradigm to optimize the feature selection process for polarity classification. This allows selecting a subset of Pareto optimal features that are not dominated by any other feature. In summary, the main contributions made in this paper are:

- A set of relevant metrics for the purpose of features selection is investigated.
- Based on these metrics (or dimensions), a feature skyline is computed that significantly reduces the features space. By this way, irrelevant features are ruled out and then reducing the noisy in the classification process.
- A set of experiments are conducted to show the interest of our method with regard to dimensionality reduction rate.

The paper is structured as follows. Section 2 recalls some basics notions on feature selection, skyline queries and presents the problem of interest. In Sect. 3, we provide some related work. In Sect. 4, we present our skyline-based method to features selection. Section 5 discuss the experimental results and finally we conclude the paper.

2 Preliminaries and Problem Statement

This section describes the feature selection problem and some notions on skyline queries that are necessary to our proposal. Finally, we present the problem statement.

2.1 Feature Selection Problem

Feature selection is a search process that aims at finding a relevant feature subset from an initial set. The relevance of a feature subset always depends on the objectives and criteria of the problem to be solved. The problem can be defined formally as follows:

Definition 1. *Let X be the original set of features, with cardinality $|X| = n$, and Let J(.) be an evaluation measure to be optimized defined as J: $X' \subseteq X \to \Re$. The purpose of the feature selection is to find $X' \subset X$, such that $J(X')$ is maximum* [8].

To achieve sentiment classification using machine learning, three major steps are needed (as shown in Fig. 1). The first step is to extract an initial set of features which will serve as an input to the classifier, so, it is difficult to learn good classifiers without removing a number of irrelevant features. Thus, feature selection attempts to select the minimally sized subset in a way that the classification accuracy does not decrease. The work presented in this paper seeks to improve this step.

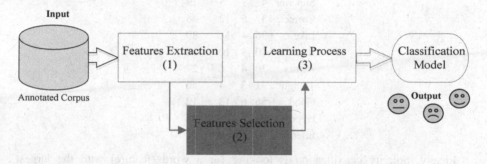

Fig. 1. Overview of sentiment classification steps.

2.2 Skyline Queries

Skyline queries [9] are example of preference queries that can help users to make intelligent decisions in the presence of multidimensional data where different and often conflicting criteria must be taken. They rely on Pareto dominance principle which can be defined as follows:

Definition 2. *Let U be a set of d-dimensional points and u_i and u_j two points of U. u_i is said to dominate in Pareto sense u_j (denoted $u_i \succ u_j$) if u_i is better than or equal to u_j in all dimensions and (strictly) better than u_j in at least one dimension. Formally, we write:*

$$u_i \succ u_j \Leftrightarrow (\forall k \in \{1,\ldots,d\}, u_i[k] \geq u_j[k]) \wedge (\exists l \in \{1,\ldots,d\}, u_i[l] > u_j[l]) \quad (1)$$

where each tuple $u_i = (u_i[1], u_i[2], u_i[3], \ldots, u_i[d])$ with $u_i[k]$ stands for the value of the tuple u_i for the attribute A_k.

Definition 3. *The skyline of U, denoted by S, is the set of points which are not dominated by any other point.*

$$u \in S \Leftrightarrow \nexists u' \in U, u' \succ u \quad (2)$$

Example 1. To illustrate the skyline, let us consider a database containing information about supposedly subjective words (features) as shown in Table 1, and for each word, four weighting metrics are used to indicate its importance in documents. The list of words includes the following information: term, term frequency (*tf*), Relevance Frequency (*rf*), Term Frequency-Inverse Document Frequency (*tf-idf*) and Multinomial Z Score (*zd*).

Table 1. List of words

Term	tf	rf	tf-idf	zd
Wrong	32	5	10	35
Superior	41	7	5	19
Correct	37	5	12	45
Like	36	4	11	39
Bad	40	8	10	18
Actors	30	4	6	27
Tonight	31	3	4	56
Good	36	6	13	12
Hope	33	6	6	95
Inferior	40	7	9	20

Ideally, polarity classification is looking for a word (feature) with the largest Relevance Frequency and Term Frequency-Inverse Document Frequency (Max *rf* and Max *tf-idf*), ignoring the other pieces of information. Applying the traditional skyline on the word list shown in Table 1 returns the following terms: Bad, Good. As can be seen, such results are the most relevant terms (see Fig. 2).

2.3 Problem Statement

Given that it is not obvious with classical features selection methods to consider several and conflicting weighting metrics in order to refine the selection for sentiment analysis. We are interested here in the way of identifying a discriminant features subset using the skyline paradigm for improving the performance of the classification step in terms of f-measure.

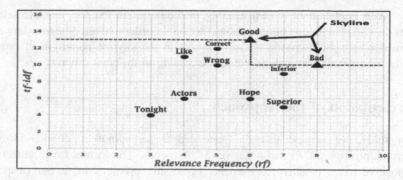

Fig. 2. Skyline of words

3 Related Work

We review here the main approaches proposed in the literature about features selection for the sentiment analysis context. In approaches based on machine learning techniques [10, 11], the performance of the algorithms strongly depends on the features used in the learning task. Thus, the presence of redundant or irrelevant features can have a significant impact on the performance of the approaches. In [10] the authors propose the use of an Entropy Weighted Genetic Algorithm that incorporates the information gain and heuristics. The results show that this combination strongly improves the accuracy of the sentiment analysis. The study in [11] evaluates a set of feature selection and weighting methods for sentiment analysis. It introduces two new methods, Senti-WordNet Subjectivity Score and SentiWordNet Proportional Difference. The former allows distinguishing between objective and subjective terms, while the latter can incorporate distinctive class capabilities for features selection.

The work done in [12] presents a new feature selection schemes that use content and syntax model to automatically learn a set of features in a review document. The features obtained yield competitive results in a maximum entropy classifier. In the same vein, in [13] explicit topic models are established to filter words and extract the training attributes of each product. Finally, several SVM classifiers are constructed to train the selected attributes to detect the corresponding implicit features for Chinese reviews. A subjectivity classification by considering improved Fisher's discriminate ratio based on feature selection method is performed in [14].

More recently, [15] emphasizes the problem of sentiment classification on imbalanced data and proposes a boundary region cutting algorithm that is only suitable for two categories. The feature selection adjustment corrects the classifier bias by assigning a higher weight to features that is more important in minority class to solve the imbalance problem. In [16] several individual feature lists obtained by different feature selection methods are aggregated to improve sentiment classification. Let us also mention the work of [17] which proposes an optimized feature reduction that incorporates information gain and genetic algorithm as feature selection. The results show a clear improvement in terms of precision.

The main limitation of almost existing methods lies in the fact that the selected features depend on a unique optimization criterion. In order to overcome this drawback, we consider here multi-criteria decision based on the skyline paradigm in order to highlight the most discriminating features according to different criteria.

4 Presentation of the Approach

In this study, we assume that the annotated target corpus is divided into two sub-corpus: positive (*SCP*) and negative (*SCN*) corpus. The former (resp. latter) contains positive (resp. negative) documents expressing opinions in favour (resp. disfavour) of the matter of interest. By this way, one can perform a better pruning on the search space related to the features documents.

The selection approach relies on three metrics (i.e., *G_idf*, *Odds Ratio* and *POS weight*). The two first metrics are computed for both positive and negative sub-corpus in order to characterize which feature best reflects the specificity of each class in regard to the other. Each feature will be then associated with a vector of five measures (G_Idf^+, G_Idf^-, Orr^+, Orr^-, Pos) where *Orr* stands for Odds Ratio metric. Based on this vector, one can compute the skyline to keep only the most interesting features (in the sense of Pareto). The overview architecture of the proposed approach is illustrated in Fig. 3.

4.1 Weighting Metrics

We describe first the set of metrics used in for the features selection method.

Global tf.idf Weight

For the first metric (i.e., Global tf.idf denoted *G_idf*), we rely on the Term Frequency-Inverse Document Frequency weight (tf-idf). The principle is to give importance to the more specific features of a document. So, to evaluate the representativeness of a feature in the corpus and not only in a document, we have to apply an aggregation function (e.g., the arithmetic mean) over the set of individual tf-idf weights of features computed on each document. Let C be a corpus of n documents and $f \in F$ a set of features, one can write:

$$G\_idf_f = \frac{1}{n}\sum_{i=1}^{n} tfidf_{f,i} \tag{3}$$

In the following, G_idf^+ (resp. G_idf-) stands for the aggregate score G_idf on SCP (resp. SCN) sub-corpus.

Odds Ratio Weight

Odds Ratio (*Orr*) gives a positive score to features that occur more often in one class than in the other, and a negative score if it occurs more in the opposite class [18]. Formally, *Orr* measure is given by (where $f \in F$):

$$\text{Orr}_f = \log\left(\frac{p(1-q)}{q(1-p)}\right) \tag{4}$$

p (resp. q) the probability of class to be predicted (resp. the opposite class) given the feature f. Following [18], a *null* value of the *Orr* parameter means that the feature belongs to the neutral polarity. As above, Orr^+ (resp. Orr^-) stands for the *Orr* weight on SCP (resp. SCN) sub-corpus.

POS Weight

As stressed in [12], an in-depth linguistic analysis based on syntactic relations in a sentence can be useful for sentiment analysis model. For instance, collocations that contain mostly adjectives-adverbs or verbs-negation features are considered more

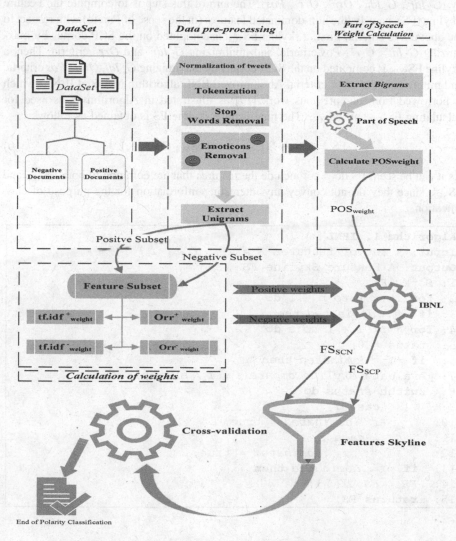

Fig. 3. Architecture of the skyline-based feature selection

sentiment bearing. Thus, to be more precise in our process, we have chosen to exploit this semantics by checking and accounting for linguistic collocation based on Part Of Speech (POS) using *Stanford Pos Tagger*. To this end, we introduce the following formula for measuring the POS weight (where *n* represents the number of documents in the target corpus):

$$POS_f = \log\left[1 + \sum_{i=1}^{n} term\_frequency_f\right] \tag{5}$$

4.2   Feature Skyline

As mentioned above, each feature is associated with a vector of five weights expressed by (G_Idf^+, G_Idf^-, Orr^+, Orr^-, Pos). The aim of this step is to compute the Feature Skyline (FS), namely, the non-dominated features (in the sense of Pareto) with respect to the above five criteria. First, a skyline FS_{SCP} is computed on the SCP corpus by maximizing G_Idf^+, Orr^+, Pos criteria, and minimizing G_Idf^- and Orr^- criteria. Then, a skyline FS_{SCN} is computed on the SCN corpus by maximizing G_Idf, Orr^-, Pos criteria, and minimizing the other criteria. An improved BNL algorithm (denoted IBNL) which is borrowed from our previous work [19] as illustrated in Algorithm 1, is used for calculating FS_{SCP} and FS_{SCN}. The final feature skyline FS is obtained as follows:

$$FS = (FS_{SCP} \cup FS_{SCN}) - (FS_{SCP} \cap FS_{SCN}) \tag{6}$$

As it can be seen, FS does not include the features that are common to both FS_{SCP} and FS_{SCN} since they do not convey any interesting information for the purpose of classification.

```
Algorithm 1. IBNL
Input: A set of features F = {f₁, f₂,…, fₙ}
Output: A Feature Skyline FS
1: Sort (F)
2: for i := 1 to n - 1 do
3: if ¬fᵢ .dominated then
4: for j := i + 1 to n do
5: status = 0;
6:   if ¬fⱼ .dominated then
7:   evaluate SkylineCompare(fᵢ, fⱼ, status);
8: switch status do
9:        case 1
10:        fᵢ .dominated = true;
11:         case 2
12:             fⱼ .dominated = true;
13:   if ¬fᵢ .dominated then
14:   FS = FS ∪{fᵢ};
15:   returns FS;
```

SkylineCompare is a function that evaluates the dominance, in the sense of Pareto, between f_i and f_j on all skyline dimensions and returns the result in *status*. It may be equal to: 0 if $f_i = f_j$, 1 if $f_i > f_j$, 2 if $f_i < f_j$ and 3 if they are incomparable.

5   Experimental Study

5.1   Dataset and Experimental Setup

The experiments are conducted on two different data sets to evaluate the proposed approach. We focus mainly on the two following criteria: (i) dimensionality reduction rate and (ii) classification rate. In Table 2, we present the distribution of documents in datasets. The First dataset "Polarity Dataset V2.0"[1] which was extracted from the Internet Movie Database (IMDB), contains 1000 positive and 1000 negative movie reviews. The second dataset "The Twitter Sentiment Analysis Dataset" contains at the origin 1 578 627 classified tweets; each row is marked as 1 for positive sentiment and 0 for negative sentiment. A reduced version was proposed in Lightside[2] workbench and contains 10 662 classified tweets distributed uniformly between the two classes. A 10-fold cross-validation procedure is utilized, which means that the original data set is randomly divided into 10 equal-sized subsamples. For each time, a single sub-sample is used for validation, whereas the rest are kept for training. All algorithms were implemented with Java 1.8 64 bits. The Weka API [20] has been used with default settings for classifiers algorithm and all experiments were conducted on a Windows 8.1 system with Intel core i7 6500U @2.5 GHz CPU 8 GB RAM.

In order to evaluate the effectiveness of the proposed feature selection method, the performances of Information Gain (IG) [21] and Pairwise Mutual Information (PMI) [22] are taken into account for the purpose of comparison. IG has been used in sentiment analysis in many works [10, 23–25] and its effectiveness has been proven. It is measured by the reduction of the uncertainty in identifying the class attribute when the value of the feature is known. It tries to find out how well each single feature separates the given dataset. On the other hand, PMI is a measure of association used in information theory and statistics, it measures the strength of association between a feature and positive or negative documents in sentiment analysis [5, 8, 23, 26]. Usually, it is used to calculate the semantics orientation of a given feature (word) by comparing its similarity to a positive/negative reference word using co-occurrence [4].

In order to measure the strength of the proposed methods, several classifiers implemented in Weka [20] were tried for the experiments. Finally, two classification algorithms were selected in relation to the stability of their results. The first is Naïve Bayes (NB) [7], which is based on the probabilistic information about text features. The second algorithm is Support Vector Machine (SVM) [6], which represents the features as points in space (feature space).

---

[1] https://www.cs.cornell.edu/people/pabo/movie-review-data/.
[2] http://ankara.lti.cs.cmu.edu/side/download.html.

The well-known precision and recall are used to compute the classification effectiveness in terms of F-measure for each category. It is computed as follows:

$$F - measure = 2 \times \frac{precision \times recall}{presicion + recall} \qquad (7)$$

Table 2. Distribution of documents in dataset

# No	Dataset description	# Documents	# Class	# Words
D1	Polarity dataset V2.0	2000	2	7 956
D2	The twitter sentiment analysis dataset	10662	2	20 392

5.2 Experimental Results

The first conducted experiments focus on reduction ratio parameter. Table 3 shows the results of dimensionality reduction rate obtained by the proposed Skyline-Based Feature Selection method (called SBFS).

Table 3. Dimensionality reduction rate with SBFS

Dataset	# Initial features	# Features after selection	Reduction rate (%)
D1	7 956	3160	60.28
D2	20 392	7 053	65.41

In order to study the impact of the reduction rate on the classification performance, we have chosen to align the reduction rate of IG and PMI methods with the rate obtained using SBFS method. Indeed, a reduced list of features can really be obtained with our method contrary to the IG and PMI which only a ranked feature list can be provided. For example in D1 dataset, a reduction rate of around 60% was obtained with our approach, so to evaluate its effectiveness we applied the same reduction rate to the top ranked features of IG and PMI methods. Figure 4 shows the classification accuracy according to the precision rate applied with the three features selection (IG, PMI, SBFS) using SVM and NB classifiers over the two corpora D1, D2.

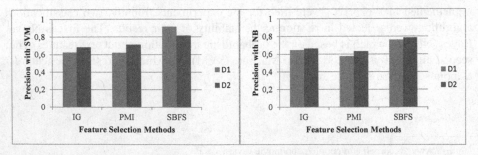

Fig. 4. Precision rate with IG, PMI, and SBFS methods

As shown in Fig. 4, the highest classification performance is achieved by SBFS with all datasets. The likely reason for this result is that SBFS is based on efficient multi-criteria optimization technique (i.e., Pareto optimality) applied to each polarity class.

For a better evaluation of the results, we propose to extend the experiments by considering different subsets of the initial features sets (Table 2). Different proportions ranging from 20% to 80% are taken into account in the evaluation as shown in Fig. 5. The predictive performance and usefulness of each feature subset are evaluated by NB classifier in term of F-measure. We select the subset having obtained the best result and will therefore deduct its reduction rate.

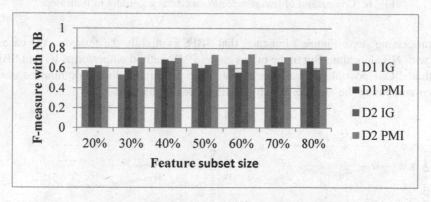

Fig. 5. Comparison of f-measure for IG and PMI on various subsets size

Figure 5 shows the overall efficiency of the each method (IG, PMI) during the variation of the size in D1 and D2 datasets. As it can be seen, the best F-measure rates across different subsets are achieved at 50% and 60% of saved features respectively for D1 and D2 with IG feature selection. For PMI method, the best F-measure is achieved when 40% and 60% of features are selected respectively for D1 and D2. These results correspond to reduction rates equivalent to 50% and 60% respectively for IG and PMI with D1 and 40% for both methods with D2. One can observe that our method SBFS is still better than IG and PMI methods.

As shown in Fig. 6, the best classification rates across the two dataset are achieved with SBFS. However, with small dataset, it is slightly difficult for SBSF to achieve a high reduction rate; this is noticeable in negative impact on classification performance, but with the larger dataset, the reduction rate achieved by SBFS is significantly improved especially with SVM algorithm.

As pointed out in [10], it is possible to obtain good classification accuracies while using less than 36% of the initial features space. This applies to SBSF as illustrated in Fig. 6 (for D2 corpus).

On the other hand, we have also conducted some experiments to analyze the required time in order to perform the learning phase that is represented in our case by the Cross-Validation including features selection step (without taking account the

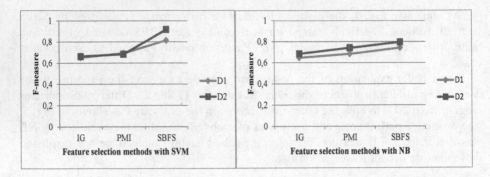

Fig. 6. Comparison of feature selection methods according to F-measure

preprocessing step). Figure 7 indicates that SBFS is slightly the fastest in the case of D2 with NB algorithm. But in the others cases, PMI method outperforms IG and SBFS methods. One possible explanation is that SBFS requires a double dominance check when computing the skyline, which is a time consuming task.

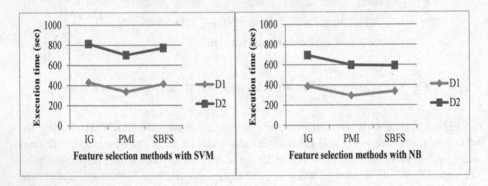

Fig. 7. Execution time (sec) using SVM and NB algorithms

6 Conclusion

In this paper, we proposed an approach that deals with the feature selection for polarity classification in sentiment opinions. The key concept of the approach is the feature skyline. Namely, the subset of features that are the most interesting (in the dominance Pareto sense) with respect to a set of relevant criteria suitably chosen. The feature skyline computed allows improving the classification performance in most cases, as shown via the conducted experiments.

As for future work, we plan first to improve the approach by using some advanced optimizations when computing the skyline. One way to improve better the classification is to take into account more criteria (metrics). This can however lead to very large feature skylines which could be inefficient for our purpose. As second perspective, we plan to investigate the feature skyline refinement by introducing an order relation between them thanks to a fuzzy counterpart of Pareto dominance relationship.

References

1. Maks, I., Vossen, P.: A lexicon model for deep sentiment analysis and opinion mining applications. Decis. Support Syst. **53**, 680–688 (2012)
2. Neviarouskaya, A., Prendinger, H., Ishizuka, M.: SentiFul: a lexicon for sentiment analysis. IEEE Trans. Affect. Comput. **2**, 22–36 (2011)
3. Boiy, E., Moens, M.F.: A machine learning approach to sentiment analysis in multilingual Web texts. Inf. Retrieval **12**, 526–558 (2009)
4. Turney, P.D.: Mining the Web for synonyms: PMI-IR versus LSA on TOEFL. In: Raedt, L., Flach, P. (eds.) ECML 2001. LNCS, vol. 2167, pp. 491–502. Springer, Heidelberg (2001). doi:10.1007/3-540-44795-4_42
5. Turney, P.D.: Thumbs up or thumbs down? Semantic orientation applied to unsupervised classification of reviews. In: 40th Annual Meeting on Association for Computational Linguistics, pp. 417–424. ACL, Stroudsburg (2002)
6. Cortes, C., Vapnik, V.: Support-vector networks. Mach. Learn. **20**, 273–297 (1995)
7. Murphy, K.P.: Naive Bayes Classifiers. University of British Columbia (2006)
8. Kumar, V., Minz, S.: Feature selection. SmartCR **4**, 211–229 (2014)
9. Börzsönyi, S., Kossmann, D., Stocker, K.: The skyline operator. In: 17th International Conference on Data Engineering, pp. 421–430. IEEE, New York (2001)
10. Abbasi, A., Chen, H., Salem, A.: Sentiment analysis in multiple languages: feature selection for opinion classification in web forums. ACM Trans. Inf. Syst. **26**, 12 (2008)
11. O'Keefe, T., Koprinska, I.: Feature selection and weighting methods in sentiment analysis. In: Proceedings of the 14th Australasian Document Computing Symposium, Sydney, pp. 67–74 (2009)
12. Duric, A., Song, F.: Feature selection for sentiment analysis based on content and syntax models. Decis. Support Syst. **53**, 704–711 (2012)
13. Xu, H., Zhang, F., Wang, W.: Implicit feature identification in Chinese reviews using explicit topic mining model. Knowl.-Based Syst. **76**, 166–175 (2015)
14. Wang, S., Li, D., Song, X., Wei, Y., Li, H.: A feature selection method based on improved fishers discriminant ratio for text sentiment classification. Expert Syst. Appl. **38**, 8696–8702 (2011)
15. Wang, Y., Li, Z., Liu, J., He, Z., Huang, Y., Li, D.: Word vector modeling for sentiment analysis of product reviews. In: Zong, C., Nie, J.Y., Zhao, D., Feng, Y. (eds.) NLPCC 2014. CCIS, vol. 496, pp. 168–180. Springer, Heidelberg (2014). doi:10.1007/978-3-662-45924-9_16
16. Onan, A., Korukoğlu, S.: A feature selection model based on genetic rank aggregation for text sentiment classification. J. Inf. Sci. **43**, 25–38 (2015)
17. Kalaivani, P., Shunmuganathan, K.L.: Feature reduction based on genetic algorithm and hybrid model for opinion mining. Sci. Program.-Neth. **12**, 15–26 (2015)
18. Shaw, J.: Term-relevance computations and perfect retrieval performance. Commun. Comput. Inf. Sci. **31**, 491–498 (1995)
19. Belkasmi, D., Hadjali, A.: MP2R: a human-centric skyline relaxation approach. In: Andreasen, T., et al. (eds.) Flexible Query Answering Systems 2015. AISC, vol. 400, pp. 227–241. Springer, Cham (2016). doi:10.1007/978-3-319-26154-6_18

20. Hall, M., Frank, E., Holmes, G., Pfahringer, B.: The WEKA data mining software: an update. ACM SIGKDD Explor. Newslett. J. **11**, 10–18 (2009)
21. Yang, Y., Pedersen, J.O.: A comparative study on feature selection in text categorization. In: ICML, vol. 97, pp. 412–420 (1997)
22. Church, K.W., Hanks, P.: Word association norms, mutual information, and lexicography. Comput. Linguist. **16**, 22–29 (1990)
23. Mukras, R., Wiratunga, N., Lothian, R.: Selecting bi-tags for sentiment analysis of text. In: Bramer, M., Coenen, F., Petridis, M. (eds.) Research and Development in Intelligent Systems XXIV, pp. 181–194. Springer, London (2008). doi:10.1007/978-1-84800-094-0_14
24. Banea, C., Mihalcea, R., Wiebe, J.: Sense-level subjectivity in a multilingual setting. Comput. Speech Lang. **28**, 7–19 (2014)
25. Deng, Z.W., Luo, K.H., Yu, H.L.: A study of supervised term weighting scheme for sentiment analysis. Expert Syst. Appl. **41**, 3506–3513 (2014)
26. Liu, Y., Jin, J., Ji, P., Harding, J.A., Fung, R.Y.: Identifying helpful online reviews: a product designer's perspective. Comput. Aided Des. **45**, 180–194 (2013)

Group Top-k Spatial Keyword Query Processing in Road Networks

Hermann B. Ekomie[1], Kai Yao[1], Jianjun Li[1(✉)], Guohui Li[1], and Yanhong Li[2]

[1] School of Computer Science and Technology,
Huazhong University of Science and Technology, Wuhan, China
ekomie@yahoo.fr, {kaiyao,jianjunli,guohuili}@hust.edu.cn
[2] Department of Computer Science, South-Central University for Nationalities,
Wuhan, China
yhli@hust.edu.cn

Abstract. With the proliferation of geo-positioning and geo-tagging, spatial keyword query (SKQ) processing is gaining great concern recently. Typically, a top-k spatial keyword query returns the k best spatio-textual objects ranked according to their proximity to the query location and relevance to the query keywords. While there has been much work devoted to top-k spatial keyword processing, most of them are either focused on single query or only suitable for Euclidean space. In this paper, we take the first step to study the problem of multiple query points (or group) top-k spatial keyword query processing in road networks. We first propose a basic group query-processing algorithm by using three distinct index structures. Next, we propose another more efficient algorithm based on the concept of Minimum Bounding Rectangle (MBR), which can significantly reduce the objects to be examined and thus achieve higher performance. Finally, extensive experiments using real data sets are conducted to validate the effectiveness of the proposed algorithms.

1 Introduction

With the increasing popularization of geo-positioning technologies, there is a rapidly growing amount of spatio-textual objects collected in many applications such as location based services and social networks, in which an object is described by its spatial location and a set of keywords. This development calls for techniques that enable the indexing of data that contains both textual descriptions and geo-locations, so as to support the efficient processing of spatial keyword queries (SKQ) that consider both spatial proximity and textual relevance between the query client and geo-textual objects.

The work was partially supported by the State Key Program of National Natural Science of China under Grant No. 61332001, the National Natural Science Foundation of China under Grant Nos. 61300045, 61572215, 61672252, 61602197, and the Fundamental Research Funds for the Central Universities, HUST-2016YXMS076, HUST-2016YXMS085.

© Springer International Publishing AG 2017
D. Benslimane et al. (Eds.): DEXA 2017, Part I, LNCS 10438, pp. 395–408, 2017.
DOI: 10.1007/978-3-319-64468-4_30

As one of the most important kind of SKQ queries, top-k spatial keyword queries return the k best spatio-textual objects ranked in terms of both spatial proximity to the query location and textual relevance to the query keywords [1–3]. The current approaches for processing top-k spatial keyword queries are restricted to the one query point scenario. However, in many cases, multiple query points need to be considered and the result should take into account each query location and textual description. For example, consider a company of travelers who visit a city and lodge in different hotels. These people may want to visit an attraction of the city, which is not far from their locations and also relevant to their personal different preferences (e.g., someone want to visit "museum" and "library", while others would like to visit "shopping mall" and "amusement park"). This problem is called the multiple query points (or group) top-k spatial keyword query processing. The most relevant study related to this problem is GLkT [4] (Group Location-aware top-k Text retrieval), which enables multiple users to get the k spatio-textual objects that best match their arguments. However, GLkT only considers query processing in Euclidean space. In real life situations, objects are usually limited in road networks (also called *constraint-based* [5–7] environment), where the distance between objects is determined by the connectivity of the network, rather than the objects' coordinates in Euclidean space. Hence, it is desirable to study the problem of processing group top-k spatial keyword queries in road networks, and this problem is quite different from and more difficult than that in Euclidean space.

In this paper, we address, for the first time, the challenging problem of processing Group Top-k Spatial Keyword Queries in road Networks (GTkSKQN). Specifically, given a set of spatio-textual objects and several query points having different locations and arguments, a GTkSKQN query returns the k best objects in terms of both spatial proximity and textual relevance to the query points. For example, given the road network depicted in Fig. 1 and the

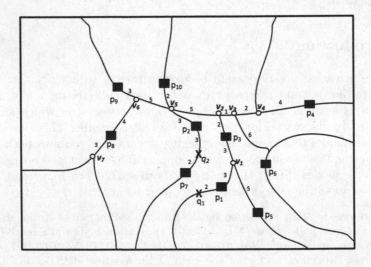

Fig. 1. A road network

corresponding data presented in Table 1, suppose two friends are making a query to meet at a restaurant, q_1 adds *"pizza"* and *"friendly"* to the description of the restaurant, while q_2 adds *"steak"* and *"friendly"* to the description of his query. Assuming that both queries were made and processed in Euclidean space, by neglecting the road network in Fig. 1 and considering the objects in a 2D space, object p_7 will be the best suit for this group query. But if we consider this group query in the road network environment, everything changes since users can only reach objects through predefined roads.

Table 1. Objects textual description

Set P	Set D	Objects with their descriptions
P1 (restaurant)	d1 (pizza)	P1 (d3, d8)
P2 (restaurant)	d2 (steak)	P2 (d3, d4)
P3 (supermarket)	d3 (friendly)	P3 (d6)
P4 (cinema)	d4 (western)	P4 (d4, d5)
P5 (cafe)	d5 (family)	P5 (d4)
P6 (bar)	d6 (shopping)	P6 (d9, d10)
P7 (restaurant)	d7 (calm)	P7 (d3, d10)
P8 (shopping mall)	d8 (affordable)	P8 (d8)
P9 (game store)	d9 (cozy)	P9 (d5)
P10 (hotel)	d10 (cheap)	P10 (d10)

To address the GTkSKQN query processing problem, we first propose a basic algorithm, which firstly computes the spatio-textual relevance score of each object located between the query points and then the remaining part of the network. After that, we combine all the results, from which we select the best k objects as the final results. Next, in order to achieve a higher performance, we propose another more efficient algorithm, based on the concept of Minimum Bounding Rectangle (MBR), which can significantly reduce the objects to be examined. In summary, the main contributions of this paper are:

- We take, to our best knowledge, the first step to address the GTkSKQN query processing problem and propose a basic algorithm by using a classical selecting method to choose the best k objects.
- Based on the concept of Minimum Bounding Rectangle, we propose a more efficient algorithm for processing GTkSKQN queries. By treating the whole query set as a query unit, we can increase the performance of the algorithm and reduce the number of node accesses significantly.
- Extensive experiments using real datasets demonstrate that our approaches are efficient in terms of both runtime and I/O cost.

The remainder of the paper is organized as follows. Section 2 gives a formal definition of the GTkSKQN query processing problem. Sections 3 and 4 present our two solutions: a basic algorithm and an enhanced algorithm, respectively. Section 5 provides the experimental results. Section 6 briefly describes related work and Sect. 7 draws a conclusion.

2 Preliminaries

We consider a road network presented as a graph $G = (V, E, W)$, where V is the set of vertices, E is the set of edges and W is the set of distance cost associated with each edge. A vertex $v \in V$ is the intersection of any two or more roads (lines) in the network. An edge $(u, v) \in E$ is any road segment made of two adjacent vertices u and v. The distance cost $w \in W$ represented by $\|u, v\|$ is the cost or the length of edge (u, v). We use $P = \{p_1, p_2, \ldots, p_m\}$ to denote a set of points of interest (PoI) and $D = \{d_1, d_2, \ldots\}$ to denote the set of textual description, P and D are related such that every element $p_j \in P$ is assigned at least one element $d_k \in D$ which becomes the textual description of p_j and eventually be part of the relevant criteria to consider during query processing. Each $p_j \in P$ is defined by its location $p_j.l$ on the graph and its textual relevance $p_j.d$. The set of spatio-textual queries is $Q = \{q_1, q_2, \ldots, q_n\}$, where each $q_i \in Q$ is represented by $q_i = (q_i.l, q_i.d)$, with $q_i.l$ being the query location and $q_i.d$ the query keywords. We assume that all queries and points of interest are on the set of edges E of the network. Moreover, to avoid confusion, neither q_i nor p_j shall be located on a vertex.

Problem Statement: Given a set of spatio-textual query Q on the road network G, and a set of spatio-textual objects P, our goal is to return the k best spatio-textual objects, in terms of shortest path and textual relevance, in decreasing order of score.

Definition 1 (Network proximity δ). *Given a query q_i with location $q_i.l$ on the graph and spatio-textual object p_j with location $p_j.l$. The network proximity between these two points is defined as:*

$$\delta(p_j.l, q_i.l) = \|p_j.l, q_i.l\| \tag{1}$$

Since we are considering a graph and we want the closest object possible in terms of distance, the network proximity is going to be the shortest path between a query point and the point of interest.

Definition 2 (Textual relevance θ) [6]. *It gives the textual importance of an object of interest to a query. Given a query textual description $q_i.d$ and an object textual description $p_j.d$. We define the textual relevance as:*

$$\theta(p_j.d, q_i.d) = \frac{\sum_{t \in q_i.d} w_{t,p_j.d}.w_{t,q_i.d}}{\sqrt{\sum_{t \in p_j.d}(w_{t,p_j.d})^2 . \sum_{t \in q_i.d}(w_{t,q_i.d})^2}} \tag{2}$$

where the weight $w_{t,p_j.d} = 1 + \ln(f_{t,p_j.d})$ with $f_{t,p_j.d}$ being the number of occurrences (frequency) of t in $p_j.d$; and the weight $w_{t,q_i.d} = \ln(\frac{|P|}{df_t})$, where $|P|$ is the number of objects in the collection and df_t is the number of objects that contains t in their description (document frequency). Many methods can be used to determine the textual relevance of $q_i.d$ and $p_j.d$. In our work we choose to use the cosine similarity between the vectors composed by the weight of the description words in $q_i.d$ and $p_j.d$, which is proposed by Zobel and Moffat [8].

Definition 3 *(Spatio-textual score τ). This is the network proximity and textual relevance computed together to determine a numerical score that will help evaluate the importance of a specific object to a query as compare to another object in view of making the best choice, in our case the object with the highest spatio-textual score is selected. Given a query q_i and an object p_j the spatio-textual relevance score τ is defined as:*

$$\tau(p_j, q_i) = \frac{\theta(p_j.d, q_i.d)}{1 + \alpha \times \delta(p_j.l, q_i.l)} \tag{3}$$

where $\alpha \in \mathbb{R}^+$ is a parameter used to define the importance between spatial and textual components. For example, if $\alpha = 0$ only textual relevance is considered, if $\alpha > 1$ the network proximity importance increases over textual relevance and the greater α the more important network proximity gets.

Definition 4 *(Group spatio-textual score T). This is the overall score of the set of query Q relative to a single point in the set of points of interest P. It is used to determine the most relevant point to the set Q, and is defined as:*

$$T(p_j, Q) = \sum_{q_i \in Q} \tau(p_j, q_i) \tag{4}$$

3   The Basic Algorithm

In this section, we present a basic algorithm for processing the GTkSKQN queries. The idea of our basic algorithm is as follows. For each query point $q_i \in Q$, we first compute the ranked list of all objects according to the ranking function τ. Then, we combine them to gain the comprehensive relevance score. Afterwards, we use a classical selecting algorithm to get the k most relevant spatio-textual objects from the dataset.

Rocha-Junior and Nørvåg [6] proposed an efficient algorithm to process top-k spatial keyword queries in road networks. We can utilize it to gain the spatio-textual relevance score for each query point. The basic indexing architecture combines a spatio-textual index, such as IR-tree, and the road network framework proposed by Papadias et al. [9]. The framework proposed by Papadias et al. permits starting a query at any location of the road network, while the spatio-textual index permits finding the objects in a given spatial region relevant for the query keywords. Figure 2 presents our indexing architecture, which consists

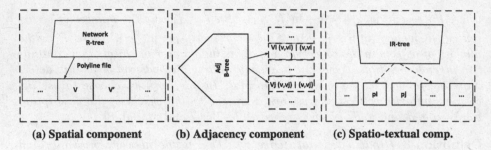

Fig. 2. Indexing architecture

of three components: (1) The spatial component is used to identify the road segment in which the query point lies; (2) The adjacency component permits retrieving the adjacent vertices of a given vertex, and is used to traverse the network; and (3) The spatio-textual component stores the objects. Moreover, we use the Dijkstra algorithm [10] to compute the shortest path between query points and different points of interest.

Algorithm 1 gives the pseudo-code of our basic algorithm, which takes the set of queries Q as input, and outputs the top-k spatio-textual objects in decreasing order of score T. At the beginning, a heap that stores the k best element is created. E stores the score T of the k-th best elements and is set to be 0 initially.

Algorithm 1. The Basic Algorithm

---

1 **Input:** A query set on road network $Q = (q_1, q_2, \ldots, q_n)$
2 **Output:** Top-k objects in decreasing order of score T
3 **begin**
4   Maxheap $H^{Q.k} = \varnothing$;
5     $E = 0$; // stores k scores in $H^{Q.k}$;
6   **foreach** q_i **do**
7        $(u, v) \leftarrow$ network edge in which q_i lies;
8   Find all vertices between any two adjacent query points;
9   **foreach** q_i **do**
10        $C \leftarrow findcandidates((u, v), q_i.d, E))$;
11   mark (u, v) as visited;
12   compute $\tau(p_j, q_i)$;
13   compute $T(p_j, Q) = \sum_{q_i \in Q} \tau(p_j, q_i)$;
14   update $H^{Q.k}$ (and E) with p_j;
15   **while** $(\sum_{q_i \in Q} \tau(p_j, q_i) > E)$ **do**
16   **foreach** *non visited adjacent edge* $(u, v) \in V$ **do**
17           $C \leftarrow findcandidates((u, v), q_i.d, E)$;
18   mark (u, v) as visited;
19   **return** $H^{Q.k}$;

The algorithm starts by finding the route that joins all the query location points $q_i.l$, then locates all the vertices on that route (line 6–8) by using the B-tree adjacency index component. The road segment on which each $q_i.l$ query lies is identified by the R-tree in the spatial component, and then the closest point p_j located on the route is found and τ is computed for each q_i (line 12). The spatio-textual objects are stored by the spatio-textual component using an IR-tree and objects of interest are maintained in a heap where the best objects are the ones with the highest score T. The algorithm stops when the remaining objects cannot have a better score than the scores of the k-th object already found, or the entire network has been expanded.

Now, we use an example to illustrate our basic algorithm, based on the road network given in Fig. 1 and the data presented in Table 1. To simplify presentation, we assume $k = 1$ and for each occurrence of a query keyword in the description of an object $p.d$, the textual relevance θ should be incremented by 0.25, with $\theta_{max} = 1$ and $\alpha = 1$. Our basic algorithm starts by locating all the query points by accessing the R-tree of the network, in our case two query points, q_1 with textual description $q_1.d$ ("pizza", "calm") and q_2 with textual description $q_2.d$ ("steak", "friendly"), are returned. Then, by using the B-tree adjacency component, all the vertices between q_1 and q_2 are located, which are (v_1, v_2, v_5). After that, we first consider q_1. The closest object to q_1 is p_1 and $||q_1.l, p_1|| = 2$. According to Table 1, $p_1.d = ($ "friendly", "affordable") and $q_1.d$ query keywords are ("pizza", "calm"), the edge in which p_1 lies is marked as visited and the score $\tau(q_1, p_1)$ is computed as $\frac{0.25}{1+2} = 0.083$. The same process is conducted for p_3. However, the query keywords $q_1.d$ do not appear in $p_3.d$, we have $\theta = 0$ and $\tau(q_1, p_3) = 0$. The algorithm proceeds to the next point p_2, where $\tau(q_1, p_2) = \frac{0.25}{1+18} = 0.013$. After that, q_2 is selected. Following a similar way, we have $\tau(q_2, p_2) = \frac{0.25}{1+3} = 0.062$, $(q_2, p_3) = \frac{0}{1+13} = 0$ and $(q_2, p_1) = \frac{0.25}{1+19} = 0.012$. The aggregate score of p_1 is then computed as $T(p_1, Q) = 0.083 + 0.012 = 0.095$, E is updated to $E = 0.095$ and p_1 inserted into $H^{Q.k}$. Similarly, the aggregate score of p_2 is computed as $T(p_2, Q) = 0.013 + 0.062 = 0.075$, which is smaller than the current value of E. After having expanded all the objects located on the route between q_1 and q_2, the algorithm scans for other points. The point p_7 has one word in its description that occurs in $q_1.d$ and $q_2.d$. The algorithm computes $\tau(q_2, p_7) = \frac{0.25}{1+2} = 0.083$ and $\tau(q_1, p_7) = \frac{0.25}{1+23} = 0.010$, the aggregate score of p_7 is $T(p_7, Q) = 0.083 + 0.010 = 0.093 < E$. Therefore, p_1 is maintained in $H^{Q.k}$. Since there is no other possible object with an aggregate score $T > E$, the algorithm stops and returns p_1 as the top-1 object.

4   The MBR-Based Algorithm

Granting the fact that the basic algorithm can process GTkSKQN queries, it is inefficient and may produce a considerable amount of candidates. For each $q_i \in Q$, it requires retrieving all the spatio-textual objects and computing relevance score (may result in high I/O cost), without any pruning. Also the search space of each query point expands towards all directions. This may include too many candidate objects and cause unnecessary network distance computation. In order to address these limitations, we propose a novel algorithm based on the concept of minimum bounding rectangle, which can solve the problem by a single traversal. The main idea is to treat the whole query set as a query unit, and then we can use it to prune the search space. As shown in Fig. 3, we firstly compute the minimum bounding rectangle for query set, and then use this rectangle to proceed the query process. Specifically, starting from the root of the IR-tree for dataset D, we first visit those nodes that may contain candidate objects. For example, if the distance between $R1$ and MBR is too far, $R1$ might be discarded without being truly visited. In order to control expansion directions, the network distances to each query point from those spatio-textual objects can be computed by using the A^* algorithm [11].

Definition 5 ($\delta'(e,MBR)$). *Given an object e and the MBR of a query set Q. The spatial proximity between e and MBR is defined in the following equation:*

$$\delta'(e, MBR) = 1 - \frac{D_\varepsilon(e.l, MBR.l)}{D_{\max}} \tag{5}$$

where MBR is the minimum bounding rectangle of Q, $D_\epsilon(e.l, MBR.l)$ is the Euclidean distance between $e.l$ and $MBR.l$, and D_{max} is the maximum distance in the location space.

According to the properties of minimum bounding rectangle, we have $D_\varepsilon(e.l, MBR.l) \leq D_\varepsilon(e.l, q_i.l), q_i \in Q$. Therefore, we have $\delta'(e, MBR) \geq \delta(e, q_i), q_i \in Q$. Note that e could be an intermediary-node N of IR-tree.

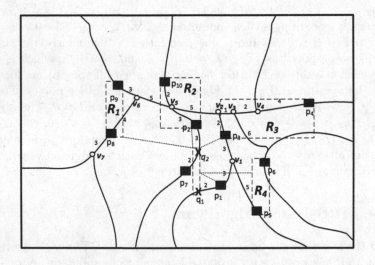

Fig. 3. Example of the MBR-based algorithm

Definition 6 ($\theta'(e, MBR)$). *Given an object e and the MBR of a query set Q. The textual relevance between e and MBR is defined in the following equation:*

$$\theta'(e, MBR) = MAX\{\theta(e, q_i)\}, q_i \in Q \tag{6}$$

That is, $\theta'(e, MBR)$ is the largest textual relevance score of e with q_i that belongs to Q. Therefore, it is easy to see that $\theta'(e, MBR) \geq \theta(e, q_i), q_i \in Q$.

Definition 7 *(pseudo relevance score: $\tau'(e, MBR)$).* *Given an object e and the MBR of a query set Q. The spatio-textual relevance score between e and MBR is defined as:*

$$\tau'(e, MBR) = \alpha \cdot \delta'(e, MBR) + (1 - \alpha) \cdot \theta'(e, MBR) \tag{7}$$

According to Definitions 5 and 6, it is easy to know that the pseudo relevance score is larger than or equal to the real relevance score. Since the query q is enclosed in the MBR of Q, the minimum Euclidean distance between MBR and e is no larger than the Euclidean distance between q and e: $\delta'(e, MBR) \geq \delta(e, q_i), q_i \in Q$. Then, the textual relevance between e and MBR is larger than the textual relevance between e and q_i: $\theta'(e, MBR) \geq \theta(e, q_i), q_i \in Q$ So, we finally have: $\tau'(e, MBR) \geq \tau(e, q_i), \forall q_i \in Q$. We can use this character to prune the search space.

Algorithm 2 shows the pseudo code of the MBR-based algorithm, where $Result\_k$ is used to store the k most relevant objects found so far, and ϵ is the relevance score of the k-th most relevant object in $Result\_k$. Note that if $Result\_k$ contains fewer than k members, ϵ is set to zero. When an object is inserted into $Result\_k$, an existing member is replaced if it already contains k members. $List$ is a priority queue, and the objects in this queue are in decreasing order of relevance score.

Firstly, we compute the minimum bounding rectangle of query set and the root of index structure (e.g., IR-tree) is inserted into $List$. Then, $Result\_k$ is set to empty and ϵ is set to zero. Next, we get the top element e in $List$ and compute its pseudo relevance score $\tau'(e, MBR)$. If e is an intermediate node in the index structure, we compute the pseudo relevance score for each entries in e (line 12). The entry whose pseudo relevance score is smaller or equal to the score of the k-th object already found should be discarded (lines 13–14). This is guaranteed by the fact that pseudo relevance score is an upper bounder of the real relevance score. If e is an object in the index structure, we compute the real group relevance score of it and compare it with ϵ (lines 16–18). If it is a better one, we insert this object into $Result\_k$ and update ϵ (lines 19–20). Of course, if $Result\_k$ contains fewer than k, ϵ is set zero.

In this paper, we use the A^* algorithm to find the network distance between query point and the object. In order to improve the efficiency of distance computing, we maintain the heap contents for each query point. Note that all queries have a common set of source points (i.e., the query points Q). Therefore, we can reuse information about network nodes visited by previous queries. In particular, the network nodes discovered by previous queries are stored in a hash table

---

Algorithm 2. The MBR-based Algorithm

---

1 **Input:**   a query set on road network $Q = \{q_1, q_2, ..., q_n\}$;
2 **Output:**   top-k most relevant spatio-textual objects;
3 **begin**
4 $\quad$ compute MBR of the query set;
5 $\quad$ $Result\_k \leftarrow \varnothing$; $\quad$ //k best objects found so far
6 $\quad$ $List \leftarrow$ Index.root; $\quad$ //objects in decreasing order of $\tau'(object, MBR)$;
7 $\quad$ $\epsilon \leftarrow 0$; $\quad$ //k_{th} score in $Result\_k$;
8 $\quad$ $e \leftarrow List.pop()$;
9 $\quad$ **while** $(e \neq \varnothing$ **and** $(\tau'(e, MBR) > \epsilon)$ **do**
10 $\quad\quad$ **if** e *is an intermediate node* **then**
11 $\quad\quad\quad$ **foreach** *entry in e* **do**
12 $\quad\quad\quad\quad$ compute $\tau'(entry, MBR)$;
13 $\quad\quad\quad\quad$ **if** $\tau'(e, MBR) \leq \epsilon$ **then**
14 $\quad\quad\quad\quad\quad$ continue; $\quad$ //skip this step;
15 $\quad\quad\quad\quad$ $List \leftarrow entry$;
16 $\quad\quad$ **else**
17 $\quad\quad\quad$ compute real group spatio-textual relevance score of e: $\tau(e, Q)$;
18 $\quad\quad\quad$ **if** $\tau(e, Q) > \epsilon$ **then**
19 $\quad\quad\quad\quad$ $Result\_k \leftarrow e$;
20 $\quad\quad\quad\quad$ update ϵ; $\quad$ //while$|Result\_k| < k$, $\epsilon = 0$;
21 $\quad\quad$ $e \leftarrow List.pop()$;

---

and we maintain heap contents for each query points. Therefore, we can use the previous state of the A^* heap to resume search.

5 Experiments

In this section, we evaluate the performance of the proposed two algorithms: the basic algorithm (denoted by "Basic Algorithm (BA)") and the MBR-based algorithm (denoted by "Enhanced Algorithm (EA)") by using two real datasets: EURO[1] and Australia[2]. Table 2 summarizes these two datasets. They are two real datasets that contain points of interest (e.g., restaurant, hotel, park) in Europe and Australia, respectively. Each point of interest, which can be regarded as a spatial web object, contains a geographical location and a short description (name, features, etc.). Both algorithms were implemented in Java, and an Intel(R) Core(TM) i5-5200u CPU @2.20 GHz with 4 GB RAM was used for the experiments. The index structure is disk resident, and the page size is 4 KB and the maximum number entries in internal nodes is set to 100.

In order to make the query Q resemble what users would like to use, each $q_i \in Q$ is distributed uniformly in the MBR of Q and has 3 tokens which are

---

[1] http://www.pocketgpsworld.com.
[2] http://www.openstreetmap.org/.

Table 2. Dataset properties

Property	Euro	Australia
Total number of vertices	200,793	1,274,323
Total number of edges	255,279	1,957,436
Total number of objects	35,490	97,032
Total number of keywords	105,737	457,719
Average number of keywords per object	7.79	5.63
Data size	57M	490M

randomly generated from datasets. Unless stated explicitly, parameters are set as follows by default: $\alpha = 0.5$, $k = 10$. In all experiments, we use workloads of 100 queries and report average costs of the queries.

The first experiment shows the effect of the cardinality of Q. We fix MBR of Q to 5% of the workspace of dataset and vary $|Q|$ from 1 to 8. The results are shown in Fig. 4. The CPU cost increases in both algorithms when the cardinality of Q increases from 1 to 8. This is because the distance and textual relevance computations for qualifying objects increase with the number of query points. However, the performance of EA is better than BA when $|Q|$ is larger than 1, due to the high pruning power of the enhanced algorithm. On the other hand, the cardinality of Q has little effect on the node accesses of EA, because it does not influence the pruning power.

To evaluate the effect of the MBR size of Q, we set $|Q|$ to 2 and vary MBR from 5% to 30% of the datasets. The results are shown in Fig. 5. We can see that the MBR size of Q does not influence node accesses metric of BA significantly, because it does not play an important role in the process of BA. However, the cost of EA increases with the MBR size. The reason is that the pruning power of

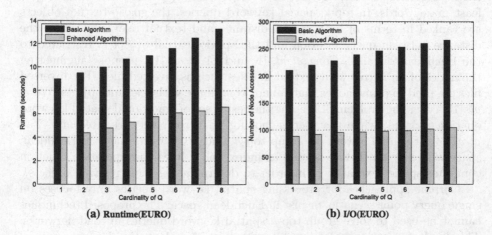

(a) **Runtime(EURO)**   (b) **I/O(EURO)**

Fig. 4. Results for varying cardinality of Q

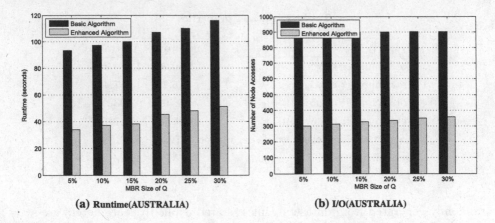

(a) Runtime(AUSTRALIA)   (b) I/O(AUSTRALIA)

Fig. 5. Results for varying MBR size of Q

EA degrades with the increase of the MBR size of Q, but EA still outperforms BA remarkably.

6   Related Work

In the context of spatial keyword query, De Felipe *et al.* [12] propose a new data structure that integrates signature files and R-trees. Each node of the R-tree employ a signature to indicate the keywords present in the node sub-tree. The hybrid index structure enables to utilize both spatial information and text information to prune the search space at query time. However, this proposal is limited by its use of signature files. A hybrid index structure that combines the R*-tree and bitmap indexing is developed to process a new type of query called the m-closest keyword query [13] that returns the closest objects containing at least m keywords. In top-k spatial keyword queries, the spatio-textual objects are ranked in terms of both spatial distance and textual relevance, where the distance between query location and the spatio-textual object is restricted to the Euclidean distance. Cong *et al.* [1] and Li *et al.* [14] propose augmenting the nodes of an R-tree with textual indexes such as inverted files. The inverted files are used to prune nodes that cannot contribute with relevant objects. Wu *et al.* [15] cover the problem of keeping the result set of traditional spatial keyword queries updated, while the user is moving on a road network. Recently, Yao *et al.* [4] presented a work on group top-k spatial keyword query in Euclidean space where by an IR-tree is used as index and a linear interpolation function computes spatio-textual score in order to determine the most relevant objects.

Current approaches for processing spatial keyword queries only focuses on single query point scenario, or just in Euclidean space. The proposed techniques cannot be used to solve group top-k spatial keyword queries in road networks. This kind of query takes as input a query set Q and returns k spatio-textual objects which have highest relevance score based on all users' arguments in road networks. Our problem definition and algorithms differ significantly from previously studied spatial keyword search problems and their solutions.

7 Conclusion

In this paper, we studied the problem of group top-k spatial keyword query processing in road networks. In an attempt to efficiently solve this problem, we proposed two algorithms. We first proposed the basic algorithm by first computing a comprehensive relevance score and then using a classical selecting algorithm to get the k most relevant spatio-textual objects from the dataset. Next, based on the minimum bounding rectangle (MBR) of query set, we proposed another more efficient algorithm, which can utilize a pruning heuristic to prune the search space and thus reduce the cost significantly. Finally, extensive experimental results demonstrate the effectiveness and efficiency of our approaches.

This work serves as the first attempt to address the group top-k spatial keyword query processing problem in road networks. For future work, we plan to improve the algorithm performance by employing more efficient index structures for spatial search on road networks.

References

1. Cong, G., Jensen, C.S., Wu, D.: Efficient retrieval of the top-k most relevant spatial web objects. Proc. VLDB Endow. **2**(1), 337–348 (2009)
2. Cao, X., Chen, L., Cong, G., Jensen, C.S., Qu, Q., Skovsgaard, A., Wu, D., Yiu, M.L.: Spatial keyword querying. In: Atzeni, P., Cheung, D., Ram, S. (eds.) ER 2012. LNCS, vol. 7532, pp. 16–29. Springer, Heidelberg (2012). doi:10.1007/978-3-642-34002-4_2
3. Chen, L., Cong, G., Jensen, C.S., Wu, D.: Spatial keyword query processing: an experimental evaluation. Proc. VLDB Endow. **6**, 217–228 (2013)
4. Yao, K., Li, J., Li, G., Luo, C.: Efficient group top-k spatial keyword query processing. In: Li, F., Shim, K., Zheng, K., Liu, G. (eds.) APWeb 2016. LNCS, vol. 9931, pp. 153–165. Springer, Cham (2016). doi:10.1007/978-3-319-45814-4_13
5. Gao, Y., Zhao, J., Zheng, B., Chen, G.: Efficient collective spatial keyword query processing on road networks. IEEE Trans. Intell. Transp. Syst. **17**(2), 469–480 (2016)
6. Rocha-Junior, J.B., Nørvåg, K.: Top-k spatial keyword queries on road networks. In: Proceedings of the 15th International Conference on Extending Database Technology, pp. 168–179. ACM (2012)
7. Zhang, C., Zhang, Y., Zhang, W., Lin, X., Cheema, M.A., Wang, X.: Diversified spatial keyword search on road networks. In: EDBT, pp. 367–378 (2014)
8. Zobel, J., Moffat, A.: Inverted files for text search engines. ACM Comput. Surv. (CSUR) **38**(2), 6 (2006)
9. Papadias, D., Zhang, J., Mamoulis, N., Tao, Y.: Query processing in spatial network databases. In: Proceedings of the 29th International Conference on Very Large Data Bases, vol. 29, pp. 802–813. VLDB Endowment (2003)
10. Dijkstra, E.W.: A note on two problems in connexion with graphs. Numer. Math. **1**(1), 269–271 (1959)

11. Hart, P.E., Nilsson, N.J., Raphael, B.: A formal basis for the heuristic determination of minimum cost paths. IEEE Trans. Syst. Sci. Cybern. **4**(2), 100–107 (1968)
12. De Felipe, I., Hristidis, V., Rishe, N.: Keyword search on spatial databases. In: IEEE 24th International Conference on Data Engineering, ICDE 2008, pp. 656–665. IEEE (2008)
13. Zhang, D., Chee, Y.M., Mondal, A., Tung, A.K.H., Kitsuregawa, M.: Keyword search in spatial databases: towards searching by document. In: IEEE 25th International Conference on Data Engineering, ICDE 2009, pp. 688–699. IEEE (2009)
14. Li, Z., Lee, K.C.K., Zheng, B., Lee, W.-C., Lee, D., Wang, X.: IR-tree: an efficient index for geographic document search. IEEE Trans. Knowl. Data Eng. **23**(4), 585–599 (2011)
15. Wu, D., Yiu, M.L., Jensen, C.S., Cong, G.: Efficient continuously moving top-k spatial keyword query processing. In: 2011 IEEE 27th International Conference on Data Engineering (ICDE), pp. 541–552. IEEE (2011)

Geo-Social Keyword Top-k Data Monitoring over Sliding Window

Shunya Nishio$^{(\boxtimes)}$, Daichi Amagata, and Takahiro Hara

Department of Multimedia Engineering,
Graduate School of Information Science and Technology,
Osaka University, Yamadaoka 1-5, Suita, Osaka, Japan
nishio.syunya@ist.osaka-u.ac.jp

Abstract. Recently, in many applications, points of interest (PoIs) have generated data objects based on Publish/Subscribe (Pub/Sub) model, and users receive their preferable data objects. Due to the prevalence of location based services and social network services, in addition, locations, keywords, and social relationships are considered to be meaningful for data retrieval. In this paper, we address the problem of monitoring top-k most relevant data objects over a sliding window, w.r.t. distance, keyword, and social relationship. If we have a lot of queries, it is time-consuming to check the result update of all queries. To solve this problem, we propose an algorithm that maintains queries with a Quadtree and accesses only queries with possibilities that a generated data object becomes top-k data. Moreover, we utilize k-skyband technique to quickly update query results. Our experiments using real datasets verify the efficiency of our algorithm.

Keywords: Pub/Sub · Social network · Continuous top-k query

1 Introduction

Due to the prevalence of GPS-enabled mobile devices, data retrieval based on location information and keywords has been becoming essential for many applications which employ Publish/Subscribe (Pub/Sub) model [11,15,17]. That is, users receive only data objects that are relevant to their preferences registered in advance, when points of interest (PoIs) generate data objects [2,4]. In this model, many data objects are continuously generated, so a top-k query that retrieves only k most relevant data objects to users' preferences is useful [2,9]. Moreover, due to the development of social networking service (SNS), such as Facebook and Twitter, data retrieval based on social relationships attracts much attention [1,13]. In particular, by extracting users' preferences with social filtering, it becomes easier to retrieve user requiring data [3]. For example, users have social relationships with their interests on PoIs, such as "like" of Facebook. Figure 1 shows social relationships (described by edges) between u_1, u_2, and PoIs. In this example, u_2 has social relationship with PoI p_1, hence when p_1 generates a data

© Springer International Publishing AG 2017
D. Benslimane et al. (Eds.): DEXA 2017, Part I, LNCS 10438, pp. 409–424, 2017.
DOI: 10.1007/978-3-319-64468-4_31

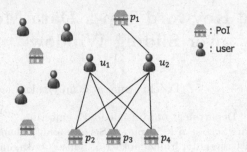

Fig. 1. Social relationships between users and PoIs

object, u_2 receives the data object. It can be seen that the social relationships of u_1 and u_2 are similar, meaning that their preferences would be similar. Although u_1 does not have social relationship with p_1, it is probable that u_1 is interested in p_1. This example implies that Pub/Sub systems should not eliminate chances that u_1 obtains data objects generated by p_1. If u_1 can receive the data objects, u_1 would find new observations, while p_1 can succeed in its advertisements.

In this paper, we consider top-k queries that rank data objects based on distance, keywords, and social relationships. More specifically, we address a problem of top-k data monitoring over a sliding window. Users can specify a query location and keywords and have different social relationships, thus each query has different top-k result. Besides, Pub/Sub systems normally have a huge number of queries, so it is challenging to update the top-k result of each query with realtime. It is also challenging to alleviate top-k re-evaluations when some top-k data objects expire triggered by window sliding.

To solve the above challenges, we propose an algorithm that maintains queries with a Quad-tree [14,16] and accesses only queries with possibilities that a generated data object becomes top-k data. Our Quad-tree supports pruning queries whose top-k results do not change. Furthermore, our algorithm employs k-skyband technique [10] to reduce top-k re-evaluations. Our experiments using real datasets verify the efficiency of our proposed algorithm.

Contribution. We summarize our contributions as follows.

- We address the problem of monitoring top-k data objects based on locations, keywords, and social relationships over a sliding window. To the best of our knowledge, this is the first work which addresses this problem.
- We propose an algorithm that maintains queries with a Quad-tree and accesses only queries with possibilities that a generated data object becomes top-k data. Moreover, we utilize k-skyband technique to quickly update the query result when a top-k data object expires.
- Through experiments using real datasets, we show that our algorithm reduces update time and is more efficient than baseline algorithms.

Organization. In Sect. 2, we define the problem. Section 3 describes our proposed algorithm, and we show our experimental results in Sect. 4. We review related works in Sect. 5. Finally, Sect. 6 concludes the paper.

2 Preliminaries

In this paper, P denotes a set of PoIs, O denotes a set of data objects generated by PoIs, and Q denotes a set of queries. Normally, user are not interested in old data. We thereby avoid providing such data objects. To this end, we employ a sliding window defined below.

Definition 1 (Sliding window). *Given a stream of data objects arriving in time order, a sliding window W over the stream with size $|W|$ consists of the most recent $|W|$ data objects.*

Data Object Model. A data object is defined as $o = (id, loc, key, t)$, where $o.id$ is a data identifier (id), $o.loc$ is the location of PoI that has generated o, $o.key$ is a set of keywords, and $o.t$ is the generation time of o.

Query Model. A query issued by a user u is defined as $q_u = (id, loc, key, k)$, where $q_u.id$ is a query id, $q_u.loc$ is a query location, $q_u.key$ is a set of query keywords, and $q_u.k$ is the number of data objects that u is willing to receive.

Our system delivers each data object to its relevant queries. To measure the relevance between a query q_u and a data object o, we define a scoring function. Before introducing the function, we define location score, keyword score, and social score.

Definition 2 (Location score). *The location score, $dist(q_u, o)$, between q_u and o is defined as follows.*

$$dist(q_u, o) = 1 - \frac{d(q_u.loc, o.loc)}{MAXloc} \qquad (1)$$

where $d(q_u.loc, o.loc)$ is the Euclidean distance between q_u and o, and $MAXloc$ is the maximal possible distance in the space.

Definition 3 (Keyword score). *The keyword score, $key(q_u, o)$, between q_u and o is defined as follows.*

$$key(q_u, o) = \frac{2|q_u.key \cap o.key|}{|q_u.key| + |o.key|} \qquad (2)$$

Example 2.1. *If $q_u.key = \{w_1, w_2\}$ and $o.key = \{w_2, w_3, w_4\}$, $key(q_u, o) = \frac{2 \times 1}{2 + 3} = 0.4$.*

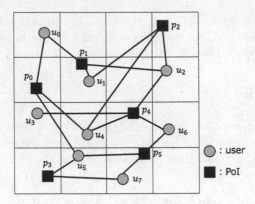

Fig. 2. A social graph

The social score is calculated based on social relationships between users and PoIs. Figure 2 shows a social graph that illustrates social relationships (described by edges) between users and PoIs. Let E denote a set of edges in a social graph, and let $e_{u,p}$ denote the edge between u and p.

Definition 4 (Social score). *The social score, $socio(q_u, p)$, between q_u and p is defined as follows.*

$$socio(q_u, p) = \begin{cases} 1 & (e_{u,p} \in E) \\ \max \frac{2|P_u \cap P_{u'}|}{|P_u| + |P_{u'}|} & (e_{u,p} \notin E) \end{cases} \tag{3}$$

where $P_u = \{\forall p' \in P | \exists e_{u,p'} \in E\}$ and $P_{u'} = \{\forall p' \in P | (\exists e_{u',p} \in E) \wedge (\exists e_{u',p'} \in E)\}$.

Example 2.2. *In Fig. 2, if we assume that p_0 generates o, $socio(q_{u_0}, p_0) = 1$ since $e_{u_0,p_0} \in E$. Next, we assume that p_4 generates o. Since $e_{u_1,p_4} \notin E$, we calculate the similarity of edges between u_1 and u' ($e_{u',p_4} \in E$). For example, $P_{u_1} = \{p_1, p_2\}$ and $P_{u_2} = \{p_1, p_2, p_4\}$, thus the similarity of edges between u_1 and u_2 is computed as $\frac{2 \times 2}{2 + 3} = 0.8$. Also, the similarities between u_1 and u_3, u_1 and u_4, and u_1 and u_6 are respectively 0, 0.4, and 0. Consequently, $socio(q_{u_1}, p_4) = \max\{0.8, 0, 0.4, 0\} = 0.8$.*

In a real environment, social relationships may be updated (i.e., edge insertions and deletions). It is thus possible to update $socio(q_u, p)$. However, the update frequencies of social relationships between users and PoIs are low, so we do not consider the update of social relationships in this paper, and dealing with such situations is a part of our future work.

Now we are ready to define our scoring function and geo-social keyword top-k queries.

Definition 5 (Scoring function). *The scoring function $s(q_u, o)$ is defined as follows.*

$$s(q_u, o) = dist(q_u, o) + key(q_u, o) + socio(q_u, p) \tag{4}$$

Definition 6 (Geo-social keyword top-k query). *Given a set of data objects* $O \in W$, *a geo-social keyword top-k query* q_u *retrieves* $q_u.k$ *data objects with the highest score, and its query result set* $q_u.D$ *satisfies that* $\forall o_i \in q_u.D$, $\forall o_j \in O \setminus q_u.D$, $score(q_u.o_i) \geq score(q_u, o_j)$.

Problem Statement. Given a massive number of geo-social keyword top-k queries and data objects, we aim to monitor the top-k result of each query with real-time over a sliding window W.

3 Proposed Algorithm

3.1 Overview

First, we describe the overview of the proposed algorithm. We maintain queries with a Quad-tree based on their locations, and then each node stores the query ids, a set of keywords, and the j-th ($j \geq k$) highest scores of the queries in the node. For each leaf node, we store the node id, the query ids in the node, and a digit sequence that shows the path from the root node in a node list (L_N). After construction of Quad-tree, we create a social score list (L_{ss}^i) for each PoI p_i. For each node $n \in L_N$, L_{ss}^i stores the node id and the node social score ss_n, which is the maximum value of social score between $q_u \in n$ and p_i. The construction of Quad-tree and the creation of L_{ss} are preprocessed because queries have been registered in advance.

Given a new data object o, a key challenge is to access only queries such that o becomes their top-k data. First, we create a social score table T_{ss} by combining L_N and L_{ss} of the PoI that has generated o. T_{ss} is used to calculate node social scores. Next, from the root node, we check whether the node is pruned based on the upper bound score. In other words, we prune queries whose top-k results surely do not change. Moreover, we utilize a k-skyband technique to quickly update the query result when a top-k data object expires.

In the following, Sect. 3.2 describes k-skyband. Section 3.3 describes the Quad-tree construction method, and Sect. 3.4 describes the L_{ss} creation method. Finally, Sect. 3.5 describes our algorithm that updates the top-k result of each query.

3.2 k-Skyband

We utilize k-skyband to reduce top-k re-evaluations when a top-k data object expires. Followings are formal definitions of dominance and k-skyband.

Definition 7 (Dominance). *A data object* o_1 *is dominated by another data object* o_2 *w.r.t. a query* q_u *if both* $s(q_u, o_1) \leq s(q_u, o_2)$ *and* $o_1.t < o_2.t$ *hold.*

Definition 8 (k-skyband). *The k-skyband of a query* q_u *contains a set of data objects which are dominated by at most* $(k-1)$ *other data objects.*

k-skyband contains data objects which may become top-k results when some of the current top-k data objects expire triggered by window sliding. Moreover, the top-k result of q_u belong to the k-skyband of q_u [6]. In our algorithm, k-skyband is updated only when a data object, whose score is higher than the j-th ($j \geq k$) highest score. In the rest of this paper, $q_u.S$ denotes k-skyband of q_u, and $o.dom$ denotes the number of data objects that dominant o.

3.3   Quad-Tree Construction

We adopt the liner Quad-tree structure to maintain a huge number of queries. A Quad-tree is a space partitioning tree structure in which a d-dimensional space is recursively subdivided into 2^d regions. As shown in Fig. 3a, each node resulting from a split is named in the order of NW, NE, SW, and SE. For example, Fig. 3b and c show the space partition and the corresponding tree structure of a simple Quad-tree for a given set of queries $\{q_0, ..., q_3\}$. As shown in Fig. 3c, in this paper, we use a circle and square to denote non-leaf node and leaf node respectively. A leaf node is set *black* if it is not empty, i.e., it contains at least one query. Otherwise, it is a *white* leaf node.

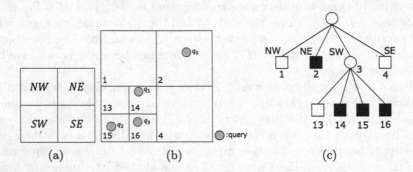

(a)   (b)   (c)

Fig. 3. Illustration of a Quad-tree

Next, we describe the Quad-tree construction method. Each query q_u is assigned into a leaf node based on its location $q_u.loc$. First, the root node is the entire space where all PoIs exist, thus the root node contains all queries. If the number of queries in a node is larger than an integer m, we partition the node into four children nodes and queries are distributed to each node based on their locations, until a node has at most m queries. For example, we assume that users in Fig. 2 issue queries whose locations are their current locations. If $m = 2$, the Quad-tree is constructed as shown in Fig. 4a.

Each node stores the query ids, a set of keywords, and the j-th highest scores of queries in the node. Let $q_u.score_j$ be the j-th highest score of q_u and its initial value is 0. If $q_u.score_j$ is updated, the information stored in nodes that contain q_u is updated. As shown in Fig. 4a, the query id and its $q_u.score_j$ are illustrated by (id, $q_u.score_j$). Node 4 contains q_6 and q_7, thus their ids, the j-th highest

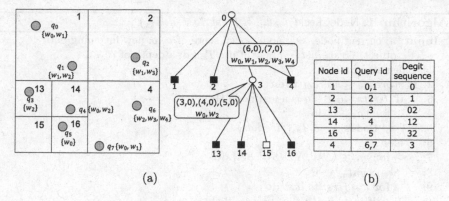

(a) (b)

Fig. 4. Construction of a Quad-tree ($m = 2$) and L_N

scores, and a set of their keywords are stored. Node 3 stores the information of q_3, q_4, and q_5 because they are in the subtree rooted at node 3.

Moreover, we create a node list L_N. For each non-empty leaf node, L_N stores the node id, the query ids in the node, and a digit sequence (ds) that shows the path from the root node. ds is calculated from the node id. In a liner Quad-tree, if a node id is x and its parent node id is y, $y = \lfloor \frac{x-1}{4} \rfloor$ and $z = x - 4y$, where z is a remainder. Moreover, z indicates that a node is NW, NE, SE, or SE of its parent node. For example, if $z = 0$, the node is NW of its parent node. By repeating this calculation, the remainder obtained in order is stored as a digit sequence in L_N. For example, since the remainder is obtained in the order of 1 and 2 for node 14, a digit sequence of 12 is stored. As shown in Fig. 4, node 15 does not contain any queries, thus node 15 is not stored in L_N.

3.4 Social Score List

Next, we describe the creation of a social score list (L_{ss}^i) for each PoI p_i. For ease of presentation, we denote a node social score between node n and p as ss_n, and

$$ss_n = \max_{q_u' \in n} socio(q_u', p). \tag{5}$$

For each node $n \in L_N$, we store the node id and ss_n in L_{ss}. It is however cost-prohibitive that each node stores all ss_n. Thus, if $ss_n = 0$, L_{ss} does not store n.

3.5 Top-k Data Update

Finally, we describe our algorithm that updates the top-k result of each query. We first describe how to process the generated data and then how to process the expired data.

T_{ss} **Creation.** First, given a new data object, we create a social score table T_{ss} by combining L_N and L_{ss} of the PoI that has generated o. T_{ss} is used to

---

Algorithm 1. NodeCheck($n, ss_n, o, first, last, d, T_{ss}$)

Input: n: current node, ss_n: node social score of n, o: new incoming data
object, $first \cdot last$: a range of use T_{ss}, d: a depth of n, T_{ss}

1: Calculate $score_{ub}$
2: **if** $score_{min} \leq score_{ub}$ **then**
3:    **if** $first = last$ //n is a leaf-node **then**
4:       **for** $\forall q_u \in n$ **do**
5:          Update($q_u, o, T_{ss}[first].ds$)
6:    **else**
7:       **for** $\forall n_c \in [NW, NE, SW, SE]$ **do**
8:          $count_{n_c} \leftarrow 0$
9:       **for** $i = first$ to $last$ **do**
10:          **if** d-th digit of $T_{ss}[i].ds = 0$ **then**
11:             $x \leftarrow NW$
12:          **else if** d-th digit of $T_{ss}[i].ds = 1$ **then**
13:             $x \leftarrow NE$
14:          **else if** d-th digit of $T_{ss}[i].ds = 2$ **then**
15:             $x \leftarrow SW$
16:          **else**
17:             $x \leftarrow SE$
18:          $count_x \leftarrow count_x + 1$
19:          **if** $ss_x < T_{ss}[i].ss_n$ **then**
20:             $ss_x \leftarrow T_{ss}[i].ss_n$
21:       $d \leftarrow d + 1$
22:       **for** $\forall n_c \in [NW, NE, SW, SE]$ //in order of $NW, NE, SW,$ and SE **do**
23:          **if** $count_{n_c} > 0$ **then**
24:             $last \leftarrow first + count_{n_c} - 1$
25:             NodeCheck($n_c, ss_{n_c}, o, first, last, d, T_{ss}$)
26:             $first \leftarrow first + count_{n_c}$
27: **else**
28:    **for** $\forall q_u \in n$ **do**
29:       **if** $q_u.s_p \leq score_{ub}$ **then**
30:          $q_u.s_p \leftarrow score_{ub}$

---

calculate the node social score. T_{ss} stores the node ids, the node social scores (ss_n), and the digit sequences (ds) of all nodes in L_N. That is, each entry of T_{ss} is $<n, ss_n, ds>$. For each node n, if $n \in L_{ss}$, T_{ss} stores $ss_n \in L_{ss}$. Otherwise, T_{ss} stores 0.

Node Check. After T_{ss} creation, from the root node, we check nodes and access only queries with possibilities that the score of o is not less than their j-th highest scores. Note that ss_n of the root node is 1. Algorithm 1 shows our algorithm. For each node n, we compare $score_{min}$ and the upper bound score ($score_{ub}$), where $score_{min}$ is the minimum value of $q_u.score_j$ ($q_u \in n$) and $score_{ub}$ of o is

$$score_{ub} = dist(n, o) + key(n, o) + ss_n. \tag{6}$$

Note that $dist(n, o)$ is calculated by using the minimum distance from o to n. If $o \in n$, however, $dist(n, o) = 1$. $key(n, o)$ is calculated as follows:

$$key(n, o) = \frac{2|n.key \cap o.key|}{|n.key \cap o.key| + |o.key|}, \tag{7}$$

where $n.key$ is a set of keywords stored in n. For each query $q_u \in n$, $dist(n, o) \geq dist(q_u, o)$ and $key(n, o) \geq key(q_u, o)$ hold. The top-k result of q_u is updated, if $score(q_u, o) \leq q_u.score_k$. Then the following theorem holds.

Theorem 1. *Given a node n, if $score_{min} > score_{ub}$, we can safely prune all the queries in n.*

Proof. The top-k result of $q_u \in n$ is updated, if $score(q_u, o) \leq q_u.score_k$. If $score_{min} > score_{ub}$, $score_{min} > dist(n, o) + key(n, o) + ss_n$. Moreover $q_u.score_k \geq q_u.score_j \geq score_{min}$ and $dist(n, o) + key(n, o) + ss_n \geq dist(q_u, o) + key(q_u, o) + socio(q_u, p) = s(q_u, o)$. Thus if $score_{min} > score_{ub}$, $q_u.score_k > s(q_u, o)$. Theorem 1 therefore holds. □

Next, we compare $score_{ub}$ and $score_{min}$ of the current node (Line 2). If $score_{ub} < score_{min}$, we prune all the queries in the current node because their k-skyband is not updated. At that time, for each query q_u, if $q_u.s_p < score_{ub}$, we assign $score_{ub}$ to $q_u.s_p$, where $q_u.s_p$ is the maximum value of $score_{ub}$ when q_u is pruned. (We describe the detail of $q_u.s_p$ later.) If $score_{ub} \geq score_{min}$, we iteratively check the children nodes of the current node. For each child node, we count the number of leaf nodes in T_{ss} and the subtree rooted at the child node. We simultaneously calculate ss_n of the children nodes (NW, NE, SW, and SE) (Lines 6–21). Next, for each child node that contains at least one leaf node, we similarly check nodes in order of NW, NE, SW, and SE (Lines 22–26). If $score_{ub} \geq score_{min}$ at a leaf node, for each query in the node, we update its k-skyband. That is, we execute Algorithm 2 that updates k-skyband, top-k result, and the information stored in Quad-tree (Lines 3–5). (We describe the detail of the algorithm later.)

Example 3.1. *In Fig. 4, we assume that $q_6.score_j = 1.8$ and $q_7.score_j = 2.2$. We furthermore assume that p_0 in Fig. 2 generates a data object o where $o.key = \{w_1, w_2, w_5\}$. We then check node 4 in Fig. 4. Since $n.key = \{w_0, w_1, w_2, w_3, w_4\}$ and $o.key = \{w_1, w_2, w_5\}$, $n.key \cap o.key = \{w_1, w_2\}$. Thus, $key(n, o) = \frac{2 \times 2}{2+3} = 0.8$. Now, we assume that $dist(n, o) = 0.6$ and $ss_n = 0.8$. In this situation, $score_{ub} = 0.6 + 0.8 + 0.8 = 2.2$. Since $score_{ub}(2.2) > score_{min}(1.8)$, we access queries in node 4 and update their k-skyband and top-k results.*

k-Skyband and Top-k Result Update. Algorithm 2 shows our update algorithm. If $score_{ub} \geq score_{min}$ at a leaf node, for each query in the node, we update k-skyband, top-k result, and the information stored in Quad-tree. First, we calculate $score(q_u.o)$ and update $o'.dom$ ($\forall o' \in q_u.S$) (Lines 1–4). If $o'.dom \geq k$, we delete o' from $q_u.S$. Next, if $score(q_u, o) \geq q_u.score_k$, we update the top-k result $q_u.D$ from $q_u.S$ (Lines 8–9). In our algorithm, the information stored in Quad-tree is updated triggered by k-skyband update. For each node that contains q_u whose k-skyband is updated, we update $q_u.score_j$ (Lines 10–12).

Algorithm 2. Update(q_u, o_{in}, ds)

Input: q_u: a query, o_{in}: a new incoming data, ds: a digit sequence
1: Calculate $s(q_u, o_{in})$
2: **for** $\forall o \in q_u.S$ **do**
3:     **if** $s(q_u, o_{in}) \geq s(q_u, o)$ **then**
4:         $o.dom \leftarrow o.dom + 1$
5:         **if** $o.dom \geq q_u.k$ **then**
6:             $q_u.S \leftarrow q_u.S - \{o\}$
7: $q_u.S \leftarrow q_u.S + \{o_{in}\}$
8: **if** $s(q_u, o_{in}) \geq q_u.score_k$ **then**
9:     Update $q_u.D$ from $q_u.S$
10: $N \leftarrow$ A set of nodes that are calculated by ds
11: **for** $\forall n \in N$ **do**
12:     Update $q_u.score_j$

Algorithm 3. Topk-Reevaluation(o)

Input: o: an expired data object
1: **for** each query q_u such that $o \in q_u.S$ **do**
2:     $q_u.S \leftarrow q_u.S - \{o\}$
3:     **if** $o \in q_u.D$ **then**
4:         **if** $|q_u.S| \geq q_u.k$ **then**
5:             Extract $q_u.D$ from $q_u.S$
6:             **if** $q_u.s_p \geq q_u.score_k$ **then**
7:                 Extract $q_u.D, q_u.S$ from W
8:         **else**
9:             Extract $q_u.D, q_u.S$ from W

Data Expiration. Algorithm 3 shows our top-k re-evaluation algorithm. First, for each query whose k-skyband contains the expired data object o, we delete o from $q_u.S$. We then update the top-k result of query q_u where $o \in q_u.D$. The k data objects with the highest score in $q_u.S$ become top-k result. If $|q_u.S| < k$ or a data object with score less than $q_u.s_p$ becomes top-k result, we need to re-evaluate its top-k result from W. Recall that $q_u.s_p$ is the maximum value of $score_{ub}$ when q_u is pruned. Thus, if a data object with score less than $q_u.s_p$ becomes top-k result, it is possible that a data object pruned before becomes top-k result. We compute $score(q_u, o')$ of a data object $o' \in W$, and then update $q_u.D$ and $q_u.S$.

4   Experiments

In this section, we provide an experimental evaluation on the performance of our proposed algorithm. Since this is the first work addressing the problem of monitoring top-k data objects based on locations, keywords, and social relationships over a sliding window, we compare our algorithm with following three baseline algorithms.

- **baseline 1.** This algorithm does not utilize our query index. Given a new data object, we update k-skyband and top-k results of all queries. When a top-k data object expires, we update the top-k result from k-skyband.
- **baseline 2.** This algorithm utilizes our query index, but does not utilize the k-skyband technique. Given a new data object, we access only queries with possibilities that the score of the data object is not less than the j-th highest score of their queries. When a top-k data object expires, we check all data objects in W to update the top-k result.
- **baseline 3.** This algorithm utilizes our query index, but does not utilize the k-skyband technique. The difference with baseline 2 is that, when we calculate the score of a data object o, we utilize L_{ss}. Specifically, when we calculate the social score of o, if the leaf node n containing q_u is in L_{ss} of the PoI that has generated o, we assign $ss_n \in L_{ss}$ to the social score. Otherwise, we assign 0 to the social score. We then calculate the approximate score of o and compare the score with $q_u.score_k$. If the score is higher than $q_u.score_k$, we calculate the accurate score of o and update the top-k result.

4.1 Experimental Settings

All algorithms were implemented in C++, and all experiments were conducted on a PC with 3.47 GHz Intel Xeon processor and 192 GB RAM.

Dataset. Our experiments were conducted on two real datasets: Yelp[1] and Brightkite[2]. The statistics of the two datasets are summarized in Table 1. Each dataset contains PoIs with location and check-ins, thus we utilize the check-ins as social relationships between users and PoIs. For $o.key$, we randomly select a keyword set from Yelp reviews. Each keyword set contains 1 to 11 keywords. Figure 5 shows the histogram of the number of keywords in a keyword set.

Table 1. Dataset statistics

	Yelp	Brightkite
Number of users	$366, 715$	$50, 687$
Number of PoIs	$60, 785$	$772, 631$
Number of check-ins	$1, 521, 160$	$1, 072, 965$

Parameters. Table 2 summarizes the parameters and the default values (bold ones) used in the experiments. We used 10 and $2k$ as default values of m and j. We generated a top-k query for each user. For $q_u.loc$, we assign a random location in the space where PoIs exist. We randomly selected a keyword set from Yelp reviews as $q_u.key$ and $q_u.k$ was a random number between 1 to k_{max}.

[1] http://www.yelp.com/dataset_challenge/.
[2] http://snap.stanford.edu/data/loc-brightkite.html.

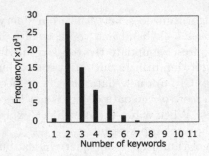

Fig. 5. Histogram of the number of keywords in a keyword set

Table 2. Configuration of parameters

Parameter	Value
Maximum value of $q_u.k$, k_{max}	**5**, **10**, 15, 20, 25, 30
$\|Q\|$ $[\times 10^3]$	**50**, 100, 150, 200, 250, 300, 350 (Yelp)
	10, 20, 30, 40, 50 (Brightkite)
$\|W\|$ $[\times 10^6]$	**0.5**, 1, 1.5, 2, 2.5, 3

4.2 Experimental Result

We report the average update time (the average time to finish updating the top-k result of each query triggered by window sliding [sec]) of each algorithm.

Effect of k_{max}. We examine the impact of k_{max}, and Fig. 6 shows the result. As shown in Fig. 6, our algorithm updates the top-k results faster than the baseline algorithms. All the algorithms need longer update time as k_{max} increases. The difference in the update time between baselines 2 and 3 in Yelp case is larger than that in Brightkite case. In Yelp dataset, it can be seen that the number of POIs is smaller than the number of users, meaning that users would check in the

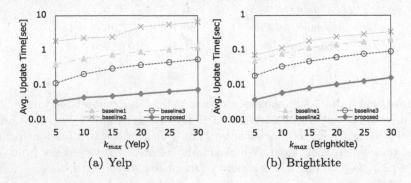

(a) Yelp   (b) Brightkite

Fig. 6. Effect of k_{max}

same PoIs. It thus takes much time to calculate social score, and the performance of baseline 2 that calculates social scores of all data objects decreases.

Effect of $|Q|$. We next examine the impact of $|Q|$, and Fig. 7 shows the result. As shown in Fig. 7, all the algorithms need longer update time as $|Q|$ increases and our algorithm updates the top-k results faster than the baseline algorithms. Specifically, the update time of our algorithm is at least four times faster than those of the baseline algorithms.

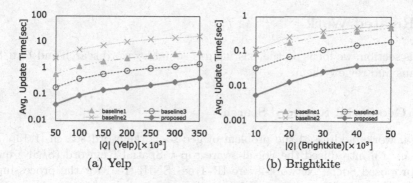

(a) Yelp   (b) Brightkite

Fig. 7. Effect of $|Q|$

Effect of $|W|$. We finally examine the impact of $|W|$, and Fig. 8 shows the result. When $|W|$ increases, the update time of each baseline algorithm is constant or increases. However, the update time of our algorithm decreases. The update time includes generated data processing time (i.e., GDP) and expired data processing time (i.e., EDP). When we increase $|W|$, the GDP time of baselines 2 and 3, and our algorithm decreases, due to the fact that a large window size usually leads to top-k results with higher score. Thus, a new data object affects less queries, resulting in shorter GDP time. On the other hand, the GDP time of baseline

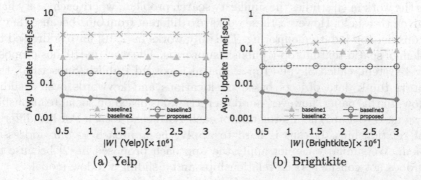

(a) Yelp   (b) Brightkite

Fig. 8. Effect of $|W|$

1 is constant because baseline 1 always checks all queries. With increasing $|W|$, the possibility that an expired data object is contained in the top-k results of queries becomes low. The number of top-k re-evaluations hence decreases. On the other hand, when a top-k data object expires, baselines 2 and 3 check all data objects in W. When $|W|$ increases, the EDP time of baselines 2 and 3 therefore increases. Recall that, we utilize k-skyband technique in baseline 1 and our algorithm, thus we can quickly update the top-k result when a top-k data object expires.

5 Related Work

In this section, we focus on the existing geo-social keyword search and Pub/Sub systems and review them.

5.1 Geo-Social Keyword Search

Several works addressed the problem of geo-social keyword search [1,13]. The work in [13] introduced the Social-aware top-k Spatial Keyword (SkSK) query and proposed Social Network-aware IR-Tree (SNIR-tree) for the processing of SkSK queries. SNIR-tree is an R-Tree, where each node also contains a set of users relevant to the POIs indexed by the subtree rooted at the node. In this paper, however, the definition of social relationships is different from ours and this tree structure does not support efficient top-k data monitoring. Also, the work in [1] considers snapshot queries over static data while our problem focuses on continuous query over streaming data.

5.2 Pub/Sub System

Numerous works studied Pub/Sub systems. Nevertheless, most of the existing Pub/Sub systems, e.g., [7,8,12,15], do not consider spatial information. Recently, spatial-keyword Pub/Sub system has been studied in a line of work (e.g., [2,4,5, 10,11]). Among them, the works in [5,11] study the boolean matching problem while the work in [4] studies the similarity search problem, where each query has a pre-given threshold. However, these works are different from ours because they do not consider top-k data monitoring. The only works in [2,10] have addressed the problem of spatio-textual top-k publish/subscribe. The work in [10] has proposed a novel system, called Skype (Top-k Spatial-Keyword Publish/Subscribe), that monitors top-k data objects based on locations and keywords over a sliding window. In this paper, however, our data retrieval is based on social relationships and we need to store the information of each PoI because each PoI has different social relationships. It thus is hard to apply the approach to our addressing problem. We similarly cannot apply the approach proposed in [2] because the work does not consider social relationships and a sliding window model.

6   Conclusion

In this paper, we addressed the problem of monitoring top-k most relevant data objects, w.r.t. distance, keyword, and social relationship, with sliding window setting. To quickly update the top-k result of each query, we proposed an algorithm that maintains queries with a Quad-tree and accesses only queries with possibilities that a generated data object becomes the top-k data. Moreover, we utilize k-skyband technique to quickly update the query result when a top-k data object expires. We evaluated our proposed algorithm by experiments on real data, and the results show that our algorithm reduces the update time and is more efficient than the three baseline algorithms.

In this paper, we did not consider the update of social relationships. As part of our future work, we plan to design an algorithm which can deal with updates of social relationships.

Acknowledgement. This research is partially supported by the Grant-in-Aid for Scientific Research (A)(26240013) of the Ministry of Education, Culture, Sports, Science and Technology, Japan, and JST, Strategic International Collaborative Research Program, SICORP.

References

1. Ahuja, R., Armenatzoglou, N., Papadias, D., Fakas, G.J.: Geo-social keyword search. In: Claramunt, C., Schneider, M., Wong, R.C.-W., Xiong, L., Loh, W.-K., Shahabi, C., Li, K.-J. (eds.) SSTD 2015. LNCS, vol. 9239, pp. 431–450. Springer, Cham (2015). doi:10.1007/978-3-319-22363-6_23
2. Chen, L., Cong, G., Cao, X., Tan, K.-L.: Temporal spatial-keyword top-k publish/subscribe. In: ICDE, pp. 255–266 (2015)
3. Groh, G., Ehmig, C.: Recommendations in taste related domains: collaborative filtering vs. social filtering. In: SIGGROUP, pp. 127–136 (2007)
4. Hu, H., Liu, Y., Li, G., Feng, J., Tan, K.-L.: A location-aware publish/subscribe framework for parameterized spatio-textual subscriptions. In: ICDE, pp. 711–722 (2015)
5. Li, G., Wang, Y., Wang, T., Feng, J.: Location-aware publish/subscribe. In: KDD, pp. 802–810 (2013)
6. Mouratidis, K., Bakiras, S., Papadias, D.: Continuous monitoring of top-k queries over sliding windows. In: SIGMOD, pp. 635–646 (2006)
7. Mouratidis, K., Pang, H.: Efficient evaluation of continuous text search queries. IEEE TKDE **23**(10), 1469–1482 (2011)
8. Sadoghi, M., Jacobsen, H.-A.: BE-Tree: an index structure to efficiently match Boolean expressions over high-dimensional discrete space. In: SIGMOD, pp. 637–648 (2011)
9. Shraer, A., Gurevich, M., Fontoura, M., Josifovski, V.: Top-k publish-subscribe for social annotation of news. PVLDB **6**(6), 385–396 (2013)
10. Wang, X., Zhang, Y., Zhang, W., Lin, X., Huang, Z.: Skype: top-k spatial-keyword publish/subscribe over sliding window. PVLDB **9**(7), 588–599 (2016)
11. Wang, X., Zhang, Y., Zhang, W., Lin, X., Wang, W.: AP-Tree: efficiently support continuous spatial-keyword queries over stream. In: ICDE, pp. 1107–1118 (2015)

12. Whang, S.E., Garcia-Molina, H., Brower, C., Shanmugasundaram, J., Vassilvitskii, S., Vee, E., Yerneni, R.: Indexing Boolean expressions. PVLDB **2**(1), 37–48 (2009)
13. Wu, D., Li, Y., Choi, B., Xu, J.: Social-aware top-k spatial keyword search. In: MDM, pp. 235–244 (2014)
14. Zhang, C., Zhang, Y., Zhang, W., Lin, X.: Inverted linear quadtree: efficient top k spatial keyword search. In: ICDE, pp. 901–912 (2013)
15. Zhang, D., Chan, C.-Y., Tan, K.-L.: An efficient publish/subscribe index for e-commerce databases. PVLDB **7**(8), 613–624 (2014)
16. Zhang, D., Tan, K.-L., Tung, A.K.: Scalable top-k spatial keyword search. In: EDBT, pp. 359–370 (2013)
17. Zheng, K., Su, H., Zheng, B., Shang, S., Xu, J., Liu, J., Zhou, X.: Interactive top-k spatial keyword queries. In: ICDE, pp. 423–434 (2015)

Geo-Social Keyword Skyline Queries

Naoya Taguchi[1]([✉]), Daichi Amagata[2], and Takahiro Hara[2]

[1] University of Tokyo, Tokyo, Japan
guchio@logos.t.u-tokyo.ac.jp
[2] Osaka University, Suita, Japan
{amagata.daichi,hara}@ist.osaka-u.ac.jp

Abstract. Location-Based Social Networking Services (LBSNSs) have been becoming increasingly popular. One of the applications provided by LBSNSs is a PoI search based on spatial distance, social relationships, and keywords. In this paper, we propose a novel query, Geo-Social Keyword Skyline Query (GSKSQ), which returns the skyline of a set of PoIs based on a query point, the social relationships of the query owner, and query keywords. Skyline is the set of data objects which are not dominated by others. We also propose an index structure, Social Keyword R-tree, which supports efficient GSKSQ processing. The results of our experiments on two real datasets Gowalla and Brightkite demonstrate the efficiency of our solution.

1 Introduction

With the wide spread of mobile devices such as smart phones and tablets, people use Location-Based Services (LBS) and Social Networking Services (SNS) in their daily lives [4,8]. Due to this fact, Location-Based Social Networking Services (LBSNS), e.g., Facebook and Yelp, are also prevalent [3]. In a LBSNS, we can search points of interests (PoIs) based on three criteria, spatial distance, social relationships, and keywords [1].

In this paper, we propose a novel query, Geo-Social Keyword Skyline Query (GSKSQ), which returns the skyline [2] of a set of PoIs based on a query point, the social relationships of the query owner, and user-specified keywords. Skyline is the set of data objects which are not dominated by others. Informally, a data object o is dominated by another data object o' if o is worse than o' for all attributes. The result of a GSKSQ thus supports multi-criteria decision making in LBSNS applications. We show a practical example of a GSKSQ in Example 1.

Example 1. *Assume that a user u_a searches a restaurant on a LBSNS, and u_a can choose it based on his/her current position, the number of check-ins of his/her friends, and the number of keywords matched with a query. In this case, u_a can find such a restaurant easily by using a GSKSQ. Figure 1a represents an example situation where u_a specifies his/her position as a query point and "Chinese restaurant" as a query keyword. Figure 1b represents the corresponding social graph. As we can see in Figs. 1a and b, restaurant B is worse than restaurant C in the three attributes, so B is dominated by C. In the same manner,*

© Springer International Publishing AG 2017
D. Benslimane et al. (Eds.): DEXA 2017, Part I, LNCS 10438, pp. 425–435, 2017.
DOI: 10.1007/978-3-319-64468-4_32

we can see that restaurants D, E, F, G, H, I, and J are dominated by C. As a result, u_a obtains restaurants A and C as the query result, and he/she can easily choose preferable one from these two restaurants.

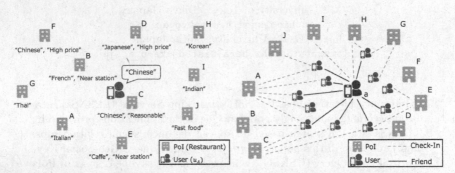

(a) An example of user-specified keyword (Chinese) and a set of restaurants

(b) A social graph example

Fig. 1. A practical example of a GSKSQ

A naive way to calculate a Geo-Social Keyword Skyline (GSKS) is to check whether or not a given PoI is dominated by any other PoIs. However, this approach is computationally expensive particularly when the number of PoIs is large [2]. In addition, because the result of a GSKSQ is dependent on a query, we cannot pre-compute the results of any queries. In this paper, we propose an index structure called Social Keyword R-tree (SKR-tree), which is a kind of aR-tree [5], and supports efficient GSKSQ processing. With the SKR-tree, we can retrieve the result while pruning unpromising PoIs. Our contributions in this paper are summarized as follows: (1) We propose a novel query, GSKSQ, which returns the skyline of a set of PoIs based on a query point, the social relationships of the query owner, and user-specified keywords. (2) We propose an index structure called SKR-tree to calculate a GSKS efficiently. (3) We propose a scoring function to support efficient retrieval on SKR-tree. We design this function based on an idea about user behaviors, and show that this function is optimal w.r.t. the number of node accesses of the SKR-tree. (4) The results of our experiments on real datasets Gowalla[1] and Brightlite[2] demonstrate the efficiency of our solution.

The organization of this paper is as follows. Section 2 defines the problem of this paper. We describe our algorithm in Sect. 3, and Sect. 4 presents our experimental results. Finally, we conclude this paper in Sect. 5.

---

[1] https://snap.stanford.edu/data/loc-gowalla.html.

[2] https://snap.stanford.edu/data/loc-brightkite.html.

2 Problem Definition

2.1 Geo-Social Keyword Skyline Query (GSKSQ)

We define a PoI $p_i \in P$ as $p_i = \langle loc, key \rangle$, where i is the identifier of p_i, loc is the location of p_i, and key is the set of keywords held by p_i. When a user u retrieves a GSKS, he/she issues a GSKSQ by specifying a query point and query keywords. A GSKSQ is defined as $q_u = \langle loc, key \rangle$. In response to q_u, each $p_i \in P$ obtains attribute values $p_i.G(q_u)$, $p_i.S(q_u)$, and $p_i.K(q_u)$, which respectively correspond to a spatial distance, social relationships, and keywords. $p_i.G(q_u)$ is the Euclidean distance between $q_u.loc$ and $p_i.loc$, which is defined as follows.

$$p_i.G(q_u) = dist(q_u.loc, p_i.loc)$$

Note that smaller $p_i.G(q_u)$ is better. $p_i.S(q_u)$ is the number of users who are friends of u and have checked-in to p_i. So, $p_i.S(q_u)$ is defined as

$$p_i.S(q_u) = |u.friends \cap p_i.checkin|,$$

where $u.friends$ is the set of users who are friends of u, and $p_i.checkin$ is the set of users who have checked-in to p_i. For example, in Fig. 1b, $A.S(q_u)$ is 4, $C.S(q_u)$ is 3, and $F.S(q_u)$ and $I.S(q_u)$ are 0. Larger $p_i.S(q_u)$ is better. $p_i.K(q_u)$ is the number of the common keywords in $q_u.key$ and $p_i.key$, and is defined as follows.

$$p_i.K(q_u) = |q_u.key \cap p_i.key|$$

As same as $p_i.S(q_u)$, larger $p_i.K(q_u)$ is better. Based on these three attribute values, we define dominance below.

Definition 1 (Dominance). *Let q_u be a geo-social keyword skyline query. If p_i and p_j satisfy that $(p_i.G(q_u) \geq p_j.G(q_u)) \wedge (p_i.K(q_u) \leq p_j.K(q_u)) \wedge (p_i.S(q_u) \leq p_j.S(q_u))$, we represent this condition as follows.*

$$p_i \preceq p_j$$

Furthermore, if p_i and p_j satisfy that $(p_i \preceq p_j) \wedge [((p_i.G(q_u) > p_j.G(q_u)) \vee (p_i.K(q_u) < p_j.K(q_u)) \vee (p_i.S(q_u) < p_j.S(q_u))]$, p_i is dominated by p_j, and we represent this condition as follows.

$$p_i \prec p_j$$

We now define a GSKS as follows.

Definition 2 (Geo-Social Keyword Skyline (GSKS)). *Let P be a set of PoIs, and $q_u = \langle loc, key \rangle$ be a geo-social keyword skyline query. The Geo-Social Keyword Skyline of P calculated on q_u is the subset P' of P, which satisfies that*

$$\nexists p_j \in P \text{ such that } p_i \prec p_j, \tag{1}$$

for $\forall p_i \in P'$.

Our objective is to achieve efficient processing of a GSKSQ.

Algorithm 1. BASELINE ALGORITHM

---

1   for $\forall p_i \in P$ do
2   Calculate each attribute value of p_i

3   for $\forall p_i \in P$ do
4   for $\forall p_j \in P$ do
5   if $p_i \succ p_j$ then
6        $P \leftarrow P \backslash \{p_j\}$
7   if $p_i \prec p_j$ then
8        $P \leftarrow P \backslash \{p_i\}$
9   break

10 return P

2.2   Baseline

We show a baseline algorithm for processing a GSKSQ in Algorithm 1. In Algorithm 1, we calculate all attribute values of all PoIs at first (lines 1–2). We then check whether or not a given PoI dominates or is dominated by other PoIs (lines 3–9), and finally we obtain a GSKS (line 10).

Algorithm 1 is computationally expensive because of two reasons. First, we have to calculate all attribute values of all PoIs. In particular, it increases computational cost w.r.t. $p_i.S(q_u)$. This is because to calculate this attribute value, we have to check how many friends of the query owner checked-in for all PoIs. Second, the baseline algorithm executes dominance check for all PoIs, and this operation is the main overhead of skyline computation [7]. We propose a more efficient algorithm which alleviates this cost.

3   Proposed Solution

3.1   Social Keyword R-Tree (SKR-Tree)

We propose an index structure called SKR-Tree to achieve efficient GSKSQ processing. SKR-tree is a kind of a R-tree as illustrated in Fig. 2. Nodes of SKR-tree store information on the three attributes, and are classified to two kinds of nodes, leaf nodes and internal nodes. A leaf node n_i corresponds to a PoI p_i, and contains:

- a pointer to p_i,
- $n_i.loc$, which is the location of p_i,
- $n_i.key$, which is the set of keywords held by p_i,
- $n_i.S$, which is the upper-bound social value of p_i.

We detail $n_i.S$, which is the upper bound of $p_i.S(q_u)$ for any q_u, in Sect. 3.2. For example, in Fig. 2, leaf node n_A (n_B) contains a pointer to p_A (p_B), $n_A.key = \{k_1\}$ ($n_B.key = \{k_2\}$) is the set of keywords, and $n_A.S = 1$ ($n_B.S = 3$) is the

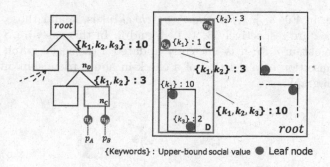

{Keywords} : Upper-bound social value   ● Leaf node

Fig. 2. An example of SKR-tree

upper-bound social value. An internal node n_i is a Minimum Bounding Rectangle (MBR) which consists of its all child nodes, and contains:

- $n_i.key$, which is the union of the keyword sets which are held by its all child nodes,
- $n_i.S$, which is the maximum upper-bound social value among all its child nodes' ones.

For example, in Fig. 2, internal node n_C is an MBR which consists of its child nodes n_A and n_B, a set of keywords $\{k_1, k_2\} = n_A.key \cap n_B.key$, and 3, which is the maximum of $n_A.S$ and $n_B.S$.

Given a GSKSQ q_u, we can calculate the three attribute values for all n_i, and we represent them as $n_i.G(q_u)$, $n_i.S$, and $n_i.K(q_u)$. If n_i is a leaf node, $n_i.G(q_u)$ is $p_i.G(q_u)$. If n_i is an internal node, $n_i.G(q_u)$ is the Euclidean distance between $q_u.loc$ and the nearest neighbor point in the corresponding MBR of n_i. $n_i.S$ is an upper-bound social value of n_i, and $n_i.K(q_u)$ is $|q_u.key \cap n_i.key|$. By using these attribute values, we can execute dominance check between a node n_i and a PoI p_j. If n_i is dominated by p_j, PoIs which are pointed by n_i's descendant nodes are not in the GSKS, so we can prune the subtree rooted at n_i. This reduces attribute value calculation, node accesses, and dominance checks. Therefore, we can accelerate query processing performance. Note that once SKR-tree is constructed, it is updated efficiently by incremental manner [5].

3.2 Upper-Bound Social Value Calculation

To construct SKR-tree, we have to calculate upper-bound social values, which are defined below, for all leaf nodes.

Definition 3 (Upper-bound social value). *Given a leaf node n_i and a set of all users U, $n_i.S$ is calculated as follows.*

$$n_i.S = max_{\forall u_j \in U} |u_j.friends \cap p_i.checkin|$$

For example, in Fig. 3, $|u_a.friends \cap p_A.checkin|$ is 3, and those for u_b, u_c, u_d, u_e, and u_f are respectively 3, 1, 1, 4, and 2. In this case, $n_A.S = 4$. Note that once we obtain $n_i.S$, it is efficiently updated if the social graph is updated because an insertion or a deletion of a check-in and a friendship affects just a part of social graph.

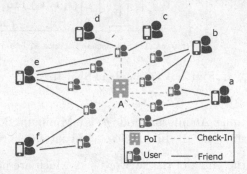

Fig. 3. An example social graph to calculate an upper-bound social value

3.3   GSKSQ Processing Algorithm

We propose Algorithm 2 to calculate a GSKS efficiently on a SKR-tree.

In Algorithm 2, we traverse nodes based on scores. These scores are obtained by a scoring function F, and higher scores mean higher priorities. We achieve this priority by using a priority queue Q_F, and this queue is initialized by the root node n_{root} (line 1). At the first of each iteration, we pop and get the top node of Q_F, n_{temp}, and check whether or not n_{temp} is dominated by PoIs in P_{GSKS}^{temp}, which is an intermediate result set (lines 3–8). If n_{temp} is dominated by $p_i \in P_{GSKS}^{temp}$, we prune the subtree rooted at n_{temp}, and proceed to the next iteration. Otherwise, if its child nodes $n_i \in n_{temp}.children$ are internal nodes, we calculate the three attribute values of n_i, and push them to Q_F (lines 9–13). If $n_i \in n_{temp}.children$ are leaf nodes, we calculate $p_i.G(q_u)$, $p_i.S(q_u)$, and $p_i.K(q_u)$ where p_i is pointed by n_i. Then, we execute dominance checks for all p_i against the PoIs in P_{GSKS}^{temp}. After that, we add the PoIs pointed by nodes in $n_{temp}.children$ which are not dominated by $p_j \in P_{GSKS}^{temp}$ to P_{GSKS}^{temp}, and remove non-skyline PoIs from P_{GSKS}^{temp} (lines 14–23). We continue this iteration while Q_F is not empty (line 24).

3.4   Scoring Function Design

The node traversal order in Algorithm 2 is dependent on a scoring function F. We design F based on two criteria. First one is the number of node accesses. To achieve fast query processing, the number of node accesses should be minimized.

Algorithm 2. PROPOSED ALGORITHM

1 $Q_F.\textbf{push}(n_{root})$
2 **while** $Q_F \neq \emptyset$ **do**
3      $n_{temp} = Q_F.top$
4      $Q_F.\textbf{pop}$
5   **for** $\forall p_i \in P_{sky}^{temp}$ **do**
6   **if** $n_{temp} \prec p_i$ **then**
7             $n_{temp}.children = \emptyset$
8   **break**

9   **if** $n_{temp}.children$ *are not leaf* **then**
10   **for** $\forall n_i \in n_{temp}.children$ **do**
11   Calculate each attribute value of n_i
12             $n_i.score = F(n_i, q_u)$
13             $Q_F.\textbf{push}(n_i)$

14   **else**
15   **for** $\forall n_i \in n_{temp}.children$ **do**
16   Calculate each attribute value of $n_i.p$
17             $P_{sky}^{temp} \leftarrow P_{sky}^{temp} \backslash \{\forall p_j \in P_{sky}^{temp} | p_j \prec n_i.p\}$
18   **if** $\nexists p_j \in P$ *such that* $p_i \prec p_j$ **then**
19               $P_{sky}^{temp} \leftarrow P_{sky}^{temp} \cup \{n_i.p\}$

20 **return** P_{sky}^{temp}

Second one is the number of dominance checks. It is intuitively known that if P_{GSKS}^{temp} has a skyline PoI that dominates many PoIs, we can reduce unnecessary checks. Therefore, we should obtain such a PoI as soon as possible.

Now we address the first requirement and introduce Theorem 1. Proof is omitted due to space limitation.

Theorem 1. *Assume that we process a GSKSQ by using Algorithm 2. Let n_i and n_j be arbitral nodes of SKR-tree, which satisfy that $n_i \prec n_j$. If F satisfies that*

$$F(q_u, n_i) < F(q_u, n_j), \tag{2}$$

the number of node accesses is minimum for retrieving the GSKS on the SKR-tree.

Based on Theorem 1, we address the second requirement. Our F consists of three sub-scoring functions f_G, f_S, and f_K, which correspond to the three attributes. As well as F, higher scores of these sub-scoring functions mean higher priorities. Note that, PoIs which dominate larger space potentially dominate may nodes and PoIs. Here, we design F and the three sub-scoring functions as follows.

$$F(q_u, n_i) = \begin{cases} f_G(q_u, n_i) \cdot f_S(q_u, n_i) \cdot f_K(q_u, n_i) & (f_G(q_u, n_i) \neq 0) \\ -\frac{1}{f_S(q_u, n_i) \cdot f_K(q_u, n_i)} & (f_G(q_u, n_i) = 0) \end{cases}, \tag{3}$$

$$f_G(q_u, n_i) = dist_{max} - n_i.G(q_u), \tag{4}$$

$$f_S(q_u, n_i) = \begin{cases} n_i.S(q_u) & (n_i.S(q_u) \neq 0) \\ \alpha & (n_i.S(q_u) = 0) \end{cases}, \tag{5}$$

$$f_K(q_u, n_i) = \begin{cases} n_i.K(q_u) & (n_i.K(q_u) \neq 0) \\ \alpha & (n_i.K(q_u) = 0) \end{cases}, \tag{6}$$

where $dist_{max}$ is the farthest distance among the ones between $q_u.loc$ and all corners of n_{root}, which is an MBR, and α is a real number which satisfies $0 < \alpha < 1$. The reason why F, f_S, and f_K have 2 cases is that the most natural definitions of them, which is the first cases, do not satisfy Eq. (2) in case that one or more of f_G, f_S, and f_K is 0.

In addition to this, we refine f_S by considering a practical characteristic of social relationships. Assume that a user issues a GSKSQ. It can be expected that he/she tends to issue this query in his/her friends' living area. This is because users on SNSs are more likely to connect with people they already know, or have some offline basis for the connection [6]. In this case, his/her friends tend to check-in to near PoIs to the query point. This suggests that distance between attribute and social attribute values have correlation. To this end, we design a sub-scoring function f_S', which employs f_S of Eq. (5) and is employed instead of f_S in Eq. (3), as follows.

$$f_S'(q_u, n_i) = \begin{cases} f_S(q_u, n_i) \cdot f_G{}^d(q_u, n_i) & (n_i.G(q_u) \neq 0) \\ f_S(q_u, n_i) & (n_i.G(q_u) = 0) \end{cases}, \tag{7}$$

where d is a real number which provides the influence of distance attribute. Note that d is selected by some empirical studies.

Based on the above discussion, we finally redesign a scoring function F which we propose in this paper and is described below.

$$F(q_u, n_i) = \begin{cases} f_G{}^{d+1}(q_u, n_i) \cdot f_S(q_u, n_i) \cdot f_K(q_u, n_i) & (f_G(q_u, n_i) \neq 0) \\ -\frac{1}{f_S(q_u, n_i) \cdot f_K(q_u, n_i)} & (f_G(q_u, n_i) = 0) \end{cases}$$

This scoring function is based on Eqs. (3) and (7).

4   Experimental Evaluation

4.1   Experimental Setup

We used two real datasets of Gowalla and Brightkite detailed in Table 1. These datasets include: (1) PoIs which hold IDs and locations, (2) users who hold logs of friendships and check-ins. For each PoI in a given dataset, we assigned five synthetic keywords, and the number of distinct keywords in the dataset

Table 1. Datasets

Dataset	#Users	#PoIs	#Friendships	#Check-ins
Gowalla	196,591	1,280,969	1,900,655	3,981,334
Brightkite	58,228	772,965	428,157	1,072,965

is 10,000. To simulate semantic proximities (e.g., the keyword of "restaurant" tends to appear with the keyword of "lunch"), we used continuous five integers calculated by modulo 10,000 as the keywords (e.g., 9,999, 0, 1, 2, 3). For these datasets, we run 100 GSKSQs by the following way. We first pick a random user and choose 1–5 query keywords in the same way employed for PoIs. We calculate the MBR of PoIs which the user has checked-in, and choose a random location from the MBR as a query point. For α in Eqs. (5) and (6), we use 0.001. All algorithms were implemented in C++, and executed on Intel Xeon 3.47 GHz with 192 GB RAM.

4.2 Experimental Result

Pre-processing. The calculation times of upper-bound social values are only 67.7 s for Gowalla, and 35.2 s for Brightkite. These are trivial times in practice. The calculation for Gowalla takes more time because Gowalla has more users, check-ins, and friendships. In addition, the construction times of SKR-tree are 4.38×10^4 s for Gowalla, and 2.38×10^4 s for Brightkite. These are reasonable times in practice because we can update them once we construct them. Gowalla takes more time because it has more PoIs.

Results with Different Number of Query Keywords. We run GSKSQs with different number of query keywords to demonstrate the efficiency of our algorithm, and to investigate the impact of the number of query keywords. The result is shown in Fig. 4. We used the best d shown in Table 2, which are acquired from empirical experimental results. We can see that our algorithm significantly outperforms the baseline. It is also seen that both algorithms need longer processing times as the number of query keywords increases. When the number of query keywords is large, the number of PoIs which have the common keywords with query keywords is also large. Hence, the number of dominance checks increases.

Table 2. The best d for each number of query keywords

Dataset	Number of query keywords				
	1	2	3	4	5
Gowalla	101	116	122	122	122
Brightkite	119	119	119	120	120

We here show the pruning rate of our algorithm in Fig. 5. We define pruning rate as a rate of pruned nodes to the entire nodes. According to Fig. 5, the pruning rates for Gowalla are lower than those of Brightkite. This difference is explained by Table 1. The differences of the numbers of users, check-ins, and friendships are larger than the difference of the number of PoIs, and that causes differences of attribute values of social relationships. These pruning rates bring the results that the processing times of the proposed algorithm on Gowalla are in 43%–49% compared to that of the baseline one, while those on Brightkite are in 12%–16%.

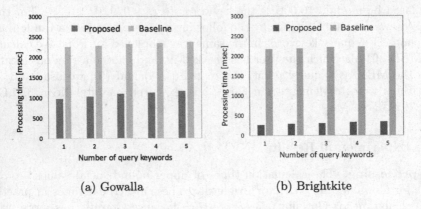

(a) Gowalla   (b) Brightkite

Fig. 4. Impact of the number of query keywords

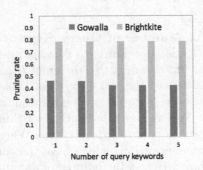

Fig. 5. Pruning rates for the two datasets

5   Conclusion

In this paper, we introduced a novel query, GSKSQ. Although this query supports decision making in LBSNS, it requires high processing cost when we use a naive algorithm. To overcome this problem, we proposed an index called SKR-tree and an algorithm which uses this index. Our algorithm optimizes a number of node accesses, and achieves effective dominance checks by considering a practical characteristic of social relationships. The experimental results demonstrate that the performance of our algorithm is better than that of the baseline algorithm.

References

1. Ahuja, R., Armenatzoglou, N., Papadias, D., Fakas, G.J.: Geo-social keyword search. In: Claramunt, C., Schneider, M., Wong, R.C.-W., Xiong, L., Loh, W.-K., Shahabi, C., Li, K.-J. (eds.) SSTD 2015. LNCS, vol. 9239, pp. 431–450. Springer, Cham (2015). doi:10.1007/978-3-319-22363-6_23
2. Borzsony, S., Kossmann, D., Stocker, K.: The skyline operator. In: ICDE, pp. 421–430 (2001)
3. Emrich, T., Franzke, M., Mamoulis, N., Renz, M., Züfle, A.: Geo-social skyline queries. In: Bhowmick, S.S., Dyreson, C.E., Jensen, C.S., Lee, M.L., Muliantara, A., Thalheim, B. (eds.) DASFAA 2014. LNCS, vol. 8422, pp. 77–91. Springer, Cham (2014). doi:10.1007/978-3-319-05813-9_6
4. Mouratidis, K., Li, J., Tang, Y., Mamoulis, N.: Joint search by social and spatial proximity. IEEE TKDE 27(3), 781–793 (2015)
5. Papadias, D., Kalnis, P., Zhang, J., Tao, Y.: Efficient OLAP operations in spatial data warehouses. In: Jensen, C.S., Schneider, M., Seeger, B., Tsotras, V.J. (eds.) SSTD 2001. LNCS, vol. 2121, pp. 443–459. Springer, Heidelberg (2001). doi:10.1007/3-540-47724-1_23
6. Steinfield, C., Ellison, N., Lampe, C., Vitak, J.: Online social network sites and the concept of social capital. Front. New Media Res. 15, 115–131 (2012)
7. Tan, K.-L., Eng, P.-K., Ooi, B.C., et al.: Efficient progressive skyline computation. In: VLDB, vol. 1, pp. 301–310 (2001)
8. Tao, Y., Sheng, C.: Fast nearest neighbor search with keywords. IEEE TKDE 26(4), 878–888 (2014)

FTGWS: Forming Optimal Tutor Group for Weak Students Discovered in Educational Settings

Yonghao Song[1,2], Hengyi Cai[1,2], Xiaohui Zheng[1,2], Qiang Qiu[1], Yan Jin[1], and Xiaofang Zhao[1(✉)]

[1] Institute of Computing Technology, Chinese Academy of Sciences, Beijing, China
{songyonghao,caihengyi,zhengxiaohui,qiuqiang,zhaoxf}@ict.ac.cn,
jinyan@ncic.ac.cn
[2] University of Chinese Academy of Sciences, Beijing, China

Abstract. The task of experts discovering, as one of the most important research issues in social networks, has been widely studied by many researchers in recent years. However, there are extremely few works considering this issue in educational settings. In this work, we focus on the problem of forming tutor group for weak students based on their knowledge state. To solve this problem, a novel framework based on Student-Skill Interaction (SSI) model and set covering theory is proposed, which is called FTGWS. The FTGWS framework contains three major steps: firstly, building SSI models for each student and each skill he or she has encountered; then, discovering the top-k weak students based on their knowledge state; finally, forming the optimal tutor group for each weak student. We evaluate our framework on a real-word dataset which contains 28834 students and 244 skills. The experiments show that the framework is capable of producing high-quality solutions (for 93% of weak students, the size of the optimal tutor group can be decreased up to 2 students).

Keywords: Tutor group · Grouping students · Weak student · Cooperative learning · Student-Skill Interaction Model (SSI)

1 Introduction

With the booming popularity of web-based educational settings, such as Coursera, Khan Academy, and ASSISTment, e-learning has attracted much attention of educators, governments and the general public [1]. E-learning aims to make high quality online learning resources to the world, and has attracted a diverse population of students from a variety of age groups, educational backgrounds and nationalities [3]. Despite these successes, providing high quality online education is a multi-faceted and complex system [2]. Two particular problems that have vexed researchers and educators for a long time are how to identify students who are at risk of poor performance early and how to create tutor groups for

© Springer International Publishing AG 2017
D. Benslimane et al. (Eds.): DEXA 2017, Part I, LNCS 10438, pp. 436–444, 2017.
DOI: 10.1007/978-3-319-64468-4_33

these weak students so that they can augment their knowledge with cooperative learning from each other [3–6].

In this research work, we explore how to identify weak students and how to form the optimal tutor group for weak students based on their knowledge state which is described as two skill sets. The formal definition of these two problems will be given in preliminary section. If a set of students that together have all of the required skills which a weak student has not mastered, through the cooperative learning between the weak student and this set of students can improve the performance of this weak student [5,8]. This set of students is defined as a tutor group of one specific weak student. Based on above idea, a FTGWS framework is proposed, where weak students are discovered based on their interaction records in system and the optimal tutor group will be generated based on students knowledge state. Our main contributions can be summarized as follows:

(1) We give the formal definitions of difficulty of skills and learning rate of students, then the concept of weak students is defined, and a algorithm called FKWS to discover top-k weak students has been designed.
(2) We introduce the formal definition of tutor group and convert the problem of forming optimal tutor group to the minimum set cover problem (SCP) which has been proved a NP-hard problem; a heuristic algorithm based on genetic algorithm is implemented to solve this problem.
(3) Extensive experiments on the real world data set[1] which contains 28834 students and 244 skills are carried out. The experimental results are capable of producing high-quality solutions (for 93% weak students, the size of the optimal tutor group can be decreased up to 2 students).

2 Preliminaries

2.1 Notations and Definitions

The mathematical denotations throughout this paper are listed in Table 1.

Definition 1 (Difficulty Coefficient of Skill). *Given a skill k_j, a set of students $S^j = \{s_1^j, s_2^j, \ldots, s_m^j\}$ who have exercised k_j and the matrix $SK_{M \times N}$. The difficult coefficient of skill k_j is*

$$d_j = 1 - \frac{\sum_{s_i^j \in S^j} P_{i,j}(T)}{||S^j||} \tag{1}$$

Definition 2 (Learning Ability of Student). *Given a student s_i, a set of skills $K^i = \{k_1^i, k_2^i, \ldots, k_n^i\}$ which s_i has exercised and the matrix $SK_{M \times N}$. The learning ability of student s_i is*

$$l_i = \frac{\sum_{k_j^i \in K^i} P_{i,j}(T)}{||K^i||} \tag{2}$$

---

[1] https://sites.google.com/site/assistmentsdata/home/2012-13-school-data-with-affect.

Table 1. Mathematical notations used in this paper

Notation	Description
M, N, D	Number of students, number of skills and data set, respectively
$S = \{s_1, s_2, \ldots, s_i, \ldots, s_M\}$	Set of students where s_i is the student i
$W = \{s_1, s_2, \ldots, s_i, \ldots, s_k\}$	Set of students where s_i is a poor performance student, $W \subseteq S$
$K = \{k_1, k_2, \ldots, k_j, \ldots, k_N\}$	Set of skills where k_j is the skill j
$K^i = \{k_1^i, k_2^i, \ldots, k_n^i\}$	Set of skills which s_i has exercised
$S^j = \{s_1^j, s_2^j, \ldots, s_m^j\}$	Set of students who have exercised k_j
$R^{i,j} = r_1^{i,j} r_2^{i,j} \ldots r_l^{i,j}$	Response sequence of s_i on k_j, e.g. 01001011111
$P_{i,j}(L_0)$	Probability that s_i masters skill k_j initially
$P_{i,j}(T)$	Probability that s_i transforms k_j from unlearned state to learned
$P_{i,j}(G)$	Probability that s_i guesses correctly on k_j
$P_{i,j}(S)$	Probability that s_i slips (make a mistake) on k_j
$s_i k_j$	SSI model of s_i for k_j, it is a four-tuple: $\{P_{i,j}(L_0), P_{i,j}(T), P_{i,j}(G), P_{i,j}(S)\}$
$SK_{M \times N}$	Matrix formed based on SSI model where $SK_{ij} = s_i k_j$
$DMap\langle k_j, d_j \rangle$	Collection contains coefficient of difficulty d_j of skill k_j
$LMap\langle s_i, l_i \rangle$	Collection contains learning rate l_i of student s_i
MS_{s_i}	Mastered skill set of student s_i
TS_{s_i}	Target skill set of student s_i
TG_{s_i}	Optimal tutor group for s_i

Definition 3 *(Mastered Skill). Given a student s_i, a skill k_j, a SSI model $s_i k_j = \{P_{i,j}(L_0), P_{i,j}(T), P_{i,j}(G), P_{i,j}(S)\}$, a response sequence $R^{i,j} = r_1^{i,j} r_2^{i,j} \ldots$ and a determining factor e. Let $n = ||R^{i,j}||$. If the following condition is satisfied, then k_j is a mastered skill of s_i.*

$$P_{i,j}(L_{n-1}|r_{n-1}^{i,j} = 1) = \frac{P_{i,j}(L_{n-1}) * (1 - P_{i,j}(S))}{P_{i,j}(L_{n-1}) * (1 - P_{i,j}(S)) + (1 - P_{i,j}(L_{n-1})) * P_{i,j}(G)}$$

$$P_{i,j}(L_{n-1}|r_{n-1}^{i,j} = 0) = \frac{P_{i,j}(L_{n-1}) * P_{i,j}(S)}{P_{i,j}(L_{n-1}) * P_{i,j}(S) + (1 - P_{i,j}(L_{n-1})) * (1 - P_{i,j}(G))} \quad (3)$$

$$P_{i,j}(L_n) = P_{i,j}(L_{n-1}) + (1 - P_{i,j}(L_{n-1})) * P_{i,j}(T)$$

and

$$P_{i,j}(L_n) \geq e \quad (4)$$

Definition 4 *(Target Skill). Given a student s_i, a skill k_j, and a determining factor ε, obtaining the coefficient of difficulty d_j of skill k_j from $DMap\langle k_j, d_j \rangle$, and obtaining the learning rate l_i of student s_i from $LMap\langle s_i, l_i \rangle$. If the following condition is satisfied, then k_j is a target skill of s_i.*

$$l_i \geq \varepsilon d_j \quad (5)$$

2.2 Problems Formulation

The major tasks of this research are discovering weak students who are at risk of poor learning performance, and seeking out the optimal tutor group for each of them based on students interaction records in e-learning system. Based on the notations and definitions provided, the problems to be solved in this paper are formulated as follows.

Problem 1 (Discovering Top-K Poor Performance Students, FKWS). Given a student s_i, the mastered skill set MS_{s_i} of s_i, and the target skill set TS_{s_i} of s_i, the function $f(s_i, Performance)$ is used to calculate the performance score of s_i.

$$f(s_i, Perf) = \frac{||MS_{s_i}||}{||TS_{s_i}||} \tag{6}$$

Based on Eq. 6, the top-k poor performance student can be sought out. Before the definition of forming tutor groups for weak students, the tutor skill set is defined as follows.

Definition 5 *(Tutor Skill Set). Given a weak student w_i and the other students set $S = \{s_1, s_2, \ldots, s_i, \ldots, s_{M-1}\}$ where $w_i \notin S$. Given mastered skill set MS_{w_i} and target skill set TS_{w_i} of w_i, and other students mastered skill sets $\{MS_{s_1}, MS_{s_2}, \ldots, MS_{s_i}, \ldots, MS_{s_{M-1}}\}$. Let $UMS_{w_i} = TS_{w_i} - MS_{w_i}$ and $I_i = TS_{w_i} \cap MS_{s_i}$, the tutor skill set of w_i is*

$$TutorSet_{w_i} = \cup_{i=1}^{M-1} I_i \ (I_i \in UMS_{w_i}) \tag{7}$$

Problem 2 (Forming the optimal tutor group for weak students, FTGWS). Given a weak student w_i and the other students $S = \{s_1, s_2, \ldots, s_i, \ldots, s_{M-1}\}$ where $w_i \notin S$. Given mastered skill set MS_{w_i} and target skill set TS_{w_i} of w_i, and other students mastered skill sets $\{MS_{s_1}, MS_{s_2}, \ldots, MS_{s_i}, \ldots, MS_{s_{M-1}}\}$. Based on Definition 5, the problem of forming the optimal tutor group for weak student w_i is to find a student set $S_{w_i} \subset S$, where

$$\cup_{s_i \in S_{w_i}} MS_{s_i} = TutorSet_{w_i} \ and \ S_{w_i} = argmin|S_{w_i}| \tag{8}$$

3   FTGWS Framework

In this section, the FTGWS framework is presented in detail. The design of our algorithm is inspired by a simple idea: each skill has an inherent difficulty and each student has an inherent learning ability, based on these hypotheses, the target skill set and the mastered skill set of each student can be obtained from this student's interaction records on skills. The criterion of poor performance students is defined according to these skill sets; then, the optimal tutor group can be formed for each weak student which is formulated by Problem 2 that is converted to minimum set cover problem (SCP). Based on the above idea, the FTGWS framework that employs SSI model and genetic algorithm is proposed, which contains three major steps (the pseudo code is shown in Algorithm 1).

To understand the work mechanism of FTGWS scheme, we give an illustrative example in Fig. 1, and each step of FTGWS is introduced in the following subsections.

Algorithm 1. FTGWS framework of forming tutor group for weak students

Initial Step: initialing the variables will be used in the following steps
    Obtain the set of students S and the set of skills K from D
    Obtain the response sequence $R^{i,j}$ from D
Step 1: Learning SSI Model for Each Student and Each Skill $SK_{M \times N}$
    $SK_{M \times N} \leftarrow LSSI(S, D, d)$ // d is the threshold of stopping learning
Step 2: Discovering Top-K Poor Performance Student W
    $W \leftarrow FKWS(SK_{M \times N}, S, D, k)$ // k is the number of weak students who are top-k
Step 3: Forming the optimal tutor group for each weak student
    $TG \leftarrow FOTG(S, W, MS, TS)$
Finalized Step:
    return TG

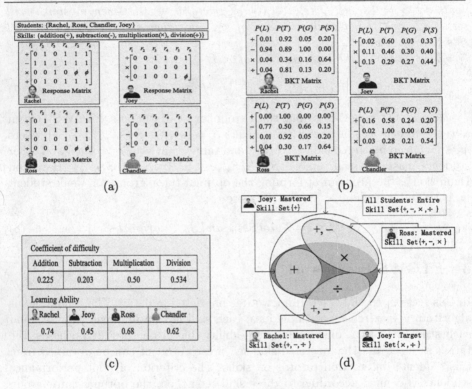

Fig. 1. The illustrative example of FTGWS framework

3.1 Learning SSI Model for Each Student and Each Skill

The Student-Skill Interaction (SSI) model proposed by Pardos & Heffernan is expanded based on standard BKT model which is a simple hidden markov model (HMM) [7,10]. The first step of SSI model is to learn student specific parameters by training all skill data of an individual student. The second step is to embed all students' specific parameter information which obtained from first step into

SSI model. The classical Baum-Welch algorithm is used to find the unknown parameters of a HMM.

Figure 1(a) shows that the entire skill set of student Rachel is $\{+, -, \times, \div\}$, for each of them, Rachel has a response sequence which obtains from interaction records by chronological order. For instance, the response sequence of Rachel on addition skill is $r_1 r_2 \ldots r_6 = [010111]$. The individual initial knowledge of Rachel is $15/22$ for all skills. As showed in Fig. 1(b), the learning rate, guess rate and slip rate on addition is 0.92, 0.05, 0.20 respectively, which are learnt by SSI model.

3.2   Discovering Top-K Poor Performance Student

According to the definitions in the preliminary section, we utilize algorithm FTWS to discover top-k poor performance students. Firstly, the learning rate l_i for a specific student s_i and the difficulty of a specific skill k_j can be calculated. Next, the mastered skill set MS_{s_i} and the target skill set TS_{s_i} can be obtained based on Definitions 3 and 4. Lastly, we calculate the score of performance for each student and find the top-k weak students.

As shown in Fig. 1(c), the difficulty of $\{+, -, \times, \div\}$ is $\{0.225, 0.203, 0.50, 0.534\}$ in which division is the hardest skill, and the learning rate of $\{$Rachel, Joey, Ross, Chandler$\}$ is $\{0.74, 0.45, 0.68, 0.62\}$ where Rachel has the best learning ability. Figure 1(d) illustrates that the top-1 weak student is Joey with mastered skill set $\{+\}$ and target skill set $\{+, \times, \div\}$, whose score of performance is 0.33 derived from Eq. 6.

3.3   Forming the Optimal Tutor Group for Weak Students

Now that the weak students have been discovered, the optimal tutor group needs to be formed for each weak student to augment their knowledge. Generally, for a weak student w_i, if $(TS_{w_i} - MS_{w_i}) \subset \cup_{s_i \in S'} MS_{s_i}$ where $S' \subset S$, the student set S' is a tutor group of w_i, we define the tutor group with the minimal size as the optimal tutor group. Based on the formal description of Problem 2 in preliminary section, the FTGWS problem is a minimum set cover problem which has been proved to be a NP-hard problem. In this paper, we employ a heuristic algorithm proposed by Beasley & Chu which is based on genetic algorithm to solve FTGWS problem [9]. The result of experiment shows that this heuristic algorithm is capable of producing high-quality solutions.

Figure 1(d) shows that the optimal tutor group obtained from FOTG algorithm is $\{$Rachel, Ross$\}$ for weak student Joey. The mastered skill sets of Rachel and Ross are $\{+, -\div\}$ and $\{+, -, \times\}$, the tutor skill set of Joey is $TS_{Joey} - MS_{Joey} = \{+, \times, \div\}$ which can be covered by the mastered skill sets of Rachel and Ross.

4   Experiments

In this section, the proposed FTGWS framework is evaluated on the real-world data set assistments_2012_2013 published by ASSISTment platform. Specifically, we show and analyze the result of every step of FTGWS framework, which includes the coefficient of difficulty of skills, the learning rate of students and the optimal tutor group.

4.1 Experimental Data

The ASSISTments data set contains 46674 students, 265 skills and 4 problem types which are choose_1, algebra, fill_in and open_response. We preprocessed the data set by deleting the records in which skill_id is null and problem type is open_response, for the reason that open response problem is always marked as correct. The final experimental data set contains 28834 students and 244 skills.

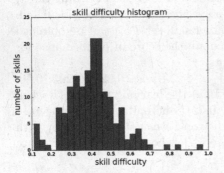

Fig. 2. DIST of skills difficulty   **Fig. 3.** DIST of students learning rate

4.2 Experimental Results and Analysis

Coefficient of Difficulty of Skills. In this group of experiments, the coefficient of difficulty for all 244 skills in the dataset were calculated based on Definition 1. Figure 2 shows that difficulty coefficients of most skills are less than 0.7, which represents these skills are relatively simple; the difficulty of skills follow the normal distribution which verified the rationality of Definition 1.

Learning Rate of Students. In this group of experiments, the learning rates of all 28834 students were calculated based on Definition 2. The student with higher learning rate tends to have better learning ability. Figure 3 shows that the mean learning rate of most students is 0.6 which indicates that most students has a normal learning ability, overall, the distribution of students learning rate fits the normal distribution represents that there are fewer prominent students or backward students.

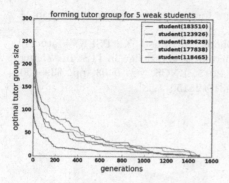

Fig. 4. Forming optimal tutor group

Optimal Tutor Group. The convergence and the stability of FOTG algorithm are evaluated and optimal tutor groups for top-100 weak students have been formed. Figure 4 demonstrates the iteration processes of FOTG algorithm, for all 5 weak students the size of their optimal tutor group can be converged to less than 3, which means the mastered skill sets of 3 students can cover the target skill set of one weak student.

5 Conclusion and Future Work

This paper proposed a novel FTGWS framework to form the optimal tutor group for weak students discovered in educational settings, which is based on BKT model and SCP theory. There are several possibilities to extend the research in the future. First, due to the high complexity of FOTG algorithm, a more effective substitutable algorithm needs to be designed to reduce the complexity of forming tutor group. Second, the FTGWS framework is not sufficiently sophisticated, an excellent student who is good at many skills maybe appears in every tutor group, this unbalance problem will be solved in the future work.

References

1. Anderson, A., Huttenlocher, D., Kleinberg, J.: Engaging with massive online courses. In: WWW, pp. 687–698 (2014)
2. Gillies, J., Quijada, J.: Opportunity to learn: a high impact strategy for improving educational outcomes in developing countries. In: USAID EQUIP (2008)
3. He, J., Bailey, J., Rubinstein, B., Zhang, R.: Identifying at-risk students in massive open online courses. In: AAAI, pp. 1749–1755 (2015)
4. Lakkaraju, H., Aguiar, E., Shan, C.: A machine learning framework to identify students at risk of adverse academic outcomes. In: KDD, pp. 1909–1918 (2015)
5. Agrawal, R., Golshan, B., Terzi, E.: Grouping students in educational settings. In: KDD, pp. 1017–1026 (2014)
6. Kim, B.W., Chun, S.K., Lee, W.G., Shon, J.G.: The greedy approach to group students for cooperative learning. In: Park, J., Yi, G., Jeong, Y.S., Shen, H. (eds.) UCAWSN & PDCAT 2016. LNEE, vol. 368, pp. 83–89. Springer, Singapore (2016). doi:10.1007/978-981-10-0068-3_10
7. Pardos, Z.A., Heffernan, N.T.: Modeling individualization in a Bayesian networks implementation of knowledge tracing. In: De Bra, P., Kobsa, A., Chin, D. (eds.) UMAP 2010. LNCS, vol. 6075, pp. 255–266. Springer, Heidelberg (2010). doi:10.1007/978-3-642-13470-8_24
8. Compton, J.I., Forbes, G.R.: Modeling success: using preenrollment data to identify academically at-risk students. In: Education Publications, No. 37 (2015)

9. Beasley, J.E., Chu, P.C.: A genetic algorithm for the set covering problem. Eur. J. Oper. Res. **94**, 392–404 (1996)
10. Song, Y., Jin, Y., Zheng, X., Han, H., Zhong, Y., Zhao, X.: PSFK: a student performance prediction scheme for first-encounter knowledge in ITS. In: Zhang, S., Wirsing, M., Zhang, Z. (eds.) KSEM 2015. LNCS, vol. 9403, pp. 639–650. Springer, Cham (2015). doi:10.1007/978-3-319-25159-2_58

Data Mining and Big Data

Mining Cardinalities from Knowledge Bases

Emir Muñoz[1,2(✉)] and Matthias Nickles[2]

[1] Fujitsu Ireland Limited, Dublin, Ireland
[2] Insight Centre for Data Analytics, National University of Ireland, Galway, Ireland
Emir.Munoz@ie.fujitsu.com

Abstract. Cardinality is an important structural aspect of data that has not received enough attention in the context of RDF knowledge bases (KBs). Information about cardinalities can be useful for data users and knowledge engineers when writing queries, reusing or engineering KBs. Such cardinalities can be declared using OWL and RDF constraint languages as constraints on the usage of properties over instance data. However, their declaration is optional and consistency with the instance data is not ensured. In this paper, we address the problem of mining cardinality bounds for properties to discover structural characteristics of KBs, and use these bounds to assess completeness. Because KBs are incomplete and error-prone, we apply statistical methods for filtering property usage and for finding accurate and robust patterns. Accuracy of the cardinality patterns is ensured by properly handling equality axioms (owl:sameAs); and robustness by filtering outliers. We report an implementation of our algorithm with two variants using SPARQL 1.1 and Apache Spark, and their evaluation on real-world and synthetic data.

1 Introduction

The Resource Description Framework (RDF) is a widely used framework for representing knowledge (bases) on the Web. RDF is schema-less, which means that it gives freedom to data publishers in describing entities and their relationships using facts without the need to observe some specified unique global data schema. Most RDF knowledge bases (KBs) avoid to include domain, range or cardinality restrictions because of the contradictions that they can generate [6,21]. For instance, having two different properties (e.g., from different ontologies) to represent the population of a country, or using two or more labels to refer to the same entity. However, the lack of a central schema causes a series of difficulties in the consumption of such data (e.g., [1,9,11,14]), e.g., having two different population numbers in the same KB. For instance, data users and knowledge engineers need an understanding of what information is available in order to write queries, and to reuse or engineer KBs [15,26]. In data management, *cardinality* is an important aspect of the structure of data. We show that the aforementioned problem can be overcome partially by mining cardinalities from instance data, and complement other methods (e.g., [30]) towards the generation of a central schema. In RDF, the cardinality of a property limits the number of values that may have for

© Springer International Publishing AG 2017
D. Benslimane et al. (Eds.): DEXA 2017, Part I, LNCS 10438, pp. 447–462, 2017.
DOI: 10.1007/978-3-319-64468-4_34

a given entity, and can be declared using the Web Ontology Language (OWL) or RDF constraint languages; however, such declarations are hand-crafted and application dependent. Therefore, ontologies rarely include cardinality declarations in practice, highlighting the need for cardinality mining methods.

In this work, our main goal is to discover such structural patterns using a bottom-up or extensional approach for mining cardinality bounds from instance data. Cardinality bounds are hidden data patterns that unveil the structure of data, and in some cases they might not even be intuitive for data creators [14]. Our approach produces *accurate* cardinalities by taking into consideration the semantics of `owl:sameAs` [1] equality axioms in KBs; and *robust* ones by not assuming completely correct data, i.e., instance data can have errors. By doing so, the output of this work can serve users to analyse completeness and consistency of data, and thus contribute towards higher levels of quality in KBs [7,16,26].

Recently, RDF constraint languages (e.g., Shape Expressions [19], Shapes Constraint Language (SHACL) [2]) have been defined to satisfy the latent requirements for constraints definition and validation of RDF in real-world use cases. They build upon SPARQL or regular expressions to define so-called *shapes* that perform for RDF the same function as XML Schema, DTD and JSON Schema perform for XML or JSON: delimit the boundaries of instance data. Here, we consider that properties in KBs are constrained by multiplicities, which cover a critical aspect of relational data modelling, referring to the number of values that a property can have. Specifically, we consider a set of cardinalities or multiplicities as part of the internal structure of an entity type in a KB. In databases, internal structure-based methods are also referred to as constraint-based approaches, and have been used for schema matching [20]. Unlike databases, RDF and OWL assume the open-world semantics, and absence of the unique name assumption (nUNA). This makes the problem of extracting cardinalities more complex than a simple application of SPARQL queries using the `COUNT` operator. Take as example the constraint "a person must have two parents": if the data contain an entity of type `person` with only one parent, *this does not cause a logical inconsistency, it just means it is incomplete, and in RDF/OWL incomplete is different from inconsistent.* To deal with these specificities, we propose a method which tackles two important challenges: (1) *KB Normalisation*, where we must deal with `owl:sameAs` (or alike) axioms representing equality between entities, and (2) *Outliers Filtering*, where we account for the probability of noise in the data, in order to extract robust cardinality patterns.

This paper is organised as follows: In Sect. 2 we review the related work about cardinality, consistency, and schema discovery in KBs. Section 3 introduces some preliminaries about the RDF model. Section 4 provides a definition and semantics for cardinality bounds in KBs considering existing languages. In Sect. 5 we define an algorithm for mining cardinality bounds in an accurate and robust manner, and propose one implementation with two normalisation

---

[1] Henceforth, we use prefixes for namespaces according to http://prefix.cc/.

[2] https://www.w3.org/TR/shacl/ (accessed on February 13, 2017).

variants. Finally, we evaluate our algorithm over different datasets in Sect. 6, and present our conclusions and outlook in Sect. 7.

2 Related Work

Cardinality Constraints/Bounds. Cardinality constraints in RDF have been defined for data validation in languages such as OWL [13], Shape Expressions (ShEx) [19], OSLC Resource Shapes [24], and Dublin Core Description Set Profiles (DSP)[3]. OSLC integrity constraints include cardinality of relations which are more similar to UML cardinality for associations (i.e., exactly-one, one-or-many, zero-or-many, and zero-or-one). However, the expressivity of OSLC is limited compared to the definitions proposed in OWL[4], DSP, Shapes Constraint Language (SHACL), and Stardog ICV[5]. All of them define flexible boundaries for cardinality constraints: a lower bound in $\mathbb{N}$, and an upper bound in $\mathbb{N} \cup \{\infty\}$. SPIN[6] Modelling Vocabulary is yet another language that based on SPARQL to specify rules and logical constraints, including cardinality. It is worth pointing out that due to the bottom-up approach taken here we do not refer to our cardinalities as constraints but as *bounds*. They can be considered as constraints only after a user assessment and application over a given dataset. Despite this, our work builds upon existing approaches for cardinality constraints in RDF and other data models such as XML [3] and Entity-Relationship [10,28].

Consistency in RDF Graphs. Consistency is a relevant dimension of data quality, and many researchers have investigated the checking and handling of inconsistencies in RDF. However, to the best of our knowledge, this is the first work focused on the extraction and study of cardinalities to detect inconsistencies in RDF, through the application of outlier detection techniques. The concept of outliers or anomaly detection is defined as "finding patterns in data that do not conform to the expected normal behaviour" (see [2] for a survey). Under the assumption that KBs are likely to be noisy and incomplete [18], there exist several approaches that aim to enhance or refine their quality and completeness (see Paulheim [16] for a recent survey). Among the most relevant to our work are: [31] which applies unsupervised numerical outlier detection methods to DBpedia for detecting wrong values that are used as literal objects of a property; and [4] that builds upon [31] by identifying sub-populations of instances where the outlier detection works more accurately, and by using external datasets accessible from the `owl:sameAs` links. Our work differs from theirs in that they focus on missing property values, not on their cardinality or multiplicity.

RDF Schema Discovery. Our goal falls into the broader area of schema discovery. Völker and Niepert in [30] introduce a statistical approach where association rule mining is used to generate OWL ontologies from RDF data, but

---

[3] http://dublincore.org/documents/dc-dsp/.

[4] OWL allows the expression of cardinalities through the `minCardinality`, `maxCardinality`, and `cardinality` restrictions.

[5] http://docs.stardog.com/icv/icv-specification.html.

[6] http://spinrdf.org/.

consider cardinality restrictions (upper bounds) only as future work. Similarly, in [8] the authors extract type definitions described by profiles which consist of a property vector, where each property is associated to a probability. A further analysis of semantic and hierarchical links between types is performed to extract a global schema. Since most KBs are generated from semi- or un-structured data, [29] analysed the case of DBpedia enrichment with axioms identified during the extraction process from Wikipedia. Such axioms are identified with methods from Inductive Logic Programming (ILP), like in [30]. Despite their bottom-up or extensional approach, similar to ours, such works aim to build or enrich ontologies with missing relations, not considering any notion of cardinality nor their use to analyse completeness and/or consistency. A related approach to detect cardinality in KBs is presented by Rivero et al. in [21], which uses SPARQL 1.1 queries to automatically discover ontological models. Such models include types and properties, subtypes, domain, range, and minimum cardinalities of these properties. However, the approach presented in [21] is not able to deal with the semantics of data: both the existence of `owl:sameAs` axioms and outliers or errors in the data are ignored. For these reasons, our work is orthogonal and complementary to all aforementioned works.

3   Preliminaries

Below we first provide some preliminaries regarding RDF and its semantics.

RDF Model. Let $\mathcal{R}$ be the set of *entities*, $\mathcal{B}$ the set of *blank nodes*, $\mathcal{P}$ the set of *predicates*, and $\mathcal{L}$ the set of *literals*. A finite knowledge base $\mathcal{G}$ is a set of *triples* $t := (s, p, o) \in (\mathcal{R} \cup \mathcal{B}) \times \mathcal{P} \times (\mathcal{R} \cup \mathcal{B} \cup \mathcal{L})$, where s is the *subject*, p is the *predicate*, and o is the *object* of t. We define the functions $\text{PRED}_\mathcal{G}(\tau) = \{p \mid \exists s, o \ (s, p, o) \in \mathcal{G} \wedge (s, \text{rdf:type}, \tau) \in \mathcal{G}\}$ that returns a set of predicates appearing with instances of entity type τ; and $\text{TRIPLES}_\mathcal{G}(p) = \{(s, p, o) \mid \exists s, o \ (s, p, o) \in \mathcal{G}\}$ that returns the triples in $\mathcal{G}$ with property p.

UNA 2.0. The *unique name assumption* (UNA) is a simplifying assumption made in some ontology languages and description logics. It means that two different names always refer to different entities in the world [23]. On the one hand, OWL default semantics does not adopt the UNA, thus two different constants can refer to the same individual—a desirable behaviour in an environment such as the Web. On the other hand, validation checking approaches in RDF usually adopt a *closed world assumption* (CWA) with UNA, i.e., inferring a statement to be false on the basis of failure to prove it, and if two entities are named differently, they are assumed to be different entities. To deal with this, SHACL defines the so-called *UNA 2.0* which is a simple workaround where all entities are treated as different, unless explicitly stated otherwise by `owl:sameAs`. From a practical point of view, it is a desirable feature for mining algorithms to consider the semantics of RDF graphs avoiding misinterpretations of the data. Figure 1 (left) shows an example where the adoption of normal UNA will lead to a count of five different entities (i.e., `ex:A`, `ex:B`, `ex:C`, `ex:D`, `ex:E`), and

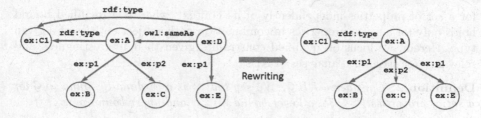

Fig. 1. Example of UNA 2.0 as defined in SHACL.

for `ex:A` the property `ex:p1` would have cardinality 1. While adopting UNA 2.0 (Fig. 1, right) changes the counts: now we have four different entities (i.e., `ex:A`, `ex:B`, `ex:C`, `ex:E`), and the cardinality of `ex:p1` in `ex:A` is 2. Here, we call *rewriting* the process of applying UNA 2.0 to an unnormalised KB.

Since the cardinality of a property is severely affected when `owl:sameAs` axioms are not considered, hereafter, we adopt UNA 2.0 and satisfy this requirement, allowing us to correctly interpret the KB semantics.

4   Cardinality Bounds in RDF

In this section we introduce a definition of cardinality bounds in KBs that generalises the semantics of the definitions discussed in Sect. 3.

Cardinality (also known as multiplicity) covers a critical aspect of relational data modelling, referring to the number of times an instance of one entity can be related with instances of another entity. A *cardinality bound* is a restriction on the number of elements in the relation. In particular, in KBs we have relations between entities of a given type through properties, and we want to specify bounds for such relationships. For example, we would like to express that a drug has only one molecular formula, but can be associated to a finite set (of known or unknown size, the latter denoted as ∞) of adverse drug reactions.

Definition 1. *A cardinality bound in RDF knowledge bases restricts the number of property values related with an entity in a given context (i.e., a particular type, or the whole KB). Formally, a cardinality bound φ is an expression of the form $card(P, \tau) = (min, max)$ where $P \subseteq \mathcal{P}$, τ is an entity type, and where $min \in \mathbb{N}$ and $max \in \mathbb{N} \cup \{\infty\}$ with $min \leq max$. Here $|P|$ denotes the number of properties in φ, min is called the* lower bound, *and max the* upper bound *of φ. If τ is defined ($\tau \neq \varepsilon$), we say that φ is* qualified; *otherwise we say that φ is* unqualified.

The semantics of this definition of cardinality bounds limits the maximum and minimum count that a given set of properties can have in a given context as in SHACL, DSP, ICV and OWL. The lower bound of a cardinality may take on values in $\mathbb{N}$, whilst upper bounds can be ∞ to represent that there is an unknown upper limit. An unqualified bound is independent of a type (context), i.e., it holds for a set of properties independently of its context, whereas a qualified bound holds only for a set of properties in combination with subject entities of a given type. Herein, we focus on qualified constraints given their interestingness and relevance for structural analyses of KBs.

Definition 2. *Consider a KB $\mathcal{G}$. We say that φ is a cardinality bound in $\mathcal{G}$ for a set of properties $P_\varphi \subseteq \mathcal{P}$, a lower bound min_φ, and upper bound max_φ, if*

$$\forall s \in (\mathcal{R} \cup \mathcal{B}) \ (min_\varphi \leq |\{p : p \in P_\varphi \wedge \exists o \ (s, p, o) \in \mathcal{G}\}| \leq max_\varphi).$$

If φ is qualified to τ then to satisfy φ, $\mathcal{G}$ also needs to satisfy the condition that $\forall s \in (\mathcal{R} \cup \mathcal{B}) \ (s, \mathrm{rdf : type}, \tau) \in \mathcal{G}$.

Although our approach is able to compute an upper bound cardinality, this limit is uncertain when considering RDF's open world assumption (OWA). For instance, even when the data show that an entity person has maximum two children, this might be wrong when considering other entities. More certain cardinality bounds can be mined from reliable or complete graphs usually existent within specific domains. Therefore, we refer to cardinality bounds as "patterns" when they are automatically extracted from raw KBs, and as "constraints" when normatively assessed by a user and applied in order to restrict a KB.

```
1  SELECT $this
2  WHERE {
3          $this $PROPERTY ?value .
4      }
5  } GROUP BY $this
6  HAVING (COUNT(?value) < $minCount)
```

Fig. 2. SPARQL 1.1 definition of a minimum cardinality constraint.

In practice, cardinality bounds can be used to validate KBs using SPARQL 1.1 queries. For instance, Fig. 2 shows the SPARQL query proposed to validate a lower bound min_φ ($minCount). The query represents restrictions on the number of values, ?value, that the $this node may have for the given property. A validation result must be produced if the number of value nodes is less than $minCount. Similarly, to validate an upper bound (max_φ) restriction for a property, we can change the HAVING condition to '>'. Note that SHACL, ShEx, and other constraint languages only allow the definition of one condition at a time per property. Therefore, to validate our cardinalities with multiple properties, one must apply an SPARQL 1.1 query like the one in Fig. 2 independently for each property and bound. In Sect. 5 we will show how a single SPARQL 1.1 query can be used to extract both minimum and maximum bounds at once.

Example 1. *The following expressions define cardinality bounds for different entity types in different domains.*

1. $card(\{ \mathrm{mondial : name}, \mathrm{mondial : elevation} \}, \mathrm{mondial : Volcano}) = (1, 1)$,

2. $card(\{\,\mathtt{mondial:hasCity}\,\},\mathtt{mondial:Country}) = (1, \infty)$,
3. $card(\{\,\mathtt{dcterms:contributor}\,\},\mathtt{bibo:Book}) = (0, \infty)$,
4. $card(\{\,\mathtt{dcterms:language}\,\},\mathtt{bibo:Book}) = (1, 2)$.

As suggested in the previous example, when the upper bound is unclear we use ∞ in the cardinality bound to express that uncertainty.

5 Mining Cardinality Patterns

In the following, we introduce our main algorithm for mining cardinality patterns from KBs. We also present two different implementations: one based on SPARQL 1.1 that uses a graph databases approach to normalise and extract cardinalities; and another based on Apache Spark that applies a MapReduce or divide-and-conquer strategy to divide the data and run the steps in parallel.

5.1 Algorithm

We present Algorithm 1 as an efficient solution to mine accurate and robust cardinality patterns from any KB. This algorithm is designed to mine qualified cardinalities, i.e., there is a context type; however, it can be easily extended to mine unqualified cardinalities. From a data quality perspective, it is desirable that the mined cardinalities bounds (see Sect. 4) are accurate and robust. Algorithm 1 outputs a set of cardinality patterns, which are called "accurate" because we consider the semantics of owl:sameAs axioms, and "robust" because we perform an outliers detection and filtering over noisy cardinality counts.

Our mining algorithm has three main parts: (1) <u>KB normalisation:</u> represented by the NORMALISE($\cdot$) function, receives an (unnormalised or with multiple equal entities) KB as input and applies an on-the-fly rewriting process to consider the semantics of owl:sameAs (or other alike relation), where one can build cliques from grouping equal entities. This function could be considered optional though in cases where users want information about unnormalised bounds—at the cost of accuracy. (2) <u>Cardinalities extraction:</u> performed by the function CARDPATTERNS($\cdot$), it is called to retrieve cardinality pairs (entity, cardinality) from the data for a given property. The cardinalities for a fixed property are stored in a map which is after used to identify noisy values. (3) <u>Outliers filtering:</u> represented by the FILTEROUTLIERS($\cdot$) function, receives a map of (entity, cardinality) pairs and applies unsupervised univariate statistical methods to identify and remove noisy or outside of a range values to ensure robustness.

Next, we present an example for the application of Algorithm 1, and describe each of its part in more details in Sects. 5.2 to 5.4.

Example 2. *Let us consider a KB with entities* ex:s1 *and* ex:s2*, and properties* ex:p1 *and* ex:p2*. The* NORMALISE *function takes all triples in the KB and replaces duplicates by one representative element equivalence type induced by* owl:sameAs*-cliques. Then, for each property it extracts the cardinality values using the function* CARDPATTERNS *to get values:* $[1, 1, 1]$ *for* ex:p1*, and*

Algorithm 1. CARDBOUNDS: Extraction of cardinality bounds.

Input: a knowledge base $\mathcal{G}$; and a context τ
Output: a set Σ of cardinality bounds
 1: $\mathcal{G}' \leftarrow$ NORMALISE$(\mathcal{G}, \tau)$
 2: $P \leftarrow$ PRED$_{\mathcal{G}'}(\tau)$ ◁ *Retrieve all predicates for entities of type τ*
 3: **for all** $p \in P$ **do**
 4:     $\mathcal{D} \leftarrow$ TRIPLES$_{\mathcal{G}'}(p)$ ◁ *Retrieve all triples with property p*
 5:     $\mathcal{M} \langle u, v \rangle \leftarrow$ CARDPATTERNS$(\mathcal{D})$ ◁ *u is an entity, and v a cardinality*
 6:     *inliers* $\leftarrow$ FILTEROUTLIERS$(\mathcal{M}.v)$
 7:     $\Sigma.add(card(\{p\}, \tau) = (MIN(inliers),$ MAX$(inliers)))$
 8: **end for**

$[3, 3, 25]$ *for* `ex:p2` *. Next, the function* FILTEROUTLIERS *determines that there are no outliers for property* `ex:p1` *, but that a cardinality of 25 is an outlier for* `ex:p2` *. Thus, 25 is removed from the patterns leaving* $[3, 3]$ *as robust cardinalities for* `ex:p2` *. Finally, the cardinality bounds (min, max) are extracted from the remaining inlier cardinalities by using simple* MIN *and* MAX *functions.*

5.2 Knowledge Bases Equality Normalisation

Knowledge bases contain different types of axioms, being `owl:sameAs` and alike the most important when computing cardinalities. Regardless of the approach, by not considering these axioms a method loses its accuracy and cannot ensure that the cardinality bounds are consistent with the data. Unlike [21], we perform an on-the-fly normalisation of the graph in order to capture the semantics of `owl:sameAs` axioms without having to modify the underlying data. A simple SPARQL query using the COUNT operator will return two instances of `ex:C1` instead of the expected count 1 for the example in Fig. 1 (left). To overcome this issue, we propose an axiomatisation with two rules, namely, *subject-equality* and *object-equality* (Table 1), where duplicated elements are replaced by one representative element equivalence type induced by `owl:sameAs`. This normalisation can be done replacing the underlying data [12] or on-the-fly (without modification) when needed. However, if the underlying data is modified, the links to other KBs stated by the `owl:sameAs` axioms are overwritten and lost. Instead, here we follow an on-the-fly overwrite (line 1 in Algorithm 1) which performs the modifications in memory. This was also used by Schenner et al. in [25]. The representatives are selected from the so-called `owl:sameAs`-*cliques*, which are sets of entities all of which are equal to each other [12]. In practice, the axiomatisation of Table 1 can be implemented on-the-fly either by using SPARQL 1.1 or programmatically. We briefly introduce these two options as follows:

SPARQL Rewrite. We make use of the nested SPARQL 1.1 query in Fig. 4, which contains three sub-queries and aims to obtain (entity, cardinality) pairs for a given property and entity type (line 3). Embedded in `SQ-1` are `SQ-2` and `SQ-3` performing the clique generations for subject and object, respectively. Sub-query

Table 1. An axiomatisation for reduction on equality

$(s,p,o) \wedge (s',p',o') \wedge$ $(s',\texttt{owl:sameAs},s)$	$(s,p,o) \wedge (s',p',o') \wedge$ $(o',\texttt{owl:sameAs},o)$
$(s,p,o),(s,p',o')$	$(s,p,o),(s',p',o)$
(subject-equality)	(object-equality)

SQ-2 applies the subject-equality rule, and sub-query SQ-3 applies the object-equality rule. In a clique generation, a graph search is done in all directions of the graph to find equal entities, which incurs in a high complexity. For each clique found, a representative is selected, thus omitting all "clone" entities. Intuitively, the query used here can be seen as complex and resource demanding. Hence, we also propose a faster solution that works outside of a SPARQL endpoint.

Programmatic Rewrite. We can also frame the extraction of cardinality patterns as the well-known words count problem. So, we can easily parallelise the algorithm using frameworks such as Apache Spark[7]. By using Spark and the `filer` and `map` operations, we implemented a parallel rewrite, where the `owl:sameAs`-cliques are generated and used to normalise the KB triple by triple (see Fig. 3, left). We generate the `owl:sameAs`-cliques as follows: for each $(s,\texttt{owl:sameAs},o)$ triple we lexically compare s and o and select the minimum (e.g., s), which becomes the *representative*; add a mapping from the minimum to the other (e.g., from s to o); if the no-minimum (e.g., o) was the representative of other entities, then we update their mappings in cascade with the new representative (e.g., s). We then apply a `map` operation over each initial triple and overwrite it according to the `owl:sameAs`-cliques to obtain $\mathcal{G}'$ in $O(1)$.

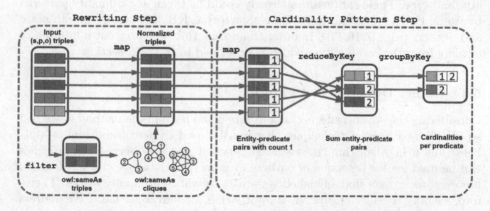

Fig. 3. Cardinality patterns extraction using Apache Spark.

---

[7] http://spark.apache.org/ (version 2.1.0).

```
1  PREFIX rdf: <http://www.w3.org/1999/02/22-rdf-syntax-ns#>
2  PREFIX owl: <http://www.w3.org/2002/07/owl#>
3  SELECT ?first_subj (COUNT(DISTINCT ?first_obj) AS ?nbValues) WHERE {
4      { SELECT DISTINCT ?first_subj ?first_obj WHERE {            % (SQ-1)
5          ?subj $property ?obj .
6          { SELECT ?subj ?first_subj WHERE {                      % (SQ-2)
7              ?subj a $type .
8              ?subj ((owl:sameAs|^owl:sameAs)*) ?first_subj .
9                  ?notfirst ((owl:sameAs|^owl:sameAs)*) ?first_subj .
10                 FILTER (STR(?notfirst) < STR(?first_subj))
11             } FILTER(!BOUND(?notfirst))
12         }}
13         { SELECT ?obj ?first_obj WHERE {                        % (SQ-3)
14             ?obj ((owl:sameAs|^owl:sameAs)*) ?first_obj .
15                 ?notfirst ((owl:sameAs|^owl:sameAs)*) ?first_obj .
16                 FILTER (STR(?notfirst) < STR(?first_obj))
17             } FILTER(!BOUND(?notfirst))
18         }}
19     }}
20 } GROUP BY ?first_subj
```

Fig. 4. Query the cardinality of a property for every entity of a given type.

5.3 Detection of Cardinality Patterns

After the normalisation step, cardinalities can be collected for each property (line 5 in Algorithm 1) ensuring their accuracy, which is a major difference w.r.t. previous approaches such as Rivero et al. [21]. In the SPARQL-based approach, Fig. 4 shows a query which performs both the normalisation of $\mathcal{G}$ and the detection of cardinality patterns in one place. Complex SPARQL queries are hard to evaluate and optimise [27], making this approach very inefficient and poorly scalable. On the other hand, the Spark-based approach can make use of multiple machines to scale and process the KB in splits. We show a comparison of both approaches latter in Sect. 6. Regardless of the approach, the output of the cardinality extraction is a map of (entity, cardinality) pairs for a given property and type. These cardinalities already could be taken as cardinality patterns by users. However, several works have shown that KBs frequently contain noise and outliers (e.g., [7,16,17]). In order to address this, we carry out a filtering of outliers from the cardinalities, which is described in the next section.

5.4 Outlier Detection and Filtering

Considering the adverse effects that outliers could have in the method described so far, we now present techniques that can be used to detect and remove outliers (line 6 in Algorithm 1). Several supervised and unsupervised approaches can be used for the detection of outliers in numerical data (see [18] for details); however, we did not find any labelled dataset for valid cardinality values. Therefore, we only consider unsupervised approaches for univariate data. We address the detection of outliers in a sequence of numbers as a statistical problem. Interestingly, outlier detection approaches determine a lower and upper bound on the range of data, similarly to the semantics of a cardinality bound. The *extreme studentized deviation* (ESD) identifier [22] is one of the most popular approaches and computes the mean μ and standard deviation σ values and considers as outlier any value outside of the interval $[\mu - t \cdot \sigma, \mu + t \cdot \sigma]$, where $t = 3$ is usually used. The problem with ESD is that both the mean and the standard deviation are themselves sensitive to the presence of outliers in the data. *Hampel* identifier [18] appears as an option, where the mean is replaced by the median *med*, and the standard deviation by the median absolute deviation (*MAD*). The range for outliers is now: $[med - t \cdot MAD, med + t \cdot MAD]$. Since the median and MAD are more resistant to the influence of outliers than the mean and standard deviation, Hampel identifier is generally more effective than ESD. Although, Hampel sometimes could be considered too aggressive, declaring too many outliers [18]. Boxplot appears as a third option, and defines the range: $[Q1 - c \cdot IQD, Q3 + c \cdot IQD]$, where $Q1$ and $Q3$ are the lower and upper quartiles, respectively, and $IQD = Q3 - Q1$ is the interquartile distance—a measure similar to the standard deviation. The parameter c is similar to t in Hampel and ESD, and is commonly set to $c = 1.5$. Boxplot is better suited for distributions that are moderately asymmetric, because it does not depends on an estimated "centre" of the data. Thus, in our evaluation we use boxplot rule to determine cardinality outliers.

6 Evaluation

Next, we evaluate the application of our mining algorithm in its two variants against real-world and synthetic KBs. After, we use the mined cardinality bounds to analyse the notions of completeness and consistency of KBs.

6.1 Settings

Datasets. We used five datasets with different number of triples and owl:sameAs axioms. The chosen datasets are diverse in domain, features, and represent real-world and synthetic data. We present their characteristics in Table 2, and describe them as follows:

Table 2. Datasets characteristics

DATASET	№ TRIPLES	№ TYPES	№ PROP.	№ SAMEAS
OpenCyc	2,413,894	7,613	165	360,014
UOBM	2,217,286	40	29	0
British National Library	210,820	24	45	14,761
Mondial	186,534	27	60	0
New York Times People	103,496	1	20	14,884

- OpenCyc[8] is a large general KB released in 2012 that contains hundreds of thousands of terms in the domain of human knowledge covering places, organisations, business related things, people among others.
- UOBM[9] is a synthetic dataset that extends the Lehigh University Benchmark (LUMB), a university domain ontology, that contains information about faculties and students.
- British National Library[10] is a dataset published by the National Library of the UK (second largest library in the world) about books and serials.
- Mondial[11] is a database compiled from geographical Web data sources such as CIA World Factbook, and Wikipedia among others.
- New York Times People[12] is a compilation of the most authoritative people mentioned in news of the New York Times newspaper since 2009.

Test Settings. We implemented our cardinality mining Algorithm 1 using Python 3.4 and Apache Spark 2.1.0. We use a Intel Core i7 4.0 GHz machine with 32 GB of RAM running Linux kernel 3.2 to run experiments on different KBs. Although Spark can run on multiple machines, we only tested it on a single machine using multiple parallel processes—one per core using 8 cores in total.

6.2   Results

First, we quantitatively compare the runtime of Algorithm 1 with both implementations, SPARQL and Spark, to normalise (rewrite) KBs, retrieve, and filter cardinality bounds. We then analyse qualitatively the use of the identified cardinality bounds to assess two crucial notions of KBs and databases: completeness and consistency.

Quantitative Evaluation. Intuitively, based on the scalability of Spark, one can foresee that the parallelised variant of our algorithm (Fig. 3) will outperform the other that uses SPARQL. To test this we ran both implementations on the British National Library (BNL) and Mondial datasets, where only the first contains `owl:sameAs` axioms. For BNL, we considered the type $\tau =$ Book with 7 predicates and obtained average runtime of 253.908 s for the SPARQL implementation, and 15.634 s for the Spark one. This shows that the Spark implementation is 16x faster than using SPARQL, while performing the same task on BNL dataset. For Mondial, we considered the type $\tau =$ River with 8 predicates and obtained average runtime of 117.739 s for the SPARQL implementation, and 2.948 s for the Spark one. This shows that the Spark implementation is 40x faster than using SPARQL, while performing the same task on Mondial dataset. It is worth pointing out that the times on Mondial dataset are lower than for BNL because of the lower number of instances and the absence of `owl:sameAs`

---

[8] http://www.cyc.com/platform/opencyc.
[9] https://www.cs.ox.ac.uk/isg/tools/UOBMGenerator/.
[10] http://www.bl.uk/bibliographic/download.html.
[11] http://www.dbis.informatik.uni-goettingen.de/Mondial/.
[12] https://datahub.io/dataset/nytimes-linked-open-data.

axioms. Finally, our experiments show that the outlier detection method (i.e., boxplot) does not add a significant overhead to the whole process and scales well for different data sizes.

Qualitative Evaluation. The characteristics of the datasets range between 1 up to 7,613 types and 20 up to 165 properties. To keep our study manageable, we selected randomly one entity type per dataset (5 in total) and five properties per type (25 in total). For each type, we show (see Table 3) the number of `owl:sameAs`-cliques generated, and the number of triples before and after the rewriting process. To show the benefits of studying cardinality constraints derived from automatically discovered bounds in KBs, we bring to the fore their use on the realm of completeness and consistency. We evaluate each entity type in these two dimensions from a common sense point of view. Because the consideration of cardinality bounds is application dependent, here we try to abstract (without loss of generality) from individual use cases. The cardinalities presented herein are considered robust bound assessed to be a constraint by a knowledge engineer. We consider that a property p in the context of a type τ is *complete* given a cardinality constraint if every entity s of type τ has the 'right number' of triples (s, p, o), and *incomplete* otherwise. For example, a constraint might be that all books must have at least one property `title`, but the same it is not true for property `comment`. Also, we consider that a property p in the context of a type τ is *consistent* if the triples with predicate p and subject s (of type τ) comply with the cardinality bounds, and *inconsistent* otherwise. For example, a constraint might be that all books must have always 1 `title`; however, we found five books which violate this constraint having 2 titles. Based on a set of verified discovered robust bounds, in Table 3 we show the ratios of completeness and consistency found in the 5 properties per type. For example, 2/5 completeness ratio in the entity type Book indicates that 2 out of 5 properties presented complete data, and the rest was incomplete. We did the same to measure consistency. In particular, we noticed a strong consistency on synthetic datasets, where it is normal to define an ontology that all instances generated will satisfy.

Table 3. Evaluation of completeness and consistency per dataset: one type and five random properties per type.

Class	№ sA-Cliques	№ Triples Befo./After	Complet. Ratio	Consist. Ratio
Fashion Model	118	1060/928	2/5	5/5
Research Assistant	0	135197/135197	4/5	5/5
Book	4515	97101/83556	2/5	3/5
Country	0	21766/21766	1/5	4/5
Concept	4979	58685/48780	2/5	5/5

Figure 5 shows a subset of the cardinality bounds extracted for the entity type Volcano in Mondial KB. At a first glance, the extracted patterns correspond to what a knowledge engineer could expect: all volcanoes have a name and latitude/longitude coordinates, not all have information about their last eruption, and they have 1 to 3 locations for those that are in the intersection of different countries. Notice that even when we obtain a cardinality for `mondial:locatedIn` with upper bound 3, this cannot be considered as a constraint and used for validation until it is assessed by a user. We found a similar situation with the property `mondial:hasCity` in the type `mondial:Country`, where the robust cardinality identified was $card(\{mondial:hasCity\}, mondial:Country) = (1,31)$. However, based on that upper limit, there are 25 identified outliers, which are not real outliers. For instance, we have China with 306 cities, USA with 250 cities, Brazil with 210 cities, Russia with 171 cities, and India with 99 cities, which are outside of the range considered as robust. Based on this information, data users and knowledge engineers could be able to determine whether a given cardinality pattern should be promoted to become a constraint (i.e., verified discovered robust bounds) or not. We argue that by detecting cardinality inconsistencies and incompleteness we can determine structural problems at the instance level. This can be used to guide repair methods and move towards better quality of KBs.

```
1  card({mondial:name}, mondial:Volcano)=(1,1)
2  card({mondial:elevation}, mondial:Volcano)=(1,1)
3  card({mondial:longitude}, mondial:Volcano)=(1,1)
4  card({mondial:latitude}, mondial:Volcano)=(1,1)
5  card({mondial:locatedIn}, mondial:Volcano)=(1,3)
6  card({mondial:lastEruption}, mondial:Volcano)=(0,1)
```

Fig. 5. Cardinality bounds subset for entity type `mondial:Volcano`.

7   Conclusions and Outlook

KBs contain implicit patterns and constraints not always stated in ontologies or schemata neither clear for data creators and consumers. In this paper, we have introduced an extensional approach to the discovery of so-called cardinality bounds in KBs. Cardinality bounds are proposed as a solution for the need to (partially) cope with the lack of explicit structure in KBs. We presented two implementations of our approach, namely SPARQL- and Spark-based, and evaluated them against five different datasets including real-world and synthetic data. Our analysis shows that cardinality bounds can be mined efficiently, and are useful to understand the structure of data at a glance, and also to analyse the completeness and consistency of data.

Future work in this area could go into various directions. First, we notice that cardinality bounds give an indication about the completeness of data, and thus can be used to guide methods for knowledge completion or link prediction [5]. Second, further research is required on the assessment of the discovered cardinality bounds by users, and whether they can be promoted as normative constraints that could be used for validation [19]. Finally, the structure provided by cardinality patterns could serve to generate schema graphs that represent the characteristics that KBs naturally exhibit, unlocking management problems such as query optimisation [27].

Acknowledgements. This work has been supported by TOMOE project funded by Fujitsu Laboratories Ltd., Japan and Insight Centre for Data Analytics at National University of Ireland Galway, Ireland.

References

1. Bosch, T., Eckert, K.: Guidance, please! Towards a framework for RDF-based constraint languages. In: Proceedings of the International Conference on Dublin Core and Metadata Applications (2015)
2. Chandola, V., Banerjee, A., Kumar, V.: Anomaly detection: a survey. ACM Comput. Surv. **41**(3), 15 (2009)
3. Ferrarotti, F., Hartmann, S., Link, S.: Efficiency frontiers of XML cardinality constraints. Data Knowl. Eng. **87**, 297–319 (2013)
4. Fleischhacker, D., Paulheim, H., Bryl, V., Völker, J., Bizer, C.: Detecting errors in numerical linked data using cross-checked outlier detection. In: Mika, P., et al. (eds.) ISWC 2014. LNCS, vol. 8796, pp. 357–372. Springer, Cham (2014). doi:10. 1007/978-3-319-11964-9_23
5. Galárraga, L., Razniewski, S., Amarilli, A., Suchanek, F.M.: Predicting completeness in knowledge bases. In: WSDM, pp. 375–383. ACM (2017)
6. Glimm, B., Hogan, A., Krötzsch, M., Polleres, A.: OWL: yet to arrive on the web of data? In: LDOW, CEUR Workshop Proceedings, vol. 937. CEUR-WS.org (2012)
7. Hogan, A., Harth, A., Passant, A., Decker, S., Polleres, A.: Weaving the pedantic web. In: LDOW, CEUR Workshop Proceedings, vol. 628. CEUR-WS.org (2010)
8. Kellou-Menouer, K., Kedad, Z.: Evaluating the gap between an RDF dataset and its schema. In: Jeusfeld, M.A., Karlapalem, K. (eds.) ER 2015. LNCS, vol. 9382, pp. 283–292. Springer, Cham (2015). doi:10.1007/978-3-319-25747-1_28
9. Lausen, G., Meier, M., Schmidt, M.: SPARQLing constraints for RDF. In: EDBT, pp. 499–509 (2008)
10. Liddle, S.W., Embley, D.W., Woodfield, S.N.: Cardinality constraints in semantic data models. Data Knowl. Eng. **11**(3), 235–270 (1993)
11. Motik, B., Horrocks, I., Sattler, U.: Bridging the gap between OWL and relational databases. Web Seman.: Sci. Serv. Agents World Wide Web **7**(2), 74–89 (2009)
12. Motik, B., Nenov, Y., Piro, R.E.F., Horrocks, I.: Handling Owl:sameAs via rewriting. In: AAAI, pp. 231–237. AAAI Press (2015)
13. Motik, B., Patel-Schneider, P.F., Parsia, B.: OWL 2 Web Ontology Language structural specification and functional-style syntax, 2nd edn (2012). http://www.w3. org/TR/2012/REC-owl2-syntax-20121211/
14. Muñoz, E.: On learnability of constraints from RDF data. In: Sack, H., Blomqvist, E., d'Aquin, M., Ghidini, C., Ponzetto, S.P., Lange, C. (eds.) ESWC 2016. LNCS, vol. 9678, pp. 834–844. Springer, Cham (2016). doi:10.1007/978-3-319-34129-3_52
15. Neumann, T., Moerkotte, G.: Characteristic sets: accurate cardinality estimation for RDF queries with multiple joins. In: ICDE, pp. 984–994. IEEE Computer Society (2011)
16. Paulheim, H.: Knowledge graph refinement: a survey of approaches and evaluation methods. Semant. Web **8**(3), 489–508 (2017)

17. Paulheim, H., Bizer, C.: Improving the quality of linked data using statistical distributions. Int. J. Semant. Web Inf. Syst. **10**(2), 63–86 (2014)
18. Pearson, R.K.: Mining Imperfect Data: Dealing with Contamination and Incomplete Records. Society for Industrial and Applied Mathematics, Philadelphia (2005)
19. Prud'hommeaux, E., Gayo, J.E.L., Solbrig, H.R.: Shape expressions: an RDF validation and transformation language. In: SEMANTICS, pp. 32–40. ACM (2014)
20. Rahm, E., Bernstein, P.A.: A survey of approaches to automatic schema matching. VLDB J. **10**(4), 334–350 (2001)
21. Rivero, C.R., Hernández, I., Ruiz, D., Corchuelo, R.: Towards discovering ontological models from big RDF data. In: Castano, S., Vassiliadis, P., Lakshmanan, L.V., Lee, M.L. (eds.) ER 2012. LNCS, vol. 7518, pp. 131–140. Springer, Heidelberg (2012). doi:10.1007/978-3-642-33999-8_16
22. Rosner, B.: Percentage points for a generalized ESD many-outlier procedure. Technometrics **25**(2), 165–172 (1983)
23. Russell, S., Norvig, P.: Artificial Intelligence: A Modern Approach, 2nd and 3rd edn. Pearson Education, London (2009)
24. Ryman, A.G., Hors, A.L., Speicher, S.: OSLC resource shape: a language for defining constraints on linked data. In: Proceedings of the WWW 2013 Workshop on Linked Data on the Web (2013)
25. Schenner, G., Bischof, S., Polleres, A., Steyskal, S.: Integrating distributed configurations with RDFS and SPARQL. In: Configuration Workshop, CEUR Workshop Proceedings, vol. 1220, pp. 9–15. CEUR-WS.org (2014)
26. Schmidt, M., Lausen, G.: Pleasantly consuming Linked Data with RDF data descriptions. In: COLD. CEUR-WS.org (2013)
27. Schmidt, M., Meier, M., Lausen, G.: Foundations of SPARQL query optimization. In: ICDT, pp. 4–33. ACM (2010)
28. Thalheim, B.: Fundamentals of cardinality constraints. In: Pernul, G., Tjoa, A.M. (eds.) ER 1992. LNCS, vol. 645, pp. 7–23. Springer, Heidelberg (1992). doi:10.1007/3-540-56023-8_3
29. Töpper, G., Knuth, M., Sack, H.: DBpedia ontology enrichment for inconsistency detection. In: I-SEMANTICS, pp. 33–40. ACM (2012)
30. Völker, J., Niepert, M.: Statistical schema induction. In: Antoniou, G., Grobelnik, M., Simperl, E., Parsia, B., Plexousakis, D., Leenheer, P., Pan, J. (eds.) ESWC 2011. LNCS, vol. 6643, pp. 124–138. Springer, Heidelberg (2011). doi:10.1007/978-3-642-21034-1_9
31. Wienand, D., Paulheim, H.: Detecting incorrect numerical data in DBpedia. In: Presutti, V., d'Amato, C., Gandon, F., d'Aquin, M., Staab, S., Tordai, A. (eds.) ESWC 2014. LNCS, vol. 8465, pp. 504–518. Springer, Cham (2014). doi:10.1007/978-3-319-07443-6_34

Incremental Frequent Itemsets Mining
with IPPC Tree

Van Quoc Phuong Huynh[1]([⊠]), Josef Küng[1], and Tran Khanh Dang[2]

[1] Faculty of Engineering and Natural Sciences (TNF),
Institute for Application Oriented Knowledge Processing (FAW),
Johannes Kepler University (JKU), Linz, Austria
{vqphuynh, jkueng}@faw.jku.at
[2] Faculty of Computer Science and Engineering,
HCMC University of Technology, Ho Chi Minh City, Vietnam
khanh@hcmut.edu.vn

Abstract. Frequent itemsets mining plays a fundamental role in many data
mining problems such as association rules, correlations, classifications, etc. The
current era of Big Data, however, has been challenging mining techniques for its
large scale datasets. To tackle this problem, we propose a solution for incrementally mining frequent itemsets through an algorithm named IFIN, a prefix
tree structure IPPC-Tree and related algorithms. Instead of using a global order
of frequent items in a dataset like FP-Tree, the IPPC-Tree maintains a local and
changeable order of items in a path of nodes from the root to a leave. This allows
the IFIN not to waste computational overhead for the previously processed part
of data when a new one is added or the support threshold is changed. Therefore,
much processing time is saved. We conducted experiments to evaluate our
solution against the well-known algorithm FP-Growth and other two
state-of-the-art ones FIN and PrePost+. The experimental results show that IFIN
is the most efficient in time and memory consuming, especially the mining time
with different support thresholds.

Keywords: Incremental · Frequent itemsets mining · Data mining · Big Data ·
IPPC-Tree · IFIN

1 Introduction

Frequent itemsets mining can be briefly described as follows. Given a dataset of n
transactions $D = \{T_1, T_2, \ldots, T_n\}$, the dataset contains a set of m distinct items
$I = \{i_1, i_2, \ldots, i_m\}$, $T_i \subseteq I$. A k-itemset, IS, is a set of k items $(1 \leq k \leq m)$. Each itemset
IS possesses an attribute, *support*, which is the number of transactions containing IS.
The problem is featured by a support threshold ε which is the percent of transactions in
the whole dataset D. An itemset IS is called frequent itemset iff $IS.support \geq \varepsilon * n$. The
problem is to discover all frequent itemsets existing in D.

Discovering frequent itemsets in a large dataset is an important problem in data
mining. Nowadays, in Big Data era, this problem as well as other mining techniques
has been being challenged by very large volume and high velocity of datasets. To

© Springer International Publishing AG 2017
D. Benslimane et al. (Eds.): DEXA 2017, Part I, LNCS 10438, pp. 463–477, 2017.
DOI: 10.1007/978-3-319-64468-4_35

confront with this challenging, we introduce a prefix tree structure IPPC-Tree (Incremental Pre-Post-Order Coding Tree), mining algorithm IFIN (Incremental Frequent Itemset Nodeset) and relevant algorithms aiming at providing an efficient method for discovering frequent itemsets with incremental mining ability. IPPC-Tree is a combination of (1) the idea of flexible and local order of items in a path from the root to a leave node in CATS-Tree [3] and (2) the PPC-Tree [5] which each node in PPC-Tree is identified by a pair of codes: *pre-order* and *post-order*. The tree construction takes only one time of dataset scanning and is independent to the support threshold. This allows a constructed IPPC-Tree to be built up with an additional dataset. The IFIN algorithm is an enhanced version of the state-of-the-art algorithm FIN [4] through providing the ability of incremental mining and adapting to be fitting with the IPPC-Tree structure. By that solution, the running time is significantly saved.

The rest of this paper is organized as follows. In Sect. 2, some related works are presented. Section 3 introduces the IPPC-Tree structure, relevant algorithms and preliminaries. The algorithm IFIN is mentioned in Sect. 4, and followed with experiments in Sect. 5. Finally, conclusions are given in Sect. 6.

2   Related Works

Problem of mining frequent itemsets was started up by Agrawal and Srikant with algorithm Apriori [1]. This algorithm generates candidate $(k + 1)$-itemsets from frequent k-itemsets at the $(k + 1)^{th}$ pass and then scans dataset to check whether a candidate $(k + 1)$-itemsets is a frequent one. Many previous works were inspired by this algorithm. Algorithm Partition [8] aim at reducing I/O cost by dividing dataset into non-overlapping and memory-fitting partitions which are sequentially scanned in two phases. In the first phase, local candidate itemsets are generated for each partition, and then they are checked in the second one. DCP [9] enhances Apriori by incorporating two dataset pruning techniques introduced in DHP [10] and using direct counting method for storing candidate itemsets and counting their support. In general, Apriori-like methods suffer from two drawbacks: a deluge of generated candidate itemsets and/or I/O overhead caused of repeatedly scanning dataset. Two other approaches, which are more efficient than Apriori-like methods, are also proposed to solve the problem: (1) frequent pattern growth adopting divide-and-conquer with FP-Tree structure and FP-Growth [2], and (2) vertical data format strategy in Eclat [11]. FP-Growth and algorithms based on it such as [12, 13] are efficient solutions as unlike Apriori, they avoid many times of scanning dataset and generation-and-test. However, they become less efficient when datasets are sparse. While algorithms based on FP-Growth and Apriori use a horizontal data format; Eclat and some other algorithms [8, 14, 15] apply vertical data format, in which each item is associated a set of transaction identifiers, Tids, containing the item. This approach avoids scanning dataset repeatedly, but a huge memory overhead is expensed for sets of Tids when dataset becomes large and/or dense. Recently, two remarkably efficient algorithms are introduced: FIN [4] with POC-Tree and PrePost+ [5] with PPC-Tree. These two structures are prefix trees and similar to FP-Tree, but the two mining algorithms use additional data structures, called Nodeset and N-list respectively, to significantly improve mining speed.

To better deal with the challenge of high volume in Big Data, in addition to the ideas of parallel mining for existing algorithms such as [16] for Eclat, incremental mining approaches are also considered as a potential solution. Some typical algorithms in this approach are algorithm FELINE [3] with CATS-Tree structure and IM_WMFI [17] for mining weighted maximal frequent itemsets from incremental datasets. These methods are both based on the well-known FP-Tree for its efficiency.

3 IPPC Tree

3.1 IPPC Tree Construction

IPPC-Tree is a prefix tree and possesses two properties, Property 1 and 2. IPPC-Tree includes one root labeled "*root*" and a set of prefix sub trees as its children. Each node in the sub trees contains the following attributes:

- *item-name*: the name of an item in a transaction that the node registered.
- *support* (or *local support* of an item): the number of transactions containing the node's *item-name*. Conversely, *global support* of an item, without concerning nodes, is the number of transactions containing the item.
- *pre-order* and *post-order*: two global identities in the IPPC-Tree which are sequent numbers generated by traversing the tree with pre and post order respectively.

For a clearer overview, Table 1 provides a comparison among the four similarity tree structures FP-Tree, CATS-Tree, PPC-Tree and IPPC-Tree.

Table 1. Comparison among FP-Tree, CATS-Tree, PPC-Tree and IPPC-Tree

	FP-Tree	CATS-Tree	PPC-Tree	IPPC-Tree
Items building the tree	Frequent items	All items	Frequent items	All items
Node attributes	- Item-name - Support	- Item-name - Support	- Item-name - Support - Pre-order - Post-order	- Item-name - Support - Pre-order - Post-order
Header table and node chains of the same item	Yes	Yes	No	No
Order of items in paths from the root to leaves	Global support	Local support	Global support	Local support
Local order of items in a path are flexible	No	Yes	No	Yes
Order of child nodes with the same parent node	No	Descending of support	No	No

Property 1. For a given IPPC-Tree, there exist no duplication nodes with the same item in a path of nodes from the root to a leave node.

Property 2. In a given IPPC-Tree, the *support* of a parent node must be greater than or equal to the sum of all its children's *support*.

The construction of the IPPC-Tree does not require a given support threshold. The tree is a compact and information-lossless structure of the whole items of all trans-actions in a given dataset *D*. Local order of items in a path of nodes from the root to a leave is flexible and can be changed to improve compression while remaining Property 2. To guarantee this, two conditions for swapping are as follows.

Child Swapping. A node can be swapped with its child node if it has only one child node, its *support* is equal to its child's *support*, and the number of child nodes of its child is not greater than one.

Descendant Swapping. Given a path of k nodes $N_1 \rightarrow N_2 \rightarrow \cdots \rightarrow N_k (k > 2)$, N_i is parent node of N_j $(i < j)$; if every node N_i $(i < k)$ satisfies the Child Swapping condition, node N_1 can be swapped with descendant node N_k.

To demonstrate the building process of an IPPC-Tree, the Fig. 1 records transaction by transaction in Table 2 inserted into an empty IPPC-Tree. Initially, the tree has only the root node, and transaction 1(b, e, d, f, c) is inserted as it is in Fig. 1(a). The Fig. 1(b) is of the tree after transaction 2 (d, c, b, g, f, h) is added. The item b in transaction 2 is merged with node b in the tree. Although transaction 2 does not contain item e, but its common items d, f and c can be merged with the corresponding nodes. The item d is found common, so it is merged with node d after node d is swapped[1] with node e to guarantee the Property 2. Similarly, items f and c are merged with node f and c respectively; and the remaining items g and h are inserted as a child branch of node c. In Fig. 1(c), transaction 3 (f, a, c) is processed. Common item f is found that can be merged with node f, so node f is

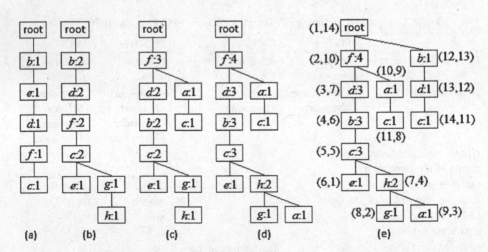

Fig. 1. An illustration for constructing an IPPC-Tree on example transaction dataset

---

[1] Swapping two nodes is simply exchanging one's item name to that of the other.

Table 2. Example transaction dataset

ID	Items in transactions	ID	Items in transactions
1	b, e, d, f, c	4	a, b, d, f, c, h
2	d, c, b, g, f, h	5	b, d, c
3	f, a, c		

swapped with node b. Item c is also a common one, but it is not able to be merged with node c as node d does not satisfy the Descendant Swapping condition with node c. Then the items a and c are added as a branch from node f. When transaction 4 (a, b, d, f, c, h) is added in Fig. 1(d), common items f, d, b and c are merged straightforwardly with corresponding nodes f, d, b and c. The remaining items a and h are then inserted into the sub tree having root node c. The item h is found common with node h in the second branch. Node h and item h, therefore, are merged together after node h is swapped with node g. The last item a is then inserted as a new child branch from node h. Insertion of transaction 5 (b, d, c) is depicted in Fig. 1(e). All items in transaction 5 are common, but they cannot be merged with nodes b, d and c as node f does not guarantee the Child Swapping condition. Thus, transaction 5 is added as a new child branch of root node.

After the dataset has been processed, each node in the IPPC-Tree is attached with a pair of sequent numbers (*pre-order*, *post-order*) by scanning the tree with pre order and post order traversals through procedure **AssignPrePostOrder**. For an example, node $(4, 6)$ is identified by *pre-order* = 4 and *post-order* = 6, and it registers item b with *support* = 3. Above is all concepts of IPPC-Tree construction; for a formal and detail description, algorithm **BuildIPPCTree**, is presented as follows.

Algorithm 1: BuildIPPCTree
Input: Dataset D, root node R, item list $\mathcal{L}$
Output: An IPPC-Tree with root R, sorted item list $\mathcal{L}$
1. **For Each** transaction $T \in D$
2.   Update items and their frequencies in $\mathcal{L}$ from items in T;
3.   **InsertTransaction**(T, R);
4. **End For**
5. **AssignPrePostOrder**(R);
6. Sort items in $\mathcal{L}$ increasingly based on their frequencies;

Procedure InsertTransaction(Transaction T, Node R)
1. $subNode \leftarrow R$; $notMerged$;
2. **While**($T \neq \emptyset$)
3. $notMerged \leftarrow$ **true**;
4.   **For Each** child node N of $subNode$
5.   **If**($N.item\text{-}name \in T$) {$notMerged \leftarrow$ **false**; $N.support$++;
6. $subNode \leftarrow N$; $T \leftarrow (T \setminus N.item\text{-}name)$; **break**;}
7.   **End For**
8.   **If**($notMerged$) **break**;
9. **End While**
10. **If**($T = \emptyset$) **Return**;
11. **For Each** child node N of $subNode$
12.   **If**(**MergeDescendants**(T, N)) **Return**;

```
13. End For
14. Insert T as a new branch from subNode (added nodes are ini-
    tialized at 1 for their supports);
Function MergeDescendants(Transaction T, Node N)
 1.  subNode ← N; mrgNode ← N; merged ← false;
 2.  While(subNode satisfies the Child Swapping condition)
 3.     descendant ← subNode.child;
 4.     If(descendant.item-name ∈ T){
 5.        T ← (T \ descendant.item-name); merged ← true;
 6.        Exchange item names of mrgNode and descendant;
 7.        mrgNode.support++; mrgNode ← mrgNode.child;}
 8.     subNode ← descendant;
 9.  End While
10.  If(merged) Insert T as a new branch from mrgNode.parent
     (added nodes are initialized at 1 for their supports);
11.  Return merged;


Procedure AssignPrePostOrder (Node R)
     // PreOrder and PostOrder are initialized at 1.
 1.  R.pre-order ← PreOrder; PreOrder++;
 2.  For Each child node N of R Do AssignPrePostCode(N);
 3.  R.post-order ← PostOrder; PostOrder++;
```

As the IPPC-Tree construction is independent to the support threshold and the global order of items in a dataset, a built IPPC-Tree from a dataset D is reusable for different support thresholds and changed dataset $D' = D \pm \Delta D$. To complete providing incremental ability for the IPPC-Tree, methods of storing and loading for the tree and item list $\mathcal{L}$ must be proposed, in which the data format and algorithms are their two features. For the simplicity of storing and loading for $\mathcal{L}$, this detail will not be mentioned here for concision. Besides *item-name*, *support*, etc., the important information for loading a node is its parent's information to identify where the node was in the built tree. By utilizing the *pre-order* (or *post-order*), the global identity, the requirement is resolved. The data format for a single node record is as follows.

<*parent's pre-order*>:<*pre-order*>:<*post-order*>:<*item-name*>:<*support*>

We employ Breadth-First-Search traversal to store the IPPC-Tree. In fact, the storing phrase can utilize other strategies such as pre order traversal, but the sequence of node records generated by Breadth-First-Search traversal is more convenient in loading phrase. The reason is that the records of all child nodes with the same parent node are continuous together. By storing the data record of each single node on a line, the stored data for the example tree in Fig. 1(e) is in right column of Table 3. The algorithm for loading the IPPC-Tree, procedure **LoadIPPCTree**, is presented in Table 3.

Table 3. Loading algorithm and data format for IPPC-Tree

```
Procedure LoadIPPCTree(File F, Root R, L)      <No. Trans.>
1.   Load item list L; TransNum ← 0;           -1:1:14:root:0
2.   Load sequentially TransNum and R from F;  1:2:10:f:4
3.   ParentNode ← R; NodeList ← Ø;             1:12:13:b:1
                                               2:3:7:d:3
4.   For Each line L in data file F            2:10:9:a:1
5.     Create a node N from L;                 12:13:12:d:1
6.     parentID ← <parent's pre-order>;        3:4:6:b:3
7.     Add N into the end of NodeList;         10:11:8:c:1
8.     While(parentID <> ParentNode.pre-order){ 13:14:11:c:1
9.       ParentNode ← NodeList[0];             4:5:5:c:3
10.      Remove ParentNode from NodeList;}     5:6:1:e:1
11.    Add N as a child of ParentNode;         5:7:4:h:2
12.  End For                                   7:8:2:g:1
                                               7:9:3:a:1
```

3.2 Preliminaries

In this subsection, some IPPC-Tree related definitions and lemmas are introduced as preliminaries for IFIN algorithm. In addition to the IPPC-Tree, another output of Algorithm 1 is the increasingly ordered list of items based on their frequencies $L = \{I_1, I_2, \ldots, I_n\}$. For the convenience of expressing the relative order between two items, we denote $I_i \prec I_j$ to indicate that I_i is in front of I_j in $L (1 \leq i < j \leq n)$. There are two premises of traversing a tree with pre order and post order as follows:

Premise 1. Traversing a tree to process a work at each node with pre order, it must be that (1) N_1 is an ancestor of N_2 or (2) N_1 and N_2 stay in two different branches (N_1 in the left and N_2 in the right) iff the work is done at N_1 before N_2.

Premise 2. Traversing a tree to process a work at each node with post order, it must be that (1) N_1 is an ancestor of N_2 or (2) N_1 and N_2 stay in two different branches (N_1 in the right and N_2 in the left) iff the work is done at N_2 before N_1.

By applying a work which assigns an increasingly global number at each node on Premises 1 and 2, two following lemmas are directly deduced.

Lemma 1. For any two different nodes N_1 and N_2 in the IPPC-Tree, N_1 is an ancestor of N_2 iff $N_1.pre\text{-}order < N_2.pre\text{-}order$ and $N_1.post\text{-}order > N_2.post\text{-}order$.

Lemma 2. For any two nodes N_1 and N_2 in two different branches of the IPPC-Tree, N_1 is in the left branch and N_2 in the right one iff $N_1.pre\text{-}order < N_2.pre\text{-}order$ and $N_1.post\text{-}order < N_2.post\text{-}order$.

Definition 1 (nodeset of an item). Given an IPPC-Tree, the *nodeset* of an item I, denoted by NS_I, is a set of all nodes in the IPPC-Tree with ascending order of *pre-order* and *post-order* in which all the nodes register the same item I.

In case N_1 and N_2 register the same item, N_1 and N_2 must be in two different branches because of Property 1. By traversing the IPPC-Tree with pre order, all nodes with the same item I, sequentially from the left-most branch to the right-most one, are added into the end of the list of nodes reserved for the item I. Hence, according to Lemma 2, the increasing orders of both *pre-order* and *post-order* are guaranteed. Finally, we will have *nodesets* for all items in $\mathcal{L}$. For an instance, the *nodeset* for item c in the example IPPC-Tree Fig. 1(e) will be $NS_c = \{(5, 5, 3), (11, 8, 1), (14, 11, 1)\}$. Here, each node N is depicted by a triplet of three numbers (N.*pre-order*, N.*post-order*, N.*support*).

Lemma 3. Given an item I and its nodeset is $NS_I = \{N_1, N_2, \ldots, N_l\}$, the *support* (or . *global support*) of item I is $\sum_{i=1}^{l} N_i$.*support*.

Rationale. According to Definition 1, NS_I includes all nodes registering the item I, and each node's *support* is the *local support* of the item I. Hence, the *global support* is the sum of all nodes' supports in NS_I. ∎

Definition 2 (nodeset of a k-itemset, $k \geq 2$). Given two $(k-1)$-itemsets $P_1 = p_1 p_2 \cdots p_{k-2} p_{k-1}$ with *nodesets* NS_{P_1} and $P_2 = p_1 p_2 \ldots p_{k-2} p_k$ with *nodeset* NS_{P_2} ($p_1 \prec p_2 \prec \ldots \prec p_k$), the *nodeset* of k-itemset $P = p_1 p_2 \ldots p_{k-2} p_{k-1} p_k$, NS_P, is defined as follows.

$$NS_P = \left\{ D_k \middle| \begin{bmatrix} D_k = Descendant(N_i, M_j) \ with \ N_i \in NS_{P_1} \wedge M_j \in NS_{P_2} \\ D_k \in NS_{P_1} \wedge D_k \in NS_{P_2} \end{bmatrix} \right\}$$

Function $Descendant(N_i, M_j)$ means that there has been an ancestor-descendant relationship between N_i and M_j, and the output is the descendant node.

Lemma 4. Given a k-itemset P and its nodeset is $NS_P = \{N_1, N_2, \ldots, N_l\}$, the *support* of the itemset P is $\sum_{i=1}^{l} N_i$.*support*.

Proof. By inductive method, the proof begins with *nodesets* for 2-itemsets.
According to the Definition 2, the *nodeset* of 2-itemset $p_1 p_2$ ($p_1 \prec p_2$) is a set of all descendant nodes from all pairs of nodes $N_i \in NS_{p_1}$ and $M_j \in NS_{p_2}$ that both N_i and M_j stay in the same path of nodes from root to a leave. Following the Property 2, descendant nodes' *supports* are lesser than or equal to that of the corresponding ancestor . nodes; so the *supports* of descendant nodes are the local *supports* of all 2-itemsets $p_1 p_2$ distributed in the IPPC-Tree. Consequently, Lemma 4 holds in case $k = 2$.

Assume Lemma 4 holds for case k. We need to proof Lemma 4 also holds for case $k + 1$. We have the assumptions:

1. k-itemset $P_1 = p_1 p_2 \ldots p_{k-1} p_k$ with its nodeset $NS_{P_1} = \{N_1, N_2, \ldots, N_{l1}\}$
2. k-itemset $P_2 = p_1 p_2 \ldots p_{k-1} p_{k+1}$ with its nodeset $NS_{P_2} = \{M_1, M_2, \ldots, M_{l2}\}$
3. For each N_i, there will be a path of k nodes from the root to N_i (except the root node); and N_i is the bottom node in the path visualized in Fig. 2(a). Each node in the path registers a certain item in $\{p_1, p_2, \ldots, p_{k-1}, p_k\}$. Each node M_j is similar to N_i, see Fig. 2(b).
4. $(k + 1)$-itemset $P = p_1 p_2 \ldots p_{k-1} p_k p_{k+1}$ with its *nodeset* $NS_P = \{D_1, D_2, \ldots, D_{l3}\}$

According to the Definition 2, each D_i will fall into one of two cases:

Case 1 (Fig. 2(c)): $D_i = Descendant(N_i, M_j)$, without loss of generality, assume that N_i is the descendant node of M_j. In fact, two paths of nodes $1 \rightarrow 2 \rightarrow \cdots \rightarrow (k-1) \rightarrow k$ and $1' \rightarrow 2' \rightarrow \cdots \rightarrow (k-1)' \rightarrow k'$ must be in the same path from the root to a leave node as each child node has only one parent node. Because two paths of nodes share the common items in $\{p_1, p_2, \ldots, p_{k-1}\}$, and no duplicate nodes exist in the same path (Property 1); there will be $(k - 1)$ pairs of identical nodes in which each pair includes one node from $1 \rightarrow 2 \rightarrow \cdots \rightarrow (k - 1)$ and the other from $1' \rightarrow 2' \rightarrow \cdots \rightarrow (k - 1)' \rightarrow k'$. This derives that there are only one path of $(k + 1)$ unique nodes registering $(k + 1)$ items in the item list $(p_1, p_2, \ldots, p_{k+1})$. Therefore, the *support* of N_i or D_i is a local *support* of the $(k + 1)$-itemset $P = p_1p_2\cdots p_{k-1}p_kp_{k+1}$.

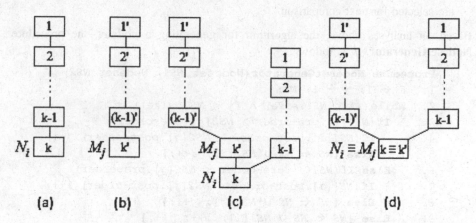

Fig. 2. Cases in Definition 2

Case 2 (Fig. 2(d)): $N_i \equiv M_j$. This means that the items of N_i and M_j are the same and must register one in $\{p_1, p_2, \ldots, p_{k-1}\}$. Hence, $(k - 2)$ remaining common items are shared by two paths of nodes $1 \rightarrow 2 \rightarrow \cdots \rightarrow (k - 1)$ and $1' \rightarrow 2' \rightarrow \cdots \rightarrow (k - 1)'$. By the same reasoning as in case 1, these two paths must be in the same path and there are $(k - 2)$ pairs of identical nodes. Consequently, the number of unique nodes in the only node path is $(1 + (k - 2) + 2 = k + 1)$, and these $(k + 1)$ nodes register $(k + 1)$ items in the list of items $(p_1, p_2, \ldots, p_{k+1})$. Thus, the *support* of N_i or D_i is a local *support* of the $(k + 1)$-itemset $P = p_1p_2\cdots p_{k-1}p_kp_{k+1}$.

Based on the two cases, Lemma 4 also holds for case $k + 1$. Hence, Lemma 4 holds. ■

Given two $(k - 1)$-itemsets $P_1 = p_1p_2\cdots p_{k-2}p_{k-1}$ and $P_2 = p_1p_2\cdots p_{k-2}p_k$ with their *nodesets* $NS_{P_1} = \{N_1, N_2, \ldots, N_{l1}\}$ and $NS_{P_2} = \{M_1, M_2, \ldots, M_{l2}\}$; at first glance, the computational complexity of generating *nodeset* NS_P for k-itemset $P = p_1p_2\cdots p_k$ is $O(l1 * l2)$. In fact this complexity can be reduced significantly to $O(l1 + l2)$, a linear cost, by utilizing Lemmas 1 and 2. For each pair of nodes N_i and M_j ($1 \leq i \leq l1$, $1 \leq j \leq l2$), there are the following five cases:

1. (N_i.*pre-order* > M_j.*pre-order*) $\wedge$ (N_i.*post-order* > M_j.*post-order*): The relationship between N_i and M_j is not an ancestor-descendant relationship, so no node is added to NS_P. Certainly, M_j also does not have this relationship with remaining nodes in NS_{P_1} as increasing orders of both *pre-order* and *post-order* in *nodesets*. Therefore, M_{j+1} is selected as the next node for the next comparison.

2. $(N_i.pre\text{-}order > M_j.pre\text{-}order) \wedge (N_i.post\text{-}order < M_j.post\text{-}order)$: N_i is added to NS_P as N_i is the descendant node of M_j. Consequently, N_{i+1} is selected as the next node for the next comparison.

3. $(N_i.pre\text{-}order < M_j.pre\text{-}order) \wedge (N_i.post\text{-}order > M_j.post\text{-}order)$: Similar to the case 2, M_j is added to NS_P, and M_{j+1} is the next node for the next comparison.

4. $(N_i.pre\text{-}order < M_j.pre\text{-}order) \wedge (N_i.post\text{-}order < M_j.post\text{-}order)$: This case is similar to the case 1; and N_{i+1}, therefore, is the next node for the next comparison.

5. $N_i \equiv M_j$: This identical node N_i is added to NS_P. Two new nodes N_{i+1} and M_{j+1} are selected for next comparison.

Based on analyses above, the algorithm for generating a *nodeset*, the procedure **NodesetGenerator**, is as follows.

```
Procedure NodesetGenerator(Nodeset NS1, Nodeset NS2)
1.   i ← 1; j ← 1; NS;
2.   While((i < NS1.size) ∧ (j < NS2.size))
3.     If(NS1[i].pre-order > NS2[j].pre-order)
4.       If(NS1[i].post-order > NS2[j].post-order) j++;
5.       Else {NS ← NS ∪ NS1[i]; i++;}
6.     Else If(NS1[i].pre-order < NS2[j].pre-order)
7.       If(NS1[i].post-order < NS2[j].post-order) i++;
8.       Else {NS ← NS ∪ NS2[j]; j++;}
9.     Else {NS ← NS ∪ NS1[i]; i++; j++;}
10.  End While
11.  Return NS;
```

It is easy to see that the increasing order of nodes in *NS* is guaranteed as these nodes are inserted to the end of *NS* in that order. Therefore, *NS* is also a *nodeset*.

Lemma 5 (superset equivalence). Given an item I and an itemset P $(I \notin P)$, if the *support* of P is equal to the *support* of $P \cup \{I\}$, the *support* of $A \cup P$ is equal to the *support* of $A \cup P \cup \{I\}$. Here $(A \cap P = \emptyset) \wedge (I \notin A)$.

Proof. As the *support* of itemset P is equal to that of $P \cup \{I\}$, any transaction containing P also contains the item I. Apparently, if a transaction contains $A \cup P$, it must contain P. This means that the numbers of transactions containing $A \cup P$ and $A \cup P \cup \{I\}$ are the same. Therefore, Lemma 5 holds. ∎

4   Algorithm IFIN

In this section, we present the algorithm IFIN based on the preliminaries introduced in the previous section. There are three running modes in algorithm IFIN: (1) **Just-Building-Tree**, just build an IPPC-Tree from a dataset D; (2) **Incremental**, load an IPPC-Tree from a previously stored tree Tree-D and build up the loaded IPPC-Tree with an incremental dataset D; (3) **Just-Loading-Tree**, just load an IPPC-Tree from a previously stored tree Tree-D. Each mode can be performed with different support thresholds (lines 5–32) with only one time of constructing the IPPC-Tree (lines 1–4). Lines 9–16 generate list of candidate 2-itemsets $C2$ as well as their *supports*. This task is ignored if the current running session performs for following times of mining with other support thresholds. Lines 20–28 create the corresponding *nodeset* for each frequent 2-itemsets in $L2$. From the second time of mining, just new frequent 2-itemsets' *nodesets* are generated. In lines 30–31, each frequent 2-itemset in $L2$ will be extended by the recursive procedure **GenerateFrequentItemsets** to discover longer frequent itemsets.

```
Algorithm 2: IFIN
Input: Stored tree Tree-D, incremental dataset D, ε
Output: Set of frequent k-itemsets L
1.   Create the root node R; L ← ∅;
2.   If(Tree-D <> null) LoadIPPCTree(Tree-D, R, L);
3.   If(D <> null) BuildIPPCTree(D, R, L);
4.   HasMap<itemset, support> C2 ← ∅;
5.   LOOP:
6.   Ask for a new support threshold ε or exit;
7.   Filter frequent items in L based on ε and add to L1;
8.   If(C2 <> ∅) Goto SKIP;
9.   Scan Each node N in IPPC-Tree by pre order traversal
10.      I_N ← N.item-name;
11.      For Each ancestor A of N
12.         I_A ← A.item-name;
13.         If(I_N < I_A) C2.add(I_N I_A, I_N I_A.support + N.support);
14.         Else C2.add(I_A I_N, I_A I_N.support + N.support);
15.      End For
16.   End Scan

17.  SKIP:
18.  L2' ← L2; L2 ← ∅;
19.  Filter frequent itemsets in C2 based on ε and add to L2;
20.  Scan Each node N in IPPC-Tree by pre order traversal
21.      I_N ← N.item-name;
22.      For Each ancestor A of N
23.         I_A ← A.item-name;
24.         If(I_N < I_A) IS ← I_N I_A;
25.         Else IS ← I_A I_N;
26.         If((IS ∈ L2)∧(IS ∉ L2')) nodeset_IS.add(N);
27.      End For
28.   End Scan
29.  F ← F ∪ F1; F ← F ∪ F2;
30.  For Each I_i I_j ∈ F2
31.      GenerateFrequentItemsets(I_i I_j, {I|I ∈ F1, I_j < I}, ∅);
32.  Goto LOOP;
```

Like FIN and PrePost[+], the procedure **GenerateFrequentItemsets** searches on a space of itemsets which is demonstrated by a set-enumeration tree [6] constructing from the list of ordered frequent items *L1*. An example of the search space for the dataset in Table 1 with support threshold $\varepsilon = 0.6$ is visualized in Fig. 3. The procedure employs two pruning strategies to greatly narrow down the search space. The first strategy is that if P is not a frequent itemset, its supersets are not either, and the second one is the superset equivalence introduced in Lemma 5. There are three input parameters for procedure **GenerateFrequentItemsets**: (1) *FIS* is a frequent itemset which will be extended; (2) *CI* is a list of candidate items used to expand the *FIS* with one more item; (3) *Parent_FISs* is the set of frequent itemsets generated at the parent of *FIS* in the set-enumeration tree. The detail procedure is as follows.

```
Procedure GenerateFrequentItemsets(FIS, CI, Parent_FISs)
1.   nextCI ← ∅; eqItems ← ∅; extFISs ← ∅;
2.   For Each item I ∈ CI
3.     IS = (FIS\{FIS.last_item}) ∪ {I};
4.     extIS = FIS ∪ {I};
5.     extIS.nodeset ← NodesetGenerator(FIS.nodeset,IS.nodeset);
6.     If(extIS.support = FIS.support) eqItems.add(I);
7.     Else If(extIS is an frequent itemset){
8.       nextCI.add(I); extFISs.add(extIS); F.add(extIS);}
9.   End For
10.  If(eqItems <> ∅)
11.    SoS ← set of all subsets of eqItems, excluding ∅;
12.    For Each IS ∈ SoS Do F.add(FIS ∪ IS);
13.    If(Parent_FISs <> ∅)
14.      Production ← {P| P = P1∪P2, P1 ∈ SoS, P2 ∈ Parent_FISs};
15.      For Each IS ∈ Production Do F.add(FIS ∪ IS);
16.      Parent_FISs ← Parent_FISs ∪ Production;
17.    End If
18.    Parent_FISs ← Parent_FISs ∪ SoS;
19.  End If
20.  If(Parent_FISs <> ∅)
21.    Production ← {P| P = P1∪P2, P1 ∈ extFISs, P2 ∈ Parent_FISs};
22.    F ← F ∪ Production;
23.  End If;
24.  For Each itemset IS ∈ extFISs
25.    GenerateFrequentItemsets(IS, nextCI, Parent_FISs);
```

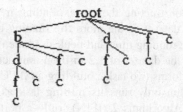

Fig. 3. Set-enumeration tree for example dataset Table 1, support threshold $\varepsilon = 0.6$

5 Experiments

All experiments were conducted on a 1.86 GHz processor and 4 GB memory computer with Window 8.1 operating system. To evaluate the performance, we used the Market-Basket Synthetic Data Generator [7], based on the IBM Quest, to prepare a dataset of 1.2 million transactions. The average transaction length and number of distinguishing items are 10 and 1000 respectively. The dataset was divided into six equal parts, 200 thousand transactions for each one, for emulating incremental dataset. The algorithm IFIN was compared with two state-of-the-art algorithms FIN and Pre-Post[+] and the well-known one FP-Growth. All the four algorithms were implemented in Java.

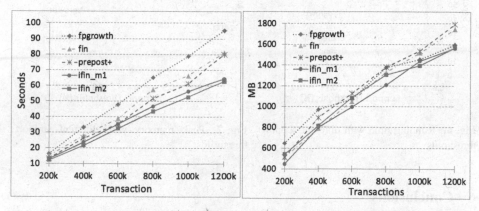

Fig. 4. Running time on incremental datasets **Fig. 5.** Peak memory on incremental datasets

Figures 4 and 5 sequentially demonstrate the running time and peak used memory for the four algorithms on incremental datasets at the support threshold $\varepsilon = 0.1\%$. Two running modes are performed by IFIN algorithm: **Incremental** (ifin_m1) and **Just-Loading-Tree** (ifin_m2). For all four algorithms, both running time and peak memory increase linearly when the dataset is accumulated. Follow the increasing of the dataset size, while there is not much difference in used memory of the four algorithms; the running time of IFIN becomes more discrepant compared with that of the remaining algorithms. The key reason is that with the same dataset, loading a stored built

IPPC-Tree is faster than constructing the corresponding trees in FIN, PrePost$^+$ and FP-Growth. The larger the dataset is, the more the running time difference is.

In the Figs. 6 and 7, the running time and peak used memory are visualized for the four algorithms mining on the dataset of 1.2 million transactions with different ε values. At $\varepsilon = 0.6\%$, IFIN performs two tasks: building an IPPC-Tree and mining; but for other ε values, the algorithm only runs its mining task as the built tree is reused. Besides, according to the Algorithms 2 (IFIN), only a portion of its mining is performed. Consequently, with $\varepsilon \neq 0.6\%$, both running time and used memory of IFIN take an overwhelming advantage against that of the three remaining algorithms. The IFIN algorithm's mining tasks at all ε values are in the same running session; therefore, fairly, the peak used memory in the cases $\varepsilon \neq 0.6\%$ and the case $\varepsilon = 0.6\%$ are the same. The algorithm FP-Growth uses memory more efficient than the two algorithms FIN and PrePost$^+$. However, its running time is considerably longer than that of FIN and PrePost$^+$. Algorithm PrePost$^+$ is more efficient than FIN in both running time and used memory, but this dominance of PrePost$^+$ is not significant.

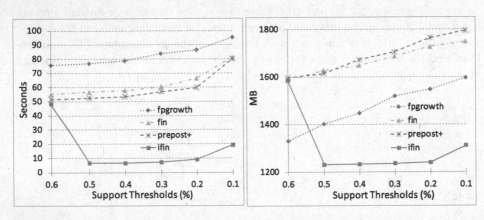

Fig. 6. Running time with different support thresholds

Fig. 7. Peak memory with different support thresholds

6   Conclusions

In this paper, we propose a solution for incrementally mining frequent itemsets through IPPC-Tree and the algorithm IFIN. The tree structure IPPC-Tree is independent to support threshold and the global order of items. Thus, this allows a previously constructed tree to be built up with a new additional dataset without wasting time to rebuild the old one. IFIN is based on the IPPC-Tree. However, in the same running session, the algorithm also possesses its own incremental property which some portions in mining task are skipped when mining with other support thresholds. Besides, to complete the ability of incremental mining, a stored data format, and corresponding storing and loading algorithms for IPPC-Tree were also proposed.

For the future work, we are interested in conducting experiments on other real datasets and improving IFIN's performance with parallel solutions on both local and distributed environment to better confront with inherent challenges of Big Data as our next steps.

Acknowledgements. We highly appreciate the reviewer's comments which are really helpful for us to improve the manuscript and our future works as well.

References

1. Agrawal, R., Srikant, R.: Fast algorithms for mining association rules. In: Proceedings of 20th International Conference on VLDB, pp. 487–499 (1994)
2. Han, J., Pei, J., Yin, Y.: Mining frequent itemsets without candidate generation. ACM Sigmod Rec. **29**(2), 1–12 (2000)
3. Cheung, W., Zaïane O.R.: Incremental mining of frequent patterns without candidate generation or support constraint. In: Proceedings of the 7th International Database Engineering and Applications Symposium, pp. 111–116. IEEE (2003)
4. Deng, Z.-H., Lv, S.-L.: Fast mining frequent itemsets using nodesets. Expert Syst. Appl. **41**(10), 4505–4512 (2014)
5. Deng, Z.-H., Lv, S.-L.: PrePost$^+$: an efficient N-lists-based algorithm for mining frequent itemsets via children-parent equivalence pruning. Expert Syst. Appl. **42**(13), 5424–5432 (2015)
6. Rymon, R.: Search through systematic set enumeration. In Proceeding of the 1st International Conference on Principles of Knowledge Representation and Reasoning, pp. 539–550 (1992)
7. Market-Basket Synthetic Data Generator. https://synthdatagen.codeplex.com/
8. Savasere, A., Omiecinski, E., Navathe, S.: An efficient algorithm for mining association rules in large databases. In: VLDB, pp. 432–443 (1995)
9. Perego, R., Orlando, S., Palmerini, P.: Enhancing the *Apriori* algorithm for frequent set counting. In: Kambayashi, Y., Winiwarter, W., Arikawa, M. (eds.) DaWaK 2001. LNCS, vol. 2114, pp. 71–82. Springer, Heidelberg (2001). doi:10.1007/3-540-44801-2_8
10. Park, J.S., Chen, M.S., Yu, P.S.: Using a hash-based method with transaction trimming and database scan reduction for mining association rules. IEEE Trans. Knowl. Data Eng. **9**(5), 813–825 (1997)
11. Zaki, M.J.: Scalable algorithms for association mining. IEEE Trans. Knowl. Data Eng. **12**(3), 372–390 (2000)
12. Grahne, G., Zhu, J.: Fast algorithms for frequent itemset mining using FP-Trees. Trans. Knowl. Data Eng. **17**(10), 1347–1362 (2005)
13. Liu, G., Lu, H., Lou, W., Xu, Y., Yu, J.X.: Efficient mining of frequent itemsets using ascending frequency ordered prefix-tree. DMKD J. **9**(3), 249–274 (2004)
14. Shenoy, P., Haritsa, J.R., Sudarshan, S.: Turbo-charging vertical mining of large databases. In: 2000 SIGMOD, pp. 22–33 (2000)
15. Zaki, M.J., Gouda, K.: Fast vertical mining using diffsets. In: 9th SIGKDD, pp. 326–335 (2003)
16. Liu, J., Wu, Y., Zhou, Q., Fung, B.C.M., Chen, F., Yu, B.: Parallel Eclat for opportunistic mining of frequent itemsets. In: Chen, Q., Hameurlain, A., Toumani, F., Wagner, R., Decker, H. (eds.) DEXA 2015. LNCS, vol. 9261, pp. 401–415. Springer, Cham (2015). doi:10.1007/978-3-319-22849-5_27
17. Yun, U., Lee, G.: Incremental mining of weighted maximal frequent itemsets from dynamic databases. Expert Syst. Appl. **54**, 304–327 (2016)

MapFIM: Memory Aware Parallelized Frequent Itemset Mining in Very Large Datasets

Khanh-Chuong Duong[1,2(✉)], Mostafa Bamha[2], Arnaud Giacometti[1], Dominique Li[1], Arnaud Soulet[1], and Christel Vrain[2]

[1] Université Francois Rabelais de Tours, LI EA 6300, Blois, France
{Arnaud.Giacometti,Dominique.Li,Arnaud.Soulet}@univ-tours.fr
[2] Université d'Orléans, INSA Centre Val de Loire, LIFO EA 4022, Blois, France
{Khanh-Chuong.Duong,Mostafa.Bamha,Christel.Vrain}@univ-orleans.fr

Abstract. Mining frequent itemsets in large datasets has received much attention, in recent years, relying on MapReduce programming models. Many famous FIM algorithms have been parallelized in a MapReduce framework like *Parallel Apriori*, *Parallel FP-Growth* and *Dist-Eclat*. However, most papers focus on work partitioning and/or load balancing but they are not extensible because they require some memory assumptions. A challenge in designing parallel FIM algorithms is thus finding ways to guarantee that data structures used during mining always fit in the local memory of the processing nodes during all computation steps.

In this paper, we propose MapFIM, a two-phase approach for frequent itemset mining in very large datasets relying both on a MapReduce-based distributed Apriori method and a local in-memory method. In our approach, MapReduce is first used to generate local memory-fitted prefix-projected databases from the input dataset benefiting from the Apriori principle. Then an optimized local in-memory mining process is launched to generate all frequent itemsets from each prefix-projected database. Performance evaluation shows that MapFIM is more efficient and more extensible than existing MapReduce based frequent itemset mining approaches.

Keywords: Frequent itemset mining · MapReduce programming model · Distributed file systems · Hadoop framework

1 Introduction

Frequent pattern mining [2] is an important field of Knowledge Discovery in Databases. This task aims at extracting a set of events (called itemsets) that occur frequently within database entries (called transactions). For more than 20 years, a large number of algorithms have been proposed to mine frequent patterns as efficiently as possible [1]. In big data era, proposing efficient algorithms that handle huge volumes of transactions remains an important challenge due to the memory space required to mine all frequent patterns. To tackle this issue, several proposals have been made to work in distributed environments where

© Springer International Publishing AG 2017
D. Benslimane et al. (Eds.): DEXA 2017, Part I, LNCS 10438, pp. 478–495, 2017.
DOI: 10.1007/978-3-319-64468-4_36

the major idea is to distinguish two phases: a global one and a local one. A first global phase uses MapReduce distributed techniques for mining the most frequent patterns whose calculation requires a large part of the data that does not fit in memory. Then, a second local phase mines on a single machine all the supersets of a pattern obtained at the previous phase. Indeed, these supersets can be mined using only a part of the data that can fit in the memory of a single machine. Intuitively the first phase guarantees the possibility of working on a large volume of data while the second phase preserves a reasonable execution time. Unfortunately the current proposals fail to be fully extensible, *i.e.* mining becomes intractable as soon as the number of transactions is too large or the minimum frequency threshold is too low.

Indeed, a major difficulty consists in determining the balance between the global phase and the local phase. If an approach relies too heavily on the local phase, it can only deal with high minimum frequency thresholds where the amount of candidate patterns or the projected database fit in memory. For instance, Parallel FPF algorithm in [7] where the projected databases are distributed cannot deal with low minimum thresholds when at least one projected database does not fit in the memory of a machine. Conversely, if an approach relies too heavily on the global phase, it will be very slow because the cost of communication is high. For instance, Parallel Apriori [8] is quite slow for low thresholds because all patterns are extracted in the global phase. In BigFIM [14], the user sets a minimum length k below which the itemsets are mined globally while the larger itemsets are mined locally as they cover a smaller set of transactions that can fit in memory. The problem is that this length is difficult to determine for the user as it varies, depending on the dataset and on the available memory. To illustrate this issue encountered with the threshold k, Fig. 1 plots the maximum length of frequent itemsets with the dataset Webdocs (see Sect. 5 for details) varying the minimum frequency threshold. In [14], it is suggested to use a global phase with itemsets of size $k = 3$ and for larger itemsets, it is assumed that the conditional databases will fit in the memory. However, from Fig. 1, it is easy to see that 3 is not a sufficiently high threshold because there is at least one itemset of size 4 that covers more than 40% of transactions. Moreover, it does not take into account the fact that two patterns of the same size may have very different frequencies. In this paper, we propose a fine-grained method depending

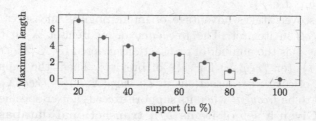

Fig. 1. Maximum length of frequent itemsets in dataset Webdocs

on the frequency of each itemset for determining whether it is possible to switch from the global phase to the local phase.

Contributions of the Paper. We propose the algorithm MapFIM (Memory aware parallelized Frequent Itemset Mining) which is, to the best of our knowledge, the first algorithm extensible with respect to the number of transactions. The advantage of this extensibility is that it is possible to process large volumes of data (although the addition of machines does not necessarily improve run-time performance as it is the case with scalability). The key idea is to introduce a maximum frequency threshold β above which frequency counting for an itemset is distributed on several machines. We prove that there exists at least one setting of β for which the algorithm is extensible under the conditions that the FIM algorithm used locally takes a memory space bounded with respect to the size of a projected database and that the set of items holds in memory. We show how to empirically determine this parameter in practice. Indeed, the higher this threshold, the faster the mining (because more patterns are mined locally). Finally, an experimental section illustrates the extensibility and the efficiency of MapFIM compared to the state-of-the-art algorithms.

Section 2 formulates the problem of frequent itemset mining in an extensible way in order to deal with huge volumes of transactions. Section 3 shows that existing proposals in literature are not extensible. In Sect. 4, we present how our algorithm MapFIM works and in particular, we detail the two phases (global and local ones) and the switch between the two. In Sect. 5, we empirically evaluate MapFIM against the state-of-the-art methods by comparing execution times and memory consumption. Section 6 briefly concludes.

2   Problem Formulation

2.1   Frequent Itemset Mining Problem

Let $\mathcal{I} = \{i_1 < i_2 < \ldots < i_n\}$ be a set of n ordered literals called *items*. An itemset (or a pattern) is a subset of $\mathcal{I}$. The language of itemsets corresponds to $2^{\mathcal{I}}$. A transactional database $\mathcal{D} = \{t_1, t_2, \ldots, t_m\}$ is a multi-set of itemsets of $2^{\mathcal{I}}$. Each itemset t_i, usually called a *transaction*, is a database entry. For instance, Table 1 gives a transactional database with 10 transactions t_i described by 6 items $\mathcal{I} = \{a, b, c, d, e, f\}$.

Pattern discovery takes advantage of interestingness measures to evaluate the relevancy of an itemset. The *frequency* of an itemset X in the transactional database $\mathcal{D}$ is the number of transactions covered by X [2]: $freq(X, \mathcal{D}) = |\{t \in \mathcal{D} : X \subseteq t\}|$ (or $freq(X)$ for sake of brevity). Then, the *support* of X is its proportion of covered transactions in $\mathcal{D}$: $supp(X, \mathcal{D}) = freq(X, \mathcal{D})/|\mathcal{D}|$. An itemset is said to be *frequent* when its support exceeds a user-specified minimum threshold α. **Given a set of items $\mathcal{I}$, a transactional database $\mathcal{D}$ and a minimum support threshold, frequent itemset mining (FIM) aims at enumerating all frequent itemsets.**

Table 1. Original dataset

Transaction	Items	Transaction	Items
t_1	a	t_6	a, d
t_2	a, b	t_7	b, c
t_3	a, b, c	t_8	c, d
t_4	a, b, c, d	t_9	c, e
t_5	a, c	t_{10}	f

2.2 The MapReduce Programming Model

MapReduce is a simple yet powerful framework for implementing distributed applications without having extensive prior knowledge of issues related to data redistribution, task allocation or fault tolerance in large scale distributed systems.

Google's MapReduce programming model presented in [6] is based on two functions: **map** and **reduce**, that the programmer is supposed to provide to the framework. These two functions should have the following signatures:

map: $(k_1, v_1) \longrightarrow list(k_2, v_2),$
reduce: $(k_2, list(v_2)) \longrightarrow list(v_3).$

The **map** function has two input parameters, a key k_1 and an associated value v_1, and outputs a list of intermediate key/value pairs (k_2, v_2). This list is partitioned by the MapReduce framework depending on the values of k_2, with the constraint that all elements with the same value of k_2 belong to the same group.

The **reduce** function has two parameters as inputs: an intermediate key k_2 and a list of intermediate values $list(v_2)$ associated with k_2. It applies the user defined merge logic on $list(v_2)$ and outputs a list of values $list(v_3)$.

In this paper, we use an open source version of MapReduce, called Hadoop, developed by The Apache Software Foundation. Hadoop framework includes a distributed file system called HDFS[1] designed to store very large files with streaming data access patterns.

MapReduce excels in the treatment of data parallel applications, where computation can be decomposed into many independent tasks, involving large input data. However MapReduce's performance may degrade in the case of dependent tasks or in the presence of skewed data due to the fact that, in Map phase, all the emitted key-value pairs (k_2, v_2) corresponding to the same key k_2 are sent to the same reducer. This may induce a load imbalance among processing nodes and also can lead to task failures whenever the list of values corresponding to a specific key k_2 cannot fit in processing nodes available memory [3,4]. For scalability, MapReduce algorithm's design must avoid load imbalance among

---

[1] HDFS: Hadoop Distributed File System.

processing nodes while reducing disks I/O and communication costs during all stages of MapReduce jobs computation.

2.3 The Challenge of Extensibility

Guaranteeing the correct execution of a method whatever the volume of input data is a classic challenge in MapReduce framework through the notion of scalability. Scalability refers to the capacity of a method to perform similarly even if there is a change in the order of magnitude of the data volume, in particular by adding new machines (as mapper or reducer). We introduce the notion of *extensibility*, which refers to the capacity of a method to deal with an increase in the data volume but without performance guarantees.

More precisely, our goal is to efficiently process transaction databases whatever the number of transactions when the set of items remains unchanged. This situation covers many practical use cases. For instance, in a supermarket the set of products is relatively stable while new transactions are added continuously. We then formalize the notion of extensibility with respect to the number of transactions as follows:

Definition 1 (Transaction-extensible). *Given a set of items $\mathcal{I}$, a FIM method is said to be transaction-extensible iff it manages to mine all frequent itemsets whatever the number of transactions in $\mathcal{D} = \{t_1, \ldots, t_m\}$ (where $t_i \subseteq \mathcal{I}$) and the minimum support threshold α.*

This definition is particularly interesting for a pattern discovery task. Indeed, the transaction-extensible property guarantees that for a given set of items $\mathcal{I}$, the method will always be able to mine all the frequent itemsets whatever the number of transactions in $\mathcal{D}$ and the minimum frequent threshold α.

In the remainder of the paper, we aim at proposing the first transaction-extensible FIM method. This goal is clearly a challenge because it is difficult to control the amount of memory required for a frequent itemset mining. The following section reviews the shortcomings of the various literature proposals.

3 Related Work

Due to the explosive growth of data, many parallel methods of frequent pattern mining (FPM) algorithms have been proposed in the literature, mainly to extract frequent itemsets [7–9,15–17,20], but also to extract frequent sequences [5,13]. In this section, we only consider related work involving the parallelization of FPM algorithms on the MapReduce framework.

A first category of approaches includes works that are specific parallelizations of existing FPM algorithms. For example, different adaptations of Apriori on MapReduce have been proposed [8,9]. These implementations of Apriori are not *transaction-extensible* since they assume that at each level, the set of candidate itemsets can be stored in the main memory of the worker nodes (mappers or reducers). In Sect. 4.3, we show how this limitation can be overcome using

HDFS to store the set of candidates. Different implementations of FP-Growth on MapReduce also exist [7,20]. The main idea of these implementations is to distribute the conditional databases of the frequent items to the mappers. However, these proposals do not guarantee that the conditional databases can be stored in the worker nodes, and therefore, these parallelizations of FP-Growth are also not *transaction-extensible*. More recently, Makanju et al. [12] propose to use Parent-Child MapReduce (a new feature of IBM Platform Symphony) to overcome the limitations of the previous implementations of FP-Growth. The authors show that their method can provide significant speed-ups over Parallel FP-Growth [7]. However, their method requires to predict the processing loads of a FP-Tree which is a particularly difficult challenge.

A second category of approaches includes works that are independent of a specific FPM algorithm, meaning that after a data preparation and partitioning phase, they can use any existing FPM methods to locally extract patterns. In this category, we can distinguish two sub-categories of approaches. At a high-level, the methods in the first sub-category carefully partitions the original dataset in such a way that each partition can be mined independently and in parallel [13,15,16]. Once partitions have been constructed (in a first global phase), an arbitrary FPM algorithm can be used to mine each partition (in this second local phase, the partition are mined independently and in parallel). In order to maintain completeness, it is important to note that some partitions built by these approaches can overlap and that some interesting pattern can be generated several times. However, these approaches are more efficient than SON Algorithm [17] because locally frequent itemsets are necessarily globally frequent. Thus, compared to SON Algorithm, after the local phase, it is not necessary to compute the supports of the locally frequent itemsets with respect to the whole dataset. Finally, because these approaches cannot guarantee that all the partitions will fit in main memory (of the mappers or reducers), it is important to note that they are not *transaction-extensible*.

The approaches in the second sub-category do not initially partition the dataset, but the search space (the pattern language), thereby ensuring that each interesting pattern is only generated once. We can consider that Parallel FP-Growth (PFP) [8] also belongs to this second sub-category of methods. However, because PFP partitions the search space only considering single frequent items, it is not efficient. In order to overcome this type of limitation, Moens et al. [14] propose to use longer frequent itemsets as prefixes for partitioning the search space. In a first and global phase, their algorithm (called *BigFIM*) mines the frequent k-itemsets using a MapReduce implementation of Apriori. Then, in a second phase, subset of prefixes of length k are passed to worker nodes. These worker nodes use the conditional databases of prefixes to mine interesting patterns that are more specific, assuming that the conditional databases can fit in the main memory of the worker nodes. In practice, note that the choice of the parameter k can be very difficult. Indeed, if the user chooses a value of k that is too low, then BigFIM will not pass (because a conditional database will not fit in main memory). On the other hand, if the user chooses a value of k that is too high, then the first global phase of BigFIM (which computes the frequent k-itemsets) will be time consuming and not efficient. It explains why we propose in this paper a new approach that do not require the involvement of the user to fix a parameter such as k, and automatically detect when it is possible to switch from a global phase to a local phase.

4   MapFIM: A MapReduce Approach for Frequent Itemset Mining

4.1   Overview of the Approach

The key idea of our proposal is to enumerate in a breadth-first search manner all itemsets using distributed techniques (global mining phase) until one reaches a point of the search space where all its supersets can be mined on a single machine (local mining phase). This point of the search space is reached as soon as an itemset has a support sufficiently low to guarantee that the projected database (plus the amount of memory required to enumerate the itemsets) holds in memory. To do this, we introduce a maximum frequency threshold β to indicate when it is possible to switch to the local mining phase. Given a transactional database $\mathcal{D}$ and a maximum support threshold β, an itemset X is said to be *overfrequent* if its support exceeds β: $supp(X, \mathcal{D}) \geq \beta$. In the following, we denote:

- $\mathcal{L}$ the set of frequent itemsets, *e.g.* set of itemsets X such that $supp(X, \mathcal{D}) \geq \alpha$.
- $\mathcal{L}^{>\beta}$ the set of *overfrequent* itemsets, *e.g.* set of itemsets X such that $supp(X, \mathcal{D}) > \beta$.
- $\mathcal{L}^{\leq\beta}$ the set of frequent but not overfrequent itemsets, *e.g.* set of itemsets X such that $\alpha \leq supp(X, \mathcal{D}) \leq \beta$. It is clear that $\mathcal{L}^{\leq\beta} = \mathcal{L} \setminus \mathcal{L}^{>\beta}$.
- $\mathcal{L}_k$, $\mathcal{L}_k^{>\beta}$, $\mathcal{L}_k^{\leq\beta}$ are respectively the set of frequent, overfrequent, frequent but not overfrequent k-itemsets.
- $\mathcal{C}$ the set of candidates. $\mathcal{C}_k$ the set of candidate k-itemsets. $\mathcal{C}_k$ is generated by the join $\mathcal{L}_{k-1} \bowtie \mathcal{L}_{k-1}$.
- $\mathcal{D}'$ the compressed database from $\mathcal{D}$, by removing infrequent items, *e.g.* items that are not in $\mathcal{L}_1$.

More precisely, given a set of items $\mathcal{I}$, a transactional database $\mathcal{D}$, a minimum support threshold α and a maximum support threshold β, the algorithm MapFIM (for Memory aware parallelized Frequent Itemset Mining) enumerates all frequent itemsets by using three phases:

1. **Data preparation:** This phase initializes the process by compressing the transactional database based on frequent 1-itemsets. Let $\alpha = 20\%$ and $\beta = 50\%$. Considering the example given by Table 1, as only a, b, c and d are frequent, the original dataset is compressed as shown by Table 2. Transactions are updated for removing non-frequent items. Transactions t_1 and t_{10} are removed because they cannot contain an itemset of size 2 (or greater). At the end of this phase, we have $\mathcal{L}_1^{>\beta} = \{a, c\}$ and $\mathcal{L}_1^{\leq\beta} = \{b, d\}$.

Table 2. Compressed dataset

Transaction	Items
t_2	a, b
t_3	a, b, c
t_4	a, b, c, d
t_5	a, c
t_6	a, d
t_7	b, c
t_8	c, d

Table 3. Conditional datasets

Itemset	Projected datasets
b	$\mathcal{D}_b : \{\{c\}, \{c, d\}, \{c\}\}$
d	$\emptyset$
ab	$\mathcal{D}_{ab} : \{\{c\}, \{c, d\}$
ac	$\mathcal{D}_{ac} : \{\{d\}\}$
ad	$\emptyset$
cd	$\emptyset$

2. **Global mining:** This phase mines all potentially overfrequent itemsets using Apriori algorithm. An itemset is potentially overfrequent whenever at least one direct subset is overfrequent. For instance, four candidates of size 2 are generated from overfrequent items in $\mathcal{L}_1^{>\beta}$, e.g. $\mathcal{C}_2^{>\beta} = \{ab, ac, ad, cd\}$ (these candidate 2-itemsets are potentially overfrequent). The support of all candidates in $\mathcal{C}_2^{>\beta}$ are evaluated during this global phase. But, as their support is greater than α, but below β, $\mathcal{L}_2^{>\beta}$ is empty. No more candidates are generated and MapFIM moves to the next phase.

3. **Local mining:** This phase mines itemsets from non overfrequent itemsets. In our running example, the prefix-based supersets generated from $\mathcal{L}_1^{\leq\beta} = \{b, d\}$ and $\mathcal{L}_2^{\leq\beta} = \{ab, ac, ad, cd\}$ will be evaluated during this phase. Each prefix is considered individually by using a projected database as given in Table 3. Typically, abc will be generated from the prefix ab.

Of course, the maximum support threshold β is a very crucial parameter for balancing the mining process. Section 5.2 will show how to set β in practice. Note that varying this parameter enables us to unify different state-of-the-art methods. By choosing $\beta = \alpha$, MapFIM algorithm is very similar to parallel Apriori [8] as the local mining phase is ignored. At the opposite, with $\beta = 100\%$, MapFIM is similar to parallel FP-Growth algorithm [7] which only relies on a local mining phase after the data preparation. Interestingly, when the maximum support threshold β is between α and 100%, MapFIM benefits from the same ideas as BigFIM with the important difference that we are sure that all prefixes examined by the local mining phase are not overfrequent. Consequently, we are sure that they can be locally processed in-memory.

Sections 4.2, 4.3, and 4.4 detail respectively the three main phases of Map-FIM: data preparation, global mining and local mining. Finally, Sect. 4.5 demonstrates its completeness and its extensibility with respect to the number of transactions.

4.2 Data Preparation

In this phase, frequent items, *e.g.* items in $\mathcal{L}_1$ are found. This can be achieved by adapting the *Word Count* problem [6]. Each item is considered as a word and by using a MapReduce phase for Word Counting problem, we get the support of every item. Then, by using α and β parameters, $\mathcal{L}_1^{\leq\beta}$ and $\mathcal{L}_1^{>\beta}$ are constructed. Finally, the compressed data $\mathcal{D}'$ is generated and put in HDFS. This can be solved by a simple Map phase, where each mapper reads a block of data and removes items which are not in $\mathcal{L}_1$, then emits transactions with at least two frequent items.

4.3 Global Mining Based on Apriori

This phase is similar to the parallel implementation of Apriori algorithm [8]. The key difference is on the way candidates are generated. In Apriori algorithm, at each iteration, the set of candidates $\mathcal{C}_k$ is generated by the join $\mathcal{L}_{k-1} \bowtie \mathcal{L}_{k-1}$[2]. From the definition of $\mathcal{L}^{\leq\beta}$ and $\mathcal{L}^{>\beta}$, we have:

$$\mathcal{C}_k = \mathcal{L}_{k-1} \bowtie \mathcal{L}_{k-1}$$
$$= (\mathcal{L}_{k-1}^{\leq\beta} \bowtie \mathcal{L}_{k-1}) \cup (\mathcal{L}_{k-1}^{>\beta} \bowtie \mathcal{L}_{k-1})$$

We define $\mathcal{C}_k^{\leq\beta} = \mathcal{L}_{k-1}^{\leq\beta} \bowtie \mathcal{L}_{k-1}$ and $\mathcal{C}_k^{>\beta} = \mathcal{L}_{k-1}^{>\beta} \bowtie \mathcal{L}_{k-1}$ and thus: $\mathcal{C}_k = \mathcal{C}_k^{\leq\beta} \cup \mathcal{C}_k^{>\beta}$.

The idea developed in this paper is to use MapReduce framework to globally mine candidates in $\mathcal{C}_k^{>\beta}$, and then locally mine $\mathcal{C}_k^{\leq\beta}$ in a local phase. Algorithm 1 presents this algorithm.

At each iteration in Main() function, $\mathcal{C}_k^{>\beta}$ is computed by the join $\mathcal{L}_{k-1}^{>\beta} \bowtie \mathcal{L}_{k-1}$ and sent to all Mappers. The Map function counts the frequency of all the candidates belonging to $\mathcal{C}_k^{>\beta}$ in a parallel way. Let us precise that the counting of the frequency is performed at the end of each Mapper, when the combine function summarizes the emission of 1 for each itemset. Then, the Reduce function sums the frequency obtained by each mapper. If a candidate X is frequent, it is put into $\mathcal{L}_k^{>\beta}$ or $\mathcal{L}_k^{\leq\beta}$ depending on whether it is *overfrequent* or not. In case $\mathcal{C}_k^{>\beta}$ is too large to be handled by Mappers, we partition this set into, for example, l subsets and we run l MapReduce phases instead of one. Finally, the global mining achieves a good load balance because the compressed database $\mathcal{D}'$ is distributed equally among mappers and all mappers handle the same candidate set.

---

[2] In our work, in order to generate each candidate once, we use a prefix-based join operation. More precisely, given two set of k-itemsets $\mathcal{L}_{k-1}$ and $\mathcal{L}'_{k-1}$, the join of $\mathcal{L}_{k-1}$ and $\mathcal{L}'_{k-1}$ is defined by: $\mathcal{L}_{k-1} \bowtie \mathcal{L}'_{k-1} = \{(i_1,\ldots,i_k) \mid (i_1,\ldots,i_{k-2},i_{k-1}) \in \mathcal{L}_{k-1} \wedge (i_1,\ldots,i_{k-2},i_k) \in \mathcal{L}'_{k-1} \wedge i_1 < \cdots < i_{k-1} < i_k\}$.

Algorithm 1. Global Mining

---

1 **Function** Main():
2     $k = 2$;
3   **while** $|\mathcal{L}_{k-1}^{>\beta}| > 0$ **do**
4        $\mathcal{C}_k^{>\beta} = \mathcal{L}_{k-1}^{>\beta} \bowtie \mathcal{L}_{k-1}$;
5   Send $\mathcal{C}_k^{>\beta}$ to all Mappers;
6   Map phase;
7   Reduce phase;
8        $k = k + 1$;
9   **return** ;

10 **Function** Map(*String key, String value*):
11   // *key*: input name, *value*: input contents;
12   **foreach** *transaction* $t \in value$ **do**
13   **foreach** *itemset* $X \in \mathcal{C}_k^{>\beta}$ **do**
14   **if** $X \subseteq t$ **then**
15   Emit(X, 1);
16   **return** ;

17 **Function** Reduce(*String key, Iterator values*):
18   // *key*: a candidate X, *values*: a list of counts;
19     $frequency = 0$;
20   **foreach** $v \in values$ **do**
21        $frequency = frequency + v$;
22   **if** $\alpha * |D| \leq frequency$ **then**
23        $\mathcal{L}_k = \mathcal{L}_k \cup \{X\}$;
24   **if** $frequency \leq \beta * |\mathcal{D}|$ **then**
25           $\mathcal{L}_k^{\leq\beta} = \mathcal{L}_k^{\leq\beta} \cup \{X\}$;
26   **else**
27           $\mathcal{L}_k^{>\beta} = \mathcal{L}_k^{>\beta} \cup \{X\}$;
28   **return** ;

4.4 Local Mining of Frequent Itemsets

As described in the previous sections, the two-phase mining strategy guarantees the efficiency of MapFIM. Indeed, once it is estimated, through the use of the parameter β, that each projected-database with respect to a prefix generated can always be handled by a single node in the cluster, MapFIM switches to the local mining phase.

In the local mining phase, the frequent itemset enumeration is completed by using efficient algorithms (for instance, Eclat or LCM) that fit the memory constraints required by single nodes. This step is still MapReduce driven: local memory-fitted projected-databases are dispatched to each node (as Reducers) that allow to run any local FIM algorithm. The complete local mining process is shown in Algorithm 2.

Algorithm 2. Local Mining

---

1 **Function** Map(*String key, String value*):
2 | // *key*: input name, *value*: input contents;
3 | **foreach** *itemset* $X \in \mathcal{L}^{\leq \beta}$ **do**
4 | | Let $i =$ the last item in X;
5 | | **foreach** *transaction* $t \in$ *value that contains* X **do**
6 | | | Create $t' = t$;
7 | | | Remove every item j in t' such that $j \leq i$;
8 | | | Emit(X, t')
9 | **return** ;

10 **Function** Reduce(*String key, Iterator values*):
11 | // *key*: an itemset X, *values*: a list of transactions;
12 | Create an empty file f_{in} in local disk;
13 | Save values to f_{in};
14 | Run a local FIM program with input=f_{in}, output=f_{out}, support=$\alpha * \frac{|values|}{|\mathcal{D}|}$;
15 | **foreach** *frequent itemset* $X' \in f_{out}$ **do**
16 | | $X'' = X \cup X'$;
17 | | $\mathcal{L} = \mathcal{L} \cup \{X''\}$;
18 | **return** ;

In the Map phase, we consider frequent itemsets $X \in \mathcal{L}^{\leq \beta}$ as prefixes and construct their projected databases. For each $X \in \mathcal{L}^{\leq \beta}$, let i denote the last item in X. The projected-database $\mathcal{D}'$ is built by: (1) pruning every transaction $t \in \mathcal{D}'$ that does not contain X (2) pruning every item $j \leq i$ since these items cannot expand X due to the prefix-based join. As shown in Algorithm 2, each Mapper reads a block of data, then for each $X \in \mathcal{L}^{\leq \beta}$, it emits every transaction t that contains X after pruning unnecessary items.

In the Reduce phase, a local FIM algorithm is independently called to enumerate all the frequent itemsets for each projected-database. More precisely, in the Reduce phase, each *key* is a frequent itemset $X \in \mathcal{L}^{\leq \beta}$ and each list of *values* contains all transactions of the projected-database of X. They are saved to a local file so that the local FIM algorithm can work on it. For each itemset X' being frequent in the projected-database, the itemset $X'' = X \cup X'$ is frequent in $\mathcal{D}$. Notice that in the case $\mathcal{L}^{\leq \beta}$ is too large to fit in memory of Mappers, we partition this set, for example, into l subsets and repeat the local mining in l MapReduce phases until every itemset in $\mathcal{L}^{\leq \beta}$ is handled.

An algorithm adapted to the local mining phase must be able to enumerate all the itemsets corresponding to a given prefix in a bounded memory space.

Level-wise algorithms will therefore not be adapted since it is difficult to limit themselves to a given prefix and the amount of memory required is very variable. Similarly, approaches based on FP-trees do not guarantee a bounded amount of memory for tree storage. However vertical database layout based approaches such as Eclat or LCM fit well the requirement of bounded memory usage.

Due to the difference in size among projected databases, the local mining can lead to a load imbalance among reducers. In [14], the authors of BigFIM algorithm have experimented different strategies to assign the prefixes and it is shown that a random method can achieve a good workload balancing. Following this way, we decide in our implementation to assign randomly projected databases to reducers.

4.5 Completeness and Extensibility

Thanks to the complementarity of global and local mining phases, this section demonstrates that MapFIM is correct and complete, but also is transaction-extensible:

Proposition 1. *MapFIM is **correct**, i.e., all itemsets returned by the algorithm are frequent and **complete**, i.e., all frequent itemsets are returned by the algorithm.*

Idea of the Proof: The algorithm counts the support of each itemset and returns only frequent itemsets, therefore it is correct. We give here an idea of the proof of the completeness. Let I be a k-frequent itemset, $\mathcal{I} = (i_1, \ldots, i_k)$, with $i_1 < i_2 \ldots < i_k$. Let I_j denote $I = (i_1, \ldots, i_j)$.

If $k = 1$, then I is computed during data preparation. If $k > 1$, then we have two cases:

- $supp(I_{k-1}, \mathcal{D}) > \beta$, $I_{k-1} \in \mathcal{L}_{k-1}^{>\beta}$. Then, since $i_k > i_{k-1}$, I is generated and evaluated during the global mining phase.
- $supp(I_{k-1}, \mathcal{D}) \leq \beta$. Let j be the smallest index such that $supp(I_j, \mathcal{D}) \leq \beta$, i.e., $I_j \in \mathcal{L}_j^{\leq\beta}$. Then frequent itemsets starting by I_j will be mined in the local mining step, from the conditional database with respect to I_j. It is built by considering all transactions in $\mathcal{D}$ containing I_j and removing from these transactions all items i with $i \leq i_j$. Since I is ordered, if I is frequent in $\mathcal{D}$ then $\{i_{j+1}, \ldots, i_k\}$ is frequent in the conditional database w.r.t I_j and will be found during the local mining phase.

The main challenge faced by MapFIM is to deal with a very large number of transactions. This is possible because the preparation and the scanning of this transactional database is distributed on several mappers and the set of generated candidates that is potentially huge is stored on the distributed file system. Therefore, in addition to being complete, MapFIM is transaction-extensible as introduced by Definition 1:

Proposition 2 (Transaction-extensible). *Assuming the distributed file system has an infinite storage capacity, the algorithm MapFIM is transaction-extensible when the set of items I holds in memory and the local frequent itemset mining method takes space $O(l \times \beta)$ where l is the length of the longest transaction.*

Idea of the Proof: The first step of data preparation is not a problem as it is similar to a word counting. The second step is also transaction-extensible because the set of frequent items holds in memory as we make the assumption that the set of all items holds in memory. Global mining phase does not raise any problem because all candidates are stored on the distributed file system (which has an infinite storage capacity) and can be partitioned into independent subsets of candidates. For local mining phase, the mining algorithm for a prefix takes a memory space proportional to the size of its projected database so there is at least one β such that each projected database holds in memory.

5   Experiments

We have chosen the dataset Webdocs [10], one of the largest commonly used datasets in Frequent Itemset Mining. It is derived from real-world data and has a size of 1.48 GB. It was obtained from the Frequent Itemset Mining Implementations Repository at http://fimi.ua.ac.be/data/. We have also generated a synthetic dataset by using the generator from the IBM Almaden Quest research group. Their program can no longer be downloaded and we have used another implementation at https://github.com/zakimjz/IBMGenerator. The command used to generate our dataset is: `./gen lit -ntrans 10000 -tlen 50 -nitems 10 -npats 1000 -patlen 4 -fname Synthetic -ascii`

The characteristics of the two datasets are given in Table 4.

Table 4. Characteristic of the two used datasets

Dataset	# Transactions	# Items	Avg length	FileSize
Webdocs	1,692,082	5,267,656	177	1.48 GB
Synthetic	10,000,000	10,000	50	2.47 GB

5.1   Performance Results

To evaluate the performance of MapFIM presented in this paper, we compared it to Parallel FP-Growth (PFP) [7] and BigFIM algorithms [14]. We believe that PFP and BigFIM are the best approaches for itemset mining in Hadoop MapReduce framework. We implemented MapFIM in Hadoop 2 and for the local mining step, we use a local program based on Eclat/LCM algorithm [18,19]. The program is implemented in C++ by Borgelt at http://www.borgelt.net/eclat.html.

PFP implementation is present in the library Apache Mahout 0.8 [11] and BigFIM implementation based on Hadoop 1 is provided by the authors at https://gitlab.com/adrem/BigFIM-sa.

All the experiments were performed on a cluster of 3 machines. Each machine has 2 Xeon Cpu E5-2650 @ 2.60 GHz with 32 cores and 64 GB of memory. MapFIM and PFP were tested in Hadoop 2.7.3 while BigFIM was experimented in Hadoop 1.2.1. We configured Hadoop environment to use up to 30 cores and 60 GB of memory for each machine[3]. We have experimented the three approaches with different values of the minimum support threshold α.

In all the experiments, the time was limited to 72 h and we report the total execution time in seconds. PFP program was tested with its default parameter and BigFIM program was configured with parameter $k = 3$ as suggested by the authors. With this configuration, BigFIM uses a parallel Apriori approach to mine all 3-frequent itemsets before switching to global mining. It is shown in [14] that with $k = 3$, BigFIM achieves good performance.

For the Dataset Synthetic. The value of the minimum support threshold α varies from 1% to 2%. In this dataset, there is no itemset whose support is greater than 30% and for that reason, MapFIM with $\beta = 30\%$ or $\beta = 50\%$ or $\beta = 100\%$ takes the same amount of execution time. All three approaches can enumerate all the frequent itemsets without running out of memory and the results of MapFIM (with $\beta = 30\%$), BigFIM and PFP are shown in Fig. 2. It is clear that PFP is the slowest while MapFIM and BigFIM are comparable. This dataset is generated randomly and there is no long frequent itemsets. Indeed, most of the frequent 3-itemset appears in only 9% of transactions. As a consequence, both BigFIM and MapFIM can achieve a good workload balancing and a good performance.

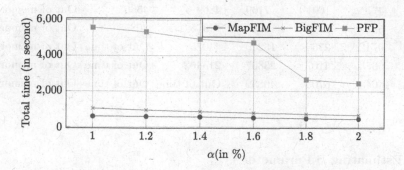

Fig. 2. Performance with the synthetic dataset

For the Dataset Webdocs. The three programs were tested with various values of the minimum support threshold α. In MapFIM, we set the value of

---

[3] In our configuration, there is no real difference of performance between Hadoop 1.2.1 and Hadoop 2.7.3.

β to $100\%, 50\%, 30\%$ and $\beta = \alpha$. As shown in Fig. 1, this dataset is expected to be hard to mine as it has long frequent itemsets as well as itemsets that are very frequent. For example, in this dataset, there exists a frequent 7-itemset that occurs in 20% of the transactions and at least one frequent 3-itemset that appears in more than 60% of transactions.

The results with the dataset Webdocs are shown in Table 5. It is surprising that PFP cannot solve the dataset Webdocs with a support below to 15%, but it requires a huge memory for the Reduce phase: with $\alpha = 15\%$, each reducer in PFP can take as much as 60 GB of memory. On the contrary, both MapFIM and BigFIM are effective in memory and never run out of memory in our setup. The results show that MapFIM outperforms both BigFIM and PFP, especially for low values of support. As expected, our algorithm works better with higher value of β. However, in case when the support α is equal or higher than 9%, MapFIM with $\beta = 30\%$ has a worse performance than with $\beta = \alpha$. Indeed, MapFIM with $\beta = \alpha$ ignores completely the Local Mining step and it is similar to parallel Apriori algorithm. With a high value of support, the Apriori approach is still effective because the number of candidates is not huge. However, when α is lower, applying the Local Mining step is efficient, as shown in our experiment.

Table 5. Performance with the Webdocs dataset

α	MapFIM				BigFIM	PFP
	$\beta = 100\%$	$\beta = 50\%$	$\beta = 30\%$	$\beta = \alpha$		
20%	162	360	641	280	421	1278
15%	211	684	1960	466	3370	3882
10%	454	1357	5183	2349	26258	Out of memory
9%	477	1691	7109	4869	45665	Out of memory
8%	581	2117	10558	12975	80858	Out of memory
7%	674	2760	15249	44075	Out of time	Out of memory
6%	967	4107	23807	215402	Out of time	Out of memory
5%	1804	6705	40832	Out of time	Out of time	Out of memory

5.2   Estimating β Parameter

In our algorithm, a good value of β is important for getting high performance. The higher the value of β is, the better performance we get in general but more memory is required. In this subsection, we present a method for estimating a good value of β.

It is proven in [18] that LCM algorithm requires an amount of memory linear to the input size. As we use a local program based on Eclat/LCM algorithm [18,19], we expect that the program requires a maximum of $f(input\_size)$ of memory, where $f()$ is a linear function. To simplify the estimation, we suppose that the maximum memory needed by the program is $\gamma \times input\_size$, where $input\_size$ is measured as the total length of all transactions in the dataset. We try to figure out the value of γ by experiments with various datasets. With each dataset, we run the program with support = 0% to report the maximum memory used during one hour by the program. Then we compute $\gamma = \frac{input\_size}{max\_memory}$ and report the result in Table 6. In these experiments, we use datasets from Frequent Itemset Mining Implementations Repository at http://fimi.ua.ac.be/data/.

From experiments, the value of γ varies from 0.017 to 0.043, with an average value of 0.023 and a standard deviation of 0.00785. For instance, the value of γ for dataset Webdocs is 0.018. Next, we test if $\gamma = 0.018$ is a good value to estimate β with dataset Webdocs. We run MapFIM with different values of β from 100% to 20% and the minimum support threshold $\alpha = 10\%$. The approximate memory required is computed by: $\gamma \times \beta \times input\_size$, where $input\_size$ is the total length of the transactions in the compressed data $\mathcal{D}'$. We report the approximate memory required $w.r.t$ $\gamma = 0.018$ and the real value of maximum memory used by the local program during the mining.

The result is expressed in Fig. 3 and as expected, the real value of max memory is always lower but not much lower than the approximate memory calculated.

From those observations, we propose to set the value of β in MapFIM by:

$$\beta = \frac{M_{Reduce} - M_{reduce\_task}}{\gamma \times input\_size} \tag{1}$$

where M_{Reduce} is the limit of memory of a Reducer, M_{reduce_task} is the memory required for a reduce task without running the local mining program[4] and $input\_size$ is the total length of transactions in the compressed dataset $\mathcal{D}'$.

Table 6. the γ value with Borgelt's implementation of Eclat/LCM

Dataset	input_size	max_memory (in Kilobyte)	γ
accidents	11500870	228400	0.020
connect	2904951	58160	0.020
kosarak	8019015	193644	0.024
pumsb	3629404	63332	0.017
retail	908576	25588	0.028
T40I10D100K	3960507	71988	0.018
T10I4D100K	1010228	25916	0.026
chess	118252	5100	0.043
pumsb_star	2475947	47424	0.019
webdocs	299887139	5422024	0.018

---

[4] In our implementation, M_{reduce_task} is around 300 MB.

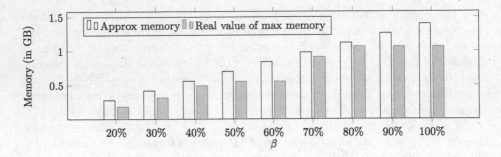

Fig. 3. Memory on the Webdocs dataset

6   Conclusion and Future Work

In this paper, we present MapFIM, a MapReduce based two-phase approach to efficiently mine frequent itemsets in very large datasets. In the first global mining phase, MapReduce is used to generate local memory-fitted prefix-projected databases from the input dataset benefiting from the Apriori principle. Then, in a local mining phase, an optimized in-memory mining process is launched to enumerate in parallel all frequent itemsets from each prefix-projected database. Compared to other existing approaches, our algorithm implements a fine-grained method to switch from global phase to the local phase. Moreover, we show that our method is transaction-extensible, meaning that given a fixed set of items, it can mine all frequent itemsets whatever the number of transactions and the minimum support threshold. To the best of our knowledge, our algorithm is the first to guarantee this property.

Our experimental evaluations show that MapFIM outperforms the best existing MapReduce based frequent itemset mining approaches. Moreover, we show how to calibrate and set the unique parameter β of our algorithm. This point is particularly important, since an optimal value of parameter β guarantees a high performance level.

Future work will be devoted to make MapFIM scalable. This can be achieved by using similar approaches, based on randomized key redistributions introduced in [3,4] for join processing, allowing to avoid the effects of data skew while guaranteeing perfect balancing properties during all the stages of join computation in large scale systems even for a highly skewed data.

Acknowledgement. This work is partly supported by the GIRAFON project funded by *Centre-Val de Loire*.

References

1. Aggarwal, A.C., Han, J.: Frequent Pattern Mining. Springer, Heidelberg (2014)
2. Agrawal, R., Srikant, R., et al.: Fast algorithms for mining association rules. In: Proceedings of VLDB 1994, vol. 1215, pp. 487–499 (1994)

3. Al Hajj Hassan, M., Bamha, M.: Towards scalability and data skew handling in groupby-joins using MapReduce model. In: Proceedings of ICCS 2015, pp. 70–79 (2015)
4. Al Hajj Hassan, M., Bamha, M., Loulergue, F.: Handling data-skew effects in join operations using MapReduce. In: Proceedings of ICCS 2014, pp. 145–158. IEEE (2014)
5. Beedkar, K., Berberich, K., Gemulla, R., Miliaraki, I.: Closing the gap: sequence mining at scale. ACM Trans. Database Syst. **40**(2), 8:1–8:44 (2015)
6. Dean, J., Ghemawat, S.: MapReduce: simplified data processing on large clusters. Commun. ACM **51**(1), 107–113 (2008)
7. Li, H., Wang, Y., Zhang, D., Zhang, M., Chang, E.Y.: PFP: parallel FP-growth for query recommendation. In: Proceedings of RecSys 2008, pp. 107–114. ACM (2008)
8. Li, N., Zeng, L., He, Q., Shi, Z.: Parallel implementation of Apriori algorithm based on MapReduce. In: Proceedings of SNDP 2012, pp. 236–241. IEEE (2012)
9. Lin, M.-Y., Lee, P.-Y., Hsueh, S.-C.: Apriori-based frequent itemset mining algorithms on MapReduce. In: Proceedings of ICUIMC 2012, pp. 76:1–76:8 (2012)
10. Lucchese, C., Orlando, S., Perego, R., Silvestri, F.: Webdocs: a real-life huge transactional dataset. In: FIMI, vol. 126 (2004)
11. Apache Mahout. Scalable machine learning and data mining (2012)
12. Makanju, A., Farzanyar, Z., An, A., Cercone, N., Hu Z.Z., Hu, Y.: Deep parallelization of parallel FP-growth using parent-child MapReduce. In: Proceedings of BigData 2016, pp. 1422–1431. IEEE (2016)
13. Miliaraki, I., Berberich, K., Gemulla, R., Zoupanos, S.: Mind the gap: large-scale frequent sequence mining. In: Proceedings of SIGMOD 2013, pp. 797–808. ACM (2013)
14. Moens, S., Aksehirli, E., Goethals, B.: Frequent itemset mining for big data. In: Proceedings of BigData 2013, pp. 111–118. IEEE (2013)
15. Salah, S., Akbarinia, R., Masseglia, F.: Data partitioning for fast mining of frequent itemsets in massively distributed environments. In: Chen, Q., Hameurlain, A., Toumani, F., Wagner, R., Decker, H. (eds.) DEXA 2015. LNCS, vol. 9261, pp. 303–318. Springer, Cham (2015). doi:10.1007/978-3-319-22849-5_21
16. Salah, S., Akbarinia, R., Masseglia, F.: Optimizing the data-process relationship for fast mining of frequent itemsets in MapReduce. In: Perner, P. (ed.) MLDM 2015. LNCS, vol. 9166, pp. 217–231. Springer, Cham (2015). doi:10.1007/978-3-319-21024-7_15
17. Savasere, A., Omiecinski, E., Navathe, S.B.: An efficient algorithm for mining association rules in large databases. In: Proceedings of VLDB 1995, pp. 432–444. Morgan Kaufmann Publishers Inc., San Francisco (1995)
18. Uno, T., Asai, T., Uchida, Y., Arimura, H.: LCM: an efficient algorithm for enumerating frequent closed item sets. In: FIMI, vol. 90. Citeseer (2003)
19. Zaki, M.J., Parthasarathy, S., Ogihara, M., Li, W., et al.: New algorithms for fast discovery of association rules. KDD **97**, 283–286 (1997)
20. Zhou, L., Zhong, Z., Chang, J., Li, J., Huang, J.Z., Feng, S.: Balanced parallel FP-growth with MapReduce. In: Proceedings of YC-ICT 2010, pp. 243–246 (2010)

Utilizing Bat Algorithm to Optimize Membership Functions for Fuzzy Association Rules Mining

Anping Song[1(✉)], Jiaxin Song[1], Xuehai Ding[1], Guoliang Xu[1], and Jianjiao Chen[2]

[1] School of Computer Engineering and Science,
Shanghai University, Shanghai, China
{apsong, dinghai}@shu.edu.cn,
{songjx, xglreal}@i.shu.edu.cn
[2] School of Medicine, Sylvester Miller Cancer Center,
University of Miami, Coral Gables, USA
jxcl665@miami.edu

Abstract. In numerous studies on fuzzy association rules mining, membership functions are usually provided by experts. It is unrealistic to predefine appropriate membership functions for every different dataset in real-world applications. In order to solve the problem, metaheuristic algorithms are applied to the membership functions optimization. As a popular metaheuristic method, bat algorithm has been successfully applied to many optimization problems. Thus a novel fuzzy decimal bat algorithm for association rules mining is proposed to automatically extract membership functions from quantitative data. This algorithm has enhanced local and global search capacity. In addition, a new fitness function is proposed to evaluate membership functions. The function takes more factors into account, thus can assess the number of obtained association rules more accurately. Proposed algorithm is compared with several commonly used metaheuristic methods. Experimental results show that the proposed algorithm has better performance, and the new fitness function can evaluate the quality of membership functions more reasonably.

Keywords: Bat algorithm · Fuzzy association rules · Membership functions

1 Introduction

Fuzzy association rules mining is used to find interesting correlations among items from quantitative data. Membership functions have a critical influence on these rules, but they are usually provided by experts. Therefore many researchers study how to automatically learn and optimize membership functions. Recently some metaheuristic algorithms have been successfully applied to the membership functions optimization. Hong proposed a framework based on genetic algorithm (GA) [1] and a parallel GA approach [2] for fuzzy association rules mining. Palacios proposed an approach to mine fuzzy association rules from low quality data based on the 3-tuples linguistic representation model [3]. Matthews provided an evolutionary algorithm to discover temporal

© Springer International Publishing AG 2017
D. Benslimane et al. (Eds.): DEXA 2017, Part I, LNCS 10438, pp. 496–504, 2017.
DOI: 10.1007/978-3-319-64468-4_37

rules [4], and a GA-based approach for web usage mining [5]. Ant colony algorithm [6] and particle swarm optimization algorithm [7] were also applied to the optimization of membership functions.

Bat algorithm (BA) is a metaheuristic algorithm proposed by Yang [8]. The advantage of rapid convergence makes it be applied in many fields [9, 10], and researchers did a lot work to improve its efficiency. In this paper, on the basis of previous studies [11, 12], a fuzzy decimal bat algorithm (FDBA) for association rules mining is proposed. The proposed algorithm can dynamically tune membership functions instead of predefining them, and FDBA is compared with uniform fuzzy partition (UFP), GA and ant colony system (ACS). In addition, a new fitness function is proposed, and the proposed fitness function is compared with Hong et al. [1].

The rest of paper is organized as follows. Membership functions representation and fitness function are stated in Sect. 2. FDBA is introduced in Sect. 3. Experimental results are shown in Sect. 4. Finally, conclusions are given in Sect. 5.

2 Membership Functions Representation and Fitness Function

2.1 Membership Functions Representation

Since isosceles triangular and trapezoidal membership function are most commonly used, in this paper the two types are used for membership functions representation, as shown in Fig. 1. Assume item I_j is divided into m linguistic terms. R_{j1} and R_{jm} are the membership functions of the first and the last linguistic term respectively. R_{jk} is the membership function of the $k-th(1 < k < m)$ linguistic term. c_{jk} is the center abscissa, w_{jk} is the half range of the spread region.

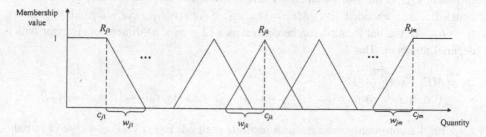

Fig. 1. Membership functions of item I_j

2.2 Fitness Function

To evaluate the membership functions, a new fitness function is proposed. It consists of two parts: (1) the *suitability* [1] of the membership functions. (2) the number of association rules obtained.

If more association rules are obtained, it is more likely to find the valuable relationships among the items. Therefore high quality membership functions are expected to generate more association rules. However, calculating the number of association rules will lead to a huge amount of computation. Hong et al. [1] considered that a larger number of 1-itemsets implied more interesting association rules. So the fitness function is defined as

$$Fit_1(I) = \frac{|L_1|}{suitability(I)} \tag{1}$$

Where $|L_1|$ is the number of frequent 1-itemsets.

In this paper another factor is considered for the fitness function, which is the average support of frequent 1-itemsets.

$$Fit_2(I) = \frac{|L_1| + \eta \times meanSupp(L_1)}{suitability(I)} \tag{2}$$

Where Fit_2 is the fitness function for items $I_1 \cdots I_j \cdots I_n$. $meanSupp(L_1)$ is the average support of frequent 1-itemsets. η is a weight which is set depending the domain size of the problem. Experiment results show that when $n - 3 \leq \eta \leq n + 3$, it has a better performance.

3 Fuzzy Decimal Bat Algorithm for Association Rules Mining

3.1 Encoding and Initialization

In this paper, the membership functions are encoded as a string of decimal numbers. As shown in Fig. 1, the two parameters c and w are used. The membership functions of item I_j are encoded as $MF_j = (c_{j1}, w_{j1}, \cdots, c_{jk}, w_{jk}, \cdots c_{jm}, w_{jm})$. For items $I_1 \cdots I_j \cdots I_n$, the bat location can be defined as a $(2 \times m \times n)$-dimensional vector with decimal numbers. That is

$$\begin{aligned} x &= MF_1 \cdots MF_j \cdots MF_n \\ &= (c_{11}, w_{11}, \cdots, c_{1m}, w_{1m}, \cdots, c_{j1}, w_{j1}, \cdots, c_{jm}, w_{jm}, \cdots, c_{n1}, w_{n1}, \cdots, c_{nm}, w_{nm}) \end{aligned} \tag{3}$$

UFP is a commonly used artificial partition method. Using UFP as a case of initial individuals can avoid the quality of initialization being too low. Set the size of bats' population as N. One of the bats is the UFP, and the remaining bats are randomly generated.

3.2   Adaptive Frequency

Frequency f is used to adjust the velocity of bats. FDBA assigns different frequencies to different dimensions.

$$f_{ij} = f_{min} + \frac{\sqrt{(D_{i,min} - D_{ij})^2}}{D_{i,max} - D_{i,min}}(f_{max} - f_{min}) \tag{4}$$

Where $D_{ij} = \sqrt{(x_{ij} - x_{*j})^2}$, x_{ij} and x_{*j} denote the $j-th$ dimension of the $i-th$ solution and the global optimal solution respectively. f_{ij} is the frequency value assigned to the $j-th$ dimension of the $i-th$ solution. f_{min} and f_{max} denote the minimum and maximum values of the frequency respectively. $D_{i,min}$ and $D_{i,max}$ respectively represent the closest and the farthest solutions to the optimal solution among all dimensions of the $i-th$ solution.

3.3   Update of Velocity

The velocity v_i^t of the $i-th$ bat at t iteration is defined as follows.

$$v_i^t = \omega v_i^{t-1} + \xi(x_i^t - x_*)f_i + (1 - \xi)(x_i^t - x_k^t)f_i \tag{5}$$

Where f_i is the frequency of the $i-th$ bat. $\omega = \frac{t_{max}-t}{t_{max}}(\omega_{max} - \omega_{min}) + \omega_{min}$ is an inertia weight. t_{max} and t are the maximum and current number of iterations respectively. ω_{min} and ω_{max} are the minimum and maximum values of ω respectively. x_* is the current best solution among the population. $x_k(i \neq k)$ is one of the solutions randomly chosen. $\xi = 1 + \left(\frac{t_{max}-t}{t_{max}}\right)^n (\xi_{init} - 1)$. ξ_{init} is the initial value of ξ, n is a nonlinear modulation index. In this paper n is 3. ξ gradually increases with iterations, and the effect of x_* gradually grows to be greater than x_k.

3.4   Hybridization with Invasive Weed Optimization

To improve the search capability of the algorithm, FDBA combines the bats algorithm with another heuristic algorithm, namely the invasive weed optimization (IWO) algorithm [13]. IWO simulates the processes of weed growth, seed dispersal and competition disappearance in nature.

4   Experiments and Results

4.1   Algorithm Experimental Platform and Problem Description

UCI machine learning data sets [14] are commonly used in data mining. Table 1 shows the characteristics of the 4 UCI datasets. In the experiment, algorithms only performed

Table 1. The specifications of datasets

Dataset	Transactions	Total items	Quantitative items
Ecoli	336	9	7
ConcreteData	1030	9	9
Yeast	1484	10	7
Stock	536	10	9

for quantitative items. All of our experiments are conducted on a 2.2 GHz PC with 8G memory, on Windows 10.

FDBA is compared with UFP, GA [1] and ACS [6]. In the experiment, linguistic terms number $m = 3$, $t_{max} = 10000$, $N = 10$. The values assigned to the parameters of GA are same as [3], and ACS are same as [9]. For each parameter of FDBA, several values had been selected. Experiment results show that the following values have better performance. $\eta = 9$, $f_{min} = 0$, $f_{max} = 0.3$, $\omega_{max} = 1$, $\omega_{min} = 0.6$, $\xi_{init} = 0.6$. In this experiment, the value of minimum support is 0.4.

4.2 Experimental Results Using Fit₁

The results obtained by UFP, GA, ACS and FDBA using Hong's fitness function are presented in Table 2. *Conf* stands for the minimum confidence. Figure 2 shows the fitness values of FDBA, GA and ACS along with different numbers of iterations.

From both Fig. 2 and Table 2, we can see that FDBA achieves the best fitness. FDBA is an effective method to learn membership functions. Hong believed that larger number of frequent 1-itemsets is likely to generate more association rules. However, Table 2 shows that for some cases, it does not conform to this rule. Therefore it is not reasonable to use only the frequent 1-itemset as the criterion to measure the number of association rules.

4.3 Experimental Results Using Fit₂

The results obtained by UFP, GA, ACS and FDBA using the proposed fitness function are presented in Table 3. Figure 3 shows the fitness values of FDBA, GA and ACS along with different numbers of iterations using Fit₂.

From both Fig. 3 and Table 3, we can see that FDBA algorithm still obtains the best fitness results as using Fit₁. In most cases, the solutions with better fitness can generate more association rules. The fitness function proposed in this paper can better evaluate the number of association rules, and FDBA algorithm can get more association rules compared with UFP, GA and FDBA.

Table 2. Results obtained by different approaches using Fit$_1$

Dataset	Algorithm	Fitness	Number of association rules			
			$Conf = 0.8$	$Conf = 0.7$	$Conf = 0.6$	$Conf = 0.5$
Ecoli	UFP	0.7142857	17	17	17	22
	GA	1.622276	278	297	349	439
	ACS	1.736977	70	85	88	95
	FDBA	2	13	14	14	14
ConcreteData	UFP	0.8888889	30	35	38	54
	GA	1.39975	86	112	142	176
	ACS	1.6406	6	7	9	12
	FDBA	1.983734	0	0	0	0
Yeast	UFP	1.285714	37	52	67	85
	GA	1.571429	197	261	284	325
	ACS	1.90167	14	18	22	22
	FDBA	2	12	12	12	12
Stock	UFP	1	985	1462	1892	1892
	GA	1.415215	277	309	366	524
	ACS	1.655768	14	20	20	20
	FDBA	2	0	0	0	0

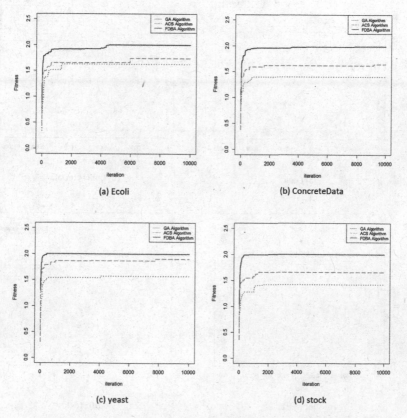

Fig. 2. The fitness values along with different numbers of iterations

Table 3. Results obtained by different approaches using Fit$_2$

Dataset	Algorithm	Fitness	Number of association rules			
			$Conf = 0.8$	$Conf = 0.7$	$Conf = 0.6$	$Conf = 0.5$
Ecoli	UFP	1.600567	17	17	17	22
	GA	2.152069	362	397	485	556
	ACS	1.918688	66	108	152	167
	FDBA	2.271544	1932	1932	1932	1932
ConcreteData	UFP	1.513087	30	35	38	54
	GA	1.923361	85	123	157	184
	ACS	1.639191	63	123	160	177
	FDBA	1.931084	5211	6087	7145	8174
Yeast	UFP	2.048874	37	52	67	85
	GA	2.18622	1856	1932	1932	1932
	ACS	2.047964	1009	1501	1811	1919
	FDBA	2.283982	1932	1932	1932	1932
Stock	UFP	1.635783	985	1462	1892	1892
	GA	1.925472	18660	18660	18660	18660
	ACS	1.764947	15037	18203	18660	18660
	FDBA	1.994329	18660	18660	18660	18660

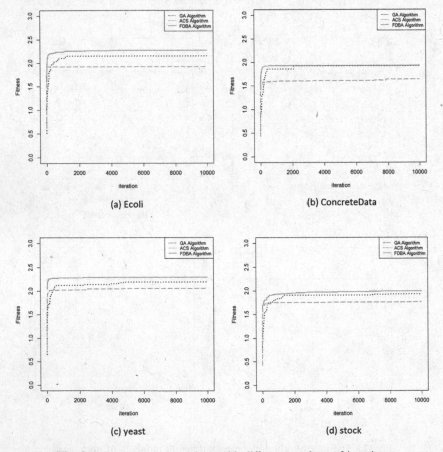

Fig. 3. The fitness values along with different numbers of iterations

5 Conclusions

In this paper, a novel fuzzy decimal bat algorithm for association rules mining is proposed to learn membership functions automatically instead of being provided by experts. As the previous fitness function sometimes can't correctly evaluate the number of association rules generated, besides the number of large 1-itemsets, we also consider the mean support of the large 1-itemsets as a factor of fitness function. In order to verify our approach, we ran the FDBA with different fitness functions on serval UCI datasets. Experimental results show that FDBA can obtain better fitness compared with UFP, GA and ACS. In addition, the proposed fitness is consistent with the number of generated rules. As each item is independent in the process of optimization and the evaluation for fitness values is time-consuming, in the future, we will consider use parallel bat algorithm to improve the speed up of the algorithm.

References

1. Hong, T.P., Chen, C.H., Wu, Y.L., Lee, Y.C.: A GA-based fuzzy mining approach to achieve a trade-off between number of rules and suitability of membership functions. Soft. Comput. **10**(11), 1091–1101 (2006)
2. Hong, T.P., Lee, Y.C., Wu, M.T.: An effective parallel approach for genetic-fuzzy data mining. Expert Syst. Appl. **41**(2), 655–662 (2014)
3. Palacios, A.M., Palacios, J.L., Sánchez, L., Alcalá-Fdez, J.: Genetic learning of the membership functions for mining fuzzy association rules from low quality data. Inf. Sci. **295**, 358–378 (2015)
4. Matthews, S.G., Gongora, M.A., Hopgood, A.A.: Evolutionary algorithms and fuzzy sets for discovering temporal rules. Int. J. Appl. Math. Comput. Sci. **23**(4), 855–868 (2013)
5. Matthews, S.G., Gongora, M.A., Hopgood, A.A., Ahmadi, S.: Web usage mining with evolutionary extraction of temporal fuzzy association rules. Knowl.-Based Syst. **54**, 66–72 (2013)
6. Wu, M.T., Hong, T.P., Lee, C.N.: A continuous ant colony system framework for fuzzy data mining. Soft. Comput. **16**(12), 2071–2082 (2012)
7. Alikhademi, F., Zainudin, S.: Generating of derivative membership functions for fuzzy association rule mining by Particle Swarm Optimization. In: 2014 International Conference on Computational Science and Technology (ICCST), pp. 1–6. IEEE (2014)
8. Yang, X.S.: A new metaheuristic bat-inspired algorithm. In: González, J.R., Pelta, D.A., Cruz, C., Terrazas, G., Krasnogor, N. (eds.) Nature Inspired Cooperative Strategies for Optimization (NICSO 2010). Studies in Computational Intelligence, vol. 284, pp. 65–74. Springer, Heidelberg (2010). doi:10.1007/978-3-642-12538-6_6
9. Pérez, J., Valdez, F., Castillo, O.: Modification of the bat algorithm using fuzzy logic for dynamical parameter adaptation. In: IEEE Congress on Evolutionary Computation (CEC) 2015, pp. 464–471. IEEE (2015)
10. Song, A., Ding, X., Chen, J., Li, M.: Multi-objective association rule mining with binary bat algorithm. Intell. Data Anal. **20**(1), 105–128 (2016)
11. Yilmaz, S., Kucuksille, E.U.: Improved bat algorithm (IBA) on continuous optimization problems. Lect. Notes Softw. Eng. **1**(3), 279 (2013)

12. Yilmaz, S., Kucuksille, E.U.: A new modification approach on bat algorithm for solving optimization problems. Appl. Soft Comput. **28**, 259–275 (2015)
13. Mehrabian, A.R., Lucas, C.: A novel numerical optimization algorithm inspired from weed colonization. Ecol. inform. **1**(4), 355–366 (2006)
14. UC Irvine Machine Learning Repository. http://archive.ics.uci.edu/ml/

A Formal Approach for Failure Detection in Large-Scale Distributed Systems Using Abstract State Machines

Andreea Buga[✉] and Sorana Tania Nemeş

Christian Doppler Laboratory for Client-Centric Cloud Computing,
Johannes Kepler University of Linz, Software Park 35, 4232 Hagenberg, Austria
{andreea.buga,t.nemes}@cdcc.faw.jku.at

Abstract. Large-scale distributed systems have been widely adopted in various domains due to their ability to compose services and resources tailored to user requirements. Such systems are characterized by high complexity and heterogeneity. Maintaining a high-level availability and a normal execution of the components implies precise monitoring and robust adaptation. Monitors capture relevant metrics and transform them to meaningful knowledge, which is further used in justifying adaptation actions. The current paper proposes an Abstract State Machine model for defining monitoring processes addressing failures and unavailability of the system nodes. The specification is simulated and validated with the aid of the ASMETA toolset. The solution is complemented with a small ontology reflecting the structure of the system. We emphasize the role of formal models in achieving the proposed requirements.

Keywords: Formal modeling · Abstract State Machines · Failure detection · Ontology · Model validation

1 Introduction

Traditional software solutions could not keep up with the user requirements. Large-scale distributed systems (LDS) use network connections to compose services and engage resources for fulfilling different tasks. LDS principles contributed to the development of grid and cloud systems. Due to high complexity and heterogeneity, LDS face various failures of components, which might trigger a chain of service unavailability. Monitoring and adaptation processes play a key role in detecting issues and restoring the system to a normal execution.

The paper addresses the robustness of monitoring solutions for failure detection in LDS and contributes with a formal model that focuses on correct behavior of monitors. We employ the Abstract State Machine (ASM) software engineering method to elaborate the specification. We perform model simulation and validation with the aid of the ASMETA toolset[1]. Thus, we analyze system behavior

---

[1] http://asmeta.sourceforge.net/.

© Springer International Publishing AG 2017
D. Benslimane et al. (Eds.): DEXA 2017, Part I, LNCS 10438, pp. 505–513, 2017.
DOI: 10.1007/978-3-319-64468-4_38

and identify design issues before the software development phase. A knowledge scheme contains the main concepts to be used by the monitoring solution.

The remainder of the paper is structured as follows. Section 2 presents the problem domain, the architecture and the knowledge base of the monitoring solution. The proposed ASM formal model is defined in Sect. 3, followed by its validation in Sect. 4. Relevant work in the area is presented in Sect. 5. We draw the conclusions and present a direction for future work in Sect. 6.

2   System Overview

2.1   System Architecture

Applications providing services from LDS to users are organized in several tiers as illustrated in Fig. 1. The client side consists of various devices that access the services. Their requests are processed by a client-provider interaction middleware that focuses on the delivery of secure and adaptable services to the client [5].

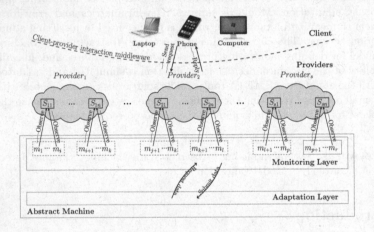

Fig. 1. Architecture of the system

Each node is assigned a set of monitors, in order to ensure a more reliable evaluation. When issues are detected, data are submitted to adapters, which construct or retrieve reconfiguration plans based on the problem details. Adapters enact repair actions and request the monitors to evaluate their efficiency. The current paper describes the behavior and properties specific to the monitors.

Monitors are also part of the LDS and can exhibit failures. We address the aspect of random behavior of the monitors and assign a measure of their trustworthiness. The confidence degree of a monitor depends on the correctness of its diagnoses in comparison with the decision taken by the leader assigned to the node. Monitors whose confidence drops below a minimum accepted value are disabled. The minimum threshold is set for each system according to its strategy.

2.2   Knowledge Representation

Monitoring is an important part of the MAPE-K cycle for self-adapting systems. The knowledge base contains essential information for choosing a plan. As shown in Fig. 2 we built an ontology, which comprises only a small part of the needed concepts, but which is sufficient for the formalization of monitors.

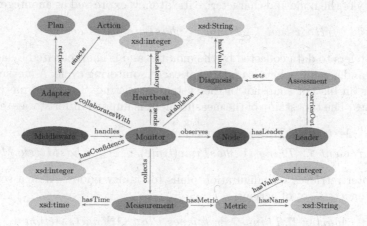

Fig. 2. Ontology of the monitoring solution

The monitor observes a node by sending heartbeats and collecting measurements to establish a diagnosis. Each monitor has a confidence degree reflecting its accuracy. Monitors collaborate with adapters by reporting issues or evaluating the efficiency of an adaptation plan.

$$Monitor \sqsubseteq Thing \sqcap \exists collects.Measurement \sqcap = 1observes.Node$$
$$\sqcap \exists collaboratesWith.Adapter \sqcap \exists establishes.Diagnosis$$
$$\sqcap \exists sends.Heartbeat \sqcap \exists hasConfidence(integer)$$

A monitor has initially a confidence value of 100, which decreases whenever its diagnosis is different than the one proposed by the majority of monitors assigned to the node as shown by Eq. 1. The reduction is proportional to a penalty, which depends on how critical the system is.

$$conf\_degree(m)- = \frac{|diagn| - |similar\_diagn|}{|diagn|} \cdot *penalty, \in Monitors \quad (1)$$

Nodes have a leader which executes an assessment by requesting input from all the monitors appointed to the node. The leader is a different component than the monitor and it performs an evaluation when a monitor reports a problem. The assessment contains the diagnosis established by the majority of the monitors.

$$Node \sqsubseteq Thing \sqcap \exists hasLeader.Leader$$
$$Leader \sqsubseteq Thing \sqcap \exists carriesOut.Assessment$$
$$Assessment \sqsubseteq Thing \sqcap \exists sets.Diagnosis$$
$$Diagnosis \sqsubseteq Thing \sqcap \exists hasValue(String)$$

Heartbeats are ping requests sent by the monitor to the node. They reflect the availability of the node and characterize its latency, expressed as an integer value.

$$Heartbeat \sqsubseteq Thing \sqcap \exists hasLatency(integer)$$

A metric refers to data collected by the monitor and is characterized by a unique identifier and a value, which is updated at each monitoring cycle. A measurement represents an instant snapshot of the system consisting of a set of metrics and their values. The timestamp of the measurement is important for storage cleanup.

$$Metric \sqsubseteq Thing \sqcap \exists hasName(String) \sqcap \exists hasValue(integer)$$
$$Measurement \sqsubseteq Thing \sqcap \exists hasTime(time) \sqcap \exists \geq 1 \ hasMetric.Metric$$

The adapter retrieves reconfiguration plans for faulty nodes. After a scheme is found, it enacts a set of actions.

$$Adapter \sqsubseteq Thing \sqcap \exists retrieves.Plan \sqcap \exists enacts.Action$$

3 Formal Specification of the Monitoring Solution

3.1 Background on ASM Theory

ASMs extend the Finite State Machines by allowing input and output states to have data structures. An ASM machine has a finite set of transitions **if** *Condition* **then** *Updates*, where *Update* is a finite set of assignments $f(t_1, ..., t_n) := t$ [4]. ASM rules allow expressing parallelism, sequentiality, concurrency and non-determinism and their functions emphasize separation of concerns.

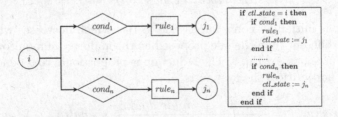

Fig. 3. Structure of a control state ASM

Definition 1. *A* control state ASM *follows the rule structure from Fig. 3: any control state i verifies at most one true guard, $cond_k$, triggering $rule_k$ and moving from state i to state s_k. If no guard is fulfilled, the machine executes no action.*

3.2   Ground Model

Control state ASMs capture natural-language requirements of the system. A middleware agent initializes the system and assigns monitors to each node. Monitors execute rules to evaluate the nodes. Each node has a leader that collects diagnoses from the corresponding monitors and performs an assessment.

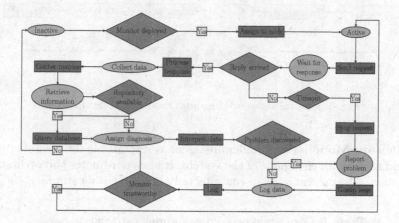

Fig. 4. Control state ASM ground model of monitor module

Monitor Model. After deployment by the middleware, the monitor passes to the *Active* state, where it sends a heartbeat request to its assigned node. The monitor advances then to the *Wait for response* state where it checks two guards. First, it verifies if a request response is received. If so, it verifies if the delay of the response is acceptable. If this condition is satisfied, the monitor moves to the *Collect data* state. If no response is received or if the response has a big delay, the monitor moves to the *Report problem* state. In the *Collect* data state, the monitor gathers low level metrics and then moves to the *Retrieve information* state, where it fetches past monitoring data if the repository is available. The monitor moves to the *Assign diagnosis* state where it interprets the available data. If a problem is discovered, it passes to the *Report problem*, otherwise it goes to the *Log data* state. When an issue is found, the monitor triggers the leader of the node to inquire all his monitoring counterparts and carry out a collaborative diagnosis. After reporting the issue, the monitor moves to the *Log data* state. Here, the confidence degree of the monitor is checked and if the monitor is still trustworthy it starts a new monitoring cycle. Alternatively, it moves to the *Inactive* state and waits to be deployed again in the system. The behavior of the monitor is captured by its corresponding ASM module in Fig. 4.

Leader Model. The middleware assigns to each node a leader module, which is responsible for collaborative decisions. If an issue is reported the leader moves to the *Evaluate* state, where it requests diagnoses from the monitors assigned to the node and goes to the *Assess* state. During assessment, the diagnosis voted by the majority is chosen. The structure of the diagnosis is reflected by Table 1. After evaluation, the leader moves to the *Idle leader* state and clears previous data. At a new assessment, the leader moves to the *Evaluate* state and restarts the cycle as illustrated in Fig. 5. The current model assumes that the leader is reliable. However, we aim to address its possible failure in a future refinement.

Fig. 5. Control state ASM ground model of leader module

Middleware Model. The middleware agent is responsible for assigning monitors and a leader for each node of the system. It also coordinates and orchestrates the system. The current version contains only one *Executing* state.

Table 1. Correlation between monitoring data and diagnosis

DIAGNOSIS	Latency	CPU usage	Memory usage	Storage usage	Work capacity
NORMAL	<10	<40	<40	<40	>85
CRITICAL	<10	>40	>40	>40	<85
FAILED	>10	NA	NA	NA	NA

3.3   Translation of the Models to AsmetaL Language

Monitors are the core components and their functions reflect the structure of the proposed knowledge base. Instead of defining a function *observes: Monitor* → *Node*, we stored the sequence of all monitors assigned to a node. Measurements of a monitor are stored in a sequence of *(Metric, Integer)* pairs.

Leader module consists of three rules to request and clear data, and assess the node status as shown in Code 1. The assessment establishes the diagnosis voted by the majority of the monitors. By analyzing decisions from different counterparts we reduce the side-effects of random failures of the monitors. In a future refinement we aim to add weights equal to confidence degree of monitors. Thus, monitors with a lower accuracy have a smaller impact on the evaluation.

The middleware agent assigns monitors and a leader for each node. The number of monitors to draft is constant and depends on how critical the system is. The reader is advised to consult the full AsmetaL specification[2].

---

[2] http://cdcc.faw.jku.at/staff/abuga/dexa_specs.rar.

```
module Leader                                               rule r_ClearData ($l in Leader) = skip
signature:                                                  rule r_LeaderProgram =
    enum domain Leader_States = {IDLE_LEADER|EVALUATE|ASSESS}   par
    controlled leader_state : Leader —> Leader_States             if (leader_state (self) = EVALUATE) then
    controlled assessment: Leader —> Diagnosis                       par
definitions:                                                            r_RequestData [self]
    rule r_AssessNode ($l in Leader) =                                  leader_state(self) := ASSESS
        if (max(failed_diagnoses($l), critical_diagnoses($l))        endpar
            = failed_diagnoses($l)) then                         endif
        if (max(failed_diagnoses($l), normal_diagnoses($l))      if (leader_state (self) = ASSESS) then
            = failed_diagnoses($l)) then                             par
            assessment ($l) := FAILED                                   r_AssessNode [self]
        else                                                            leader_state(self) := IDLE_LEADER
            assessment ($l) := NORMAL                                endpar
        endif                                                    endif
    else                                                         if (leader_state (self) = IDLE_LEADER) then
        if(max(critical_diagnoses($l),normal_diagnoses($l))         seq
            = critical_diagnoses($l)) then                             r_ClearData [self]
            assessment ($l) := CRITICAL                                if (is_evaluation_needed(self)) then
        else                                                               leader_state (self) := EVALUATE
            assessment ($l) := NORMAL                                   endif
        endif                                                        endseq
    endif                                                        endif
    rule r_RequestData ($l in Leader) = skip                 endpar
```

Code 1. Leader ASM module

4   Validation of the Model

Scenarios capture execution flows given specific function values of the system. We check if the leader correctly assesses a node, given the diagnosis of each monitor and if monitors assign a correct evaluation based on Table 1. For validation we defined a *Node* instance and a basic set of monitoring data. AsmetaV tool permits validation of scenarios defined with the Avalla language presented by [6].

We assigned five *Monitors* to the node and values for the collected metrics that correspond to different diagnoses, such that *Monitor1, Monitor4* should assess that the node failed, *Monitor2, Monitor5* should assess that the node is in a critical state and *Monitor3* should find the execution of the node to be normal. According to the model, the *Leader* must choose the diagnosis established by the majority of the monitors at the moment when the evaluation is carried out. The scenario to validate the model and its simulation trace are shown in Code 2. At the time the leader executes the assessment, three monitors did not carry out a diagnosis. Thus, the leader evaluation relies on a fragmented view on the system (of the first and fourth monitor).

5   Related Work

LDS monitoring has been developed together with the expansion of cloud and grid systems. In comparison to our ASM approach, CloudML uses an UML extension to express Quality of Service (QoS) models, monitoring rules and data, and adaptation concerns of multi-cloud systems [2].

```
set assigned_monitors(node_1) := [mon_1, mon_2, mon_3, mon_4, mon_5] and has_leader(node_1) := leader_1 and leader_state(leader_1) := IDLE_LEADER;
step
set heartbeat_response_arrived(heartbeat_1):=false and heartbeat_response_arrived(heartbeat_2):=true and .. heartbeat_response_arrived(heartbeat_5):=true;
set heartbeat_latency(heartbeat_2) := 5 and heartbeat_latency(heartbeat_3) := 7 and heartbeat_latency(heartbeat_4) := 25 and heartbeat_latency(
       heartbeat_5) := 2;
step
set heartbeat_response_arrived(heartbeat_1) := true and heartbeat_latency(heartbeat_1) := 21;
set monitor_measurements(mon_2) := [("Latency", 35), ("CPU_Usage", 10), ("Storage_Usage", 15), ("Memory_Usage", 20), ("Bandwidth", 20)];
set monitor_measurements(mon_3) := [("Latency", 5), ("CPU_Usage", 10), ("Storage_Usage", 15), ("Memory_Usage", 10), ("Bandwidth", 50)];
set monitor_measurements(mon_5) := [("Latency", 2), ("CPU_Usage", 40), ("Storage_Usage", 15), ("Memory_Usage", 10), ("Bandwidth", 30)];
step
set is_repository_available(mon_2) := true and is_repository_available(monitor_3) := false and is_repository_available(mon_5) := true;
step step step
check assessment(leader_1) = FAILED;
```

```
<State 1 (controlled)>                                              calc_metrics(mon_3)=[("Normalized_Delay",0.1),("Work_capacity",88.33)]
leader_state(leader_1)=IDLE_LEADER                                  calc_metrics(mon_5)=[("Normalized_Delay",0.066),("Work_capacity",78.33)]
mon_state(mon_1, mon_2, mon_3, mon_4, mon_5)=WAIT_FOR_RESPONSE      diagnosis(mon_1, mon_3, mon_4)=NORMAL
<State 2 (controlled)>                                              diagnosis(mon_2, mon_5)=CRITICAL
diagnosis(mon_4)=FAILED                                             diagnosis_history(mon_1, mon_4)=[FAILED]
leader_state(leader_1)=IDLE_LEADER                                  leader_state(leader_1)=ASSESS
mon_state(mon_1)=WAIT_FOR_RESPONSE                                  mon_state(mon_1)=ACTIVE
mon_state(mon_2, mon_3, mon_5)=COLLECT_DATA                         mon_state(mon_2, mon_5)=REPORT_PROBLEM
mon_state(mon_4)=REPORT_PROBLEM                                     mon_state(mon_3)=LOG_DATA
<State 3 (controlled)>                                              mon_state(mon_4)=WAIT_FOR_RESPONSE
diagnosis(mon_1, mon_4)=FAILED                                      <State 6 (controlled)>
leader_state(leader_1)=IDLE_LEADER                                  assessment(leader_1)=FAILED
mon_state(mon_1)=REPORT_PROBLEM                                     calc_metrics(mon_2)=[("Normalized_Delay",1.75),("Work_capacity",85.0)]
mon_state(mon_2, mon_3, mon_5)=RETRIEVE_INFO                        calc_metrics(mon_5)=[("Normalized_Delay",0.066),("Work_capacity",78.33)]
mon_state(mon_4)=LOG_DATA                                           diagnosis(mon_1, mon_3)=NORMAL
trigger_gossip(mon_4)=true                                          diagnosis(mon_2, mon_5)=CRITICAL
<State 4 (controlled)>                                              diagnosis(mon_4)=FAILED
diagnosis(mon_1)=FAILED                                             diagnosis_history(mon_1, mon_4)=[FAILED]
diagnosis(mon_4)=NORMAL                                             diagnosis_history(mon_3)=[NORMAL]
diagnosis_history(mon_4)=[FAILED]                                   leader_state(leader_1)=IDLE_LEADER
leader_state(leader_1)=EVALUATE                                     mon_state(mon_1)=WAIT_FOR_RESPONSE
mon_state(mon_1)=LOG_DATA                                           mon_state(mon_2, mon_5)=LOG_DATA
mon_state(mon_2, mon_3, mon_5)=ASSIGN_DIAGNOSIS                     mon_state(mon_3)=ACTIVE
mon_state(mon_4)=ACTIVE                                             mon_state(mon_4)=REPORT_PROBLEM
trigger_gossip(mon_1)=true                                          trigger_gossip(mon_2, mon_5)=true
<State 5 (controlled)>
calc_metrics(mon_2)=[("Normalized_Delay",1.75),("Work_capacity",85.0)]    "check_succeeded:_assessment(leader_1)_=_FAILED"
```

Code 2. A validation scenario and its simulation trace

The work of [3,9] served as a good ASM representation of LDS processes. While their work focuses on the service execution, we formalize the monitoring layer. In [1], the authors propose an ASM model to analyze MAPE-K loops of decentralized self-adapting systems. Flexibility and robustness to silent node failures of the specification is validated and verified with ASMETA. While this work focuses on adaptation, we build the models for the monitoring solution. In [7], the authors discuss the relevance of ontologies and use both low- and high-level metrics for automatic reasoning in performance monitoring. mOSAIC project describes an ontology that ensures interoperability of services in multiclouds [8]. Both works are relevant for the structure of our ontology.

6   Conclusions and Future Work

The paper proposes an approach for achieving a reliable monitoring solution for LDS. By employing the ASM formal method we analyze the properties of the model. The monitoring processes are an essential part of the MAPE-K self-adaptive systems that aim to achieve high availability of services. Correct monitoring leads to an increased reliability of services. The knowledge scheme presented in the paper supports the solution and is reflected in the model. We used the AsmetaV tool to validate the behavior of the leader and the monitors.

As a future work we will refine the model and include finer level details related to the analysis of monitoring data and the collaborative diagnosis. We plan to use the AsmetaSMV tool to prove the intended properties of the solution.

References

1. Arcaini, P., Riccobene, E., Scandurra, P.: Modeling and analyzing MAPE-K feedback loops for self-adaptation. In: 2015 IEEE/ACM 10th International Symposium on Software Engineering for Adaptive and Self-Managing Systems, pp. 13–23, May 2015
2. Bergmayr, A., Rossini, A., Ferry, N., Horn, G., Orue-Echevarria, L., Solberg, A., Wimmer, M.: The evolution of CloudML and its manifestations. In: Proceedings of the 3rd International Workshop on Model-Driven Engineering on and for the Cloud (CloudMDE), Ottawa, Canada, pp. 1–6, September 2015
3. Bianchi, A., Manelli, L., Pizzutilo, S.: An ASM-based model for grid job management. Informatica **37**(3), 295–306 (2013). Slovenia
4. Börger, E., Stark, R.F.: Abstract State Machines: A Method for High-Level System Design and Analysis. Springer, New York (2003). doi:10.1007/978-3-642-18216-7
5. Bósa, K., Holom, R.M., Vleju, M.B.: A formal model of client-cloud interaction. In: Thalheim, B., Schewe, K.D., Prinz, A., Buchberger, B. (eds.) Correct Software in Web Applications and Web Services. Texts & Monographs in Symbolic Computation, pp. 83–144. Springer, Cham (2015). doi:10.1007/978-3-319-17112-8_4
6. Carioni, A., Gargantini, A., Riccobene, E., Scandurra, P.: A scenario-based validation language for ASMs. In: Börger, E., Butler, M., Bowen, J.P., Boca, P. (eds.) ABZ 2008. LNCS, vol. 5238, pp. 71–84. Springer, Heidelberg (2008). doi:10.1007/978-3-540-87603-8_7
7. Funika, W., Janczykowski, M., Jopek, K., Grzegorczyk, M.: An ontology-based approach to performance monitoring of MUSCLE-bound multi-scale applications. Procedia Comput. Sci. **18**, 1126–1135 (2013)
8. Moscato, F., Aversa, R., Di Martino, B., Fortiş, T.F., Munteanu, V.: An analysis of mOSAIC ontology for cloud resources annotation. In: 2011 Federated Conference on Computer Science and Information Systems, pp. 973–980, September 2011
9. Németh, Z.N., Sunderam, V.: A formal framework for defining grid systems. In: 2014 14th IEEE/ACM International Symposium on Cluster, Cloud and Grid Computing, p. 202 (2002)

Author Index

Printed in the United States
By Bookmasters